MOLECULAR EVOLUTION

Molecular Evolution

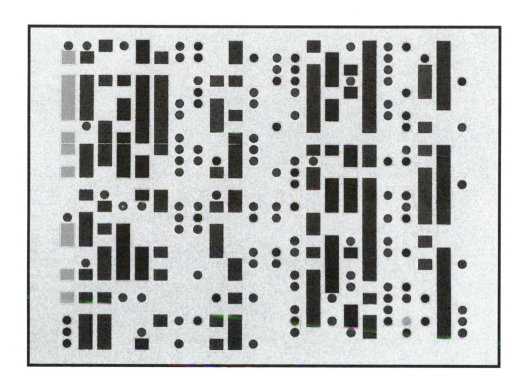

Wen-Hsiung Li

University of Texas
Health Science Center at Houston

Sinauer Associates, Inc., Publishers
Sunderland Massachusetts 01375 U.S.A.

About the Cover

Internal repeats in exons 4 of human apolipoproteins A-I, A-IV, E, A-II, C-III, C-II, and C-I. Most of the repeats are 22-mers (22 residues long), each of which is made up of two 11-mers, and the other repeats are 11-mers. The colors indicate proline (orange); aspartic acid or glutamic acid (red, acidic); arginine or lysine (purple, basic); and methionine, valine, leucine, isoleucine, phenylalanine, tyrosine, and tryptophan (green, hydrophobic). The remaining amino acids (glycine, alanine, serine, threonine, asparagine, glutamine, histidine, and cysteine) are uncolored and are referred to as indifferent. A column containing 16 or more amino acids of the same group (except for the indifferent group) is said to possess that character, and the amino acids of that group are presented as solid bars. See Figure 10.5 for more details. Image by L. Chan, Baylor College of Medicine.

MOLECULAR EVOLUTION

FAX 413-549-1118
email: publish@sinauer.com

Library of Congress Cataloging-in-Publication Data
Li, Wen-Hsiung, 1942–
    Molecular evolution / Wen-Hsiung Li.
        p.   cm.
    Includes bibliographical references and index.
    ISBN 0-87893-463-4 (cloth)
        1. Molecular evolution.   I. Title.
QH325.L655   1997   96-38032
572.8'38—dc21   CIP

Printed in Canada
5 4 3 2

To the memory of my parents

Contents in Brief

Table of Contents

CHAPTER 12

Evolution by Transposition and Horizontal Transfer 335

CHAPTER 13

Genome Organization and Evolution 379

CHAPTER 14

Roles of Mutation and Selection in Molecular Evolution 419

Preface

This book represents my effort to provide a synthesis of the exciting developments in molecular evolution in the past two decades. It describes the dynamics of evolutionary change at the molecular level, the driving forces behind the evolutionary process, novel evolutionary phenomena revealed by molecular data, the effects of various molecular mechanisms on the structure of genes and genomes, and the methodology involved in the statistical analysis of molecular data from an evolutionary perspective.

The Introduction gives a brief history of molecular evolution in the pre-DNA era. The presentation is, of course, biased by my personal perspective, and some parts may presuppose a substantial background in molecular evolution, but some knowledge of the major issues and controversies that occupied much of the research activity of molecular evolutionists before 1980 should be helpful for understanding many of the topics subsequently covered in the book. Readers may wish to revisit parts of the Introduction when they are invited to in the text.

Chapter 1 provides basic knowledge of molecular biology such as the genomic structure of prokaryotic and eukaryotic genes and genetic codes, and describes some biochemical properties of amino acids and proteins. Chapter 2 is a summary of the basic theory of population genetics that is required for understanding the effects of natural selection, mutation, and random drift, and the process of gene substitution. The background material in these two chapters prepares a student for studying evolutionary problems at the molecular level in a quantitative manner. Additional molecular biology background is presented later in relevant chapters, and further population genetics theory is provided in Chapter 9.

Chapter 3 describes models of evolutionary nucleotide change in DNA sequences. A relatively detailed presentation of models is given because a comprehension of the process of nucleotide substitution is essential for understanding the process of molecular evolution and the methodologies for estimating distances between DNA sequences and reconstructing trees. Estimation of evolutionary distances is obviously a basic subject in the study of molecular evolution and is described in some detail in Chapter 4. Because of its importance, the methodology of tree reconstruction is described in detail in Chapter 5, including statistical tests of phylogenies as well as various reconstruction methods. These three chapters provide the basic tools for studying molecular evolution. They are probably the most difficult part of the book. However, I have tried to reduce much of the mathematics involved and to use intuitive arguments and examples to help readers understand these methods. A reader who finds the material difficult may first learn the simpler methods, coming back to the more sophisticated ones later. To illustrate the power of molecular approaches to phylogenetic studies, Chapter 6 provides a number of examples where molecular studies have resolved some longstanding issues in systematics, or pointed to a new direction for research.

Chapter 7 describes nonrandom codon usage and the rate and pattern of nucleotide substitution and discusses the mechanisms of molecular evolution. Examples of adaptive evolution are given, from lysozymes in cow, langur (a monkey), and hoatzin (a bird) and hemoglobins in mammals and birds. Chapter 8 deals with the controversial issue of the molecular clock hypothesis.

Chapter 9 provides methods for analyzing DNA polymorphism data and discusses issues concerning the mechanisms underlying the maintenance of DNA polymorphism in natural populations.

Chapter 10 describes the evolutionary significance of gene duplication, which includes partial and complete gene duplication, and genome duplication. It provides examples to show how gene duplication can give rise to gene families or superfamilies that contribute to the physiological or morphological complexity of an organism. It also discusses the significance of domain (or exon) shuffling. Chapter 11 deals with the concerted evolution of multigene families by gene conversion, unequal crossing-over, and other mechanisms. Several examples of concerted evolution are given.

Chapter 12 deals with transposition and retrotransposition, and horizontal gene transfer. These processes are shown to have strong impacts on both our perception of the genome and our understanding of the evolution of genes. Chapter 13 attempts to describe the molecular evolutionary process at the genomic level, including the issues of genome size and genome composition. It also deals with the long-standing issue of the origin of introns.

Chapter 14 summarizes conclusions on the relative roles of mutation and selection in molecular evolution. I have taken the author's privilege to express my personal views on the subject.

As progress in molecular evolution has been exceedingly rapid, numerous interesting studies have been conducted in the past 20 years. So, how to select the materials for a readable textbook for graduate students? My approach has been to choose materials that would fit into a logical flow: First the background materials, then the essential methodology, then the evolutionary history of organisms, and finally various phenomena and mechanisms of molecular evolution. I have avoided the temptation to cover even more topics, so that each topic selected could receive in-depth coverage.

This book is intended for advanced students and researchers in molecular evolution or molecular biology. However, I have tried to present it in a manner that minimizes prerequisite knowledge in molecular biology, genetics, and mathematics.

Acknowledgments

Writing a book while maintaining active research has proved to be far more difficult than I expected. Fortunately, I have received much help from my colleagues, postdoctoral fellows, students, and friends. For favors great and small, I thank

Ron Adkins, Stephane Boissinot, Benny Chang, Didier Casane, Brian Charlesworth, Tommy Douglas, Yun-Xin Fu, Dan Graur, Xun Gu, David Hewett-Emmett, Y. Ina, Margaret Kidwell, Julia Krushkal, John Logsdon, Tomoko Ohta, Jeffrey Palmer, Brinda Rana, Andrei Rodin, Lawrence Shimmin, Song-Kun Shyue, Joana Silva, Arlin Stoltzfus, Hongmin Sun, Susan Wessler, Chung-I Wu, and Yi-Hong Zhou. I especially thank James Crow for his many valuable suggestions for improving the presentation of the Introduction and Chapter 14.

I greatly appreciate Amanda Ko's help in the preparation of this book and Y.-S. Chung's help for drawing many figures and for computing many tables. I thank Larry Chan for allowing me to use his color slide for the front cover design. Especially, I thank Peg Riley and Bruce Walsh for thoroughly reviewing the entire manuscript and for their valuable suggestions. I am greatly indebted to Dan Graur for allowing me to use much of the materials in Li and Graur's (1991) book.

This work was partly supported by the National Institutes of Health. I also gratefully acknowledge support from the Alfred P. Sloan Foundation.

Wen-Hsiung Li
December 1996

Molecular Evolution: A Brief History of the Pre-DNA Era

MOLECULAR EVOLUTION ENCOMPASSES TWO AREAS OF STUDY: (1) the evolution of macromolecules and (2) the reconstruction of the evolutionary history of genes and organisms. The first area includes the rates and patterns of change in the genetic material (e.g., DNA sequences) and its encoded products (e.g., proteins) during evolutionary time and the mechanisms responsible for such changes. The second area, also known as "molecular phylogeny" or "molecular phylogenetics," deals with the evolutionary history of organisms and macromolecules, as inferred from molecular data.

It might appear that the two areas of study constitute independent fields of inquiry, for the object of the first is to elucidate the causes and effects of evolutionary changes in molecules, while the second uses molecules merely as a tool to reconstruct the evolutionary history of organisms and their genetic constituents. In practice, however, the two disciplines are intimately interrelated, and progress in one area facilitates studies in the other. For instance, phylogenetic knowledge is essential for determining the order of changes in the molecular characters under study, and knowing the order of such changes is usually the first step in inferring their cause. Conversely, knowledge of the pattern and rate of change of a given molecule is crucial for attempts to reconstruct the evolutionary history of a group of organisms.

Traditionally, a third area of study, prebiotic evolution or the "origin of life," is also included within the framework of molecular evolution. This subject, however, involves a great deal of speculation and is less amenable to quantitative treatment. Moreover, the rules that govern the process of information transfer in prebiotic systems (i.e., systems devoid of replicable genes) are not known at the present time. Therefore, this book will not deal with the origin of life. Interested readers may consult Oparin (1957), Eigen and Schuster (1979), Cairns-Smith (1982), Dyson (1985), Loomis (1988), Eigen (1992), and Gesteland and Atkins (1993).

The study of molecular evolution has its roots in two disparate disciplines: population genetics and molecular biology. Population genetics provides the theoretical foundation for the study of evolutionary processes, while molecular biology provides the empirical data. Thus, to understand molecular evolution it is essential to acquire some basic knowledge of both molecular biology and the theory of population genetics.

Although as a discipline molecular evolution is still at a young stage of development, the pursuit of this subject actually began at the turn of the century. Studies in immunochemistry at the end of the last century and the beginning of this showed that serological cross-reactions were stronger for more closely related organisms than for less related ones. Seeing the evolutionary implications of this finding, Nuttall (1904) conducted precipitin tests of serum proteins to infer the phylogenetic relationships among various groups of animals. He determined, for example, that man's closest relatives were the apes, followed, in order of relatedness, by Old World monkeys, New World monkeys, and prosimians. However, Nuttall was ahead of his time and intense research in molecular evolution did not start until the mid 1950s when the methods of protein sequencing, tryptic fragment pattern analysis, and starch-gel electrophoresis as well as better immunological techniques were introduced into evolutionary studies (see, e.g., Brown et al. 1955; Anfinsen 1959; Zuckerkandl et al. 1960; Goodman et al. 1960). These techniques stimulated much interest in the molecular phylogeny of humans, apes, and other primates (Goodman 1962, 1963; Zuckerkandl 1963) and made a great impact on the study of molecular phylogeny in general.

The first complete sequence of a protein (insulin) was determined in 1952 by F. Sanger and colleagues, and by the mid-1950s a substantial amount of protein sequence information had been accumulated (see Harris et al. 1956 and references therein). The data revealed that amino acid substitutions occurred nonrandomly among different regions of a protein; for example, among the cattle, sheep, pig, horse, and whale insulins, changes were restricted to positions 8–10 of the A chain. The data also revealed that the majority of amino acid substitutions found in the same protein from different species did not seem to have significant effect on its biological activities. Yet a small number of amino acid differences may account for the large differences in biological activities of two different but related proteins; for example, cattle vasopressin and oxytocin differ by only two amino acids (Acher and Chauvet 1953; Du Vigneaud et al. 1953) but have very different biological properties.

The 1960s were a period of rapid progress and heated controversies in molecular evolution. Considerable sequence data for hemoglobins and cytochromes *c* had become available in the early 1960s and comparative studies of the data suggested that the rate of amino acid substitution in each of these proteins was approximately the same among different mammalian lineages (Zuckerkandl and Pauling 1962, 1965; Margoliash 1963). Zuckerkandl and Pauling (1965) therefore proposed that for any given protein the rate of molecular evolution is approximately constant in all evolutionary lineages or, in other words, that there is a molecular clock. This proposal immediately stimulated a great deal of interest in the use of macromolecules for evolutionary studies. Indeed, if proteins evolve at constant rates, they can be used to estimate the dates of species divergence and to reconstruct phylogenetic relationships among organisms. This proposal has had a tremendous impact on the development of molecular evolution.

The molecular clock hypothesis, however, also provoked a great deal of controversy because the concept of rate constancy is diametrically opposed to the erratic tempo of evolution at the morphological and physiological levels. The rates of morphological evolution in different evolutionary lineages and at different times in the same lineage are strikingly variable (Simpson 1944).

The controversy became particularly heated after the rate-constancy assumption was used as a basis for estimating the divergence times between man and the African apes (Sarich and Wilson 1967), leading to values much younger than traditional dates, and after it was taken as strong support for the neutral mutation hypothesis (Kimura 1969). Another controversy, stimulated by both protein sequence and DNA hybridization data, arose over whether the length of generation time has any significant effect on the rate of molecular evolution (see Laird et al. 1969; Kohne 1970; Wilson et al. 1977). In the 1970s statistical analyses of protein sequence data showed that amino acid substitution usually did not follow a simple Poisson process (Ohta and Kimura 1971; Langley and Fitch 1974). Nevertheless, the rate of substitution was roughly constant over time and the approximate constancy seemed more striking than the relatively minor deviations therefrom (Dickerson 1971; Wilson et al. 1977); also see Gillespie and Langley (1979) for a discussion of why deviations from the Poisson process may not be used to exclude the constancy of evolutionary rate. Further support for the hypothesis came from quantitative immunological comparisons of proteins (see Wilson et al. 1977). Therefore, by the late 1970s the molecular clock hypothesis was accepted by many molecular evolutionists (Kimura and Ohta 1974; Nei 1975; Wilson et al. 1977; Kimura 1979). However, some others, particularly Goodman (1976, 1981), continued to challenge it.

The 1960s also witnessed a revolution in population genetics. The availability of protein sequence data removed the species barrier in population genetic studies—traditional genetic analysis had been restricted to groups that could be crossed. For the first time, molecular studies had provided adequate data for examining the theory of gene substitution. Extrapolating the rate of nucleotide substitution estimated from several protein sequences to the entire mammalian genome, Kimura (1968a) arrived at a genomic rate that he thought was much too high to be compatible with Haldane's (1957) principle of the cost of natural selection. He therefore proposed that the majority of molecular changes in evolution are due to random drift of neutral or nearly neutral mutations. ("Nearly neutral" means that the selective differences between alleles are of the order of the reciprocal of the species population number or less, so that the effect of selection on the change in allele frequencies is similar to or weaker than that of random drift.) The same hypothesis was proposed independently by King and Jukes (1969). At the same time, the discovery that natural populations contain large amounts of protein polymorphism detectable by electrophoresis (Harris 1966; Lewontin and Hubby 1966) also raised many new issues, especially with respect to the mechanism of maintenance of high levels of polymorphism in populations. In the classical view, evolution and polymorphism were regarded as quite different processes, the former being dynamic whereas the latter is static. Kimura (1968a,b) proposed instead that most of the polymophism is due to neutral or nearly neutral mutations and is not maintained permanently in the population, that is to say, polymorphism is a transient phase of molecular evolution (Kimura and Ohta 1971a). The hypothesis proposed by Kimura and King and Jukes has been known as the neutral mutation hypothesis or the neutral theory of molecular evolution.

Under this hypothesis, evolution at the molecular level is largely driven not by natural selection but by mutation and random drift. Over a long time period the average rate of substitution is equal simply to the mutation rate of neutral mutation, if both are measured in the same units, e.g., years or generations. This provided a theoretical basis of rate constancy from the neutralist view; that is, if the majority of amino acid or nucleotide substitutions in evolution are indeed due to neutral or nearly neutral mutations and if the rate of such mutation per unit time is constant over time, then the rate of evolution will be approximately constant.

The neutral mutation hypothesis provoked a great controversy, for it contradicted the traditional view that natural selection is responsible for both within-species polymorphisms and between-species differences. A number of issues quickly arose (see the reviews by Lewontin 1974; Nei 1975; Ayala 1976; Kimura 1983; Gillespie 1991). First, there were challenges (Brues 1969; see also Van Valen 1963) to the concept that gene substitution creates a cost to the population, and there was divergence of opinion as to how to calculate the cost of natural selection when a large number of loci are involved (King 1967; Milkman 1967; Sved et al. 1967; Maynard Smith 1968; Kimura and Ohta 1971b; Nei 1971a).

Second, after the rate-constancy assumption was used to support the neutral mutation hypothesis (Kimura 1969), the validity of the assumption became a major issue and the rejection of the assumption (i.e., the molecular clock hypothesis) has sometimes been mistaken as a rejection of the neutral mutation hypothesis. After all, the mutation rate per unit time and other factors need not have remained constant.

Third, there was a debate on whether certain types of mutation are really selectively neutral. For example, synonymous mutations, which cause no amino acid changes, were then considered by neutralists as good candidates for neutral mutations (Kimura 1968b; King and Jukes 1969). This view was challenged by Clarke (1970a) and Richmond (1970), who argued that different synonymous codons for an amino acid may have different fitnesses because the tRNAs recognizing different codons may have different concentrations in the cell and because a tRNA recognizing multiple synonymous codons may have different binding affinities for different codons.

Fourth, studies on the pattern and rate of amino acid substitution in protein evolution revealed that amino acids that are similar in physicochemical properties are interchanged more frequently than dissimilar ones (Zuckerkandl and Pauling 1965; Epstein 1967; Clarke 1970b) and that functionally less important molecules or parts of a molecule evolve faster than more important ones (King and Jukes 1969; Dickerson 1971; Dayhoff 1972). These observations suggest that mutations with smaller effects on protein structure or function have greater chances of fixation (i.e., spreading through the species) than those with more radical effects. Clarke (1970a, 1971) submitted that these observations are consistent with the selectionist view because mutations with smaller effects would have higher chances of being beneficial (Fisher 1930). Kimura and Ohta (1974), however, reasoned that mutations, if their effects are very small, are likely to have very small selective advantages and therefore have correspondingly small probabilities of fixation. They as well as Jukes and King (1971) argued that the above observations can be much more simply explained under the neutralist view, namely that the smaller the effect a mutation has, the higher the probability will be that it is selectively neutral. This latter argument was apparently more acceptable to biochemists and molecular biologists. For this reason and because of the substantial support gained in the 1970s for the rate-constancy assumption, the neutral mutation hypothesis had gained considerable acceptance before 1980 (see Kimura 1983).

Fifth, Ohta's (1973, 1974) suggestion of the importance of slightly deleterious mutations in molecular polymorphism and evolution has attracted much attention—given the fact that most mutations in genes are deleterious, one would expect slightly deleterious mutations to be frequent. However, whether the slightly deleterious hypothesis can explain the patterns of molecular polymorphism and evolution is a subject of debate (see Nei 1987).

The controversy over the neutral mutation hypothesis has had two strong impacts on molecular evolution and population genetics. First, it has led to the general recognition that the effect of random drift cannot be neglected when considering the evolutionary dynamics of molecular changes (see Kimura 1983; Crow 1985; Nei 1987). Second, the fusion of molecular evolution and population genetics, which began with the availability of protein sequence and electrophoretic data, was much accelerated by the introduction of the concept that molecular evolution and polymorphism are two facets of the same phenomenon (Kimura and Ohta 1971a). Thus, an adequate theory of evolution at the molecular level must be consistent with both. Furthermore, because it is simple and testable, the neutral hypothesis has been of great value as a starting point of analysis, i.e., it can serve as a null hypothesis.

The 1960s and 1970s also were years of tremendous progress in molecular phylogenetics. The accumulation of protein sequences, which were more informative and easier to analyze than other types of molecular data then available, provided for the first time adequate data for studying long-term evolution such as evolutionary relationships among orders or beyond. These data stimulated many studies on the reconstruction of phylogenetic trees and on the development of tree-making methods (Eck and Dayhoff 1966; Fitch and Margoliash 1967; Dayhoff 1972). In particular, the finding that the tree reconstructed from the sequences of a *single* protein (cytochrome *c*) was similar to the species tree then known for many vertebrate and invertebrate species (Fitch and Margoliash 1967) revealed the power of molecular phylogenetics. Moreover, the data also stimulated much interest in the evolutionary relationships among protein sequences and in the methodology of sequence alignment (Needleman and Wunsch 1970; Dayhoff 1972, 1978; Sellers 1974; Waterman et al. 1976). On the other hand, although the method of protein electrophoresis is less accurate than that of protein sequencing, it is far less time-consuming and was therefore extensively used in the study of phylogenetic relationships among relatively closely related populations or species (e.g., Johnson and Selander 1971; Lakovaara et al. 1972; Nei and Roychoudhury 1974). The application of this method prompted the development of measures of genetic distance and methods of tree-making (e.g., Cavalli-Sforza and Edwards 1967; Nei 1971b, 1972; Rogers 1972; Felsenstein 1973); for example, Nei's (1972) genetic distance has greatly facilitated the study of evolutionary relationships among populations or closely related species. In addition, immunological techniques such as the method of microcomplement fixation were also extensively used in phylogenetic studies (e.g., Sarich and Wilson 1966, 1967; Maxon and Wilson 1975). The DNA hybridization method was also extensively used, particularly in the study of primate evolution and bird systematics (Sibley and Ahlquist 1984, 1990).

The rapid progress in the study of molecular evolution has been greatly facilitated by the development of high-speed computers. Most of the procedures used in molecular evolution and population genetic analysis were computation-intensive and could not have been routinely done without computers. Thanks to their ever-increasing speed and accessibility, analysis methods became more and more sophisticated. Only recently has it been possible to use some of the more rigorous methods such as maximum likelihood for the construction of trees.

This is a brief history of molecular evolution in "the pre-DNA era." The advent of various DNA techniques such as restriction analysis, gene cloning, polymerase chain reaction, and DNA sequencing techniques since the 1970s has brought an explosion of knowledge of molecular biology and a new era of exciting development in the study of molecular evolution. A better understanding of gene structure and genomic organization and the availability of a large body of molecular data have allowed a much more extensive and closer examination of long-standing issues and have raised numerous new ones. This book will discuss many of these issues and also provide a theoretical background for analyzing molecular data and for understanding arguments involved in many issues.

CHAPTER 1

Gene Structure, Genetic Codes, and Mutation

THIS CHAPTER PROVIDES SOME BASIC BACKGROUND in molecular biology that is required for studying evolutionary processes at the DNA level. The genomic structure of prokaryotic and eukaryotic genes and genetic codes are described, as are some basic properties of amino acids and proteins. Furthermore, various types of mutation and the pattern of point mutation inferred from nucleotide substitutions in noncoding sequences are discussed. Additional molecular biology background will be provided in the relevant chapters.

NUCLEOTIDE SEQUENCES

The hereditary information of all living organisms, with the exception of some viruses, is carried by **deoxyribonucleic acid** (**DNA**) molecules. DNA usually consists of two complementary chains twisted around each other to form a right-handed helix. Each chain is a linear polynucleotide consisting of four nucleotides. There are two **purines**: **adenine** (**A**) and **guanine** (**G**), and two **pyrimidines**: **cytosine** (**C**) and **thymine** (**T**). The two chains are joined together by hydrogen bonds between pairs of nucleotides. Adenine pairs with thymine by means of two hydrogen bonds, also referred to as the **weak bond**, and guanine pairs with cytosine by means of three hydrogen bonds, the **strong bond** (Figure 1.1).

Each nucleotide in a DNA sequence contains a pentose sugar (deoxyribose), a phosphate group, and a purine or a pyrimidine base. The backbone of the DNA molecule consists of sugar and phosphate moieties, which are covalently linked together by asymmetrical 5′—3′ phosphodiester bonds. Consequently, the DNA molecule is polarized, one end having a phosphoryl radical (—P) on the 5′ carbon of the terminal nucleotide, the other possessing a free hydroxyl (—OH) on the 3′ carbon of the terminal nucleotide. The direction of the phosphodiester bonds is what determines the molecule's character; thus, for instance, the sequence 5′—G–C–A–A–T—3′ is different from the sequence 3′—G–C–A–A–T—5′. By convention, DNA sequences are written in the order they are transcribed, i.e., from the 5′ end to the 3′ end, also referred to as the **upstream** and **downstream** directions, respectively. The double helical form of DNA has two strands in antiparallel array (Figure 1.2).

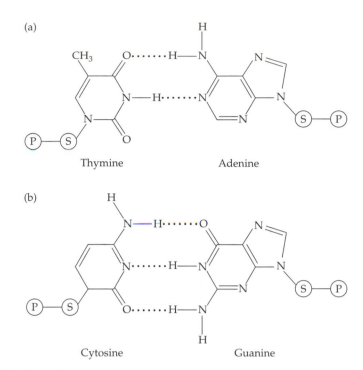

(a)

Thymine   Adenine

(b)

Cytosine   Guanine

Figure 1.1   Complementary base pairing by means of hydrogen bonds (dotted lines) between (a) thymine and adenine (weak bond), and (b) cytosine and guanine (strong bond). Ⓟ, phosphate; Ⓢ, sugar. From Li and Graur (1991).

Ribonucleic acid (RNA) is found as either a double- or a single-stranded molecule. RNA differs from DNA by having ribose as its backbone sugar moiety, instead of deoxyribose, and by using the nucleotide uracil (U) instead of thymine (T). In RNA, in addition to the canonical complementary base pairs G:C and A:U, the G:U pair is also stable; in DNA, stable pairing does not occur between G and T.

Adenine, cytosine, guanine, and thymine/uracil are referred to as the standard nucleotides. Some functional RNA molecules, most notably tRNAs, contain nonstandard nucleotides (i.e., chemical modifications of standard nucleotides) that have been introduced into the RNA after transcription.

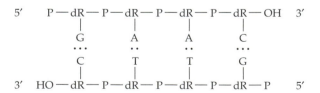

Figure 1.2   Schematic representation of the antiparallel structure of double-stranded DNA. P, phosphate; dR, deoxyribose; OH, hydroxyl; A, adenine; G, guanine; C, cytosine; T, thymine; —, covalent bond; ‥, weak bond; ⋯, strong bond. From Li and Graur (1991).

GENE STRUCTURE

Traditionally, a gene was defined as a segment of DNA that codes for a polypeptide chain or specifies a functional RNA molecule. Recent molecular studies, however, have altered our perception of genes, making a somewhat vaguer definition necessary. Accordingly, a gene is a sequence of genomic DNA or RNA that performs a specific function. Performing the function may not require the gene to be translated or even transcribed.

At present, three types of genes are recognized: (1) **protein-coding genes**, which are transcribed into RNA and subsequently translated into proteins, (2) **RNA-specifying genes**, which are only transcribed, and (3) **regulatory genes** (Cavalier-Smith 1985; Watson et al. 1987; Lewin 1994). According to a narrow definition, the third category includes only untranscribed sequences. Transcribed regulatory genes essentially belong to one of the first two categories. Protein-coding genes and RNA-specifying genes are also referred to as **structural genes**. Note that some authors restrict the definition of structural genes to include only protein-coding genes.

The transcription of both protein-coding genes and RNA-specifying genes in bacteria is carried out by one type of DNA-dependent RNA polymerase. In contrast, in nuclear genomes of eukaryotes, three types of RNA polymerases are employed (Watson et al. 1987; Lewin 1994). Ribosomal RNA (rRNA) genes are transcribed by RNA polymerase I (Pol I), protein-coding genes by RNA polymerase II (Pol II), and small cytoplasmic RNA (scRNA) genes, such as genes specifying transfer RNAs (tRNAs), by RNA polymerase III (Pol III). Some small nuclear RNA (snRNA) genes (e.g., U2) are transcribed by Pol II, others (e.g., U6) by Pol III. The snRNA gene U3 is transcribed by Pol II in vertebrates and lower eukaryotes, but by Pol III in plants (Kiss et al. 1991).

Protein-Coding Genes

A typical eukaryotic protein-coding gene consists of transcribed and nontranscribed parts (Figure 1.3a). The nontranscribed parts are designated according to their location relative to the protein-coding regions as 5' and 3' **flanking sequences**. The 5' flanking sequence contains several signals (specific sequences) that determine the initiation, tempo, and timing (or tissue specificity) of the transcription process. Because these regulatory sequences promote the transcription process, they are also referred to as **promoters**, and the region in which they reside is called the **promoter region**. The promoter region usually consists of the following signals: the **TATA box** located 19–27 base pairs (bp) upstream of the starting point of transcription, the **CAAT box** farther upstream, and one or more copies of the **GC box**, consisting of the sequence GGGCGG or its variants and surrounding the CAAT box (Figure 1.3a). The TATA box, which is usually surrounded by GC-rich sequences, controls the choice of the starting point of transcription. The CAAT and GC boxes, which may function in either orientation, control the initial binding of RNA polymerase II to DNA. Note, however, that none of the above signals is uniquely essential for promoter function. Some genes do not possess a TATA box and thus may not have a unique starting point of transcription; for example, many housekeeping-enzyme genes such as the hypoxanthine phosphoribosyl transferase and dihydrofolate reductase genes have a GC-rich region instead of the TATA box (see Dynan 1986). Other genes possess

(a)

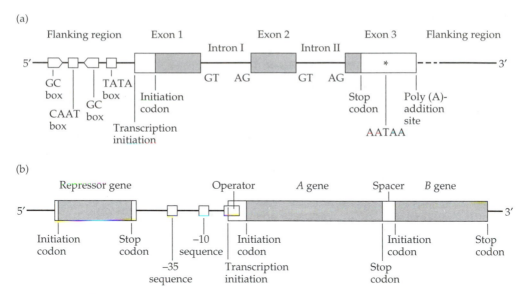

Figure 1.3 (a) Schematic structure of a typical eukaryotic protein-coding gene. Note that, by convention, the 5′ end is at the left. Rectangles denote exons; a white area in a rectangle denotes a transcribed but untranslated region, while a shaded area denotes a translated region. * denotes the site of the poly(A)-addition signal AATAA. (b) Schematic structure of an induced prokaryotic operon. Genes *A* and *B* are protein-coding genes and are transcribed into a single messenger RNA. The repressor gene encodes a repressor protein, which binds to the operator and prevents the transcription of the structural genes by blocking the movement of the RNA polymerase. The operator is a DNA region with at least 10 bases, which may overlap the transcribed region of the genes in the operon. By binding to an inducer (a small molecule), the repressor is converted to a form that cannot bind to the operator. Then RNA polymerase can initiate the transcription of the genes *A* and *B* in the operon (see Lewin 1994). In both (a) and (b), the regions are not drawn according to scale. From Li and Graur (1991).

neither a CAAT box nor any GC box, and their transcription initiation is controlled by other elements in the 5′ flanking region. The 3′ flanking sequence contains signals for the termination of the transcription process and poly(A)-addition. Presently, it is still difficult to delineate with precision the points at which a gene begins and ends.

In addition to the signals in the promoter region, there is another type of signal, called the enhancer, that is located outside the promoter region but can enhance the initiation of transcription (see Lewin 1994). An enhancer may consist of several modular elements, may function in either direction, and can be in the 5′ flanking region, in an intron, or in the 3′ flanking region. Such elements may be targets for tissue-specific or temporal regulation.

The transcription of protein-coding genes in eukaryotes starts at the **transcription initiation site** or (the **cap site** in the RNA transcript), and ends at the **termination site**, which may or may not be close to the **polyadenylation** or **poly(A) addition site** of the mature messenger RNA (mRNA) molecule. In other words, termination of transcription may occur farther downstream from the poly(A) addition site. The transcribed RNA, also referred to as **pre-messenger RNA (pre-mRNA)**, contains 5′ and 3′ **untranslated regions**, **exons**, and **introns**. Introns, or intervening sequences, are those transcribed sequences that are excised during the processing of the pre-mRNA molecule. All genomic sequences that remain in the mature mRNA following splicing are referred to as exons. Exons or parts of exons that are translated are referred to as protein-coding exons or **coding regions**. The DNA strand from which the pre-mRNA is transcribed is called the antisense strand. The untranscribed complementary strand, the sequence of which is identical to that of the pre-mRNA, is called the sense strand.

There are several types of introns, characterized by the specific mechanism with which the intron is cleaved out of the primary RNA transcript (see Lewin 1994). Here we are concerned only with the introns in the nuclear genes that are transcribed by RNA polymerase II. These introns are called **spliceosomal** introns and are enzymatically cleaved out of the pre-mRNA during the maturation of the molecule. They are different from group I and group II introns, which have self-splicing activities and are found in mitochondrial and chloroplast DNA (see Lewin 1994). The **splicing sites** or **junctions** of spliceosomal introns are probably determined by nucleotides at the 5′ and 3′ ends of each intron, known as the **donor** and **acceptor** sites, respectively. For instance, all eukaryotic nuclear introns begin with GT and end in AG (the **GT–AG rule**), and these sequences have been shown to be essential for correct excision splicing (Figure 1.3a), though there are exceptional cases. Exon sequences adjacent to introns may contribute to the determination of the splicing site. In addition, each intron contains a specific sequence, called the **TACTAAC box**, located about 30 bp upstream of the 3′ end of the intron. This sequence is well conserved in yeast, though it is more variable in higher eukaryotes. The splicing involves the cleavage of the 5′ splice junction and the creation of a phosphodiester bond between the G at the 5′ end of the intron and the A in the sixth position of the TACTAAC box. Subsequently, the 3′ splice junction is cleaved and the two exons are ligated.

The number of introns varies from gene to gene. Some genes possess dozens of introns, and some introns may be thousands of nucleotides long. Others (e.g., most histone genes) are devoid of introns altogether. Exons are not distributed evenly over the length of the gene. Some exons are clustered; others are located at great distances from neighboring exons. One such example is shown in Figure 1.4. The majority of nuclear protein-coding genes in higher eukaryotes consist mostly of introns. Not all introns interrupt coding regions. Some occur in untranslated regions, mainly in the region between the transcription initiation site and the initiation codon.

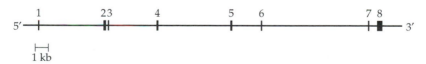

Figure 1.4   The localization of the eight exons in the human factor-IX gene. The vertical bars represent the eight exons. Only the transcribed region is shown. The exons and introns are drawn to scale. The total length of the exons is 1,386 nucleotides, as opposed to a total of length of introns of 29,954 nulecotides. From Li and Graur (1991).

Protein-coding genes in eubacteria are different from those in eukaryotes in several respects. Most importantly, they do not contain introns (Figure 1.3b). Promoters in eubacteria contain a **−10 sequence** and a **−35 sequence**, so named because they are located, respectively, 10 bp and 35 bp upstream of the initiation site of transcription. The former, also known as the **Pribnow box**, has the sequence TATAAT or its variant, while the latter has the sequence TTGACA or its variant. A prokaryotic promoter may also contain other specific sequences farther upstream from the −35 sequence.

Several structural genes in prokaryotes may be arranged consecutively to form a unit of gene expression (a polycistron) that is transcribed into one molecule of mRNA and subsequently translated into different proteins. Such a unit usually contains genetic elements that control the coordinated expression of the genes belonging to the unit. The entire arrangement of genes is called an **operon** (Figure 1.3b).

RNA-Specifying Genes

The structure of RNA-specifying genes is usually similar in eukaryotes and prokaryotes. These genes generally contain no introns. However, in some organisms, such as ciliates, slime molds, and some bacteria, RNA-specifying genes may contain introns that must be spliced out before the RNA molecule becomes functional. Sequence elements involved in the regulation of transcription of some RNA-specifying genes are sometimes included within the sequence specifying the functional end product. In particular, all eukaryotic tRNA specifying genes contain an internal transcriptional start recognized by RNA polymerase III.

Many RNA molecules are modified following transcription. Such modifications include the incorporation of standard and nonstandard nucleotides, modification of standard nucleotides into nonstandard ones, and the enzymatic addition of terminal sequences of ribonucleotides to either the 5′ or the 3′ end.

Regulatory Genes

Our knowledge of regulatory genes is much less advanced than that of the other types of genes. Traditionally, these genes would be called regulatory signals or sequences because they are not transcribed. Several such types of genes or families of genes have been tentatively identified (see Cavalier-Smith 1985). (1) **Replicator genes** specify the sites for initiation or termination of DNA replication. For example, this term may apply to the specific sequences at the origin of each replicon, which is a unit of DNA replication in a genome. These sequences may serve as binding sites for specific initiator or repressor molecules. (2) **Telomeres** are short repetitive sequences at the end of eukaryotic chromosomes; e.g., the repeat sequence is CCCTAA in humans and CCCTAAA in the plant *Arabidopsis* (see Zakian 1989 for sequences in other organisms). The telomere provides a protective "cap" for the end of a chromosome, so that it is not "sticky" and does not react with the end of another chromosome and is not susceptible to exonucleolytic degradation (see Zakian 1989; Lewin 1994). (3) **Segregator genes** provide specific sites for attachment of the chromosomes to the segregation machinery during meiosis and mitosis. This type of gene includes centromeric sequences, which have been found to consist of short repetitive sequences similar to those of telomeres (see Richards et al. 1991). (4) **Recombinator genes** provide specific recognition sites for recombination enzymes during meiosis. Many more regulatory genes may exist, some of which could be independent of structural genes in both function and location, and could be involved in complex regulatory functions, such as the ontogenetic development of multicellular organisms.

AMINO ACIDS

Amino acids are the basic structural units of proteins. All proteins in all organisms, from bacteria to humans, are constructed from the 20 amino acids listed in Table 1.1; for rare and posttranslationally modified amino acids of proteins, see Wold (1981) and Lehninger (1982).

Each amino acid has an NH_2 (amine) group and a COOH (carboxylic) group on either side of a central carbon atom, which is called the α carbon or C_α (Figure 1.5). Also attached to the α carbon are a hydrogen atom and a distinct **R group**, which is also referred to as a **side chain** (Table 1.1). The side chain varies in size, shape, charge, hydrogen-bonding capacity, and chemical reactivity, and it is the side chain that makes one amino acid different from another. In fact, classification of amino acids is usually made on the basis of their R groups.

Based on the polarity of their R groups at pH 6.0 to 7.0 (the zone of intracellular pH), the 20 amino acids can be classified into four main classes: (1) **nonpolar** or **hydrophobic** groups; (2) **neutral** (**uncharged**) **polar** groups; (3) **positively charged** groups; and (4) **negatively charged** groups. However, as we shall see, the four groups are not clear-cut and certain amino acids (e.g., glycine and tyrosine) straddle the boundary between categories. The following description is largely from Lehninger (1982), Dickerson and Geis (1983), and Stryer (1988).

Amino Acids with Nonpolar (Hydrophobic) R Groups

There are eight amino acids in this category. Alanine has a methyl R group; leucine, isoleucine, valine, and proline have an aliphatic (i.e., having an open-chain structure but no aromatic ring) hydrocarbon group; phenylalanine and tryptophan have an aromatic (benzene-like) ring (Table 1.1); and methionine contains a sulfur within the hydrocarbon chain that acts like a CH_2 group. In general, these amino acids are less soluble in water than the amino acids with polar R groups.

Hydrophobic ("water-hating") amino acids prefer to be in the internal part of a protein molecule, away from the aqueous environment of the cell. Thus, they are also called **internal** amino acids (see Dickerson and Geis 1983). Alanine is the

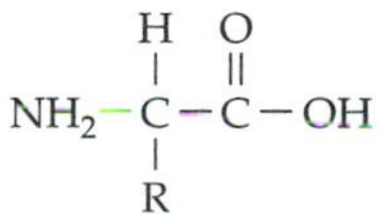

$$NH_2-\overset{\overset{\displaystyle H}{|}}{C}-\overset{\overset{\displaystyle O}{\|}}{C}-OH$$
$$\underset{\displaystyle R}{|}$$

Figure 1.5   Structure of an amino acid. It contains a central α carbon, an amine group, a carboxyl group, a hydrogen, and a side chain denoted by R.

TABLE 1.1 Primary amino acids and their three- and one-letter abbreviations, R groups, polarity, and size

Name	Three- and one-letter code	R group (side chain)	Polarity	Molecular weight
Alanine	Ala, A	$-CH_3$	Nonpolar	89.10
Leucine	Leu, L	$-CH_2CH(CH3)_2$	Nonpolar	131.18
Isoleucine	Ile, I	$-CHCH_2CH_3$ $\mid$ CH_3	Nonpolar	131.17
Valine	Val, V	$-CH(CH_3)_2$	Nonpolar	117.15
Proline	Pro, P	(structure: HOC(=O)—CH—CH₂ with HN and C—CH₂ ring) (complete structure)	Nonpolar	115.13
Phenylalanine	Phe, F	$-CH_2-$ (phenyl ring)	Nonpolar	165.19
Tryptophan	Trp, W	$-CH_2-$ (indole ring, with N–H)	Nonpolar	204.23
Methionine	Met, M	$-CH_2CH_2SCH_3$	Nonpolar	149.21
Glycine	Gly, G	$-H$	Polar(?)	75.07
Serine	Ser, S	$-CH_2OH$	Polar	105.10
Threonine	Thr, T	$-CHOH$ $\mid$ CH_3	Polar	119.12
Tyrosine	Tyr, Y	$-CH_2-$ (phenyl ring)$-OH$	Polar	181.19
Cysteine	Cys, C	$-CH_2S-H$	Polar	121.20
Asparagine	Asn, N	$-CH_2CONH_2$	Polar	132.10
Glutamine	Gln, Q	$-CH_2CH_2CONH_2$	Polar	146.15
Aspartic acid	Asp, D	$-CH_2CO_2H$	− Charged	133.10
Glutamic acid	Glu, E	$-CH_2CH_2CO_2H$	− Charged	147.10
Lysine	Lys, K	$-CH_2CH_2CH_2CH_2NH_2$	+ Charged	146.19
Arginine	Arg, R	$-CH_2CH_2CH_2NH-C(=NH)-NH_2$	+ Charged	174.20
Histidine	His, H	$-CH_2-$ (imidazole ring: HC—N, C, CH, N–H)	+ Charged(?)	155.16

least hydrophobic member of the nonpolar class and is near the borderline between nonpolar amino acids and those with uncharged polar R groups. Thus, it does not have a strong preference to be internal or **external**. Such an amino acid is said to be **ambivalent** or **indifferent** (see Dickerson and Geis 1983).

Proline differs from all the other standard amino acids in that its side chain loops back to make a second connection to its amine nitrogen (Table 1.1). This ring forces a contorted bend on the polypeptide chain. Thus, proline is especially useful in reversing the direction of a peptide backbone.

Amino Acids with Uncharged Polar R Groups

This class includes seven amino acids whose R groups contain neutral (uncharged) polar functional groups that can hydrogen-bond with water. Glycine, the borderline member of this group, is sometimes classified as a nonpolar amino acid, but its R group, a single hydrogen atom, is too small to influence the high degree of polarity of the α-amino and α-carboxyl groups. It is indifferent with respect to preference to be internal or external (see Dickerson and Geis 1983).

Serine and threonine have an aliphatic hydroxyl group. They are neither so hydrophobic nor so polar as to have a strong preference to be internal or external; they, too, are indifferent.

Tyrosine has an aromatic hydroxyl group and cysteine has an aliphatic sulfhydryl (SH) group. These R groups tend to lose protons by ionization far more readily than the R groups of other amino acids of this class. However, since they are only slightly ionized at pH 7.0, cysteine and tyrosine also belong to the indifferent group (see Dickerson and Geis 1983). Indeed, tyrosine is placed in the hydrophobic class by many authors (e.g., Fitch 1977a). In extracellular proteins cysteine usually occurs in its oxidized form cystine, in which the thiol groups of two molecules of cysteine have been oxidized to a disulfide group to provide a covalent cross-linkage between them. In intracellular proteins, the redox potential prevents disulfide bonds from forming.

Asparagine and glutamine are derived, respectively, from aspartic acid and glutamic acid by replacing the OH group with NH_2, i.e., they are the amides of aspartic acid and glutamic acid. They are easily hydrolyzed by acid or base to aspartic acid and glutamic acid, if they are not buried inside a polypepetide. Since they have very polar side chains, they are generally found on the outside of the molecule.

Amino Acids with Negatively Charged (Acidic) R Groups

The two members of this class are aspartic acid and glutamic acid, each with a carboxyl R group that is fully ionized and thus negatively charged at pH 6 to 7 (Figure 1.6). The two amino acids are hydrophilic and have a strong preference to be on the external part of a protein.

Amino Acids with Positively Charged (Basic) R Groups

Lysine has an amine-containing R group, whose nitrogen electron pair can attract a proton and acquire a positive charge (Figure 1.6). Arginine has a guanidino R group (CH_4N_2, see Table 1.1), which can acquire a proton and become positively

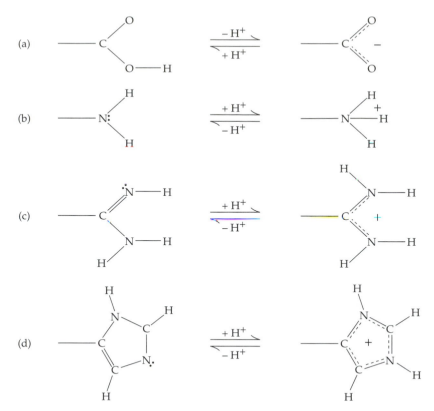

Figure 1.6   Acidic and basic groups found in protein side chains. (a) Aspartic and glutamic acids contain the carboxyl group, which can lose a hydrogen atom or proton and acquire a negative charge. (b) Lysine side chains have an amine group whose nitrogen electron pair can attract a proton and hence acquire a positive charge. (c) Arginine has a more complicated guanidino group, but the principle of acquiring an H⁺ ion is the same. (d) Histidine can attract a proton to one of its imidazole ring nitrogens. Dashed lines in these drawings represent electrons that are shared or delocalized among several atoms, lowering the overall energy of the group and making it more stable. From Dickerson and Geis (1983).

charged. Both amino acids have a net positive charge at pH 7.0. Histidine can attract a proton to one of its imidazole ($C_3H_3N_2$) ring nitrogens (Table 1.1). However, whether histidine is positively charged or neutral depends on its local environment. At pH 6.0, somewhat over 50 percent of histidine molecules possess a protonated, positively charged R group, but at pH 7.0, less than 10 percent have a positive charge. Histidine as well as lysine and arginine are hydrophilic and almost commonly external.

Among the 20 amino acids, glycine has the smallest molecular weight (75) because its R group has only a single hydrogen atom. This makes it particularly useful in tight corners in the interior of a protein where there is no room for a bulky group. In addition, like proline, it is an α-helix breaker (see next section). The largest amino acid is tryptophan, which has a molecular weight of 204. The molecular weights of the other amino acids are given in Table 1.1. The average molecular weight of amino acids in proteins is around 100–110.

Two amino acids are said to be similar if they have similar physicochemical properties. These properties include polarity and volume (size). It has been suggested that composition is also an important property for determining similarity (Grantham 1974). The "composition," c, of an amino acid is defined as the atomic weight ratio of noncarbon elements in end groups or rings to carbons in the side chain. For example, for the serine side chain —COH, $c = \frac{17}{12}$ (atomic weight of hydroxyl over that of carbon) and for lysine —CCCCNH$_2$, $c = \frac{16}{48}$ (amino over four carbons). An exchange between two amino acids are said to be conservative if the two amino acids have similar physicochemical properties (e.g., Leu and Ile) but nonconservative if otherwise (e.g., Leu and Ser).

PROTEIN STRUCTURE

A protein consists of one or more polypeptide chains, each of which has a unique, precisely defined amino acid sequence. In a polypeptide chain, the α-carboxyl group of one amino acid is joined to the α amino by a peptide bond. For example, Figure 1.7 shows the formation of a dipeptide (glycylalanine) from two amino acids (glycine and alanine) with the loss of a water molecule. Note that in this dipeptide, glycine acts as the carboxylic acid, while alanine acts as the amine. If the two roles are reversed, then the dipeptide is alanylglycine and is different from the dipeptide, glycylalanine. Thus, a polypeptide has a direction; the standard presentation of a sequence is from left to right, usually in terms of the one-letter code. For instance, glycylalanine can be presented as G.A or GA and alanyl-glycine as A.G or AG. The end containing the free amino group (i.e., the "left" end) of a polypeptide is termed the amino terminal or N-terminal end, while the end containing the free carboxyl group is termed the carboxyl terminal or C-terminal end. The amino acid sequence of a polypeptide or protein is called the **primary structure**. Polypeptides vary greatly in size, usually from 50 to 1000 amino acids long. Each amino acid unit in a polypeptide is called a residue.

A polypeptide chain rarely exists in a randomly flexed, stringlike manner. Rather, noncovalent forces (e.g., those associated with hydrogen bonds) often force regions of a polypeptide into some particular flexed conformations or structures. This level of protein structure is called the **secondary structure**. Two periodic structures, the α helix and the β-pleated sheet (or simply β sheet), are of particular interest.

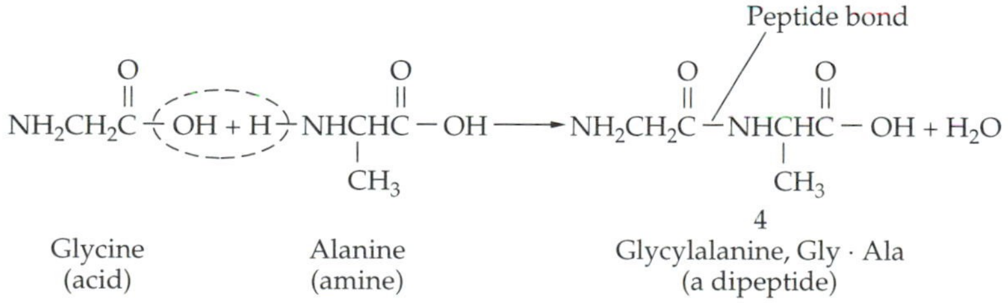

Figure 1.7 Formation of a dipeptide from two amino acids. From Holum (1978).

In the α helix (Figure 1.8), which is a rodlike structure, the main chain coils as a right-handed screw with all the side chains sticking outward in a helical array. A hydrogen bond extends from the oxygen of the CO group of each amino acid to the hydrogen of the NH group of the amino acid that is situated four residues ahead in the linear sequence. Although these hydrogen bonds are individually weak forces of attraction, they collectively add up and lead to a periodic structure. In this structure each residue is related to the next residue by a translation of 1.5 Å along the helix axis and a rotation of 100 degrees (Figure 1.8a). Because this rotation gives 3.6 amino acid residues per turn of helix, an interesting consequence is that amino acids spaced three or four apart in the linear sequence are spatially quite close to each other, whereas those spaced two apart are situated on opposite sides of the helix and so do not make contact.

In contrast to the tightly coiled structure in the α helix, the polypeptide chain in the β-pleated sheet is almost fully extended. In fact, the axial distance between adjacent amino acids is 3.5 Å, more than twice the distance (1.5 Å) in the α helix. The β sheet is formed when two or more polypeptide chains are laid down next to one another, either parallel or antiparallel, and then cross-connected by CO···HN hydrogen bonds at right angles to the chain direction (Figure 1.9). Note

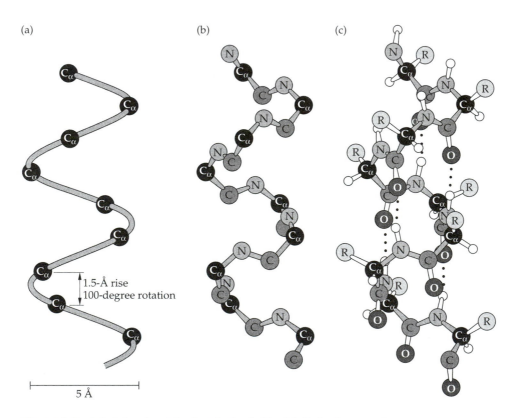

(a)

1.5-Å rise
100-degree rotation

5 Å

(b)

(c)

Figure 1.8 Models of a right-handed α helix: (a) Only the α-carbon atoms are shown on a helical thread. (b) Only the backbone nitrogen (N), α-carbon (C_α), and carbonyl carbon (C) atoms are shown. (c) The entire helix. Hydrogen bonds (denoted by•••) between NH and CO groups stabilize the helix. From Stryer (1988).

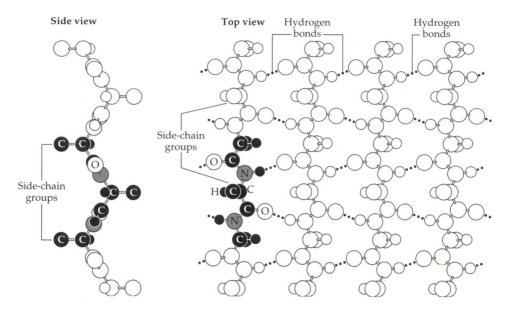

Figure 1.9   The β-pleated sheet. Hydrogen bonds (dotted lines) provide the forces keeping polypeptide strands aligned side by side. The edge view shows what is meant by the "pleats." See Holum (1978). By permission of L. Pauling and R. B. Corey.

that the hydrogen bonds are now between different polypepetide chains (or different stretches of the same chain) rather than CO and NH groups located close together in the same chain. In β sheets the side chains are pushed outward as far as possible away from the sheet surface.

Some proteins such as hemoglobin, myoglobin, and collagen have a very high α helix content, whereas others such as trypsin and chymotrypsin are virtually devoid of α helix. Two or more α helices or their variants can entwine around each other to form a coiled cable. Such cables are found in several proteins: keratin in hair, myosin in muscle, epidermin in skin, and fibrin in blood clots. They serve a mechanical role in forming stiff bundles of fibers. The β sheet is found in many enzymes and is the basic structure of silk fibers. In silk fibers the protein chains lie parallel to the fiber direction and make the fibers tough and hard to break. Also, the ability of stacks of sheets to slide over one another like pages in a book makes silk fibers flexible and supple.

The three-dimensional structure of a protein is termed its **tertiary structure**. This structure is formed from the packing of local features such as helices and sheets in a polypeptide into a three-dimensional molecule by noncovalent forces, mostly hydrogen bonds. In the process of folding, amino acid residues with hydrophobic side chains tend to be buried on the inside of the protein while those with hydrophilic side chains tend to be on the outside. This tendency is particularly strong in globular proteins, which are usually in an aqueous environment. For example, in the tertiary structure of myoglobin (Figure 1.10), the polypeptide has managed to place virtually all hydrophobic residues inside and has left hydrophilic residues outside.

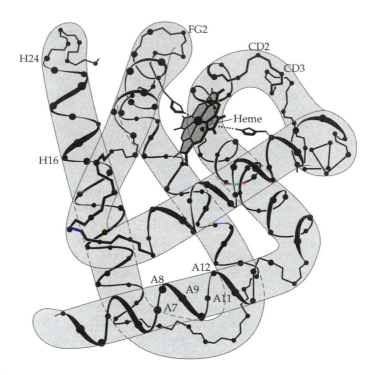

Figure 1.10   Myoglobin, the oxygen carrier in muscles. The black dots are individual α carbons of amino acid residues. Long, tubelike forms outline segments in the α-helix configuration. The identities of all residues are known. A few are lettered (to indicate a segment) and numbered (to indicate location in the segment). Hydrophilic groups are at FG2 and H16 (lysine), H24 (glutamic acid), CD2 (aspartic acid), and CD3 (arginine), to name a few. Their side chains are on exposed sites, where they can attract water molecules of the medium. Hydrophobic groups are at A7 and A9 (leucine), A8 and A11 (valine), and A12 (tryptophan), among others. Their side chains tend to be tucked inside away from the medium. See Holum (1978). Reproduced by permission, from Dickerson (1964).

Another force that can affect tertiary structure is the attraction between a positively charged and a negatively charged residue. As mentioned above, aspartic and glutamic acid residues each carry a negative charge, —COO^-, and arginine and lysine (and sometimes histidine) residues each carry a positive charge, —NH_3^+. These oppositely charged residues may attract each other if they are spatially close to each other. Such a force of attraction in a protein is called a salt bridge.

Once a polypeptide chain is formed, the molecule automatically twists itself in response to forces of potential hydrogen bonds, salt bridges, and water-avoidance or water-attraction properties into the secondary and tertiary structural features. Thus, the primary sequence of a protein determines its secondary and tertiary structures. However, our knowledge of protein folding is not yet good enough to predict accurately the secondary and tertiary structures of a protein from its primary structure. This is a challenging problem in protein chemistry today.

A protein may consist of more than one polypeptide. For example, hemoglobin, the oxygen carrier in the blood, is formed from four polypeptides: two α globins, and two β globins (Chapter 10). The final level of complexity that arises from the particular collection of polypeptides making up a protein and from the way in which the independent subunits are arranged is called the **quarternary structure**. Forces between oppositely charged sites and hydrogen bonds are involved in holding subunits together to stabilize the quaternary structure.

The above is a brief review of protein structure. For a more detailed description, readers may refer to Lehninger (1982) and Stryer (1988).

GENETIC CODES

The synthesis of proteins involves a process of decoding, whereby the genetic information carried by the mRNA is translated sequentially into amino acids through the use of transfer RNA (tRNA) mediators. Translation starts at the translation initiation site and proceeds to a stop signal. Translation involves the sequential recognition of triplets of adjacent nucleotides called **codons**. The phase in which a sequence is translated is determined by the initiation codon and is referred to as the **reading frame**. In the translational machinery, at the interface between the ribosome and the mRNA molecule, each codon is translated into a specific amino acid, which is subsequently added to the elongating polypeptide. The correspondence between the codons and the amino acids is determined by a set of rules called the **genetic code**. With few exceptions, the genetic code for nuclear protein-coding genes is **universal**, i.e., the same code is used for the translation of almost all eukaryotic nuclear genes and prokaryotic genes.

The universal genetic code is given in Table 1.2. Since a codon consists of three nucleotides and since there are four different nucleotides, there are $4^3 = 64$ possible codons. Of these, 61 code for specific amino acids and are called **sense codons**, while the remaining 3 are **nonsense** or **stop** codons that act as signals for the termination of the translation process. Since there are 61 sense codons and only 20 primary amino acids in proteins (Table 1.1), most amino acids are encoded by more than one codon. Such a code is referred to as a **degenerate code**. Different codons specifying the same amino acid are called **synonymous codons**. Synonymous codons that differ from each other at the third position only are referred to as a **codon family**. For example, the four codons for valine form a four-codon family. In contrast, the six codons for serine are divided into a four-codon family (UCU, UCC, UCA, and UCG) and a two-codon family (AGU and AGC).

The first amino acid in most eukaryotic proteins is a methionine encoded by the **initiation codon** AUG. This amino acid is usually removed in the mature protein. Most prokaryotic genes also use the AUG codon for initiation, but the amino acid initiating the translation process is a methionine derivative, namely formylmethionine.

The universal genetic code is also used in the translation process employed by the genomes of plastids, such as the chloroplasts of vascular plants. In contrast, most animal mitochondrial genomes and a few nuclear ones (e.g., *Mycoplasma* and *Tetrahymena*) use codes that are different from the universal genetic code (see Table 1.3). However, there are usually only minor differences between these codes and the universal genetic code. For example, in the vertebrate mitochondrial code, the

TABLE 1.2   The universal genetic code

Codon	Amino acid	Codon	Amino acid	Codon	Amino acid	Codon	Amino acid
UUU	Phe	UCU	Ser	UAU	Tyr	UGU	Cys
UUC	Phe	UCC	Ser	UAC	Tyr	UGC	Cys
UUA	Leu	UCA	Ser	UAA	Stop	UGA	Stop
UUG	Leu	UCG	Ser	UAG	Stop	UGG	Trp
CUU	Leu	CCU	Pro	CAU	His	CGU	Arg
CUC	Leu	CCC	Pro	CAC	His	CGC	Arg
CUA	Leu	CCA	Pro	CAA	Gln	CGA	Arg
CUG	Leu	CCG	Pro	CAG	Gln	CGG	Arg
AUU	Ile	ACU	Thr	AAU	Asn	AGU	Ser
AUC	Ile	ACC	Thr	AAC	Asn	AGC	Ser
AUA	Ile	ACA	Thr	AAA	Lys	AGA	Arg
AUG	Met	ACG	Thr	AAG	Lys	AGG	Arg
GUU	Val	GCU	Ala	GAU	Asp	GGU	Gly
GUC	Val	GCC	Ala	GAC	Asp	GGC	Gly
GUA	Val	GCA	Ala	GAA	Glu	GGA	Gly
GUG	Val	GCG	Ala	GAG	Glu	GGG	Gly

two codons AGA and AGG (denoted as AGR in Table 1.3), which specify arginine in the universal code, are used as termination codons, the codon UGA is used for tryptophan rather than for termination, and the codon AUA codes for methionine instead of isoleucine.

As many deviations from the universal code have been observed (Table 1.3), the genetic code is evolving, not frozen. For example, the codon AUA, which codes for isoleucine in the universal code, has changed to code for methionine in yeast and metazoa, but has reverted back to code for isoleucine in echinoderms. To explain changes in the code, Jukes and Osawa (1991) proposed the following hypothesis: The changes are preceded by disappearance of a codon from coding sequences in mRNA of an organism or organelle. The function of the codon that disappears is assumed by other synonynymous codons (e.g., the codon mutates to a synonymous codon), so that there is no change in amino acid sequences of proteins. The deleted codon then reappears, but it now codes for a different amino acid or serves as a stop codon. For more details about the hypothesis, see Chapter 14, Jukes and Osawa (1991), and Osawa et al. (1992).

In the mitochondrial genome of plants, the triplet CGG was once thought to code for tryptophan instead of arginine because CGG codons were often found in plant mitochondrial genes at positions corresponding to those encoding conserved tryptophans in other organisms (Fox and Leaver 1981). However, it was later found that the C residues in these codons for the conserved tryptophans were converted to U residues in the corresponding mRNA sequences, so that the CGG codons were changed to the UGG codons in the mRNA (Covello and Gray 1989). Since UGG indeed codes for tryptophan in the universal code, it appears that plant mitochondrial genomes also use the universal code. The conversion of C to U, known as a type of **RNA editing**, also occurs at some other C residues, but the specificity and structural determinants of conversion are not known at the present. For example, it is not clear why in the cytochrome *c* oxidase II gene in wheat the TCA codon at position 56 is converted to UUA in the mRNA, whereas the TCA codon at position 5 is not (Covello and Gray 1990). Because of the lack of complete knowledge, direct translation of plant mitochondrial genes into polypeptides cannot be done at the present time. There is evidence that RNA editing occurs in the mitochondria of all major groups of land plants except the Bryophyta (Hiesal et al. 1994). How such an editing system has evolved is a mystery.

MUTATION

Mutations are errors in DNA replication or errors in DNA repair. They may be classified by the length of the DNA sequence affected by the mutational event. For instance, a mutation may affect a single nucleotide (a **point mutation**), or several adjacent nucleotides. Mutations can also be classified by the type of change caused by the mutational event into (1) **substitutions**, the replacement of one nucleotide by another; (2) **recombination**, which includes crossing-over and gene

TABLE 1.3 Changes in the universal code

Codon in universal code	Changes to	Genome	
		Nuclear	**Mitochondrial**
UGA, Stop	Trp	Mycoplasma	All except plants
AUA, Ile	Met		Yeast
			Metazoa, except echinoderms
	Met back to Ile		Echinoderms, platyhelminths
AGR[a], Arg	Ser		Metazoa, except vertebrates
	Stop		Vertebrates
AAA, Lys	Asn		Flatworms, echinoderms
CUN[a], Leu	Thr		Yeasts
UAR[a], Stop	Gln	*Acetabularia*	
		Ciliated protozoa except *Euplotes*	
CUG, Leu	Ser	*Candida cylindracea*	
UGA, Stop	SeCys[b]	Vertebrates, eubacteria (in special enzymes)	

From Jukes and Osawa (1991) and Osawa et al. (1992).
[a]R, nucleotide A or G; N, any nucleotide.
[b]SeCys, selenocysteine. This occurs only at certain specific positions in the polypeptide chains of several enzymes, including glycine reductase, formate dehydrogenase in bacteria, and glutathione peroxidase of mammals and chickens.

conversion; (3) **deletions**, the removal of one or more nucleotides from the DNA; (4) **insertions**, addition of one or more nucleotides to the sequence; and (5) **inversions**, the rotation by 180° of a double-stranded DNA segment consisting of two or more base pairs (Figure 1.11).

Substitutional Mutation

Nucleotide substitutions can be classified into **transitions** and **transversions**. Transitions are substitutions between A and G (purines) or between C and T (pyrimidines). Transversions are substitutions between a purine and a pyrimidine; e.g., $A \to C$ and $A \to T$ are transversions. There are four types of transitions and eight types of transversions.

Nucleotide substitutions occurring in protein-coding regions can also be characterized by their effect on the product of translation, the protein. A substitution is said to be **synonymous** or **silent** if it causes no amino acid change (Figure 1.12a). Otherwise, it is **nonsynonymous**. Nonsynonymous, or amino acid-altering, mutations are further classified into **missense** and **nonsense** mutations. A missense mutation changes the affected codon into a codon that specifies a different amino acid from the one previously encoded (Figure 1.12b). A nonsense mutation changes a codon into one of the termination codons, thus prematurely ending the translation process and ultimately resulting in the production of a truncated protein (Figure 1.12c).

Each of the sense codons can mutate to nine other codons by means of a single nucleotide substitution. For example, CCU (Pro) can experience six nonsynonymous substitutions, to UCU (Ser), ACU (Thr), GCU (Ala), CUU (Leu), CAU (His), or CGU (Arg), and three synonymous substitutions, to CCC, CCA, or CCG. Since the universal genetic code consists of 61 sense codons, there are $61 \times 9 = 549$ possible nucleotide substitutions. If we assume that nucleotide substitutions occur at random and that all codons are equally frequent in coding regions, we

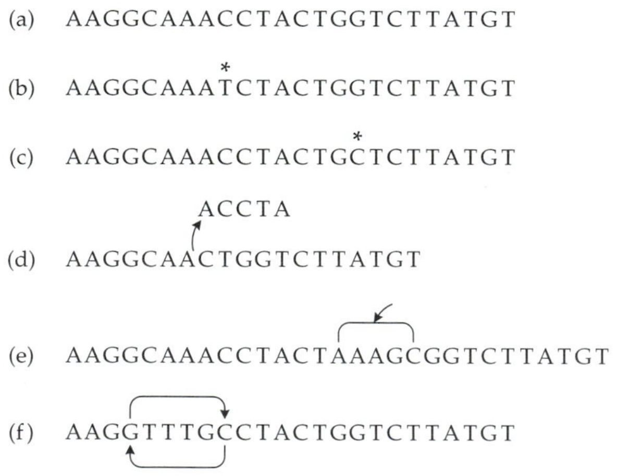

(a)   AAGGCAAACCTACTGGTCTTATGT

(b)   AAGGCAAA$\overset{*}{T}$CTACTGGTCTTATGT

(c)   AAGGCAAACCTACTG$\overset{*}{C}$TCTTATGT

(d)   AAGGCAA$\overset{\text{ACCTA}}{\curvearrowright}$CTGGTCTTATGT

(e)   AAGGCAAACCTACT$\overset{\curvearrowleft}{\overbrace{AAAGC}}$GGTCTTATGT

(f)   AAGG$\overset{\curvearrowright}{\underset{\curvearrowleft}{TTTGC}}$CTACTGGTCTTATGT

Figure 1.11 Types of mutations. (a) Original sequence; (b) a transition from C to T; (c) a transversion from G to C; (d) a deletion of the sequence ACCTA; (e) an insertion of the sequence AAAGC; (f) an inversion of 5'—GCAAAC—3' to 5'—GTTTGC—3'. From Li and Graur (1991).

Figure 1.12   Types of point mutations in a coding region: (a) synonymous, (b) missense, and (c) nonsense. From Li and Graur (1991).

can compute the expected proportion of the different types of nucleotide substitutions from the genetic code. These are shown in Table 1.4. Because of the structure of the genetic code, synonymous mutations occur mainly at the third position of codons. Indeed, almost 70% of all the possible nucleotide changes at the third position are synonymous. In contrast, all the mutations at the second position of codons are nonsynonymous, as are the vast majority (96%) of nucleotide changes at the first position.

Crossing-Over and Gene Conversion

There are two types of homologous recombination: **crossing-over** and **gene conversion**. Figure 1.13 shows a typical experiment in yeast for distiguishing between them. A diploid is constructed by crossing one haploid strain with the linked wild type alleles *A*, *B*, and *C* to a haploid containing mutant alleles *a*, *b*, and *c*. This diploid is shown at the top of the figure. The left side of Figure 1.13 shows a crossover between the first two linked loci; the segregation of the genetic markers at the two loci following the crossover will be 1*AB*:1*Ab*:1*aB*:1*ab* (tetratype) instead of 2*AB*:2*ab* (the normal parental ditype). Crossing-over is also termed **reciprocal recombination** because the tetrad that has the recombinant spore *Ab* also has the reciprocal recombinant product *aB*. The right side of Figure 1.13 shows a conversion event at the *B* locus; the segregation at the *B* locus will be 3*B*:1*b* instead of 2*B*:2*b*, although segregation at each of the other loci will be 2:2. Since it appears that one allele at the *B* locus has been converted by the other, the event is called **gene conversion**. If the *A* and *B* genes are examined together in the converted tetrad, it is noted that one spore contains the recombinant product *aB* but the other three spores contain parental combina-

TABLE 1.4   Relative frequencies of different types of mutations in a random protein-coding sequence

Type	Number	Percentage
Total in all codons	549	100
Synonymous	134	25
Nonsynonymous	415	75
Missense	392	71
Nonsense	23	4
Total in first position	183	100
Synonymous	8	4
Nonsynonymous	175	96
Missense	166	91
Nonsense	9	5
Total in second position	183	100
Synonymous	0	0
Nonsynonymous	183	100
Missense	176	96
Nonsense	7	4
Total in third position	183	100
Synonymous	126	69
Nonsynonymous	57	31
Missense	50	27
Nonsense	7	4

From Li and Graur (1991).

tions of markers. For this reason, gene conversion is also called **nonreciprocal recombination**. Gene conversion and crossing-over are often associated events; in yeast the frequency of association can be 50% or higher (see Borts and Haber 1987; Petes et al. 1991).

Figure 1.13 represents recombination at meiosis. Recombination can also occur at mitosis, but it is usually much less frequent than meiotic recombination; in yeast, for example, the former is usually 1000-fold less frequent than the latter (Esposito and Wagstaff 1981). The phenomenon of gene conversion will be discussed in more detail in Chapter 11. For the mechanism and models of recombination, readers may consult the reviews by Petes and Hill (1988), Lewin (1994), and Petes et al. (1991).

Deletions and Insertions

Deletions and insertions can occur by several mechanisms. One mechanism is **unequal crossing-over**. Figure 1.14a shows a simple model in which an unequal crossing-over between two chromosomes results in the deletion of a DNA segment in one chromosome and a reciprocal addition in the other. The chance of

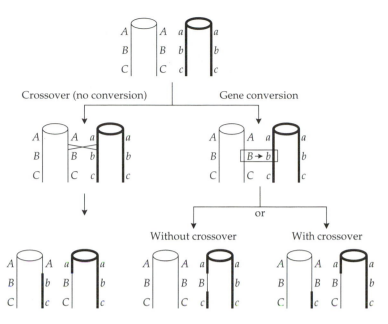

Figure 1.13   Meiotic crossing-over and gene conversion. One pair of chromosome is shown at the top; each circle represents a centromere and each line represents a chromatid. Meiotic recombination is shown between replicated nonsister homologous chromatids. The left side of the figure shows a crossover that is not associated with a detectable gene conversion event at the *B* locus. The right side shows a gene conversion event that is either accompanied or not accompanied by a crossover event. From Petes et al. (1991).

unequal crossing-over is much increased if a DNA segment is duplicated in tandem, as in Figure 1.14b, because of the increased chance of misalignment.

Another mechanism is **replication slippage** or **slipped-strand mispairing**. This type of event occurs in DNA regions containing contiguous short repeats. Figure 1.15a shows that, during DNA replication, slippage can occur because of mispairing between neighboring repeats and that slippage can result in either deletion or duplication of a DNA segment, depending on whether the slippage

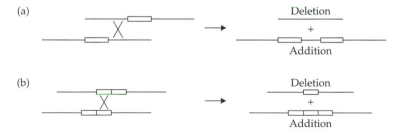

Figure 1.14   Unequal crossing-over. A box denotes a particular stretch of DNA. When a DNA segment is duplicated in tandem as in (b), the chance of misalignment increases and so does the chance of unequal crossing-over.

(a)

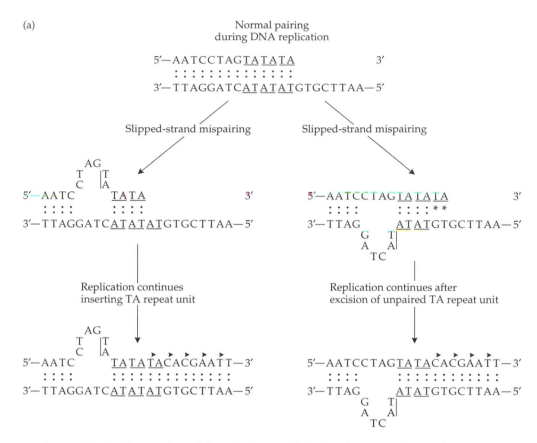

Figure 1.15 Generation of duplications or deletions by slipped-strand mispairing between contiguous repeats. Small arrows indicate direction and starting point of DNA synthesis. Dots indicate base pairing. (a) A two-base slippage in an TA repeat during DNA replication. Slippage in the $3' \to 5'$ direction results in the insertion of one TA unit (left panel). Slippage in the other direction results in the deletion of one repeat unit (right panel). The deletion shown in the right panel results from excision of the unpaired repeat unit (asterisks) at the 3' end of the growing strand,

occurs in the $5' \to 3'$ direction or in the opposite direction. Figure 1.15b shows that slipped-strand mispairing can also occur in nonreplicating DNA.

A third mechanism responsible for insertion or deletion of DNA sequences is DNA transposition, which will be dealt with in Chapter 12.

Deletions and insertions are collectively referred to as **gaps** (or indels), because when a sequence involving either an insertion or a deletion is compared with the original sequence, a gap will appear in one of the two sequences. The number of nucleotides involved in a gap event ranges from one or a few nucleotides to contiguous stretches involving thousands of nucleotides. The lengths of the gaps essentially exhibit a bimodal type of frequency distribution, with short gaps (up to 20–30 nucleotides) being mostly caused by errors in the process of DNA replication, such as the slipped-strand mispairing discussed above, and with long insertions and deletions occurring mainly because of unequal crossing-over or DNA transposition.

(b) Normal pairing
of intact chromosomal DNA

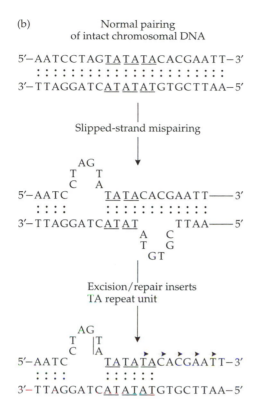

presumably by the $3' \rightarrow 5'$ exonuclease activity of DNA polymerase. (b) A two-base slippage in an TA repeat in nonreplicating DNA. Mismatched regions form single-stranded loops, which may be targets of excision and mismatch repair. The outcome (a deletion or an insertion) will depend on which strand is excised and repaired and which strand is used as template in the DNA repair process. Modified from Levinson and Gutman (1987).

In a coding region, a gap event involving a number of nucleotides that is not a multiple of three will cause a shift in the reading frame, so that the coding sequence downstream of the gap will be read in the wrong phase. Such a mutation is known as a **frameshift mutation**. Consequently, a gap not only introduces numerous amino acid changes, but may also obliterate the termination codon or bring into phase a new stop codon, thus resulting in a protein of abnormal length (Figure 1.16).

Spatial Distribution of Mutations

Mutations do not occur randomly throughout the genome. Some regions are more prone to mutate than others, and they are called **hotspots** of mutation. One such hotspot of mutation is the dinucleotide 5'—CG—3' (often called CpG), in which the cytosine is frequently methylated in higher eukaryotes and replicated with error, changing it to 5'—TG—3'. The dinucleotide 5'—TT—3' is a hotspot of mutation in prokaryotes but not in eukaryotes. In bacteria, regions within the

(a) Lys   Ala   Leu   Val   Leu   Leu   Thr   Ile   Cys   Ile   Ter
   AAG   GCA   CTG   GTC   CTG   TTA   ACA   ATA   TGT   ATA   TAA   TACCATCGCAATAGGG

               ↓

               G

   AAG   GCA   CTG   TCC   TGT   TAA   CAATATGTATATAATACCATCGCAATAGGG
   Lys   Ala   Leu   Phe   Cys   Ter

(b) Lys   Ala   Asn   Val   Leu   Leu   Thr   Ile   Cys   Ile   Ter
   AAG   GCA   AAC   GTC   CTG   TTA   ACA   ATA   TGT   ATA   TAA   TACCATCGCAATAGGG

            ↑

          G

   AAG   GCA   AAC   GGT   CCT   GTT   AAC   AAT   ATG   TAT   ATA   ATA   CCA   TCG   CAA   TAG   GG
   Lys   Ala   Asn   Gly   Pro   Val   Asn   Asn   Met   Tyr   Ile   Ile   Pro   Ser   Gln   Ter

Figure 1.16   Examples of frameshifts in reading frames caused by deletion or insertion. (a) A deletion of a G causes premature termination and (b) an insertion of G obliterates a stop codon. From Li and Graur (1991).

DNA containing short **palindromes** (i.e., sequences that read the same on the complementary strand, such as 5′—GCCGGC—3′, 5′—GGCGCC—3′, and 5′—GGGCCC—3′) were found to be more prone to mutate than other regions. In eukaryotic genomes, short tandem repeats are often hotspots for deletions and insertions, probably as a result of slipped-strand mispairing.

PATTERN OF NUCLEOTIDE SUBSTITUTION IN NONCODING SEQUENCES

Since point (substitution) mutation is one of the most important factors in the evolution of DNA sequences, molecular evolutionists have long been interested in determining the **pattern of spontaneous mutation** (e.g., Beale and Lehmann 1965; Zuckerkandl et al. 1971). This pattern provides empirical data for modeling DNA sequence evolution and can serve as a standard for inferring how far the observed frequencies of interchange between nucleotides in any given DNA sequence have deviated from the values expected under no selection, i.e., under **selective neutrality**.

One way to study the pattern of point mutation is to examine the pattern of substitution in regions of DNA that are subject to no selective constraints. Pseudogenes are particularly useful in this respect. Since they are devoid of function, all mutations occurring in pseudogenes are selectively neutral and become fixed in the population with equal probability. Thus, the pattern of nucleotide substitution in pseudogenes should reflect the pattern of spontaneous point mutation. As the pseudogene data that have been used were from mammalian nuclear pseudogenes (Table 1.5), the pattern inferred should be taken as an estimate of the mutation pattern in mammalian nuclear DNA. We shall also discuss the pattern inferred from the control (D-loop) region of human mitochondrial DNA (mtDNA), which may be taken as an estimate of the mutation pattern in mammalian mtDNA (Table 1.6).

TABLE 1.5   Pattern of nucleotide substitution in pseudogenes[a]

From	To				
	A	T	C	G	Row totals
A	—	4.7±1.3	5.0±0.7	9.4±1.3	19.1
	—	(5.3±1.4)	(5.6±0.8)	(10.3±1.4)	(21.2)
T	4.4±1.1	—	8.2±1.3	3.3±1.2	15.9
	(4.8±1.1)	—	(9.2±1.3)	(3.6±1.3)	(17.6)
C	6.5±1.1	21.0±2.1	—	4.2±0.5	31.7
	(7.1±1.3)	(18.2±2.3)	—	(4.2±0.6)	(29.5)
G	20.7±2.2	7.2±1.1	5.3±1.0	—	33.2
	(18.6±1.9)	(7.7±1.3)	(5.5±1.3)	—	(31.8)
Column totals	31.6	32.9	18.5	16.9	
	(30.5)	(31.2)	(20.3)	(18.1)	

From Gojobori et al. (1982b) and Li et al. (1984).
[a]Table entries are the inferred percentages (f_{ij}) of base changes from i to j based on 13 mammalian pseudogene sequences. Values in parentheses were obtained by excluding all CG dinucleotides from comparison.

Substitution Pattern in Pseudogenes

Figure 1.17 shows a simple method for inferring the nucleotide substitutions in a pseudogene sequence (Gojobori et al. 1982b; Li et al. 1984). Sequence 1 is a pseudogene, sequence 2 is its functional counterpart from the same species, and sequence 3 is a functional sequence that has diverged before the emergence of the pseudogene. Suppose that, at a certain nucleotide site, sequences 1 and 2 have A and G, respectively. Then we can assume that the nucleotide in the pseudogene sequence has changed from G to A if sequence 3 has G, but that the nucleotide in

TABLE 1.6   Pattern of nucleotide substitution in the control region of mtDNA[a]

From	To				
	A	T	C	G	Row totals
A	—	0.4	1.1	14.1	15.6
T	0.3	—	33.8	0.3	34.4
C	1.1	25.8	—	0.5	27.4
G	20.0	1.1	1.6	—	22.7
Column totals	21.4	27.3	36.5	14.9	

From Tamura and Nei (1993).
[a]Table entries are the inferred percentages (f_{ij}) of base changes from i to j based on 95 sequences of the control region of human mtDNA.

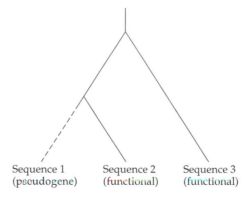

Figure 1.17 A tree for inferring the pattern of nucleotide substitution in a pseudo-gene sequence. The dashed line implies "nonfunctional." From Li and Graur (1991).

sequence 2 has changed from A to G if sequence 3 has A. However, if sequence 3 has T or C, we cannot decide the direction of change, and in this case the site is excluded from the comparison. Since the rate of substitution is usually much higher in pseudogenes than in the homologous functional gene, differences in the nucleotide sequence between a gene and a pseudogene are expected to have occurred in the pseudogene in the majority of cases.

The matrix in Table 1.5 represents the combined pattern of substitution inferred from 13 mammalian pseudogene sequences. Each entry, f_{ij}, in the matrix represents the expected number of base changes from i to j among every 100 substitutions in a random sequence (i.e., in a sequence in which the four bases are equally frequent). For example, $f_{AT} = 4.7$, that is, 4.7% of all substitutions are from A to T.

The direction of mutation is clearly nonrandom. For example, A changes more often to G than to either T or C. The four elements from the upper-right corner to the lower-left corner are the f_{ij} values for transitions, while the other eight elements represent transversions. All transitions, and in particular C → T and G → A, occur more often than transversions. The sum of the relative frequencies of transitions is 59.3% (56.3% if CG dinucleotides are excluded; see below). Note that under random mutation the expected proportion of transitions is only 33%, for there are only four types of transitions but eight types of transversions. The observed proportion is almost twice the value expected under random mutation.

Note also that some nucleotides are more mutable than others. The last column of Table 1.5 lists the relative frequencies of all mutations from A, T, C, and G to any other nucleotide. Were all four nucleotides equally mutable, we would expect a value of 25% in each of the column's elements. However, Table 1.5 shows that G mutates with a relative frequency of 33.2% (i.e., G is a highly mutable nucleotide), while T mutates with a relative frequency of 15.9 (i.e., it is less mutable than G). The bottom row of Table 1.5 shows the relative frequencies of all mutations that result in A, T, C, or G, respectively. Note that 64.5% of all mutations result in either A or T, while the random expectation is 50%. Since there is a tendency for C and G to change frequently to A or T, and since A and T are not as mutable as C and G, pseudogenes are expected to become rich in A and T. This should also be true for other noncoding regions that are subject to no functional constraint. Indeed, noncoding regions are generally found to be AT-rich.

The results in Table 1.5 are expressed in terms of the sense strand, i.e., the untranscribed strand. Thus, a change from G to A actually means that a G:C pair is replaced by an A:T pair. This change can occur as a result of either a G mutating to A in the sense strand or a C to T mutation in the complementary strand. Similarly, a change from C to T can occur as a result of either a C mutating to T in one strand or a G mutating to A in the other. If there is no difference in the pattern of mutation between the two strands, we should have $f_{GA} = f_{CT}$. Similarly, we should obtain $f_{AG} = f_{TC}$, $f_{AT} = f_{TA}$, $f_{AC} = f_{TG}$, $f_{CA} = f_{GT}$, and $f_{CG} = f_{GC}$. These equalities hold only approximately, and there may be minor asymmetries in the mutation pattern between the two strands. However, Bulmer's (1991) study of the substitution pattern in the β-globin region did not reveal any significant deviations from symmetry.

It is known that, in addition to base mispairing, the transition from C to T can also arise from conversion of methylated C residues to T residues upon deamination (Coulondre et al. 1978; Razin and Riggs 1980). The effect will elevate the frequencies of C:G $\rightarrow$ T:A and G:C $\rightarrow$ A:T; i.e., f_{CT} and f_{GA}. Since about 90% of methylated C residues in vertebrate DNA occur at 5'—CG—3' dinucleotides (Razin and Riggs 1980), this effect should be expressed mainly as changes of the CG dinucleotide to TG or CA. After a gene becomes a pseudogene, such changes would no longer be subject to any functional constraint and can therefore contribute significantly to C $\rightarrow$ T and G $\rightarrow$ A transitions if the frequency of CG is relatively high before silencing of the gene (i.e., loss of function) occurs. The substitution pattern obtained by excluding all nucleotide sites where the CG dinucleotides appear to have occurred in the ancestral sequences of these pseudogenes is given in parentheses in Table 1.5. This pattern is probably more suitable for predicting the pattern of mutations in a sequence that has not been subject to functional constraints for a long time (e.g., some parts of an intron), because in such a sequence there would exist few CG dinucleotides. The pattern with CG dinucleotides excluded is somewhat different from that obtained otherwise. In particular, the differences among the relative frequencies of the four transitions become somewhat less conspicuous, and the relative frequencies of the transversions become slightly higher, except for G $\rightarrow$ C and C $\rightarrow$ G.

Substitution Patterns in the Control Region of mtDNA

The discovery of a high ratio of transitional to transversional substitution in primate mtDNA (Brown et al. 1982) has attracted much attention, for the ratio was estimated to be close to 20, much higher than the ratio of ~1 in nuclear DNA. Although the original observation was from coding sequences and so might not reflect the sponteneous mutation pattern, the high ratio was also observed in the control region (Aquadro and Greenberg 1983; Vigilant et al. 1991), which, except for the central part, appears to be subject to very weak selective constraints and has evolved very rapidly (Horai and Hayasaka 1990; Vigilant et al. 1991; Ward et al. 1991). Recently, Kocher and Wilson (1991) and Tamura and Nei (1993) have studied the substitution pattern in the control region using extensive sequence data from humans and chimpanzees. The estimated pattern shown in Table 1.6

was based on 95 human sequences; the conservative central portion was excluded from the analysis. Obviously, all eight types of transversion occurred with very low frequencies and the average transition/transversion ratio is 15.7, similar to the value (15.0) estimated by Vigilant et al. (1991). Moreover, the transitional rate between pyrimidines (C and T) is higher than that between purines (A and G), and f_{TC} and f_{GA} are, respectively, higher than f_{CT} and f_{AG}, which suggests asymmetry in forward and backward mutation.

PROBLEMS

1. Find a complete mRNA sequence from a mammalian species. Delete the first AT you encounter in the coding region. Determine whether a premature termination codon comes into the reading frame or whether translation will extend beyond the original stop codon.

2. Explain why a frameshift mutation is likely to be more serious if it occurs in a 5′ part than if it occurs in a 3′ part of the coding region.

3. Compute the proportion of synonymous changes among mutations involving single nucleotide changes in the codon ACT under (a) the assumption that mutation is random, i.e., no difference in rate or in preference in the direction of mutation among the four nucleotides, (b) the mutation pattern in Table 1.5 (the values in parentheses), and (c) the mutation pattern in Table 1.6. Note that in parts (b) and (c), the differences in mutability among nucleotides should be taken into account.

4. Repeat problem 3 for the codon GAT.

CHAPTER 2

Dynamics of Genes in Populations

A LL EVOLUTIONARY CHANGES START WITH CHANGES WITHIN POPULATIONS. The study of genetic changes that occur in populations belongs to the domain of population genetics. This chapter reviews some basic principles of population genetics that are essential for understanding molecular evolution; further theoretical background of population genetics will be provided in Chapter 9, in which the maintenance of molecular polymorphism is discussed.

A basic problem in population genetics is to determine how the frequency of a mutant gene will change with time under the influence of various evolutionary forces. Another basic problem is how genetic variability is maintained in natural populations. In addition, from the long-term point of view, it is important to determine the probability that a new mutant will completely replace existing variants in the population and to estimate how fast the replacement will take place. Unlike morphological changes, many molecular changes are likely to have only a small effect on the phenotype of an organism, so the frequencies of molecular variants are subject to strong chance effects. Therefore, chance elements should be taken into account when dealing with molecular evolution. The role of chance effects in evolution is controversial, however, and will be discussed in this chapter and Chapter 9.

CHANGES IN ALLELE FREQUENCIES

The chromosomal or genomic location of a gene is called a **locus**, and alternative forms of the gene at a given locus are called **alleles**. In a population, more than one allele may be present at a locus, and their relative proportions are referred to as the **allele frequencies** or **gene frequencies**. For example, assume that there are two alleles with n_1 and n_2 copies at a certain locus, in a haploid population of size N. Then, their allele frequencies are equal to n_1/N and n_2/N, respectively. Note that $n_1 + n_2 = N$, and $n_1/N + n_2/N = 1$.

Evolution is a process of change in the genetic makeup of populations, with the most basic component being change in allele frequencies with time. In fact, from the evolutionary point of view, a new mutation becomes significant only if its frequency increases with time and ultimately reaches 1; the mutant gene is then said to have become **fixed** in the population. Without increasing its frequency, a mutation will have but a passing effect on the evolutionary history of the species; the only exception is that it is maintained in the population for a long time by balancing selection. For a mutant allele to increase in frequency, factors other than mutation must come into play. These factors include natural selection, random genetic drift, recombination, and migration.

To understand the process of evolution, we must study how the above factors govern the changes of allele frequencies. In this book, we discuss only natural selection and random genetic drift. In classical evolutionary studies involving morphological traits, natural selection has been considered as the major driving force of evolution. In contrast, random genetic drift is thought to have played an important role in evolution at the molecular level.

There are two mathematical approaches to studying genetic changes in populations: deterministic and stochastic. The **deterministic model** is simpler. It assumes that changes in the frequencies of alleles in a population from generation to generation occur in a unique manner and can be unambiguously predicted from knowledge of initial conditions. Strictly speaking, this approach applies only when two conditions are met: (1) the population is infinite in size and (2) the environment either remains constant with time or changes according to deterministic rules. These conditions are obviously never met in nature, and therefore a purely deterministic approach may not be sufficient to describe the temporal changes in allele frequencies in populations. Random or unpredictable fluctuations in allele frequencies must also be taken into account.

Dealing with random fluctuations requires a different mathematical approach. **Stochastic models** assume that changes in allele frequencies occur in a probabilistic manner, that is, from knowledge of the conditions in one generation we cannot predict unambiguously the allele frequencies in the next generation, but can only determine the probabilities with which certain allele frequencies will be attained. Obviously, stochastic models are preferable over deterministic ones, since they are based on more realistic assumptions. However, deterministic models are much easier to treat mathematically and, under certain circumstances, they yield sufficiently accurate approximations. In the following, we shall deal with natural selection in a deterministic fashion.

NATURAL SELECTION

Natural selection is defined as the differential reproduction of genetically distinct individuals or genotypes within a population. Differential reproduction is caused by differences among individuals in such factors as mortality, fertility, fecundity, mating success, and the viability of offspring. Natural selection is predicated on the availability of genetic variation among individuals in characters related to reproduction. It cannot occur in a population that consists of individuals that do not differ from one another in such traits. Selection leads to changes in allele frequencies over time. However, a mere change in allele frequencies from generation to generation does not necessarily indicate that natural selection is at work. Other processes, such as random genetic drift, can bring about temporal changes in allele frequencies as well (see below).

The **fitness** of a genotype, commonly denoted as w, is a measure of the individual's ability to survive and reproduce. Since the size of a population is usually constrained by the carrying capacity of the environment in which the population resides, the evolutionary success of an individual is determined not by its **absolute fitness**, but by its **relative fitness** in comparison with the other genotypes in the population. In nature, the fitness of a genotype is not expected to remain constant for all generations and under all environmental circumstances. However, by assigning a constant value of fitness to each genotype, we are able to formulate simple theories that are useful for understanding the dynamics of change in the genetic structure of populations brought about by natural selection. In the simplest class of models, we assume that the fitness of an organism is determined solely by its genetic makeup. We also assume that all loci contribute independently to the fitness of the individual, so that each locus can be treated separately.

Most new mutants arising in a population reduce the fitness of their carriers. Such mutations will be selected against and most will be eventually removed from the population. This type of selection is called **negative** or **purifying selection**. Occasionally, a new mutation may be as fit as the best allele in the population. Such a mutation is selectively **neutral**, and its fate is determined not by selection but by chance events. Rarely, a mutant that confers a selective advantage on its carriers may arise. Such a mutation will be subjected to **positive selection**. If the new mutant is advantageous only in heterozygotes but not in homozygotes, the resulting selective regime will be **overdominant selection**.

In the following, we shall consider the case of one locus with two alleles, A_1 and A_2. Each allele can be assigned an intrinsic fitness value; it can be advantageous, deleterious, or neutral. However, this assignment is only applicable to haploid organisms. In diploid organisms the fitness is determined by the interaction between the two alleles at the locus. With two alleles at a locus, there are three possible diploid genotypes: A_1A_1, A_1A_2, and A_2A_2, and their fitnesses can be denoted by w_{11}, w_{12}, and w_{22}, respectively.

Given that the frequency of allele A_1 in a population is p, and the frequency of the complementary allele, A_2, is $q = 1 - p$, we can show that under random mating, the frequencies of A_1A_1, A_1A_2, and A_2A_2 are p^2, $2pq$, and q^2, respectively. A population in which such genotypic ratios are maintained is said to be at **Hardy–Weinberg equilibrium**.

In the general case, the three genotypes are assigned the following fitness values and initial frequencies:

Genotype:	A_1A_1	A_1A_2	A_2A_2
Fitness:	w_{11}	w_{12}	w_{22}
Frequency:	p^2	$2pq$	q^2

Let us now consider the dynamics of allele frequency changes following selection. Given the frequencies of the three genotypes and their fitnesses as above, the relative frequencies of the three genotypes in the next generation will become p^2w_{11}, $2pqw_{12}$, and q^2w_{22}, for A_1A_1, A_1A_2, and A_2A_2, respectively. Therefore, the frequency of allele A_2 in the next generation will become:

$$q' = \frac{pqw_{12} + q^2w_{22}}{p^2w_{11} + 2pqw_{12} + q^2w_{22}} \tag{2.1}$$

The extent of change in the frequency of allele A_2 per generation is denoted as $\Delta q = q' - q$. We can show that:

$$\Delta q = \frac{pq[p(w_{12} - w_{11}) + q(w_{22} - w_{12})]}{p^2 w_{11} + 2pq w_{12} + q^2 w_{22}} \tag{2.2}$$

In the following, we shall assume that A_1 is the original or "old" allele in the population. We shall then consider the dynamics of change in allele frequencies following the appearance of a new allele, A_2. For mathematical convenience, we shall assign a fitness value of 1 to the $A_1 A_1$ genotype. The fitness of the newly created genotypes, $A_1 A_2$ and $A_2 A_2$, will depend on the mode of interaction between A_1 and A_2. For example, if A_2 is completely dominant to A_1, then w_{11}, w_{12}, and w_{22} can be written as 1, $1 + s$, and $1 + s$, respectively, where s is the difference between the fitness of an A_2-carrying genotype and the fitness of $A_1 A_1$. A positive value of s means an increase in fitness in comparison with $A_1 A_1$, while a negative value means a decrease in fitness. If A_2 is completely recessive, the fitnesses of the three genotypes become 1, 1, and $1 + s$, respectively.

Two common modes of interaction will be considered: (1) codominance, or genic selection, and (2) overdominance. Codominance represents a case of directional selection and is mathematically the simplest mode of interaction, while overdominance represents a type of balancing selection.

Codominance

In the **codominant mode** of selection, or **genic selection**, the two homozygotes have different fitness values, while the fitness of the heterozygote is the mean of the fitnesses of the two homozygous genotypes. The relative fitness values for the three genotypes can be written as:

Genotype:	$A_1 A_1$	$A_1 A_2$	$A_2 A_2$
Fitness:	1	$1 + s$	$1 + 2s$

From Equation 2.2 we obtain the following change in the frequency of allele A_2 per generation under codominance:

$$\Delta q = \frac{spq}{1 + 2spq + 2sq^2} \tag{2.3}$$

By iteration, this equation can be used to compute the frequency (q) of A_2 at any generation. However, the following approximation leads to a much more convenient formula. If s is small, as is usually the case, the denominator of Equation 2.3 is approximately equal to 1 and the equation becomes approximately

$$\Delta q = spq$$

This difference equation can be approximated by the following differential equation

$$\frac{dq}{dt} = spq = sq(1 - q) \tag{2.4}$$

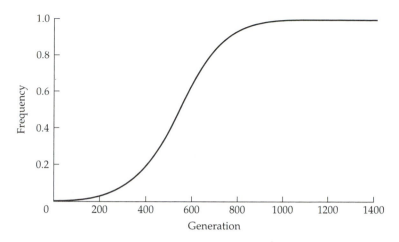

Figure 2.1 Frequency of a codominant advantageous allele with $s = 0.01$ following its appearance as a result of mutation in generation 0. From Li and Graur (1991).

The solution of this equation is given by

$$q_t = \frac{1}{1 + \left(\dfrac{1 - q_0}{q_0}\right)e^{-st}}$$

(2.5)

where q_0 and q_t are the frequencies of A_2 at time 0 and t, respectively.

Figure 2.1 illustrates the increase in the frequency of allele A_2 for $s = 0.01$. Clearly, the frequency of A_2 always increases with time. For this reason, genic selection is a type of **directional selection**. Note, however, that at low frequencies, selection for a codominant allele is not very efficient (i.e., the change in allele frequencies is slow). The reason is that, at low frequencies of A_2, the proportion of A_2 alleles residing in heterozygotes is large. For example, when the frequency of A_2 is 0.5, 50% of A_2 alleles will be carried by heterozygotes, whereas when the frequency of A_2 is 0.01, 99% of all such alleles reside in heterozygotes. Because heterozygotes, which contain both alleles, have a smaller selective advantage than do A_2A_2 homozygotes (i.e., s versus $2s$), the overall change in allele frequencies at low values of q is small.

In Equation 2.5 the frequency q_t is expressed as a function of time t. Alternatively, t can be expressed as a function of the frequency q as follows:

$$t = \frac{1}{s}\ell n \frac{q_t(1 - q_0)}{q_0(1 - q_t)}$$

(2.6)

From this equation, one can calculate the number of generations required for the frequency of A_2 to change from one value (q_0) to another (q_t).

Overdominance

In the **overdominant mode of selection**, the heterozygote has the highest fitness. Thus:

$$\begin{array}{cccc}
\text{Genotype:} & A_1A_1 & A_1A_2 & A_2A_2 \\
\text{Fitness:} & 1 & 1+s_1 & 1+s_2
\end{array}$$

In this case, $s_1 > 0$ and $s_1 > s_2$. Depending on whether the fitness of A_2A_2 is greater than, equal to, or less than that of A_1A_1, s_2 can be positive, zero, or negative. The change in allele frequencies is expressed as:

$$\Delta q = \frac{-pq(2s_1q - s_2q - s_1)}{1 + 2s_1pq + s_2q^2} \tag{2.7}$$

Figure 2.2 illustrates the changes in the frequency of an allele subject to overdominant selection. In contrast to the codominant selection regime, in which one of the alleles is eventually eliminated from the population, under overdominant selection the population sooner or later will reach an equilibrium in which the two alleles coexist. After equilibrium is reached, no further change in allele frequencies will be observed (i.e., $\Delta q = 0$). Thus, overdominant selection belongs to a class of selection regimes called **balancing** or **stabilizing selection**.

The frequency of allele A_2 at equilibrium is obtained by solving Equation 2.7 for $\Delta q = 0$:

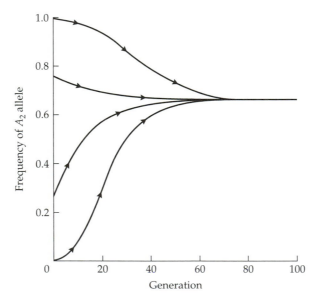

Figure 2.2   Changes in the frequency of an allele subject to overdominant selection. Initial frequencies from top to bottom: 0.99, 0.75, 0.25, and 0.01; $s_1 = 0.250$ and $s_2 = 0.125$. Since the s values are exceptionally large, the change in allele frequency is rapid. Note that there is a stable eqilibrium at $q = 0.667$. Modified from Hartl and Clark (1989).

$$\hat{q} = \frac{s_1}{2s_1 - s_2} \tag{2.8}$$

When $s_2 = 0$ (i.e., both homozygotes have identical fitness values), the equilibrium frequencies of both alleles will be 50%.

RANDOM GENETIC DRIFT

As noted above, natural selection is not the only factor that can cause changes in allele frequency. Allele frequency changes can also occur by chance, though in this case the changes are not directional but random. An important factor in producing random fluctuations in allele frequencies is the random sampling of gametes in the process of reproduction (Figure 2.3). Sampling occurs because, in

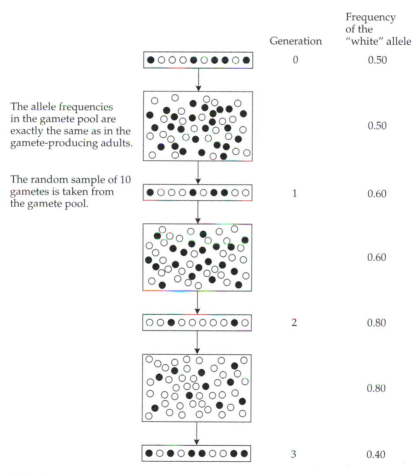

Figure 2.3   Random sampling of gametes. Allele frequencies in the gamete pools (large boxes) in each generation are assumed to reflect exactly the allele frequencies in the adults of the parental generation (small boxes). Since the population size is finite, allele frequencies fluctuate up and down. Modified from Bodmer and Cavalli-Sforza (1976).

the vast majority of cases in nature, the number of gametes available in any generation is much larger than the number of adult individuals produced in the next generation. In other words, only a minute fraction of gametes succeed in developing into adults. In a diploid population under Mendelian segregation, sampling can still occur even if there is no excess of gametes, i.e., even if each individual contributes exactly two gametes to the next generation. The reason is that heterozygotes can produce two types of gametes, but the two gametes passing on to the next generation may by chance be of the same type.

To see the effect of sampling, consider an idealized situation in which all the individuals in the population have the same fitness and selection does not operate. We further simplify the problem by considering a population with nonoverlapping generations (i.e., a group of individuals that reproduce simultaneously), such that any given generation can be unambiguously distinguished from both previous and subsequent generations. The population under consideration is diploid and consists of N individuals, so that the population contains $2N$ genes at each locus. Let us again consider the simple case of one locus with two alleles, A_1 and A_2, with frequencies p and $q = 1 - p$, respectively. When $2N$ gametes are sampled from the infinite gamete pool, the probability, P_i, that the sample contains exactly i genes of type A_1 is given by the binomial probability function:

$$P_i = \frac{(2N)!}{i!(2N - i)!} p^i q^{2N-i} \tag{2.9}$$

Since P_i is always greater than 0 for populations in which $0 < p < 1$, the allele frequencies may change from generation to generation without the aid of selection.

The frequency of allele A_1 at generation t, denoted by p_t, is a random variable. The mean and variance of p_t are given by

$$E(p_t) = p_0 \tag{2.10}$$

$$V(p_t) = p_0(1 - p_0)\left[1 - \left(1 - \frac{1}{2N}\right)^t\right] \tag{2.11}$$

$$\approx p_0(1 - p_0)(1 - e^{-t/(2N)})$$

where p_0 denotes the initial frequency and is assumed to be known (see Crow and Kimura 1970). Note that although the expectation of p_t stays the same as the intitial frequency, the variance of p_t increases with time. Thus, although random sampling of gametes produces no systematic change in allele frequency, it causes random fluctuations in allele frequency.

The process of change in allele frequency due solely to chance effects is called **random genetic drift**. One should note, however, that random genetic drift can also be caused by processes other than the sampling of gametes. For example, stochastic changes in selection intensity can also bring about random changes in allele frequencies (see Gillespie 1991).

Figure 2.4 illustrates the effects of random sampling on the frequencies of alleles in populations of different sizes. In the figure each curve represents the result of a computer simulation of the random sampling process; in the simulation, each generation is formed by random sampling of $2N$ genes (with replacement) from

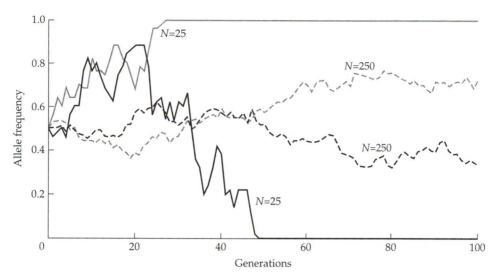

Figure 2.4   Changes in frequencies of alleles subject to random genetic drift in populations of different sizes (N). In each generation, $2N$ genes were sampled with replacement from the previous generation. For each population size, two replicates are presented. It is assumed that the effective population size N_e is equal to the actual size N.

the previous generation. The allele frequencies change from generation to generation, but the direction of change is random at any point in time. The fluctuations in allele frequency are conspicuous in the case of $N = 25$; in one simulation (replicate) the allele became fixed in the population at generation 27 and in the other the allele became lost from the population at generation 49. In the case of $N = 250$ the fluctuations in allele frequency are less pronounced and in both simulations the allele remained at an intermediate frequency at generation 100.

In the stochastic theory of allele frequency changes, one imagines an infinite array of identical populations, all of which have the same population size, are subject to the same sampling procedure, and start with the same initial frequency. At any time point t the allele frequencies in different populations represent the distribution of allele frequencies at time t. For example, the probability that the frequency of allele A_1 is in a certain frequency interval is equal to the proportion of populations in which the frequency of allele A_1 is in that interval. In computer simulation, of course, one cannot simulate an infinite array of populations. However, if the size of the array of simulated populations is large, the frequency distribution obtained will be close to the true distribution. For example, in Figure 2.5 the computer simulation was conducted with an array of 1000 populations. (In practice, it is difficult to simulate 1000 populations at the same time, and so instead one population is simulated at one time and the simulation is repeated 1000 times. Each repeat is known as one replicate.) In the simulation each population consists of $N = 10$ diploid individuals, or 20 genes, and starts with the frequencies 0.3 and 0.7 for the A_1 and A_2 alleles, respectively (a small N is used to save computer time). Since the number of genes is only 20, the exact distribution

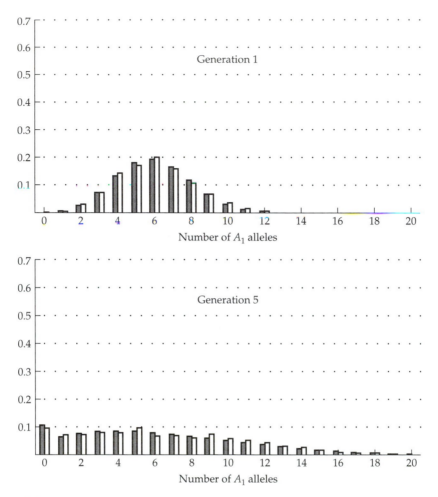

Figure 2.5   Probability distributions of allele frequencies in a diploid population of $N = 10$. The population is under random mating and is not subject to natural selection. The initial frequency of allele A_1 is 0.3, i.e., 6 A_1 alleles in the initial population. The white histograms are theoretical (exact) values and the shaded

can be obtained by the Markov chain method (see, e.g., Ewens 1979; Nei 1987). It is noted that the simulated distributions (shaded histograms) in different generations are very close to the exact distributions (white histograms).

Some interesting features emerge from Figure 2.5. At generation 1, the distribution of the frequency of A_1 follows a binomial distribution, with $p = 0.3$ and $2N = 20$. Since the variance is small, i.e., only $0.3(1 - 0.3)/20 = 0.01$, the distribution is concentrated around the mean (0.3). At generation 5, the distribution becomes flatter. The A_1 allele has become lost in 10.8% of the populations and has become fixed in 0.1% of the populations (these percentages correspond to the heights of the distribution at frequencies 0 and 1, respectively). At generation 20, the distribution becomes fairly uniform for frequencies between 0 and 1, and the probabilites of loss and fixation become 47.4% and 10.4%, respectively. At gener-

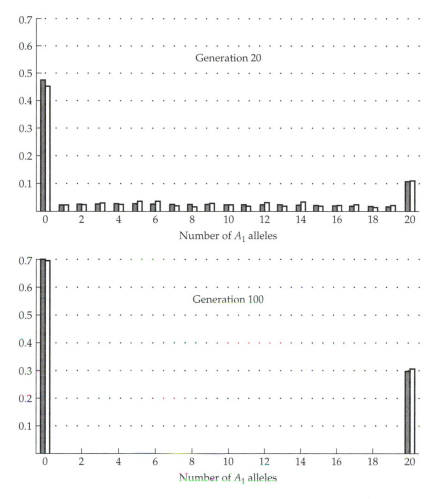

histograms are values obtained from computer simulation with 1000 replicates. The ordinate denotes the probability that the number of A_1 alleles is equal to a particular number. The heights at the two terminal classes represent the probabilities of loss and fixation of allele A_1, respectively.

ation 100, the A_1 allele has become lost in 69.7% of the populations and fixed in 29.7% of the populations; that is, it has become either lost or fixed in 99.4% (almost 100%) of the populations. If t increases further, the A_1 allele will eventually become lost in 70% of the populations and fixed in 30% of the populations; since the initial frequency of A_1 is 0.3 and since there is no factor favoring either of the two alleles, the probability that A_1 will become fixed in the population is equal to the initial frequency of A_1 (see page 47).

The time required for most of the populations to become fixed for the A_1 or A_2 allele can be roughly estimated from Equation 2.11. In this equation the only term that changes with time is $e^{-t/(2N)}$, which, for $N = 10$, reduces to only 0.007 as t increases to 100. Thus, for $t \geq 100$, $V(p_t)$ is close to $p_0(1 - p_0)$, which is the maximum value for $V(p_t)$ and is obtained when all populations have become fixed for either

A_1 or A_2. This means that when $t \geq 100$, almost all populations should have already become fixed for either A_1 or A_2. This is in agreement with the simulation result.

EFFECTIVE POPULATION SIZE

The above mathematical formulation of random genetic drift assumed an idealized population in which all individuals contribute gametes to the next generation with equal probability, generations are nonoverlapping and population size is constant over time. In practice, it is unlikely that all these conditions hold. Furthermore, as will be discussed below, there are often other factors that complicate the mathematical treatment. To simplify the mathematical formulation, Wright (1931) introduced the concept of **effective population size**, N_e, which is the size of an idealized population that would have the same effect of random sampling on gene frequency as that in the actual population. Consider a population with actual size N and assume that the frequency of allele A_1 at the present generation is p. If any of the above conditions is violated, then the variance of the frequency of allele A_1 (p') in the next generation is expected to be larger than the following binomial variance

$$V(p') = p(1-p)/(2N) \tag{2.12}$$

which can be obtained from Equation 2.11 by putting $t = 1$. The concept of effective population size is to define N_e in such a way that the actual variance is given by

$$V(p') = p(1-p)/(2N_e) \tag{2.13}$$

In general, N_e is smaller, sometimes much smaller, than N, the actual population size. Various factors can contribute to this difference. For example, in a population with overlapping generations, at any given time, part of the population will consist of individuals in either their pre- or postreproductive stage. Due to this developmental stratification, the effective size can be considerably smaller than the census size, N. For example, according to Nei and Imaizumi (1966), in humans N_e is only slightly larger than $N/3$.

Reduction in the effective population size in comparison to the census size can also occur if the number of males involved in reproduction is different from the number of females. This disparity is especially pronounced in polygamous species, such as social mammals and territorial birds, or in species in which a nonreproducing caste exists (e.g., bees, ants, and termites). If a population consists of N_m males and N_f females, N_e is given by:

$$N_e = \frac{4N_m N_f}{N_m + N_f} \tag{2.14}$$

Note that, unless the number of females equals the number of males, N_e will always be smaller than N. As an extreme example, let us assume that in a population of size N, all the females $(N/2)$ and only one male take part in the reproductive process. By using Equation 2.14, we see that $N_e = 2N / (1 + N/2)$. If N is considerably larger than 1, N_e becomes 4, regardless of the census population size.

The effective population size can also be much reduced due to long-term variations in the population size, which in turn are caused by such factors as environmental catastrophes, cyclical modes of reproduction, and local extinction and recolonization events. For example, the **long-term effective population size** in a species for a period of n generations is given by:

$$N_e = n / (1 / N_1 + 1 / N_2 + \cdots + 1 / N_n) \qquad (2.15)$$

where N_i is the population size of the ith generation. In other words, N_e equals the harmonic mean of the N_i values, and consequently it is closer to the smallest value of N_i than to the largest one. Similarly, if a population goes through a bottleneck, the effective population size is greatly reduced.

GENE SUBSTITUTION

Gene substitution is defined as the process whereby a mutant allele completely replaces the predominant or **wild type** allele in a population. In this process, a mutant allele arises in a population, usually as a single copy, and becomes **fixed** after a certain number of generations. The time it takes for a new allele to become fixed is called the **fixation time**. Not all new mutants, however, reach fixation. In fact, the majority of them are lost after a few generations. Thus, we also need to address the issue of **fixation probability** and discuss the factors affecting the chance that a new mutant allele will reach fixation in a population. New mutations arise continuously within populations. Consequently, gene substitutions occur in succession, with one allele replacing another and being itself replaced in time by a new allele. Thus, we can speak of the **rate of gene substitution**, i.e., the number of substitutions or fixations per unit time.

Fixation Probability

The probability that a particular allele will become fixed in a population depends on (1) its initial frequency, (2) its selective advantage or disadvantage, s, and (3) the effective population size, N_e. In the following, we shall consider the case of genic selection and assume that the relative fitness of the three genotypes A_1A_1, A_1A_2, and A_2A_2 are 1, $1 + s$, and $1 + 2s$, respectively.

Kimura (1962) showed that the probability of fixation of A_2 is given by

$$P = (1 - e^{-4N_e s q}) / (1 - e^{-4N_e s}) \qquad (2.16)$$

where q is the initial frequency of allele A_2. Since $e^{-x} \approx 1 - x$ when x is small, Equation 2.16 reduces to $P \approx q$ as s approaches 0. Thus, for a neutral allele, the fixation probability equals its frequency in the population. For example, in Figure 2.5 the initial frequency of allele A_1 is 30% and so it will eventually become fixed in 30% of the cases and become lost in 70% of the cases. This is intuitively understandable because in the case of neutral alleles, fixation occurs by random genetic drift, which favors neither allele.

We note that a new mutant arising as a single copy in a diploid population of size N has an initial frequency of $1/(2N)$. The probability of fixation of an individual mutant allele, P, is thus obtained by replacing q with $1/(2N)$ in Equation 2.16. When $s \neq 0$,

$$P = [1 - e^{-(2N_e s)/N}] / (1 - e^{-4N_e s}) \qquad (2.17)$$

For a neutral mutation, Equation 2.17 becomes

$$P = 1/(2N) \tag{2.18}$$

If the population size is equal to the effective population size, Equation 2.17 reduces to

$$P = (1 - e^{-2s}) / (1 - e^{-4Ns}) \tag{2.19}$$

If the absolute value of *s* is small, we obtain

$$P = 2s / (1 - e^{-4Ns}) \tag{2.20}$$

For positive values of *s* and large values of *N*, Equation 2.20 reduces to

$$P = 2s \tag{2.21}$$

Thus, if an advantageous mutation arises in a large population and its selective advantage over the rest of the alleles is small, say < 5%, the probability of its fixation is approximately twice its selective advantage. For example, if a new mutation with $s = 0.01$ arises in a population, the probability of its eventual fixation is 2%.

Let us now consider a numerical example. A new mutant arises in a population of 1000 individuals. For simplicity, we assume that $N = N_e$. The probability that this allele will become fixed in the population is $1/(2N) = 0.05\%$ if it is neutral, 2% if it confers a selective advantage of 0.01, and 0.004% if it has a selective disadvantage of 0.001 (the last two cases are computed from Equation 2.19). These results are quite noteworthy, for they mean that an advantageous mutation does not always become fixed in the population. In fact, 98% of all the mutations with a selective advantage of $s = 0.01$ will be lost by chance. On the other hand, even slightly deleterious mutations have a finite probability of becoming fixed in a population, albeit a small one. The mere fact that a deleterious allele may become fixed in a population illustrates in a powerful way the importance of chance effects in determining the fate of mutations during evolution. If the population size becomes larger, the chance effect, of course, becomes smaller. For instance, in the above example if the population size is $N = N_e = 10,000$ instead of 1,000, then the fixation probabilities become 0.005%, 2%, and $\approx 10^{-20}$, respectively. While the fixation probability for the advantageous mutation remains approximately the same, that for the deleterious allele has become very small when N_e increases from 1000 to 10,000. Therefore, in a large population, it is almost impossible for a deleterious mutation to become fixed in the population and the chance for a neutral mutation to become fixed in the population is very small; however, see below for the rate of substitution for neutral mutation.

Fixation Time

The time required for the fixation or loss of an allele depends on the frequency of the allele and the size of the population. The mean time to fixation or loss becomes shorter as the frequency of the allele approaches 1 or 0, respectively.

In terms of evolution, we are more interested in the chance of fixation of new mutations. Thus, in the following we shall deal with the mean fixation time of those mutants that will eventually become fixed in the population. This variable is called the **conditional fixation time**. In the case of a new mutation $[q = 1/(2N)]$, the mean conditional fixation time, $\bar{t}$, was calculated by Kimura and Ohta (1969). For a neutral mutation, it is approximated by

$$\bar{t} = 4N \text{ generations} \tag{2.22}$$

and, for a mutation with a selective advantage of s, it is approximated by

$$\bar{t} = (2/s)\ell n(2N) \text{ generations} \tag{2.23}$$

To illustrate the difference between different types of mutation, let us assume a mammalian species with an effective population size of about 10^6 and a mean generation time of 2 years. Under these conditions, it will take a neutral mutation, on average, $4 \times 10^6 \times 2 = 8$ million years to become fixed in the population. In comparison, a mutation with a selective advantage of 1% will become fixed in the same population in only about 5800 years. Interestingly, the conditional fixation time for a deleterious allele with a selective disadvantage s is exactly the same as that for an advantageous allele with a selective advantage s (Maruyama and Kimura 1974). This is intuitively understandable given the high probability of loss for a deleterious allele. That is, for a deleterious allele to become fixed in a population, fixation must occur very quickly.

Figure 2.6 schematically depicts the dynamics of gene substitution for advantageous and neutral mutations. Advantageous mutations are either rapidly lost or rapidly fixed in the population. In contrast, the frequency changes for neutral alleles are slow, and the fixation time is much longer than for advantageous mutants.

Rate of Gene Substitution

Let us now consider the **rate of substitution**, defined as the number of mutants reaching fixation per unit time. First, consider neutral mutations. If neutral mutations occur at a rate of u per gene per generation, then the number of mutants arising at a locus in a diploid population of size N is $2Nu$ per generation. Since the probability of fixation for each of these mutations is $P = 1/(2N)$ and since the rate of substitution is $K = 2NuP$, we have

$$K = u \tag{2.24}$$

Thus, for neutral mutations, the rate of substitution is equal to the rate of mutation, a remarkably simple result (Kimura 1968a). This result can be intuitively understood by noting that, in a large population, the number of mutations arising every generation is high, but the fixation probability of each mutation is low. In comparison, in a small population, the number of mutations arising every generation is low, but the fixation probability of each mutation is high. As a consequence, the rate of substitution for neutral mutations is independent of population size.

For advantageous mutations, the rate of substitution can also be obtained by multiplying the number of mutations arising every generation (i.e., $2Nu$) by the probability of fixation for such alleles as given in Equation 2.21. For genic selection with $s > 0$, we obtain

$$K = 4Nsu \tag{2.25}$$

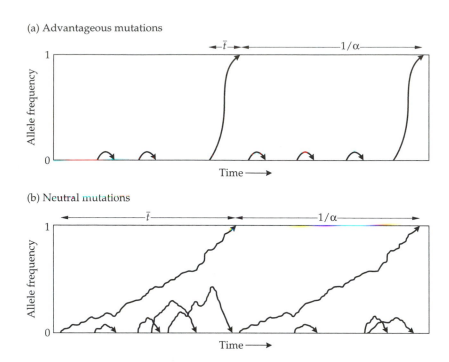

(a) Advantageous mutations

Allele frequency

Time ⟶

(b) Neutral mutations

Allele frequency

Time ⟶

Figure 2.6 Dynamics of gene substitution for (a) advantageous and (b) neutral mutations. Advantageous mutations are either quickly lost from the population or quickly fixed, so their contribution to genetic polymorphism is small. The frequency of neutral alleles, on the other hand, changes very slowly by comparison, so a large amount of transient polymorphism is generated. $\bar{t}$ is the conditional fixation time and $1/\alpha$ is the mean time between consecutive fixation events. From Nei (1987).

In other words, the rate of substitution for the case of genic selection depends on the population size (N) and the selective advantage (s), as well as on the rate of mutation (u).

EXTINCTION OF AN ALLELE UNDER MUTATION PRESSURE

At the DNA (protein) level an allele represents a nucleotide (protein) sequence. The allele will become a different allele if a mutation occurs at any nucleotide site (amino acid residue). Once it mutates to a different allele, the chance for the new allele to mutate back to the original allele is very small because the mutation must occur at the mutated site and in a specific direction. To see this, consider a gene consisting of 300 nucleotides, a fairly small gene. Suppose that a mutation occurs at the tenth nucleotide in allele A_1, changing the nucleotide from T to C, so that a new allele, A_2, is created. If a new mutation occurs in A_2, the probability for it to mutate back to A_1 is less than $1/300$ because the mutation has a probability of $1/300$ to occur at the tenth nucleotide, and even if it occurs at that site, the nucleotide may change to A or G rather than back to T. Therefore, practically every allele in a population is subject to irreversible mutation and will eventually become extinct from the population.

The question then is, "How long will it take for an allele to become extinct from a population under irreversible mutation?" Many authors have studied this problem (e.g., Ewens 1964; Crow and Kimura 1970; Nagylaki 1974; Nei 1976). The problem can be formulated as follows. Let A be the allele under consideration and put all other possible allelic forms into one single class and denote it by a. Let u be the mutation rate per gene per generation. Then, A mutates to a irreversibly at the rate of u per generation. For simplicity, let us consider only genic selection and assume that the relative fitnesses of genotypes AA, Aa, and aa are 1, $1 + s$, and $1 + 2s$, respectively. Thus, $s > 0$ ($s < 0$) means that all new mutations have a selective advantage (disadvantage) of s over A, whereas $s = 0$ means that new mutations are all neutral. Let the initial frequency of A be p and effective population size be N_e. Li and Nei (1977) developed a general formula for the mean extinction time of an allele, $T(p)$, which is not presented here because it is rather complicated in the presence of selection. For neutral mutation, if $p = 1$, the formula becomes

$$T(1) = \sum_{i=1}^{\infty} \frac{4N_e}{i(\theta + i - 1)} \tag{2.26}$$

where $\theta = 4N_e u$. If $\theta \le 0.1$, Equation 2.26 becomes approximately

$$T(1) = 4N_e \left(1 + \frac{1}{\theta}\right) = 4N_e + \frac{1}{u} \tag{2.27}$$

Table 2.1 shows the mean extinction times for neutral mutations ($S = 4N_e s = 0$), advantageous mutations ($S > 0$), and disadvantageous mutations ($S < 0$). The case

TABLE 2.1 Mean extinction time of an allele under mutation pressure[a]

		θ			
S	p	*1*	*0.1*	*0.01*	*0.001*
0	1	1.65	10.94	101.0	1001.0
	0.5	1.06	5.80	50.8	500.8
10	1	0.54	1.53	10.4	99.1
	0.5	0.29	0.32	0.39	0.99
100	1	0.10	0.20	1.05	9.61
	0.5	0.05	0.05	0.05	0.05
−5	1	9.93	251	—	—
−50	1	2×10^{18}	7×10^{20}	—	—

From Li and Nei (1977).

[a]Time is measured in units of $4N_e$ generations, where N_e is the effective population size. $\theta = 4N_e u$, $S = 4N_e s$, and p is the initial frequency of the allele (u is the mutation rate and s is the selective advantage or disadvantage). The mutant alleles are selectively neutral if $S = 0$, advantageous if $S > 0$, and disadvantageous if $S < 0$.

of $p = 1$ is of especial interest because if we consider all alleles currently existing in the population as a single allele and denote it by A, then $p = 1$ and $T(1)$ is the expected time until all presently existing alleles become extinct from the population.

First, let us consider neutral mutations. For $\theta \le 0.1$, the mean extinction time is roughly inversely proportional to $\theta = 4N_e u$ or, in other words, it is roughly proportional to $1/u$ if N_e is fixed; note that in Table 2.1, time is measured in units of $4N_e$ generations. The dependence of the mean extinction time on u is easily seen from Equation 2.27, which shows that if $4N_e < 1/u$, the mean extinction time is roughly proportional to $1/u$. For a given θ, the mean extinction time decreases with the initial frequency p. For example, for $\theta \le 0.1$, the mean extinction time for $p = 0.5$ is only about half of that for $p = 1$. This can be understood by noting that when $\theta \le 0.1$, mutations occur rarely, so that for $p = 0.5$, roughly 50% of the cases the allele will become lost in a relatively short time (less than $4N_e$ generations) and in the other 50% of the cases the allele will become fixed or nearly fixed in the population, and the mean extinction time is then similar to that for $p = 1$.

Next, let us consider advantageous mutations. Clearly, if all new mutations are advantageous, the mean extinction time for A is greatly reduced. For example, for $p = 1$ and $\theta = 0.01$, the mean extinction times (in units of $4N_e$ generations) are 101.0 for $S = 0$, 10.4 for $S = 10$, and 1.1 for $S = 100$. Note also that for the case of $p = 0.5$ and $S = 100$, the mean extinction time is independent of mutation rate. This is because the selection intensity is strong so that allele A is quickly replaced by the mutant alleles already existing in the population.

Finally, if all mutations are disadvantageous, the mean extinction time is greatly increased. For example, if $p = 1$, $\theta = 1$, and $S = -50$, the mean extinction time is $2 \times 10^{18} \times 4N_e$ generations, which is extremely long. Note that if $N_e = 12,500$, then $S = -50$ implies $s = -0.001$, which is fairly small. Thus, if the effective population size is fairly large, disadvantageous mutations have practically no chance of becoming fixed in the population. The assumption that all mutations are disadvantageous implies that the present allele is the optimal allele and that it is subject to purifying selection arising from functional or structural requirements of the sequence. The above result implies that if such requirements are stringent, then the optimal allele can persist in the population for an extremely long time. As we shall see in Chapter 7, a protein sequence with stringent structural requirements can indeed persist for millions of years without change.

GENETIC POLYMORPHISM

A locus is said to be **polymorphic** if two or more alleles coexist in the population. However, if one of the alleles has a very high frequency, say, 99% or more, then none of the other alleles is likely to be observed in a sample unless the sample size is large. Thus, for practical purposes, a locus is commonly defined as polymorphic if the frequency of the most common allele is less than 99%. This definition is obviously arbitrary and other criteria have been used in the literature.

One of the simplest ways to measure the extent of polymorphism in a population is to compute the proportion of polymorphic loci (P) by dividing the number of polymorphic loci by the total number of loci sampled. For example, if 5 of the 20 loci studied are polymorphic, then $P = 5/20 = 25\%$. This measure, however, is dependent on the number of individuals studied. A more appropriate measure of genetic variability within populations is **gene diversity**. This measure does not depend on an arbitrary delineation of polymorphism, can be computed directly from knowledge of the gene frequencies, and is less affected by sampling effects. Gene diversity at a locus is defined as:

$$h = 1 - \sum_{i=1}^{m} x_i^2 \qquad (2.28)$$

where x_i is the frequency of allele i and m is the number of alleles observed at the locus. For any given locus, h is the probability that two alleles chosen at random from the population are different from each other. In a randomly mating population, h is also the **expected heterozygosity**, i.e., the expected frequency of heterozygotes in the population for a locus with allele frequencies x_i, $i = 1,...,m$. The average of h values over all the loci studied can be used as a measure of genetic variability in the population.

The introduction of electrophoresis into population genetics in the early 1960s provided a convenient, powerful tool for studying protein polymorphism in natural populations. It was then discovered that natural populations such as humans and *Drosophila* contain a large amount of genetic variability (Harris 1966; Lewontin and Hubby 1966). In fact, early surveys revealed that the proportion of polymorphic loci is about 30% in mammalian species and can be more than 50% in *Drosophila* species (Table 2.2). The question was then how such high genetic variabilities are maintained in natural populations. This issue will be discussed in the next section and in Chapter 9.

Since the 1960s there has been much interest in developing mathematical models for studying genetic variability. A commonly used one is the infinite-allele

TABLE 2.2 **Surveys of protein polymorphism in a number of organisms**[a]

Species	Number of populations	Number of loci	P	H
Homo sapiens	1	71	0.28	0.067
Mus musculus musculus	4	41	0.29	0.091
M. m. brevirostris	1	40	0.30	0.110
M. m. domesticus	2	41	0.20	0.056
Peromyscus polionotus	7 (regions)	32	0.23	0.057
Drosophila pseudoobscura	10	24	0.43	0.128
D. persimilis	1	24	0.25	0.106
D. obscura	3 (regions)	30	0.53	0.108
D. subobscura	6	31	0.47	0.076
D. willistoni	10	20	0.81	0.175
D. melanogaster	1	19	0.42	0.119
D. simulans	1	18	0.61	0.160

From Lewontin (1974).

[a]P = proportion of polymorphic loci; H = average heterozygosity.

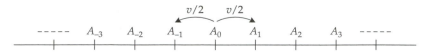

Figure 2.7   The stepwise mutation model. A_0, A_1, etc. denote the possible allelic states and v denotes the mutation rate per gene per generation. From Ohta and Kimura (1973).

model (Wright 1949; Kimura and Crow 1964), which assumes that every mutation creates a new allele or an allele that is not currently existing in the population. Under this model and the assumption of selective neutrality, the expected heterozygosity at equilibrium is given by

$$h = \frac{4N_e u}{1 + 4N_e u} \tag{2.29}$$

where N_e is the effective population size and u is the mutation rate per gene per generation (Kimura and Crow 1964). This simple formula has been heavily used as a reference point for the neutral expectation. There are of course many factors, such as selection and migration, that can cause deviations from this expectation. For example, overdominant selection can greatly increase the heterozygosity, whereas puryfing selection can reduce it (see Kimura and Crow 1964; Li 1977; Watterson 1977).

It was questioned whether the infinite-allele model is suitable for analyzing electrophoretic data, because the electrophoretic mobility of a protein may change in a stepwise manner so that recurrent and backward mutation can occur. In other words, a mutation may not create a new allele. To make the model more realistic, Ohta and Kimura (1973) proposed the stepwise-mutation model as shown in Figure 2.7. In this model the possible allelic states are visualized as the integers on a line, and a mutation causes the state of the allele to move one step either to the right or to the left. Under this model and the assumption of neutrality, the expected heterozygosity at equilibrium is given by

$$h = 1 - 1/\sqrt{1 + 8N_e u} \tag{2.30}$$

The h value under the step-mutation model can be substantially lower than that under the infinite-allele model. This model has been extended to include the possibility of two-step changes (Wehrhahn 1975; Li 1976). Although electrophoresis is now only occasionally used in polymorphism study, the stepwise model is introduced here, because it will be used when we discuss the population genetics of tamdem repeats of DNA (Chapter 13).

THE NEO-DARWINIAN THEORY AND
THE NEUTRAL MUTATION HYPOTHESIS

Darwin proposed his theory of evolution by natural selection without knowledge of the sources of variation in populations. After Mendel's laws were rediscovered and genetic variation was shown to be generated by mutation, Darwinism and

Mendelism were used as the framework of what came to be called the synthetic theory of evolution, or neo-Darwinism. According to this theory, mutation is recognized as the ultimate source of genetic variation, but natural selection is given the dominant or "creative" role in shaping the genetic makeup of populations and in the process of gene substitution.

In time, neo-Darwinism became a dogma in evolutionary biology, and selection came to be considered the only force capable of driving the evolutionary process, while other factors such as mutation and random drift were thought of as minor contributors at best. This particular brand of neo-Darwinism was called **selectionism**.

According to the selectionist or neo-Darwinian perception of the evolutionary process, gene substitutions occur as a consequence of selection for advantageous mutations. Polymorphism, on the other hand, is maintained by balancing selection. Thus, neo-Darwinists regard substitution and polymorphism as two separate phenomena driven by different evolutionary forces. Gene substitution is the end result of a positive adaptive process, whereby a new allele takes over future generations of the population if and only if it improves the fitness of the organism, while polymorphism is maintained when the coexistence of two or more alleles at a locus is advantageous for the organism or the population. Neo-Darwinian theories maintain that most genetic polymorphisms in nature are stable.

The 1960s witnessed a revolution in population genetics. The introduction of electrophoresis into population genetics studies soon led to the discovery of the existence of large amounts of genetic variability in natural populations such as human and *Drosophila* populations (Harris 1966; Lewontin and Hubby 1966). The availability of protein sequence data removed the species boundary in population genetics studies and for the first time provided adequate empirical data for examining theories pertaining to the process of gene substitution. In 1968, Kimura postulated that the majority of molecular changes in evolution are due to the random fixation of neutral or nearly neutral mutations (Kimura 1968a); it was also independently proposed by King and Jukes (1969). This hypothesis, now known as the **neutral theory of molecular evolution**, contends that at the molecular level the majority of evolutionary changes and much of the variability within species are caused neither by positive selection of advantageous alleles nor by balancing selection, but by random genetic drift of mutant alleles that are selectively neutral or nearly so. Neutrality, in the sense of the theory, does not imply strict equality in fitness for all alleles. It only means that the fate of alleles is determined largely by random genetic drift. In other words, selection may operate, but its intensity is too weak to offset the influences of chance effects. For this to be true, the absolute value of the selective advantage or disadvantage of an allele must be smaller than $1/(2N_e)$.

According to the neutral theory, the frequency of alleles is determined largely by stochastic rules, and the picture that we obtain at any given time is merely a transient state representing a temporary frame from an ongoing dynamic process. Consequently, polymorphic loci consist of alleles that are either on their way to fixation or on their way to extinction. Viewed from this perspective, all molecular manifestations that are relevant to the evolutionary process should be regarded as the result of a continuous process of a mutational input and a concomitant random extinction or fixation of alleles. Thus, the neutral theory regards substitution and polymorphism as two facets of the same phenomenon. Substitution is a long and gradual process, whereby the frequencies of mutant alleles increase or decrease randomly, until the alleles are ultimately fixed or lost by chance. At any given time, some loci will possess alleles at frequencies that are neither 0% nor 100%. These are the polymorphic loci. According to the neutral theory, most genetic polymorphism in populations is transient in nature.

The essence of the dispute between neutralists and selectionists essentially concerns the distribution of fitness values of mutant alleles. Both schools agree that most new mutations in proteins are deleterious and that these mutations are quickly removed from the population so that they contribute neither to the rate of substitution nor to the amount of polymorphism within populations. The difference concerns the relative proportion of neutral mutations among nondeleterious mutations. While selectionists maintain that very few mutations are selectively neutral, neutralists maintain that most nondeleterious mutations are effectively neutral. Of course, not all selectionists hold the same view of evolution nor do all neutralists (see reviews by Lewontin 1974; Kimura 1983; Nei 1987; Gillespie 1991). For example, Ohta's (1973, 1974) hypothesis of slightly deleterious mutation emphasizes the importance of slightly deleterious mutations in gene substitution and molecular polymorphism, whereas in Nei's (1987) view such mutations do not play an important role.

The heated controversy over the neutral-mutation hypothesis during the last two decades has had a strong impact on molecular evolution. First, it has led to the general recognition that the effect of random drift cannot be neglected when considering the evolutionary dynamics of molecular changes. Second, the synthesis between molecular biology and population genetics has been greatly strengthened by the introduction of the concept that molecular evolution and genetic polymorphism are but two facets of the same phenomenon (Kimura and Ohta 1971a). Although the controversy still continues, it is now recognized that any adequate theory of evolution must be consistent with both of these aspects of the evolutionary process at the molecular level.

PROBLEMS

1. If A_2 is completely dominant to A_1, what will be the change in the frequency of allele A_2 per generation?

2. Derive Equation 2.5 from Equation 2.4. Hint: Equation 2.4 can be rewritten as

$$\frac{dq}{q(1-q)} = sdt$$

3. Use Equation 2.6 to compute the times required for the allele frequency to increase (1) from 0.01 to 0.201, (2) from 0.3 to 0.5, and (3) from 0.8 to 0.99. From the results discuss the effect of allele frequency on the efficiency of natural selection.

4. Derive the equilibrium frequency in Equation 2.8 from Equation 2.7.

5. Use the binomial probability function (2.9) to compute the distribution of allele frequencies at generation 1 in Figure 2.5, assuming $N = 10$ and $p = 0.3$.

6. In an idealized population in which the effective size N_e is the same as the actual size N, show that the variance of allele frequency in the next generation is given by Equation 2.12.

7. What is the ratio of the effective population size to census population size in a population in which females outnumber males by 2:1?

8. A population runs through a bottleneck such that in six consecutive generations its population size is 10^4, 10^4, 10^4, 10, 10^4, and 10^4. What is its long-term effective population size?

9. What is the fixation probability of a new mutation with a selective disadvantage of 0.01 in a population in which the effective population size is 100 and $N_e = N$?

10. Discuss the differences between the neutral mutation hypothesis and the hypothesis that all mutations are neutral.

CHAPTER 3

Evolutionary Change in Nucleotide Sequences

A BASIC PROCESS IN THE EVOLUTION OF DNA SEQUENCES is the change in nucleotides with time. This process deserves a detailed consideration because it is essential for understanding the mechanism of DNA evolution and because changes in nucleotide sequences are used both for estimating the rate of evolution and for reconstructing the evolutionary history of organisms. For this reason, many models have been proposed for studying this process (e.g., Jukes and Cantor 1969; Kimura 1980, 1981a; Holmquist and Pearl 1980; Kaplan and Risko 1982; Lanave et al. 1984). We shall study first the change in a single sequence and then the divergence between two sequences. The theory presented in this chapter provides a framework for estimating the number of nucleotide substitutions between two sequences, a subject to be treated in the next chapter.

NUCLEOTIDE SUBSTITUTION IN A DNA SEQUENCE

To study the dynamics of nucleotide substitution, we must make assumptions regarding the probability of substitution of one nucleotide by another. Many such mathematical schemes have been proposed in the literature (see Zharkikh 1994). We start with the simplest models: Jukes and Cantor's (1969) one-parameter model and Kimura's (1980) two-parameter model. More general models will be discussed later.

Jukes and Cantor's One-Parameter Model

The substitution scheme of Jukes and Cantor's (1969) (the **JC**) model is shown in Figure 3.1. This model assumes no bias in the direction of change so that substitutions occur randomly among the four types of nucleotide. For example, if the nucleotide under consideration is A, it will change to T, C, or G with equal probability. In this model, the rate of substitution for each nucleotide is 3α per unit time, and the rate of substitution in each of the three possible directions of change is α. Since the model involves only one parameter, α, it is also called the **one-parameter model**.

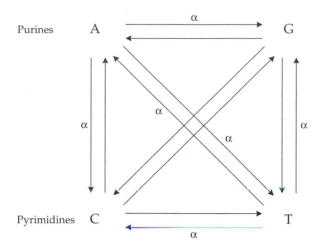

Figure 3.1   One-parameter model of nucleotide substitution. In this model, the rate of substitution in each direction is α. From Li and Graur (1991).

Let us assume that the nucleotide residing at a certain site in a DNA sequence is A at time 0. First, we consider the probability, $p_{A(t)}$, that this site will be occupied by A at time t. Since we start with A, the probability that this site is occupied by A at time 0 is $p_{A(0)} = 1$. At time 1, the probability of still having A at this site is given by

$$p_{A(1)} = 1 - 3\alpha \tag{3.1}$$

which reflects the probability, $1 - 3\alpha$, that the nucleotide has remained unchanged.

The probability of having A at time 2 is

$$p_{A(2)} = (1 - 3\alpha)p_{A(1)} + \alpha[1 - p_{A(1)}] \tag{3.2}$$

To derive this equation, we consider two possible scenarios: (1) the nucleotide has remained unchanged, and (2) the nucleotide has changed to T, C, or G but has subsequently reverted to A (Figure 3.2). The probability of the nucleotide being A at time 1 is $p_{A(1)}$, and the probability that it has remained A at time 2 is $1 - 3\alpha$. The product of the probabilities for these two events gives us the probability for the first scenario, which constitutes the first term in Equation 3.2. The probability of the nucleotide not being A at time 1 is $1 - p_{A(1)}$, and its probability of changing to A at time 2 is α. The product of these two probabilities gives us the probability for the second scenario, and constitutes the second term in Equation 3.2.

Using the above formulation, we can show that the following recurrence equation holds for any t:

$$p_{A(t+1)} = (1 - 3\alpha)p_{A(t)} + \alpha[1 - p_{A(t)}] \tag{3.3}$$

We can rewrite Equation 3.3 in terms of the amount of change in $p_{A(t)}$ per unit time as

$$p_{A(t+1)} - p_{A(t)} = -3\alpha\, p_{A(t)} + \alpha[1 - p_{A(t)}]$$

or

$$\Delta p_{A(t)} = -3\alpha\, p_{A(t)} + \alpha[1 - p_{A(t)}] = -4\alpha\, p_{A(t)} + \alpha \tag{3.4}$$

So far we have considered a discrete-time process. It is more convenient to approximate this process by a continuous-time model, by regarding $\Delta p_{A(t)}$ as the rate of change at time t. With this approximation, Equation 3.4 is rewritten as

$$\frac{dp_{A(t)}}{dt} = -4\alpha\, p_{A(t)} + \alpha \tag{3.5}$$

This is a first-order linear differential equation, and the solution is given by

$$p_{A(t)} = \frac{1}{4} + \left(p_{A(0)} - \frac{1}{4}\right)e^{-4\alpha t} \tag{3.6}$$

Since we started with A, $p_{A(0)} = 1$. Therefore,

$$p_{A(t)} = \frac{1}{4} + \frac{3}{4}e^{-4\alpha t} \tag{3.7}$$

Actually, Equation 3.6 holds regardless of the initial conditions. For example, if the initial nucleotide is not A, then $p_{A(0)} = 0$, and the probability of having A at this position at time t is

$$p_{A(t)} = \frac{1}{4} - \frac{1}{4}e^{-4\alpha t} \tag{3.8}$$

In the above, we focused on a particular nucleotide site and treated $p_{A(t)}$ as a probability. However, $p_{A(t)}$ can also be interpreted as the frequency of A in a DNA sequence. For example, if we start with a sequence made of A's only, then $p_{A(0)} = 1$, and $p_{A(t)}$ is the expected frequency of A in the sequence at time t.

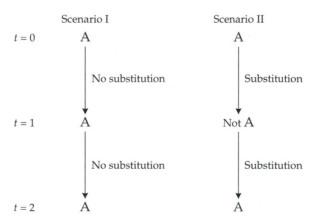

Figure 3.2   Two possible scenarios for having A at a site at time $t = 2$, given that the site had A at time 0. From Li and Graur (1991).

We can rewrite Equation 3.7 in a more explicit form to take into account the facts that the initial nucleotide is A and the nucleotide at time t is also A.

$$p_{AA(t)} = \frac{1}{4} + \frac{3}{4}e^{-4\alpha t} \qquad (3.9)$$

If the initial nucleotide is G instead of A, then from Equation 3.8 we obtain:

$$p_{GA(t)} = \frac{1}{4} - \frac{1}{4}e^{-4\alpha t} \qquad (3.10)$$

Since all the nucleotides are equivalent under the JC model, $p_{GA(t)} = p_{CA(t)} = p_{TA(t)}$. In fact, we can consider a general probability, $p_{ij(t)}$, which is the probability that a nucleotide will become j at time t, given that the initial nucleotide is i. By using this generalized notation and Equation 3.9, we obtain

$$p_{ii(t)} = \frac{1}{4} + \frac{3}{4}e^{-4\alpha t} \qquad (3.11)$$

and from Equation 3.10

$$p_{ij(t)} = \frac{1}{4} - \frac{1}{4}e^{-4\alpha t} \qquad (3.12)$$

where $i \neq j$.

Equations 3.11 and 3.12 are sufficient for describing the substitution process under the one-parameter model. For example, suppose that we wish to know the probability that the nucleotide at a given site will be A at time t. There are two possible situations. First, the nucleotide at the present time is A. In this case, $i = A$ and Equation 3.11 shows that $p_{ii(t)}$ decreases exponentially from 1 to 1/4 as t increases from 0 to ∞; see the upper dotted line denoted by p_{ii} in Figure 3.3. Second, the nucleotide at the present time is not A, but T (or C, or G). Then, $i = T$ and $j = A$ and Equation 3.12 shows that $p_{ij(t)}$ increases monotonically from 0 to 1/4 as t increases from 0 to ∞; see the lower dotted line denoted by p_{ij} in Figure 3.3. Thus, regardless of the initial condition, the probability of having A at the site will eventually become 1/4 (Figure 3.3). This also holds true for T, C, and G. Therefore, under the JC model the equilibrium frequency of each of the four nucleotides is 1/4. (The equilibrium state is also referred to as the stationary state.) This result is of course in the average sense, because in finite sequences fluctuations in nucleotide frequencies are likely to occur.

Figure 3.3 shows that the decrease in p_{ii} is linear when t is small but becomes nonlinear as t increases. The nonlinearity occurs because of back mutation, that is, the initial nucleotide may change to another nucleotide and then change back to the same nucleotide.

Kimura's Two-Parameter Model

The assumption that all nucleotide substitutions occur randomly, as in the JC model, is unrealistic in most cases. For example, transitions are generally more frequent than transversions (Chapters 1 and 7). To take this fact into account, Kimura (1980) proposed a two-parameter model (Figure 3.4). In this scheme, the rate of transitional substitution at each nucleotide site is α per unit time, whereas the rate of each of the two types of transversional substitution is β per unit time.

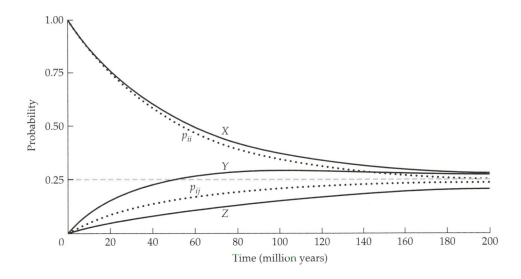

Figure 3.3 Temporal changes in the probability of having a certain nucleotide, say A, at a given nucleotide site. The two dotted lines are computed under the one-parameter model with $\alpha = 5 \times 10^{-9}$ substitutions/site/year. The line denoted by p_{ii} starts with the same nucleotide (i.e., A) while the line denoted by p_{ij} starts with a different nucleotide (i.e., T, C, or G). The three solid lines are computed under Kimura's two-parameter model with $\alpha = 10 \times 10^{-9}$ and $\beta = 2.5 \times 10^{-9}$ substitutions/site/year. The line denoted by X starts with A, the line denoted by Y starts with G (a transition), and the line denoted by Z starts with T (or C; a transversion). The dashed line denotes the equilibrium frequency (0.25).

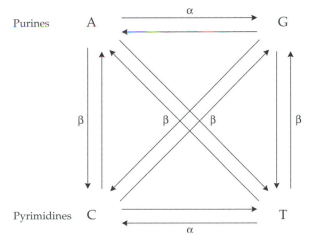

Figure 3.4 Two-parameter model of nucleotide substitution. In this model, the rate of transition (α) may not be equal to the rate of each of the two types of transversion (β). From Li and Graur (1991).

As in the one-parameter model, let $p_{A(t)}$ be the probability that the nucleotide at the site under consideration is A at time t; $p_{T(t)}$, $p_{C(t)}$, and $p_{G(t)}$ are similarly defined. The probability that the nucleotide at the site is A in the next generation is given by

$$p_{A(t+1)} = (1-\alpha-2\beta)p_{A(t)} + \beta p_{T(t)} + \beta p_{C(t)} + \alpha p_{G(t)} \tag{3.13a}$$

To derive this equation we need to consider four possible scenarios: First, the nucleotide at the site is A at time t and will remain unchanged at $t+1$. The probability that the nucleotide at the site is A at time t is $p_{A(t)}$ and the probability that it will remain unchanged at $t+1$ is $(1-\alpha-2\beta)$. The product of these two probabilities constitutes the first term in Equation 3.13a. Second, the nucleotide at the site is T at time t but will change to A at $t+1$, a transversional change. The probabilities for the two events are $p_{T(t)}$ and β, respectively, and their product constitutes the second term in Equation 3.13a. Third, the nucleotide at the site is C at time t but will change to A at $t+1$, a transversional change. The probabilities for the two events are $p_{C(t)}$ and β, respectively, and their product constitutes the third term in Equation 3.13a. Fourth, the nucleotide at the site is G at time t but will change to A at $t+1$, a transitional change. The probabilities for these two events are $p_{G(t)}$ and α, respectively, and their product constitutes the fourth term in Equation 3.13a. In the same manner, we can derive the following equations:

$$p_{T(t+1)} = \beta p_{A(t)} + (1-\alpha-2\beta)p_{T(t)} + \alpha p_{C(t)} + \beta p_{G(t)} \tag{3.13b}$$

$$p_{C(t+1)} = \beta p_{A(t)} + \alpha p_{T(t)} + (1-\alpha-2\beta)p_{C(t)} + \beta p_{G(t)} \tag{3.13c}$$

$$p_{G(t+1)} = \alpha p_{A(t)} + \beta p_{T(t)} + \beta p_{C(t)} + (1-\alpha-2\beta)p_{G(t)} \tag{3.13d}$$

To solve the above equations we need to set the initial conditions. If the initial nucleotide is A, then $p_{A(0)} = 1$ and $p_{T(0)} = p_{C(0)} = p_{G(0)} = 0$, and a solution can be obtained for the above equations. Similarly, we can obtain a solution for the initial conditions $p_{T(0)} = 1$ and $p_{A(0)} = p_{C(0)} = p_{G(0)} = 0$, and so on. For the detail of the solution, see Li (1986).

In the JC model, Equation 3.11 implies $p_{AA}(t) = p_{GG}(t) = p_{CC}(t) = p_{TT}(t)$. This equality also holds for Kimura's two-parameter model, because the four nucleotides are equivalent in this model. We shall denote this probability by $X_{(t)}$. It can be shown by using Equations (3.13a–c) that

$$X_{(t)} = \frac{1}{4} + \frac{1}{4}e^{-4\beta t} + \frac{1}{2}e^{-2(\alpha+\beta)t} \tag{3.14}$$

As in the one-parameter model, the solution is given in continuous rather than in discrete time.

Under the one-parameter (JC) model, Equation 3.12 holds regardless of whether the change from nucleotide i to nucleotide j is a transition or a transversion. In contrast, in Kimura's two-parameter model, we need to distinguish between transitional and transversional changes because the two types of change occur with different probabilities. Denote by $Y_{(t)}$ the probability that the initial nucleotide and the nucleotide at time t differ from each other by a transition. For example, if the initial nucleotide is A and the nucleotide at time t is G, then they differ by a transition and $Y_{(t)} = p_{AG(t)}$. Because of the symmetry of the substitution scheme (Figure 3.4), $Y_{(t)} = p_{AG(t)} = p_{GA(t)} = p_{TC(t)} = p_{CT(t)}$. It can be shown that

$$Y_{(t)} = \frac{1}{4} + \frac{1}{4}e^{-4\beta t} - \frac{1}{2}e^{-2(\alpha+\beta)t} \tag{3.15}$$

The probability $Z_{(t)}$ that the nucleotide at time t and the initial nucleotide differ by a specific type of transversion is given by

$$Z_{(t)} = \frac{1}{4} - \frac{1}{4}e^{-4\beta t} \tag{3.16}$$

Note that each nucleotide is subject to two types of transversion. For example, if the initial nucleotide is A, then the two possible transversional changes are A → C and A → T. Therefore, the probability that the initial nucleotide and the nucleotide at time t differ by either of the two types of transversion is twice the probability given by Equation 3.16. Note that $X_{(t)} + Y_{(t)} + 2Z_{(t)} = 1$.

Figure 3.3 shows a comparison between the one- and two-parameter models. The line denoted by X is computed from Equations 3.14. The rates used are $\alpha = 10 \times 10^{-9}$ and $\beta = 2.5 \times 10^{-9}$ substitutions per site per year; since the total rate of substitution per site is $\alpha + 2\beta = 15 \times 10^{-9}$, the rate of transition is two-thirds of the total rate. The dotted line denoted by p_{ii} is computed under the one-parameter model (i.e., Equation 3.11); the rates used are $\alpha = \beta = 5 \times 10^{-9}$, so the rate of transition is one-third of the total rate. Although the total rate of substitution per site is the same for the two curves (the two models), X decreases more slowly than p_{ii}. To understand why this difference occurs, let us assume that the initial nucleotide is A, so that X and p_{ii} represent the probabilities of having A at the site at time t under the two-parameter and one-parameter models, respectively. The probability of changing from A to non-A is the same in the two models, but the probability of changing from A to G and backward to A is higher in the two-parameter model because it has a higher rate of transition (A ↔ G) than does the one-parameter model. Because of this difference in the probability of backward substitution, X decreases more slowly than p_{ii}. However, both X and p_{ii} will approach 1/4 as $t \to \infty$. Thus, as in the one-parameter model, in Kimura's two-parameter model the equilibrium value for the frequency of each nucleotide is 1/4.

Another interesting difference between the two models is as follows. In Figure 3.3, lines Y and Z represent the probabilities that the nucleotide at time t differs from the initial nucleotide by a transition and by a specific transversion, respectively, under Kimura's two-parameter model. As in the one-parameter model (line p_{ij} in the figure), Z increases monotonically to the equilibrium value 1/4. But, surprisingly, Y is not monotonic and, in fact, can become greater than 1/4, the equilibrium value. To see why this can happen, let us study Equation 3.15. Note that $Y_{(t)}$ is > 1/4 if

$$\frac{1}{4}e^{-4\beta t} > \frac{1}{2}e^{-2(\alpha+\beta)t}$$

or

$$\frac{1}{2} > e^{-2(\alpha-\beta)t}$$

This condition holds if $\alpha > \beta$ and

$$t > \frac{1}{2}\ell n(2)/(\alpha - \beta) \approx 0.347/(\alpha - \beta) \tag{3.17}$$

Thus, if the rate of transition (α) is higher than the rate of each of the two types of transversion (β), the probability (Y) of having a transitional difference between the initial nucleotide and the nucleotide at t can become larger than 1/4 as t becomes larger than $0.693/(\alpha - \beta)$. Of course, as $t \to \infty$, Y will become 1/4.

Other Models of Nucleotide Substitution

The two-parameter model is more realistic than the one-parameter model but also involves many restrictions. For example, it requires that the four types of transition occur with the same rate, but, as seen in Chapter 1, this may not be the case. So we need to consider more general models. To do so, it is more convenient to use matrix theory. Let us number the four nucleotides A, T, C, and G as 1, 2, 3, and 4, respectively. Let $\mathbf{M}$ be a 4×4 matrix in which the i,j-th element ($i \neq j$), m_{ij}, represents the probability that the nucleotide will change to j in the next generation, given that the nucleotide is i at the present generation, and the i,i-th element, m_{ii}, represents the probability that the nucleotide will remain unchanged in the next generation. For example, for Kimura's two-parameter model, $\mathbf{M}$ is defined as

$$\mathbf{M} = \begin{bmatrix} 1-\alpha-2\beta & \beta & \beta & \alpha \\ \beta & 1-\alpha-2\beta & \alpha & \beta \\ \beta & \alpha & 1-\alpha-2\beta & \beta \\ \alpha & \beta & \beta & 1-\alpha-2\beta \end{bmatrix} \tag{3.18}$$

If we define the vector $\mathbf{p}_{(t)}$ as $\mathbf{p}_{(t)} = (p_{A(t)}, p_{T(t)}, p_{C(t)}, p_{G(t)})$, then Equations 3.13a–d can be written as

$$\mathbf{p}_{(t+1)} = \mathbf{p}_{(t)}\mathbf{M} \tag{3.19}$$

This equation can be studied by using linear algebra.

The matrix $\mathbf{M}$ in Equation 3.18 satisfies the following two conditions: (1) all elements of the matrix are nonnegative and (2) the sum of the elements in each row is 1. Such matrices are called stochastic matrices. The above formulation has in effect assumed that the substitution process is a Markov chain, the essence of which is as follows. Consider three time points in evolution, $t_1 < t_2 < t_3$. The Markovian model assumes that the status of the nucleotide site at t_3 depends only on the status at t_2 but not the status at t_1, if the status of the site at t_2 is known. In other words, if the nucleotide at the site is specified at t_2, then no additional information about the status of the site in any past generation is required (or is of help) for predicting the nucleotide at t_3. For example, if the nucleotide at time t is A, then $p_{A(t)} = 1$ and $p_{T(t)} = p_{C(t)} = p_{G(t)} = 0$, and by using Equation 3.19 one can compute the probability $p_{i(t+1)}$ that the nucleotide will be i at time $t + 1$, $i =$ A, T, C, or G; and by iteration of the same procedure, one can compute $p_{i(\tau)}$ for any future generation τ. For the theory of Markov chains, readers may consult Feller (1968).

The above description shows that Kimura's two-parameter model can be represented by the first matrix in Table 3.1; the one-parameter model is a special case of this matrix with the condition $\alpha = \beta$. By increasing the number of parameters the substitution model can be made more complex and realistic. Several such models are shown in Table 3.1. In each of these matrices the four elements on the diagonal from the upper-left corner to the lower-right corner are the probabilities that the nucleotide at the site remains unchanged in one generation, the four elements on the diagonal from the upper-right corner to the lower-left corner are the rates of transitional substitution per generation, and the other eight elements are the rates of transversional substitution per generation.

The second matrix in Table 3.1 is Blaisdell's (1985) four-parameter model. This model differs from the two-parameter model in two important respects: First, the rates of the two transitions $A \rightarrow G$ and $T \rightarrow C$ (denoted by α) may not be equal to the rates of the two transitions $G \rightarrow A$ and $C \rightarrow T$ (denoted by β). In contrast, in the two-parameter model the rates of all four types of transition are equal to α. Second, the rate of change from nucleotide i to nucleotide j may not be equal to

---

TABLE 3.1   Models of nucleotide substitution

$O\backslash S^a$	A	T	C	G
a. Two-parameter model (Kimura 1980)				
A	$1-\alpha-2\beta$	β	β	α
T	β	$1-\alpha-2\beta$	α	β
C	β	α	$1-\alpha-2\beta$	β
G	α	β	β	$1-\alpha-2\beta$
b. Four-parameter model (Blaisdell 1985)				
A	$1-\alpha-2\gamma$	γ	γ	α
T	δ	$1-\alpha-2\delta$	α	δ
C	δ	β	$1-\beta-2\delta$	δ
G	β	γ	γ	$1-\beta-2\gamma$
c. Six-parameter model (Kimura 1981a)				
A	$1-2\alpha-\gamma$	γ	α	α
T	δ	$1-2\alpha-\delta$	α	α
C	β	β	$1-2\beta-\varepsilon$	ε
G	β	β	ξ	$1-2\beta-\xi$
d. Nine-parameter model				
A	$1-g_T\beta_1-g_C\gamma_1-g_G\alpha_1$	$g_T\beta_1$	$g_C\gamma_1$	$g_G\alpha_1$
T	$g_A\beta_1$	$1-g_A\beta_1-g_C\alpha_2-g_G\gamma_2$	$g_C\alpha_2$	$g_G\gamma_2$
C	$g_A\gamma_1$	$g_T\alpha_2$	$1-g_A\gamma_1-g_T\alpha_2-g_G\beta_2$	$g_G\beta_2$
G	$g_A\alpha_1$	$g_T\gamma_2$	$g_C\beta_2$	$1-g_A\alpha_1-g_T\gamma_2-g_C\beta_2$
e. General model				
A	$1-\alpha_{12}-\alpha_{13}-\alpha_{14}$	α_{12}	α_{13}	α_{14}
T	α_{21}	$1-\alpha_{21}-\alpha_{23}-\alpha_{24}$	α_{23}	α_{24}
C	α_{31}	α_{32}	$1-\alpha_{31}-\alpha_{32}-\alpha_{34}$	α_{34}
G	α_{41}	α_{42}	α_{43}	$1-\alpha_{41}-\alpha_{42}-\alpha_{43}$

[a]O, Original nucleotide; S, substitute nucleotide.

the rate of change from j to i; for example, the rate of the transversion A → T is γ whereas the rate of the transversion T → A is δ. In contrast, in the two-parameter model the rates of forward and backward changes are equal, i.e., $m_{ij} = m_{ji}$. Obviously, having more parameters allows more flexibilities. For example, in the two-parameter model the equilibrium frequencies of the four nucleotides are the same and equal to 1/4, whereas in Blaisdell's four-parameter model the equilibrium frequencies of the four nucleotides can be different from one another and are given by

$$\hat{p}_A = 2(\beta + \gamma)\delta/(\lambda_1\lambda_2) \tag{3.20a}$$

$$\hat{p}_T = 2(\beta + \delta)\gamma/(\lambda_1\lambda_3) \tag{3.20b}$$

$$\hat{p}_C = 2(\alpha + \delta)\gamma/(\lambda_1\lambda_3) \tag{3.20c}$$

$$\hat{p}_G = 2(\alpha + \gamma)\delta/(\lambda_1\lambda_2) \tag{3.20d}$$

where $\lambda_1 = -2(\gamma + \delta)$, $\lambda_2 = -(\alpha + \beta + 2\gamma)$, and $\lambda_3 = -(\alpha + \beta + 2\delta)$.

The third matrix in Table 3.1 represents Kimura's (1981a) six-parameter model, which has been solved by Gojobori et al. (1982a). The fourth matrix represents a nine-parameter model, in which g_A, g_T, g_C, and g_G denote the equilibrium frequencies of A, T, C, and G, respectively, so the rate of substitution to a nucleotide depends on the equilibrium frequency of the nucleotide. Because the four equilibrium frequencies must sum up to 1, they actually represent only three independent parameters. In addition, α_1, α_2, β_1, β_2, γ_1, and γ_2 provide six independent parameters. So, the total number of independent parameters is nine. This model contains several submodels. For example, if $\alpha_1 = \alpha_2 = \beta_1 = \beta_2 = \gamma_1 = \gamma_2 = \alpha$, the model becomes Felsenstein's (1981) and Tajima and Nei's (1982) "equal-input model," meaning that the mutation rates to a particular nucleotide from the other three nucleotides are the same. This model, which has four independent parameters, was extended by Hasegawa et al. (1985) to a five-parameter model, which is the fourth matrix in Table 3.1 with the conditions $\alpha_1 = \alpha_2 = \alpha$ and $\beta_1 = \beta_2 = \gamma_1 = \gamma_2 = \beta$. In this model, α and β represent the rates of transitional substitution and trasnversional substitution, respectively. This model was in turn extended by Tamura and Nei (1993) to a six-parameter model by assuming that the rate of transition (α_1) between purines (A and G) is different from that (α_2) between pyrimidines (C and T), i.e., the fourth matrix in Table 3.1 with the condition $\beta_1 = \beta_2 = \gamma_1 = \gamma_2 = \beta$.

Although a 4×4 matrix can have 16 parameters, a substitution model can have at most 12 independent parameters because the sum of the elements in each row of the substitution matrix must be 1. Thus, the last matrix in Table 3.1 represents the most general model of nucleotide substitution, which has been studied by Lanave et al. (1984), Barry and Hartigan (1987), Lake (1994), Lockhart et al. (1994), and Zharkikh (1994). The general model contains any model with fewer independent parameters as a special case. For a mathematical treatment of various models of nucleotide substitution, see Zharkikh (1994). Under the general model the equilibrium frequencies of the four nucleotides can be computed by the formula of Wright (1969) and Tajima and Nei (1982).

A substitution process is said to be time-reversible if the probability of starting from nucleotide i and changing to nucleotide j in a time interval is the same as the probability of starting from j and going backward to i in the same time duration. In mathematical terms, time reversibility requires $p_{ij(t)}\hat{p}_i = p_{ji(t)}\hat{p}_j$ for all i and j and all t; $\hat{p}_i$ is the equilibrium frequency of nucleotide i. In particular, for $t = 1$, the condition becomes $\alpha_{ij}\hat{p}_i = \alpha_{ji}\hat{p}_j$ for all i and j in the last matrix of Table 3.1. For example, time-reversibility holds in Kimura's two-parameter model because $\alpha_{ij} = \alpha_{ji}$ and $\hat{p}_i = 1/4$ for all i and j. It can be shown that time-reversibility holds under the nine-parameter model in Table 3.1 but not under Blaisdell's four-parameter model or Kimura's six-parameter model. As will be seen below, time-reversibility simplifies the study of nucleotide sequence evolution.

DIVERGENCE BETWEEN DNA SEQUENCES

After the separation of two nucleotide sequences, each sequence will start accumulating nucleotide substitutions and the two sequences will diverge from each other. Initially, every substitution will increase the dissimilarity between the two sequences. However, this may no longer be the case when the degree of divergence becomes appreciable, so that multiple substitutions (also commonly dubbed as **multiple hits**) may occur at the same site. For example, if at a site the nucleotide had changed from T to A in sequence 1, then a change from T to A in sequence 2 will decrease rather than increase the degree of dissimilarity between the two sequences. This type of substitution is called **parallel substitution** (Figure 3.5). Some other types of substitution, such as back substitutions and convergent substitutions (Figure 3.5), may also decrease the divergence between the two sequences. These possibilities need to be taken into account when considering sequence divergence.

In studying the divergence between two sequences one can consider either sequence similarity or dissimilarity, using the results obtained above for a single sequence. One important point to be emphasized is that the rate of change in both quantities depends not only on the rate of substitution but also on the pattern (model) of nucleotide substitution. We shall use the simple tree shown in Figure 3.6 in which the two descendant sequences were derived from an ancestral sequence t time units ago.

Sequence Similarity

A common measure for sequence similarity is the proportion of identical nucleotides between the two sequences under study. The expected value of this proportion is equal to the probability, $I_{(t)}$, that the nucleotide at a given site at time t is the same in both sequences.

Let us start with the one-parameter model. Suppose that the nucleotide at a given site was A at time 0. At time t, the probability that a descendant sequence has A at this site is $p_{AA(t)}$, so the probability that both sequences have A at this site is $p_{AA(t)}^2$. In addition, we need to consider the possibility of parallel substitutions, that is, in both sequences A has changed to T, C, or G. The probabilities for these three cases are $p_{AT(t)}^2$, $p_{AC(t)}^2$, and $p_{AG(t)}^2$, respectively. Therefore,

$$I_{(t)} = p_{AA(t)}^2 + p_{AT(t)}^2 + p_{AC(t)}^2 + p_{AG(t)}^2 \qquad (3.21)$$

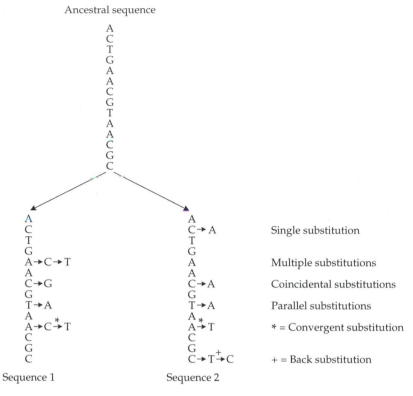

Figure 3.5 Two homologous DNA sequences that descended from an ancestral sequence and accumulated mutations since their divergence from each other. Note that although 12 mutations have accumulated, differences can be detected only at sites 2, 5, and 7, i.e., three changes. Note further that coincidental substitutions, parallel substitutions, convergent substitutions, and back substitutions all involve multiple substitutions at the same site, though perhaps in different lineages. Modified from Li and Graur (1991).

From Equations 3.11 and 3.12, we obtain

$$I_{(t)} = \frac{1}{4} + \frac{3}{4}e^{-8\alpha t} \tag{3.22}$$

This equation also holds if the initial nucleotide was T, C, or G instead of A. Therefore, $I_{(t)}$ is independent of the initial nucleotide and represents the expected proportion of identical nucleotides between two sequences that diverged t time units ago.

Next, consider Kimura's two-parameter model. As noted above, in this model, $p_{AA(t)} = X_{(t)}$, $p_{AG(t)} = Y_{(t)}$, and $p_{AT(t)} = p_{AC(t)} = Z_{(t)}$, which are given by Equations 3.14, 3.15, and 3.16, respectively. Putting these into Equation 3.21, we obtain

$$I_{(t)} = \frac{1}{4} + \frac{1}{4}e^{-8\beta t} + \frac{1}{2}e^{-4(\alpha+\beta)t} \tag{3.23}$$

As in the one-parameter model, this equation also holds for T, C, or G and so $I_{(t)}$ represents the expected proportion of identical nucleotides between two sequences that diverged t time units ago.

Finally, let us consider Blaisdell's four-parameter model. In this model the four nucleotides are no longer equivalent, and $I_{(t)}$ depends on the initial nucleotide. To take this fact into account, we shall use $I_{i(t)}$, where the subscript i signifies that the initial nucleotide was i. For example, if $i = T$, then

$$I_{T(t)} = p_{TA(t)}^{2} + p_{TT(t)}^{2} + p_{TC(t)}^{2} + p_{TG(t)}^{2}$$

The explicit formula for $I_{T(t)}$ (W.-H. Li, unpublished) will not be presented here because it is considerably more complicated than that for the two-parameter model. Since $I_{i(t)}$ now depends on i, the expected proportion, $I_{(t)}$, of identical nucleotides between the two descendant sequences depends on the nucleotide composition of the ancestral sequence. Let p_i be the proportion of nucleotide i in the ancestral sequence. Then,

$$I_{(t)} = p_A I_{A(t)} + p_T I_{T(t)} + p_C I_{C(t)} + p_G I_{G(t)} \tag{3.24}$$

If the frequencies of the four nucleotides in the ancestral sequence were at equilibrium, then $p_i = \hat{p}_i$, where $\hat{p}_i$, $i = A, T, C,$ or G, are given by Equations 3.20a–d.

We now discuss some properties of $I_{(t)}$. A comparison of Equation 3.22 with Equation 3.11 shows that in the one-parameter model $I_{(t)} = p_{ii(2t)}$. Therefore, in this model the probability for two sequences that diverged t time units ago to have the same nucleotide at a given site is the same as the probability that in a single sequence the nucleotide at a given site at time $2t$ is the same as the nucleotide at time 0. In other words, in this model we can pretend that we start from sequence 1 (or 2) in Figure 3.6 and go backward for t time units to the ancestral sequence and then go forward for t time units to sequence 2 (or 1). The one-parameter model has this property because in this model the substitution process is time reversible. Similarly, in Kimura's two-parameter model time is also reversible, and we have $I_{(t)} = X_{(2t)}$ as can be seen from Equations 3.23 and 3.14. In contrast, no such simple relationship holds for Blaisdell's four-parameter model because in that model the substitution process is not time reversible.

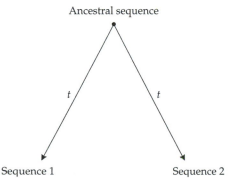

Figure 3.6   Divergence of two sequences from a common ancestral sequence t time units ago.

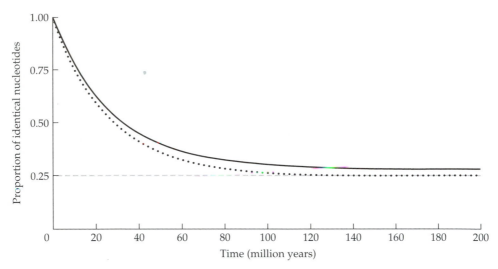

Figure 3.7   Temporal changes in the expected proportion of identical nucleotides between two sequences that diverged *t* years ago. The dotted line represents the case of one-parameter model with $\alpha = 5 \times 10^{-9}$ substitutions/site/year. The solid line represents the case of Blaisdell's four-parameter model with $\alpha = 10 \times 10^{-9}$, $\beta = 6 \times 10^{-9}$, $\gamma = 4 \times 10^{-9}$, and $\delta = 2.5 \times 10^{-9}$ substitutions/site/year.

Figure 3.7 shows the expected proportion, $I_{(t)}$, of identical nucleotides between two sequences diverged *t* years ago. The dotted line is computed under the one-parameter model with $\alpha = 5 \times 10^{-9}$ substitutions/site/year, while the solid line is computed under Blaisdell's four-parameter model with $\alpha = 10 \times 10^{-9}$, $\beta = 6 \times 10^{-9}$, $\gamma = 4 \times 10^{-9}$, and $\delta = 2.5 \times 10^{-9}$ substitutions/site/year. $I_{(t)}$ decreases faster in the one-parameter model than in Blaisdell's model. This is because in the one-parameter model substitution occurs randomly among the four nucleotides so that the chance of back substitutions in a sequence and the chance of parallel substitutions in the two sequences are lower than those in Blaisdell's model. As *t* increases, $I_{(t)}$ does not decrease to 0 but to an equilibrium value (Figure 3.7). Indeed, as can be seen from Equation 3.21, as $t \to \infty$, $I_{(t)}$ will become

$$\hat{I} = \hat{p}_A^{\,2} + \hat{p}_T^{\,2} + \hat{p}_C^{\,2} + \hat{p}_G^{\,2} \tag{3.25}$$

where $\hat{p}_i$ is the equilibrium frequency of nucleotide *i*. For the one-parameter model, $\hat{p}_i = 1/4$ for $i = A, T, C$ and G, and so $\hat{I} = 0.25$ at equilibrium. For Blaisdell's model with the parameter values used in Figure 3.7, $\hat{p}_A = 0.16$, $\hat{p}_T = 0.25$, $\hat{p}_C = 0.37$, $\hat{p}_G = 0.22$ and so $\hat{I} = 0.27$, which is only slightly larger than 0.25 (the value for the one-parameter model), although the parameter values are chosen so that the equilibrium frequencies of the four nucleotides differ considerably from equality. Since the two sequences become random sequences as $t \to \infty$, the above results show that the expected proportion of identical nucleotides between two random sequences is at least 25%.

Sequence Dissimilarity

A simple measure of sequence dissimilarity is the proportion, D, of different nucleotides between two sequences. However, since D is simply $1 - I$, it does not deserve further analysis. We shall instead consider sequence dissimilarity in terms of transitional and transversional differences, which are more informative than D.

First, consider the expected proportion, $P_{(t)}$, of transitional differences between two sequences that diverged t time units ago. Obviously, $P_{(t)} = 0$ at time 0. To see how $P_{(t)}$ increases with time, let us first use the one-parameter model. In this model $P_{(t)}$ is equal to the probability that at a given nucleotide site the two sequences differ by a transition at time t. Assume that the site was initially occupied by, say, A. To compute $P_{(t)}$ we consider all possible transitional differences at time t: (1) sequence 1 has A and sequence 2 has G or vice versa, and (2) sequence 1 has T and sequence 2 has C or vice versa. The probability for the first possibility is $2p_{AA(t)}p_{AG(t)}$ and that for the second possibility is $2p_{AT(t)}p_{AC(t)}$, where $p_{ij(t)}$ is given by Equation 3.11 if $i = j$ or by Equation 3.12 if $i \neq j$. Therefore,

$$P_{(t)} = 2p_{AA(t)}p_{AG(t)} + 2p_{AT(t)}p_{AC(t)}$$

$$= \frac{1}{4} - \left(\frac{1}{4}\right)e^{-8\alpha t} \tag{3.26}$$

A much simpler way to derive Equation 3.26 is to use the property of time-reversibility of the one-parameter model, that is, $P_{(t)} = p_{ij(2t)}$, which can readily be obtained from Equation 3.12 by replacing t by $2t$. Similarly, in Kimura's two-parameter model we have $P_{(t)} = Y_{(2t)}$, which can be obtained from Equation 3.15 by replacing t by $2t$, that is,

$$P_{(t)} = \frac{1}{4} + \left(\frac{1}{4}\right)e^{-8\beta t} - \left(\frac{1}{2}\right)e^{-4(\alpha+\beta)t} \tag{3.27}$$

Kimura (1980) obtained this formula by a different approach. Other substitution models are more complicated and are not considered here.

Next, let us consider the proportion, $Q_{(t)}$, of transversional differences between two sequences that diverged t time units ago. For Kimura's two-parameter model, we can use the property of time-reversibility and obtain $Q_{(t)} = 2Z_{(2t)}$, where the factor 2 comes from the fact that there are two types of transversion. Since $Z_{(2t)}$ is given by Equation 3.16,

$$Q_{(t)} = \frac{1}{2} - \left(\frac{1}{2}\right)e^{-8\beta t} \tag{3.28}$$

This formula was also first obtained by Kimura (1980) by a different method. For the one-parameter model, $\beta = \alpha$ in the above formula.

Figure 3.8 shows the expected proportion of transitional differences (P) and the expected proportion of transversional differences (Q) between two sequences diverged t years ago. In the one-parameter model (dotted lines) both P and Q increase monotonically to 1/4 and 1/2, respectively. In the two-parameter model (solid lines), Q increases monotonically to the equilibrium value 1/2 whereas P

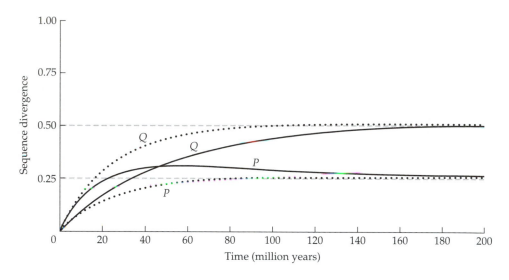

Figure 3.8   Temporal changes in the expected proportion of transitional differences (*P*) and the expected proportion of transversional differences (*Q*) between two sequences that diverged *t* years ago. Dotted lines represent the case of one-parameter model with $\alpha = 5 \times 10^{-9}$ substitutions/site/year; solid black lines represent the case of Kimura's two-parameter model with $\alpha = 10 \times 10^{-9}$ and $\beta = 2.5 \times 10^{-9}$ substitutions/site/year.

can become greater than the equilibrium value 1/4 for a long period of time. $P(t)$ can exceed 1/4 for a reason similar to that for $Y(t)$ in Figure 3.3.

NONUNIFORM RATES AMONG NUCLEOTIDE SITES

All the above treatments assumed that the rate of substitution per site (λ) is the same for all sites in the sequence, i.e., the sequence is homogeneous. This assumption is realistic only for nonfunctional sequences. For functional sequences, the substitution rate usually varies among sites. For example, some amino acid residues in cytochrome *c* appear to be invariant among species (Fitch and Margoliash 1967) and amino acids on the surface of α globin evolve 10 times faster than those residues surrounding the heme group (Chapter 7; Kimura and Ohta 1973). For more data on rate variation, see Chapter 7. Heterogeneity in substitution rate may have a strong effect on the observed degree of sequence divergence (see Palumbi 1989).

   To see how rate heterogeneity may affect the degree of sequence divergence, let us consider a simple hypothetical case in which the sequences consist of only two types of sites: the substitution rate per site per year is $\lambda_1 = 3\alpha_1$ for the first class and $\lambda_2 = 3\alpha_2$ for the second class, and the proportions of the two classes of sites are f_1 and f_2, respectively. We treat the two classes separately. For simplicity, we use the one-parameter model. For two sequences that diverged *t* years ago, the similarity for the *i*th class sites ($i = 1$ or 2) can be readily obtained from Equation 3.22 and is given by

$$I_i = \frac{1}{4} + \frac{3}{4}e^{-8\alpha_i t} \tag{3.29}$$

The average similarity over the entire sequence is

$$\bar{I} = f_1 I_1 + f_2 I_2 \tag{3.30}$$

Note that the average substitution rate per site for the above case is $3\alpha = 3(f_1\alpha_1 + f_2\alpha_2)$. For a homogeneous sequence with the substitution rate 3α, the sequence similarity (I) at time t is given by Equation 3.22, i.e.,

$$I = \frac{1}{4} + \frac{3}{4}e^{-8\alpha t} \tag{3.31}$$

Figure 3.9a shows a comparison of the two cases. The solid line represents the case of a sequence with two classes of sites with the proportions $f_1 = 0.05$ and $f_2 = 0.95$ and with the rates $\alpha_1 = 0$ and $\alpha_2 = 5 \times 10^{-9}$ substitutions per site per year. The dotted line represents the case of a homogeneous sequence with $\alpha = f_1\alpha_1 + f_2\alpha_2 = 4.75 \times 10^{-9}$. Although for any given t the two pairs of sequences have undergone the same number of substitutions, their I (sequence similarity) values differ appreciably when t is large (Figure 3.9a). For the dotted line, I will decrease to the equilibrium value 0.25. In the solid line the first class sites (5%) are invariable (i.e., $\alpha_1 = 0$), so that $I_1 = 1$ for all t's while I_2 (the I value the second class sites) will decrease to the equilibrium value 0.25; therefore, the average I value will reach the equilibrium value $0.05 + 0.95 \times 0.25 = 0.29$. Thus, as $t \to \infty$, the difference between the two curves will increase to $0.29 - 0.25 = 0.04$.

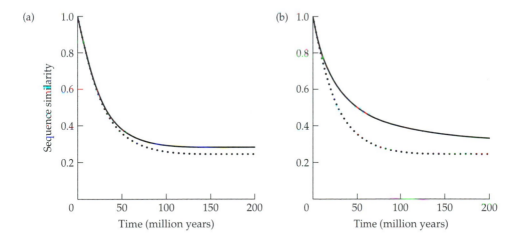

Figure 3.9 Sequence similarity between two sequences diverged t years ago. (a) The black line represents the case of a sequence with two classes of sites with the proportions of 5% and 95% and with the rates $\alpha_1 = 0$ and $\alpha_2 = 5 \times 10^{-9}$ substitutions per site per year. The dotted line represents the case of a homogeneous sequence with a rate of 4.75×10^{-9}. (b) The solid line represents the case where the rates among sites follow the gamma distribution with the parameter $a = 1$ and the mean substitution rate of 5×10^{-9}. The dotted line represents the case of a homogeneous sequence with the rate of 5×10^{-9}.

In practice, the rate variation among sites is likely to be more complicated than the above hypothetical case. However, how the rate varies among sites in a sequence has not been well studied; this question is currently a subject of much interest. Available data on amino acid substitutions in protein sequences (e.g., Uzzell and Corbin 1971; Ota and Nei 1994) and on nucleotide substitutions in DNA sequences (e.g., Kocher and Wilson 1991; Larson 1991; Wakeley 1993) suggest that, to a first approximation, the substitution rate λ varies among sites according to the gamma distribution among sites, i.e.,

$$g(\lambda) = [b^a / \Gamma(a)]e^{-b\lambda} \lambda^{a-1}$$

where $a = \bar{\lambda}^2 / V_\lambda$ and $b = \bar{\lambda} / V_\lambda$, $\bar{\lambda}$ and V_λ being the mean and variance of λ, respectively. The shape of the distribution is largely determined by the parameter a. When $a = 1$, $g(x)$ decreases exponentially from $g(0)$ to 0 as x increases from 0 to infinity (Figure 3.10). When $a < 1$, $g(x)$ approaches ∞ as x decreases to 0, meaning that a large proportion of sites in the sequence have a substitution rate close to 0. Note that $g(x)$ decreases more slowly for the case of $a < 1$ than for the case of $a = 1$ or, in other words, when $a < 1$, the distribution is more dispersed and the sequence is more heterogeneous (see the case of $a = 0.5$ in Figure 3.10). Finally, when $a > 1$, $G(0)$ is 0 and the distribution is more concentrated, implying that the sequence is less heterogeneous (Figure 3.10). More precisely, the heterogeneity is determined by the coefficient of variation, defined as $\sigma / \bar{\lambda}$, where σ is the standard deviation, i.e., the square root of $V(\lambda)$. Since $a = \bar{\lambda}^2 / V_\lambda$, the heterogeneity is

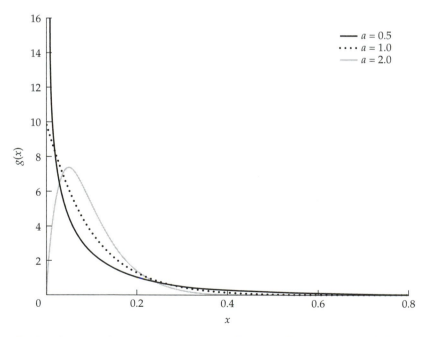

Figure 3.10   Gamma distributions with $a = 0.5$, 1, and 2. $g(x)$ is the probability density function and a is the inverse of the square of the coefficient of variation. From Tateno et al. (1994).

inversely related to a. Indeed, roughly speaking, $a = 0.5$, 1.0 and 2.0 means, respectively, strong, intermediate, and weak rate heterogeneity over sites. In the following we shall assume that λ follows the gamma distribution because there is some empirical support for this distribution and because the distribution can assume various shapes, as discussed above.

If we use the one-parameter model, the $I(\lambda)$ value for a site with substitution rate $\lambda = 3\alpha$ is given by Equation 3.22. Therefore, the mean I value over the sites between the two sequences is

$$\bar{I} = \int_0^\infty Ig(\lambda)d\lambda = \frac{1}{4} + \frac{3}{4}\left(\frac{a}{a+8\bar{\alpha}t}\right)^a \tag{3.32}$$

where $\bar{\alpha} = \bar{\lambda}/3$.

In Figure 3.9b the solid line represents the $\bar{I}$ value for the case with $\bar{\alpha} = 5 \times 10^{-9}$ and $a = 1$, and the dotted line represents the I value for the case of a homogeneous sequence with $\alpha = 5 \times 10^{-9}$. For any given t the two pairs of sequences (i.e., the two curves) should have undergone the same number of substitutions, but the solid curve decreases with t considerably more slowly than the dotted curve and the difference between the two curves becomes large as t increases. Thus, rate heterogeneity can have a strong effect on the degree of sequence divergence between two diverging sequences. However, as t approaches infinity, both curves decrease to 0.25, as can be seen from Equations 3.22 and 3.32. See Problem 9 for more discussion of the effect of rate heterogeneity on I.

PROBLEMS

1. Derive Equation 3.10 from Equation 3.6, and show that $P_{GA(t)} = P_{CA(t)} = P_{TA(t)}$ under the Jukes-Cantor model.

2. Kimura's two-parameter model should become identical to Jukes and Cantor's one-parameter model when $\alpha = \beta$. To verify this, show that when this condition is met, Formula 3.14 becomes identical to Equation 3.11, and Equations 3.15 and 3.16 both become identical to Equation 3.12.

3. Assume that $\alpha = 10 \times 10^{-9}$ and $\beta = 2.5 \times 10^{-9}$ substitutions/site/year in Kimura's two-parameter model. Use Equation 3.17 to obtain a t value and use Equations 3.15 and 3.16 to show that for this t value $Y_{(t)} > Z_{(t)}$.

4. Explain why time reversibility holds under the nine-parameter model represented by the fourth matrix of Table 3.1.

5. For the general model represented by the last matrix of Table 3.1, show that the following recurrence equation holds:

$$p_{A(t+1)} = (1 - \alpha_{12} - \alpha_{13} - \alpha_{14})p_{A(t)} + \alpha_{21}p_{T(t)} + \alpha_{31}p_{C(t)} + \alpha_{41}p_{G(t)}$$

6. Show that the proportion of identical nucleotides between two unrelated random sequences is 1/4, if the frequency of each of the four nucleotides in each sequence is 1/4.

7. Derive Equation 3.26.

8. Assume that $\alpha = 10 \times 10^{-9}$ and $\beta = 2.5 \times 10^{-9}$ substitutions/site/year. Use Equations 3.27 and 3.28 to compute the $P_{(t)}$ and $Q_{(t)}$ values for $t = 100$ and 200 million years.

9. To see the effect of rate heterogeneity on the sequence similarity (I), use Equation 3.32 to compute the $\bar{I}$ values for $a = 0.5$, 1.0 and 2.0, given $\bar{\alpha} = 5 \times 10^{-9}$ and $t = 50 \times 10^6$. Explain why the results imply that I increases with increasing rate heterogeneity.

CHAPTER 4

Estimating the Number of Nucleotide Substitutions Between Sequences

THE PRECEDING CHAPTER PROVIDES A CONCEPTUAL BASIS for understanding the process of evolutionary change in nucleotide sequences. In reality, however, with the exception of viral genes, the substitution of one nucleotide for another may take thousands or even millions of years, so that it cannot be observed within a person's life time. Therefore, to detect nucleotide substitutions in DNA sequences, we usually rely on comparison of sequences that have been derived from a common ancestral sequence. The number of nucleotide substitutions between two sequences is a quantity of foremost importance in molecular evolution because it is used to compute the rate of evolution, to estimate divergence time, and to reconstruct phylogenetic trees by many methods. This chapter is devoted to methods for estimating this number. To compare sequences, one must first align the sequences; however, this topic will be discussed last because it is more involved and not the major aim of the chapter.

As noted in Chapter 3, if multiple substitutions have occurred at any site, then the observed number of differences between the two sequences under study is smaller than the actual number of substitutions. Thus, to estimate the number of substitutions between two sequences we must correct for multiple hits. Our aim here is to study how to make such corrections. Since the pioneering work of Jukes and Cantor (1969), Holmquist (1972), and Kimura and Ohta (1972), many methods have been proposed for this purpose (e.g., Kimura 1980; Gojobori et al. 1982a; Lanave et al. 1984; Lake 1994; Lockhart et al. 1994; Zharkikh 1994). A number of methods are presented below.

The number of nucleotide substitutions between two sequences is usually expressed in terms of the number (K) of substitutions *per nucleotide site* rather than as the *total number* (N) of substitutions between the two sequences. If the number of nucleotide sites compared between the two sequences is L, then $K = N/L$. This quantity facilitates comparison of the degrees of divergence among sequence pairs that do not have the same sequence length (L). The K value is also commonly referred to as the evolutionary distance between two sequences.

Protein-coding and noncoding sequences should be treated separately, for they usually evolve at different rates. In the former case, it is advisable to distinguish between synonymous and nonsynonymous substitutions, because they may evolve at markedly different rates and because a separation of the two types of substitution provides more information for studying the mechanisms of molecular evolution (Chapter 7).

NONCODING SEQUENCES

The following methods assume that all the sites in a sequence evolve at the same rate and follow the same substitution scheme; the case of nonuniform rates will be considered later. The number of sites compared between two sequences is denoted by L. Deletions and insertions are excluded from comparison.

Jukes and Cantor's Method

This simplest method is based on the one-parameter model of nucleotide substitution. Under this model the expected proportion of identical nucleotides between two sequences that diverged t time units ago is given by

$$I_{(t)} = \frac{1}{4} + \frac{3}{4}e^{-8\alpha t}$$

(see Equation 3.22). The probability that the two sequences are different at a site at time t is $p = 1 - I_{(t)}$ and so

$$p = \frac{3}{4}\left(1 - e^{-8\alpha t}\right)$$

or

$$8\alpha t = -\ln\left(1 - \frac{4p}{3}\right) \tag{4.1}$$

The time of divergence between two sequences is usually unknown, and thus we cannot estimate α. Instead, we estimate K, the actual number of substitutions per site since the divergence between the two sequences. In the one-parameter model, $K = 2(3\alpha t)$, where $3\alpha t$ is the expected number of substitutions per site in one lineage. Using Equation 4.1 we estimate K by

$$K = -\left(\frac{3}{4}\right)\ln\left(1 - \frac{4p}{3}\right) \tag{4.2}$$

where p is assumed to be equal to the observed proportion of different nucleotides between the two sequences (Jukes and Cantor 1969). When L is large, the sampling variance of K is approximately given by

$$V(K) = p(1 - p)/\left[L(1 - 4p/3)^2\right] \tag{4.3}$$

(Kimura and Ohta 1972).

Kimura's Two-Parameter Method

In this method, the differences between two sequences are classified into transitions and transversions. Let P and Q be the proportions of transitional and transversional differences between the two sequences, respectively. Then, the number of nucleotide substitutions per site between the two sequences is estimated by

$$K = \left(\frac{1}{2}\right)\ell n(a) + \left(\frac{1}{4}\right)\ell n(b) \tag{4.4}$$

where $a = 1/(1 - 2P - Q)$ and $b = 1/(1 - 2Q)$. This formula can be obtained from Equations 3.27 and 3.28 (Kimura 1980). The sampling variance is approximately given by

$$V(K) = \left[a^2 P + c^2 Q - (aP + cQ)^2\right]/L \tag{4.5}$$

where $c = (a + b)/2$ (Kimura 1980).

Tajima and Nei's Method

Tajima and Nei (1984) proposed a method that does not require the assumption of equal frequencies of the four nucleotides. They noted that formula (4.2) can be written as

$$K = -b_1 \ell n\left(1 - p/b_1\right) \tag{4.6}$$

where $b_1 = 1 - \sum q_i^2$; q_i is the equilibrium frequency of the ith nucleotide ($i = 1, 2, 3, 4$ corresponding to A, G, T, C). In the Jukes-Cantor model, $q_i = 1/4$, so that $b_1 = 3/4$. Tajima and Nei showed that formula (4.6) also holds under the equal-input model, i.e., $\alpha_{ij} = \alpha_j$ for all $i \neq j$ (see Chapter 3). When $\alpha_j = \alpha$ for all j, this model becomes identical with the Jukes-Cantor model. If the condition $\alpha_{ij} = \alpha_j$ does not hold for all $i \neq j$, formula (4.6) tends to give an underestimate. To remove this deficiency, Tajima and Nei proposed the following formula as a general estimator of K.

$$K = -b\,\ell n(1 - p/b) \tag{4.7}$$

where $b = (b_1 + p^2/h)/2$ and

$$h = \sum_{i=1}^{3} \sum_{j=i+1}^{4} \frac{(x_{ij})^2}{2q_i q_j}$$

in which x_{ij} is the proportion of pairs of nucleotides i and j between the two homologous DNA sequences. The sampling variance for this estimator is approximately given by

$$V(K) = b^2 p(1-p)/\left[(b-p)^2 L\right] \tag{4.8}$$

Other Methods

Explicit analytic solutions have also been obtained for a number of other models (see Zharkikh 1994). Gojobori et al. (1982a) solved Kimura's (1981a) 6-parameter model and Tamura and Nei (1993) solved an equal-input related model with 6 parameters, i.e., the fourth matrix in Table 3.1 with the condition $\beta_1 = \beta_2 = \gamma_1 = \gamma_2 = \beta$. The method of Lanave et al. (1984) has been interpreted as a 12-parameter method because their transition matrix is equivalent to the last matrix in Table 3.1, but it is in effect a 9-parameter method because the estimation procedure involves some assumptions (Zharkikh 1994).

Instead of K (the number of nucleotide substitutions per site), one may consider other types of distance between sequences (Barry and Hartigan 1987; Zharkikh 1994). For example, Lake (1994) has proposed the following distance. Let N_{ij} be the number of sites where the nucleotide is i in the first sequence and j in the second sequence and let J be the determinant defined by

$$J = \begin{vmatrix} N_{AA} & N_{AT} & N_{AC} & N_{AG} \\ N_{TA} & N_{TT} & N_{TC} & N_{TG} \\ N_{CA} & N_{CT} & N_{CC} & N_{CG} \\ N_{GA} & N_{GT} & N_{GC} & N_{GG} \end{vmatrix}$$

Further, let $F_i = A_i T_i C_i G_i$, where A_i is the number of occurrences of nucleotide A in sequence i, $i = 1, 2$; T_i, C_i, and G_i are similarly defined. Then the distance between the two sequences is estimated by

$$d = -\frac{1}{4}\left[\ell n(J) - \frac{1}{2}\ell n(F_1 F_2)\right]$$

(4.9)

This distance is equal to K only if the frequencies of the four nucleotides are equal to 1/4. However, the distance is additive and is therefore called the **paralinear** distance (Lake 1994). Two other advantages of this distance is that it holds for the general (12-parameter) model of nucleotide substitution and is applicable even if the nucleotide frequencies change with time, i.e., they are nonstationary. Therefore, it may be suitable for phylogenetic reconstruction when the nucleotide composition of the sequences under study has changed with time. Lockhart et al. (1994) have proposed a distance, called the LogDet, that is similar to the paralinear distance. For more details about the statistical properties of these two distances, readers may refer to Gu and Li (1996b).

An Example

As an example of using some of the above methods to compute the number of nucleotide substitutions between two sequences (K), let us consider the first introns of human and owl monkey (a New World monkey) insulin genes. Figure 4.1 shows an alignment of the two sequences. Since we are concerned with nucleotide substitutions, we exclude the gaps from comparison; in total, 163 nucleotides are compared, i.e., $L = 163$. For the Jukes-Cantor method, we first count the number of differences between the two sequences, which is $N = 18$. We then obtain $p = N/L = 18/163 = 0.110$ as the proportion of nucleotide differences between the two sequences. Putting this p value into Equation 4.2, we obtain $K = 0.119$ as an estimate of the number of nucleotide substitutions per site between the two sequences.

For Kimura's two-parameter method, we first count the number of transitional changes (14) and the number of transversional changes (4) and then obtain the proportions of transitional and transversional differences, i.e., $P = 14/163 = 0.086$ and $Q = 4/163 = 0.025$. Putting these values into Equation 4.4, we obtain $K = 0.122$.

For Tajima and Nei's method, the computation is more complicated. First, we estimate the equilibrium frequencies of the four nucleotides A, G, T, and C, which

70
1 GTCTGTTCCAAGGGCCTTTGCGTCAGG-TGGGC-TCAGGGTT--------------CCAGGGTGGCTGG
        * # *   #   *   *   * ** *
2 GTCTGTTCCAAGGGCCTTCGAGCCAGTCTGGGCCCCAGGGCTGCCCCACTCGGGGTTCCAGAGCAGTTGG

140
1 ACCCCAGGCCCCAGCTCTGCAGCAGGGAGGACGTGGCTGGGCTCGTGAAGCATGTGGGGGTGAGCCCAGG
        * *   **   #
2 ACCCCAGGTCTCAGC---------GGGAGGGTGTGGCTGGGCTC-TGAAGCATTT--GGGTGAGCCCAGG

196
1 GGCCCCAAGGCAGGGCACCTGGCCTTCAGCCTGCCTCAGCCCTGCCTGTCTCCCAG
        *   *   #
2 GGCTC-AGGGCAGGGCACCTG-CCTTCAGC-GGCCTCAGC-CTGCCTGTCTCCCAG

Figure 4.1   Alignment of the sequences of the first introns of human (sequence 1) and owl monkey (sequence 2) insulin genes. Notation: - denotes a gap, * denotes a transitional difference, and # denotes a transversional difference.

will be denoted 1, 2, 3, and 4, respectively. Since there are 22 and 23 A nucleotides in the first and second sequences, respectively (Figure 4.1), the frequency of nucleotide A is estimated to be $q_1 = (22 + 23)/(2 \times 163) = 0.138$. Similarly, we estimate the frequencies of G, T, and C: $q_2 = 0.365$, $q_3 = 0.187$, and $q_4 = 0.310$. From the alignment we can compute $x_{ij} = y_{ij} + y_{ji}$, in which y_{ij} is the proportion of nucleotide pairs where the first and second sequences have nucleotides i and j, respectively. For example, $x_{12} = (2 + 2)/163 = 0.025$. Similarly, $x_{13} = 0.000$, $x_{14} = 0.006$, $x_{23} = 0.018$, $x_{24} = 0.000$, and $x_{34} = 0.061$. Putting the q_i and x_{ij} values into Equation 4.7, we obtain $K = 0.124$.

For the method of Lanave et al., we use a computer and obtain $K = 0.126$.

For the method of Lake (1994), we have

$$J = \begin{vmatrix} 20 & 0 & 0 & 2 \\ 0 & 24 & 5 & 1 \\ 1 & 5 & 45 & 0 \\ 2 & 2 & 0 & 56 \end{vmatrix}$$

and $A_1 = 22$, $T_1 = 30$, $C_1 = 51$, $G_1 = 60$, and $A_2 = 23$, $T_2 = 31$, $C_2 = 50$, $G_2 = 59$. Thus, $J = 1,175,566$, $F_1 = 2,019,600$, and $F_2 = 2,103,350$. Putting these values into Formula 4.9, we obtain $d = 0.147$.

It is seen that the estimate of K increases somewhat with the sophistication of the method. This change occurs because the region is rich in G and C nucleotides (68% GC) and because there are more transitional changes than transversional changes (i.e., 16 vs. 4). As a consequence, the smallest (0.119) and largest (0.126) estimates differ by about 6%. Note, however, we do not know which estimate is most accurate because we do not know the true value; the question of accuracy is discussed below. Note also that the distance obtained by Lake's (1994) method (0.147) is considerably larger than those obtained the other methods, but, as mentioned above, Lake's method was not intended for estimating K.

Comparison of Methods

It would seem that a model with a larger number of parameters is better than one with a smaller number of parameters because the former allows more flexibility and is usually more realistic than the latter. In practice, this is not always the case, for two reasons. First, in addition to assumptions on the substitution scheme, it is usually necessary to make further assumptions in the derivation of a method. For example, Tajima and Nei's method assumes that the common ancestral sequence had the equilibrium nucleotide frequencies and that these frequencies are estimated from the two extant sequences, but such assumptions are only implicit in the one-parameter and Kimura's two-parameter methods and, indeed, Formulas 4.2–4.5 do not involve nucleotide frequencies. Obviously, additional assumptions can cause estimation errors. Second, and more importantly, the effect of sampling errors, which arise because the number of nucleotides compared is finite, is usually stronger on a model with a larger number of parameters than one with a smaller number of parameters (see Gojobori et al. 1982a; Tajima and Nei 1984; Zharkikh 1994). The effect may be seen in the variance or the bias of an estimate. More seriously, sampling errors may render a method inapplicable, because all estimators of K involve logarithmic functions and the argument can become zero or negative. This chance is expected to increase with the number of parameters involved (see below). Of course, as the sequence length increases, the effect of sampling errors should decrease and a model with more parameters is likely to become better than one with fewer parameters.

Gojobori et al. (1982a), Tajima and Nei (1984), Zharkikh (1994), and others have conducted extensive simulation studies to compare the performances of various methods for estimating K. The simulation results suggest that if K is 0.5 or smaller, then estimation of K is fairly simple, and even such simple methods as the Jukes-Cantor method and the Tajima-Nei method give fairly accurate estimates of K. For more divergent sequences, the methods of Lanave et al. (1984) and Gojobori et al. are preferable to the other methods in Table 4.1, particularly for long sequences ($L \geq 1000$ bp). Tamura and Nei's (1993) 6-parameter method was not included in these studies, but it is likely to perform at least as well as that of Gojobori et al. because both methods have the same number of parameters and because the former allows more flexibility for unequal transitional and transversional rates. For example, the rates of C → A (transversion) and G → A (transition) are allowed to be unequal in Tamura and Nei's method but not in Gojobori et al. (see Table 3.1). For divergent sequences ($K > 1$), however, the methods of Lanave et al., Gojobori et al., and Tamura and Nei may not be applicable and in this case one may have to use a simpler method such as that of Tajima and Nei (1984). Table 4.1 shows Zharkikh's (1994) simulation comparison of five estimation methods for $K = 1$; six nucleotide substitution models were used to simulate sequence divergences. If the sequence length is only 100 bp, the estimates obtained by the methods of Gojobori et al. and Lanave et al. (denoted as GIN and LA) actually are, on average, slightly worse than that by the Tajima and Nei (TN) method, though they are, on average, better than those obtained by the Jukes-Cantor method and Kimura's two-parameter methods. Moreover, the proportion of inapplicable cases is much higher for GIN and LA than for the other methods. These results are due to the fact that when the sequence is short, sampling effects tend to be stronger for GIN and LA than for the one- and two-parameter meth-

TABLE 4.1 Estimates of the number of nucleotide substitutions per site (K) by different methods[a] from simulation data with sequence length L and the true value of K = 1

Models[b]	L = 100 bp					L = 1000 bp				
	JC	K2	TN	GIN	LA	JC	K2	TN	GIN	LA
JC	1.025	1.044	1.085	1.113	1.125	1.003	1.004	1.008	1.011	1.012
K2	0.935	1.050	1.091	1.094	1.107	0.901	1.006	1.011	1.016	1.017
EI	0.904	0.943	1.114	1.045	1.122	0.887	0.907	1.009	0.941	1.016
EIr	0.829	0.864	1.039	1.014	1.089	0.813	0.832	0.949	0.931	1.019
GIN	0.878	0.925	1.078	1.097	1.107	0.843	0.870	0.975	1.018	1.021
RAND	0.902	0.920	1.052	1.069	1.089	0.869	0.877	0.966	1.009	1.026
NA[c] (%)	0.0	1.1	2.2	24.4	33.9	0.0	0.0	0.3	3.7	4.8

From Zharkikh (1994).

[a]Methods: JC, Jukes and Cantor (1969); K2, Kimura's (1980) two-parameter method; TN, Tajima and Nei (1984); GIN, Gojobori, Ishii, and Nei (1982a); LA, Lanave et al. (1984).

[b]Models corresponded to the assumptions made in the simulation, so that each row simulated substitutions using the same model: JC, Jukes-Cantor (one-parameter); K2, Kimura's two-parameter model; EI, equal-input model (Tajima and Nei 1982); EIr, equal-input related models; GIN, Gojobori et al.'s (1982a) six-parameter model; RAND, random substitution parameters.

[c]NA: the proportion of inapplicable cases averaged over the six substitution models used.

ods. As the sequence length increases to 1000 bp, GIN and LA start to show better estimates than the other methods and the proportion of inapplicable cases is considerably reduced for both GIN and LA (Table 4.1). The simulation results show the difficulty of obtaining accurate estimates of K for divergent sequences. Thus, regardless of which method is used, an estimate should be taken with caution, if it is substantially larger than 1.

NONUNIFORM RATES AMONG NUCLEOTIDE SITES

All the above methods assume that the rate of substitution per site (λ) is uniform along the sequence. As noted in Chapter 3, this assumption is not realistic for functional sequences, and the decrease in similarity between two sequences may be considerably slower under nonuniform rates than under the assumption of a uniform rate. It is known that rate variation among sites can cause severe underestimation of K if the variation is not taken into account (e.g., Palumbi 1989; Shoemaker and Fitch 1989).

As noted in Chapter 3, limited data indicate that the gamma distribution may be used to describe approximately the distribution of substitution rate in a sequence. Under this distribution and under the Jukes-Cantor model, the expected number of nucleotide substitutions between two sequences ($K = 2\bar{\lambda}t$) can be obtained from Equation 3.32 and is given by

$$K = \frac{3a}{4}\left[\left(1 - \frac{4p}{3}\right)^{-1/a} - 1\right] \tag{4.10}$$

where p is the proportion of different nucleotide sites between the two sequences and $a = \bar{\lambda}^2/V(\lambda)$ (Nei and Gojobori 1986).

For Kimura's (1980) two-parameter model, Jin and Nei (1990) showed

$$K = \frac{a}{4}\left[2(1 - 2P - Q)^{-1/a} + (1 - 2Q)^{-1/a} - 3\right] \tag{4.11}$$

where P and Q are, respectively, the proportions of transitional and transversional differences between the two sequences. Tamura and Nei (1993) have solved the case of a six-parameter model and Gu and Li (1996a) have solved the nine-parameter case.

Although the gamma distribution seems to be suitable for describing the rate variation among sites in some sequences, it is unlikely to be applicable to all sequences. More empirical studies on the distribution of substitution rates among sites in sequences are needed for making better estimates of K.

PROTEIN-CODING SEQUENCES

Miyata and Yasunaga (1980), Li et al. (1985b), Nei and Gojobori (1986), and others (see Li et al. 1985a) have proposed methods for estimating the number of substitutions between two protein-coding sequences. In most of these methods, synonymous and nonsynonymous substitutions are treated separately. We shall first describe the method of Li et al. (1985b) and later the modification by Li (1993) and Pamilo and Bianchi (1993). In this method the initiation and termination codons are excluded from analysis because these two codons rarely change with time. Unless stated otherwise, the universal code is used in the following discussion.

First, classify nucleotide sites into nondegenerate, twofold degenerate, and fourfold degenerate sites. A site is **nondegenerate** if all possible changes at this site are nonsynonymous, **twofold degenerate** if one of the three possible changes is synonymous, and **fourfold degenerate** if all possible changes at the site are synonymous. For example, the first two positions of the codon TTT (Phe) are nondegenerate, while the third position is twofold degenerate (Table 1.2). In comparison, the third position of the codon GTT (Val) is fourfold degenerate. For simplicity, the third position in each of the three isoleucine codons (the universal code) is treated as a twofold degenerate site, although the degeneracy at this position is actually threefold. In mammalian mitochondrial genes, there are only two codons for isoleucine, and their third position is indeed a twofold degenerate site (Chapter 1). Using the above rules, we first count the numbers of the three types of sites in each of the two sequences compared and then compute the averages, denoting them by L_0 (nondegenerate), L_2 (twofold), and L_4 (fourfold), respectively.

Next, compare the two sequences codon by codon and infer the nucleotide differences between each pair of codons. Classify each difference according to the type of site at which it has occurred. This classification can be done by considering all possible paths between the two codons. For two codons that differ by only one nucleotide, the difference is easily inferred. For example, the difference between the two codons GTC (Val) and GTT (Val) is synonymous, while the difference between the two codons GTC (Val) and GCC (Ala) is nonsynonymous.

For two codons that differ by more than one nucleotide, we need to consider all possible minimum evolutionary pathways that can lead to the observed changes. For example, for the two codons AAT (Asn) and ACG (Thr), there are two possible pathways:

Pathway I: AAT (Asn) ↔ ACT (Thr) ↔ ACG (Thr)

Pathway II: AAT (Asn) ↔ AAG (Lys) ↔ ACG (Thr)

Pathway I requires one synonymous and one nonsynonymous change, whereas pathway II requires two nonsynonymous changes. It is known that synonymous substitutions occur, on average, considerably more often than nonsynonymous substitutions (Chapter 7), and we may assume that pathway I is more likely than pathway II. For example, if we assume a weight of 0.7 for pathway I and a weight of 0.3 for pathway II, then the number of synonymous differences between the two codons is estimated to be $0.7 \times 1 + 0.3 \times 0 = 0.7$, and the number of nonsynonymous differences is $0.7 \times 1 + 0.3 \times 2 = 1.3$. Here, the weights used are hypothetical. The weights for all possible codon pairs have been estimated by Miyata and Yasunaga (1980) from protein sequence data and by Li et al. (1985b) from DNA sequence data. If we assume that both pathways are equally likely, then for the above example the number of synonymous differences is $(1 + 0)/2 = 0.5$ and the number of nonsynonymous differences is $(1 + 2)/2 = 1.5$. Thus, the weighted and unweighted approaches may give somewhat different results. In practice, the differences in estimates between the two approaches are usually small (Nei and Gojobori 1986), but they can be important for genes coding for highly conserved proteins such as histones and actins (Li et al. 1985b).

The nucleotide differences in each class are further classified into transitional (S_i) and transversional (V_i) differences ($i = 0, 2, 4$). Note that, in the class of twofold degenerate sites, transitions are synonymous and transversions are nonsynonymous. There are no exceptions to this rule in the mammalian mitochondrial code. In the universal code, however, there are two exceptions: the first position of the arginine codons (CGA, CGG, AGA, and AGG) and the last position in the three isoleucine codons (AUU, AUC, and AUA). In all these cases, all synonymous changes are included in S_2 and all nonsynonymous changes are included in V_2.

Let $P_i = S_i/L_i$ and $Q_i = V_i/L_i$, which are, respectively, the proportions of transitional differences and transversional differences per i-fold degenerate site between the two sequences. Applying Kimura's two-parameter method to each type of sites, we can show that the mean numbers of transitional (A_i) and transversional (B_i) substitutions per ith type site are given by

$$A_i = \left(\frac{1}{2}\right)\ell n(a_i) - \left(\frac{1}{4}\right)\ell n(b_i) \tag{4.12}$$

$$B_i = \left(\frac{1}{2}\right)\ell n(b_i) \tag{4.13}$$

and their variances are given by

$$V(A_i) = \left[a_i^2 P_i + c_i^2 Q_i - (a_i P_i + c_i Q_i)^2\right] / L_i \tag{4.14}$$

$$V(B_i) = b_i^2 Q_i (1 - Q_i) / L_i \tag{4.15}$$

where $a_i = 1/(1 - 2P_i - Q_i)$, $b_i = 1/(1 - 2Q_i)$ and $c_i = (a_i - b_i)/2$. The total number of substitutions per ith type site, K_i, is given by

$$K_i = A_i + B_i \tag{4.16}$$

An approximate sampling variance of K_i can be readily obtained from Equation 4.5 by replacing P by P_i, Q by Q_i, and L by L_i. Note that A_2 and B_2 denote the numbers of synonymous and nonsynonymous substitutions per twofold degenerate site, $K_4 = A_4 + B_4$ denotes the number of synonymous substitutions per fourfold degenerate site, and $K_0 = A_0 + B_0$ denotes the number of nonsynonymous substitutions per nondegenerate site. Thus, the above formulas can be used to compare the rates of substitution at the three different types of sites.

The number of substitutions per synonymous site (K_S) is computed as follows. The estimated total number of synonymous substitutions at twofold degenerate sites is $L_2 A_2$ and that at fourfold degenerate sites is $L_4 K_4$. By following the convention to count each fourfold degenerate site as a synonymous site and to count one-third of a twofold degenerate site as synonymous and two-thirds as nonsynonymous (because two of the three possible changes are nonsynonymous), we obtain the total number of synonymous sites as $L_4 + L_2/3$. From these numbers we can show that

$$K_S = (L_2 A_2 + L_4 K_4) / (L_2 / 3 + L_4) \tag{4.17}$$

$$V(K_S) = 9\left[L_2^2 V(A_2) + L_4^2 V(K_4)\right]/(L_2 + 3L_4)^2 \tag{4.18}$$

In a similar manner, we obtain the mean and variance of K_A, the number of substitutions per nonsynonymous site.

$$K_A = (L_2 B_2 + L_0 K_0)/(2L_2/3 + L_0) \tag{4.19}$$

$$V(K_A) = 9\left[L_2^2 V(B_2) + L_0^2 V(K_0)\right]/(2L_2 + 3L_0)^2 \tag{4.20}$$

The above method is that of Li et al. (1985b). Since transitional substitutions tend to occur more often than transversional substitutions and since most transitional changes at twofold degenerate sites are synonymous changes, counting a twofold degenerate site as one-third synonymous and two-thirds nonsynonymous will tend to overestimate the rate of synonymous substitution at twofold degenerate sites. For this reason, Li (1993) and Pamilo and Bianchi (1993) proposed the following modification. Assume that the transitional rate at twofold degenerate sites is the same as that at fourfold degenerate sites. We then take the weighted average $(L_2 A_2 + L_4 A_4)/(L_2 + L_4)$ as an estimate of the average transitional rate at twofold and fourfold degenerate sites; $L_2 A_2$ is the total number of transitional substitutions at twofold degenerate sites between the two sequences, $L_4 A_4$ is the corresponding number at fourfold degenerate sites, and $L_2 + L_4$ is the total number of twofold plus fourfold degenerate sites. The number of substitutions per synonymous site and its variance are then computed as

$$K_S = (L_2 A_2 + L_4 A_4)/(L_2 + L_4) + B_4 \tag{4.21}$$

$$V(K_S) = \left[L_2^2 V(A_2) + L_4^2 V(A_4)\right]/(L_2 + L_4)^2 + V(B_4)$$
$$-2b_4 Q_4\left[a_4 P_4 - c_4(1 - Q_4)\right]/(L_2 + L_4) \tag{4.22}$$

Similarly, we have

$$K_A = A_0 + (L_0 B_0 + L_2 B_2)/(L_0 + L_2) \tag{4.23}$$

$$V(K_A) = V(A_0) + \left[L_0^2 V(B_0) + L_2^2 V(B_2)\right]/(L_0 + L_2)^2$$
$$-2b_0 Q_0\left[a_0 P_0 - c_0(1 - Q_0)\right]/(L_0 + L_2) \tag{4.24}$$

Formulas 4.22 and 4.24 correct a typographic error in Li (1993).

Table 4.2 shows a comparison of the estimates obtained by the old (Li et al. 1985b) and new (modified) methods. For the 14 pairs of mouse–rat genes, the old method gives an average K_S value of 0.183, which is 27% higher than that (0.144) obtained by the new method. For the 45 pairs of human–rodent genes, the old method gives an average K_S value of 0.690, which is 22% higher than that (0.566) obtained from the new method. To see which estimates are more reasonable, we can compare them with the number of substitutions per fourfold degenerate site estimated from the same set of genes. Because all possible changes at a fourfold degenerate site are synonymous, that site is truly synonymous. As can be seen

TABLE 4.2   Numbers of substitutions per synonymous site (K_S)
and per nonsynonymous site (K_A) between mouse
and rat genes estimated by the old and new methods

Gene	K_S		K_A	
	Old	New	Old	New
Mouse vs. rat[a]	0.183	0.144	0.0137	0.0148
Human vs. rodent[b]	0.690	0.566	0.1154	0.1248

[a]The mouse–rat comparison is based on 14 pairs of genes and is taken from Li (1993).
[b]The human–rodent comparison is based on 45 pairs of genes, the genes in Table 7.1,
excluding histones 3 and 4.

from Tables 4.2 and 4.3, the K_S values (0.144 and 0.566) estimated by the new method are close to the numbers of substitutions per fourfold degenerate site (0.137 and 0.594), whereas those estimated by the old method (0.183 and 0.690) are considerably larger. The overestimation by the old method arises because transitions occur more frequently than transversions. For example, for the 14 pairs of mouse–rat genes, the numbers of transitions and transversions per fourfold degenerate site are 0.084 and 0.053, respectively, although at each site there is only one type of transition but two types of transversion.

Table 4.2 shows that the old method gives a smaller K_A value than does the new method. However, the differences are small. The number of nondegenerate sites in a gene is usually considerably larger than that of twofold degenerate sites, so the K_A value is largely determined by the nondegenerate sites.

The new method implicitly assumes that in a gene the transitional rate at a twofold degenerate site is similar to that at a fourfold degenerate site. This assumption holds approximately for the data in Table 4.3: for the mouse–rat com-

TABLE 4.3   Numbers of transitional and transversional
substitutions per site at nondegenerate, twofold
degenerate, and fourfold degenerate sites of
codons between genes[a]

Type of substitution	Non-degenerate	Twofold degenerate	Fourfold degenerate
Mouse vs. rat			
Transition	0.009	0.097	0.084
Transversion	0.004	0.011	0.053
Total	0.013	0.108	0.137
Human vs. rodent			
Transition	0.064	0.298	0.359
Transversion	0.061	0.061	0.235
Total	0.125	0.359	0.594

[a]The genes used are the same as those in Table 4.2.

parison the number of transitional substitutions per twofold degenerate site (0.094) is somewhat higher than that at a fourfold degenerate site, while for the human–rodent comparison the situation is reversed (i.e., 0.298 vs. 0.359). The new method also assumes that the transversional rate at a twofold degenerate site is similar to that at a nondegenerate site. This assumption holds well for the human–rodent comparison (i.e., 0.061 vs. 0.061), though it does not hold well for the mouse–rat comparison (0.004 vs. 0.011).

The methods of Miyata and Yasunaga (1980) and Nei and Gojobori (1986) follow the convention to count a twofold degenerate sites one-third synonymous and two-thirds nonsynonymous and give estimates of K_S and K_A similar to those of Li et al. (1985b). Thus, they also tend to overestimate K_S.

Recently, Ina (1995) has proposed some new methods and Comeron (1995) has modified Li's (1993) methods.

SEQUENCE ALIGNMENT

Comparison of two homologous sequences involves the identification of the locations of deletions and insertions that might have occurred in either of the two lineages since their divergence from a common ancestor. This process is referred to as "sequence alignment". Note that here we assume that the two sequences under study are known to have been derived from a common ancestral sequence. This is different from the problem of testing whether two sequences are related or homologous. It is also different from the so-called homology-search problem in which one wants to know whether a sequence or part of a sequence is related to any of the sequences in a database such as GenBank. For these latter problems, readers may refer to Goad and Kanehisa (1982), Waterman (1984), Lipman and Pearson (1985), and Waterman et al. (1991).

Comparisons of two DNA or protein sequences usually cannot tell us whether an insertion has occurred in one sequence or a deletion has occurred in the other. Therefore, the outcomes of both types of events are collectively referred to as **indels** or **gaps**.

An alignment consists of a series of paired bases, one base from each sequence. There are three types of aligned pairs: (1) pairs of matched bases, (2) pairs of mismatched bases, and (3) pairs consisting of a base from one sequence and a gap (null base) from the other. Gaps are denoted by -. A matched pair implies that no substitution has occurred at the site since the divergence between the two sequences, a mismatched pair implies that at least one substitution has occurred, and a gap assumes that a deletion or an insertion has occurred at this position in one of the two sequences. Thus, an alignment represents a specific hypothesis about the evolution of the two sequences. For example, the alignment

TCAGA

TC - GT

represents the hypothesis that three of the five nucleotide sites have not undergone any change since the divergence of the two sequences, one site has undergone at least one substitution, and one site has a deletion or insertion. Note that this alignment also implies that the first, second, fourth, and fifth sites are homologous between the two sequences, while the third site might have been either inserted into the first sequence or deleted from the second sequence.

Although the process of alignment will be illustrated below by using DNA sequences, the same principle and procedure can be used to align amino acid sequences. As a matter of fact, one usually obtains more reliable alignments by using amino acid sequences than by using DNA sequences. One reason is that the former are usually better conserved than the latter. Another reason is that there are 20 amino acids but only four nucleotides, so that the probability for two non-homologous sites of the two sequences to be identical by chance is lower at the amino acid level than at the nucleotide level.

The aim here is to provide a description of the basic principles for sequence alignment rather than any actual algorithms, as many computer programs for alignment are now available (see references in Doolittle 1990); a commonly used simple alignment program is CLUSTAL V by Higgins et al. (1992).

The Dot-Matrix Method

For two sequences that are highly similar and have experienced few gap events, a reasonable alignment can be obtained either by visual inspection or by an approach called the **dot-matrix** method (Gibbs and McIntyre 1970; Maizel and Lenk 1981). In this method, the two sequences to be aligned are written out as column and row headings of a matrix (Figure 4.2). Dots are put in the matrix when the nucleotides in the two sequences are identical. If the two sequences are compeletely identical, there will be dots in all the diagonal elements of the matrix (Figure 4.2a). If the two sequences are different but can be aligned without gaps, there will be dots in most of the diagonal elements (Figure 4.2b). If a gap occurred in one of the two sequences, the alignment diagonal would be shifted vertically or horizontally (Figure 4.2c). If the two sequences differ from each other by both gaps and substitutions (Figure 4.2d), it may be difficult to identify the location of gaps and to choose between several alternative alignments. In such cases, visual inspection and the dot matrix method are not reliable, so a more rigorous method is required.

Similarity and Distance Methods

The basic principle for sequence alignment is either to maximize the number of matched pairs between the two sequences or to minimize the number of mismatched pairs, while at the same time keeping the number of gaps as small as possible. The former is called the similarity approach and was initiated in a landmark work by Needleman and Wunsch (1970); the latter is called the distance approach and was rigorously formulated by Sellers (1974) and Waterman et al. (1976).

We begin with the similarity approach. First we note that increasing the number of matches and reducing the number of gaps are two conflicting efforts. For example, consider the following two sequences:

A: TCAGACGAGTG

B: TCGGAGCTG

Since they differ in length, there must be at least one gap. One possible alignment is as follows:

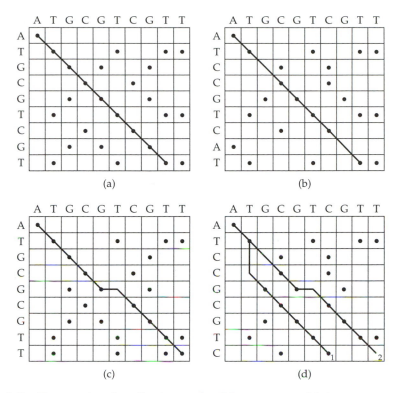

Figure 4.2   Dot matrices for aligning nucleotide sequences. (a) the two sequences are identical; (b) the two sequences differ from each other but contain no gaps; (c) the two sequences contain a gap, but are otherwise identical to each other; (d) the two sequences contain both substitutions and gaps. In (d), path 1 consists of six diagonal steps, none of which is empty, and 2 vertical steps. Path 2 contains eight diagonal steps, two of which are empty, and one horizontal step. The decision between paths 1 and 2 depends on the gap penalty, i.e., on which evolutionary path is more probable: a two-nucleotide deletion with no substitutions (path 1) or a one-nucleotide deletion and two substitutions (path 2). From Li and Graur (1991).

```
          T C A G A C G A G T G
   (I)     : :     : :         : :
          T C G G A - - G C T G
```

This alignment contains 6 matches and 1 gap of 2 nucleotides; among all the possible alignments with one gap, this alignment has the largest number of matched pairs. There are many other possible alignments, but among those with 2 gaps of one nucleotide each, the following two alignments have the largest number of matches, i.e., 7:

```
          T C A G A C G A G T G
  (II)     : :     : :   :     : :
          T C G G A - G C - T G
```

$$\text{(III)} \qquad \begin{array}{l} \text{T C A G A C G A G T G} \\ \text{: :} \quad \text{: :} \quad \text{ :} \quad \quad \text{: :} \\ \text{T C G G A - G - C T G} \end{array}$$

Which of the three alignments is the best? Obviously, we are facing a choice between having more point substitutions or having more gap events. Since comparing point substitutions with gaps is like comparing apples with oranges, we must find a common denominator with which to compare gaps and substitutions. This common denominator is called the **gap penalty**.

Let w_k be the penalty for a gap of k nucleotides. Then the Needleman-Wunsch similarity measure between two aligned sequences is defined as

$$S = x - \sum w_k z_k \qquad (4.25)$$

where x is the number of matched pairs and z_k is the number of gaps with length k. Many systems of gap penalty have been used (see Waterman 1984; McClure et al. 1994). The linear system is most commonly used because it saves computer time. In this system $w_k = a + bk$, where a and b are nonnegative parameters. However, even with this system, how to choose a and b is a difficult problem. For example, if $a = 0.5$ and $b = 2$, then the similarity for alignment I is $S = 6 - (0.5 + 2 \times 2) = 1.5$ because there are 6 matches and 1 gap of 2 nucleotides. Since $S = 2$ for both alignments II and III, they are favored over alignment I. (This example shows that there may exist more than one optimal alignment under the same criteria.) However, if $a = 1.5$ and $b = 1$, then $S = 2.5$ for alignment I, whereas $S = 2$ for alignments II and III, so that alignment I is favored. Therefore, which alignment is preferable depends on the penalty weights used. Gu and Li (1995) recently studied pseudogene sequences in humans and rodents and provided evidence that the linear penalty system is too heavy for long gaps. Their data suggested $w_k = a + b \ell n(k)$ as a penalty function, which increases with k considerably more slowly than does the linear function $w_k = a + bk$.

In general, the similarity approach is to find an alignment with a similarity measure defined by

$$S = \text{Max}\left(x - \sum w_k z_k\right) \qquad (4.26)$$

where the maximum is over all possible alignments. A commonly used algorithm for searching the S value and the optimal alignment is dynamic programming. Readers may consult Smith et al. (1981) and Waterman (1984) for this algorithm.

The distance approach is to find the alignment(s) with a distance measure defined by

$$D = \text{Min}\left(y + \sum w'_k z_k\right) \qquad (4.27)$$

where y is the number of mismatched pairs and w'_k is a gap weight analogous to w_k.

Smith et al. (1981) showed that under certain conditions the above two approaches are equivalent. Consider two DNA sequences of lengths m and n, respectively. Since the total number of nucleotides in the two sequences are $m + n$, for any alignment we should have

$$n + m = 2(x + y) + \sum k z_k \qquad (4.28)$$

Putting this relation into Equation 4.26 we obtain

$$S = \text{Max}\left[\frac{n+m}{2} - y - \sum (k/2 + w_k)z_k\right]$$

$$= \frac{n+m}{2} + \text{Max}\left[-y - \sum (k/2 + w_k)z_k\right]$$

$$= \frac{n+m}{2} - \text{Min}\left[y + \sum (k/2 + w_k)z_k\right] \tag{4.29}$$

Thus, to maximize the similarity measure defined by (4.25) is to minimize the following quantity:

$$y + \sum (k/2 + w_k)z_k$$

This is identical to minimizing the distance measure defined by Equation 4.27 if the following relation holds:

$$w'_k = k/2 + w_k \tag{4.30}$$

The above equation shows that in order to obtain the same alignment by the two approaches the penalty weights for the distance approach should be larger than those for the similarity approach. Since $w_k \geq 0$, $w'_k \geq k/2$.

Multiple Sequences

Alignment of multiple sequences by visual inspection is feasible only if the sequences under study are highly similar to one another and have not experienced many deletion or insertion events. When the similarity between sequences is not high, alignment of multiple sequences can be extremely difficult. One difficulty is to define a measure for the cost (or quality) of a multiple alignment and to choose gap costs consistent with the measure chosen (see Altschul and Lipman 1989). Sankoff (1975) proposed to use the sum of the branch lengths of the phylogenetic tree that relates the sequences; the length of a branch is the number of substitutions plus the gap costs for that branch. Another measure is the sum of the alignment costs imposed on each pair of the sequences in the multiple alignment (Altschul and Lipman 1989; Lipman et al. 1989); this is called the SP (sum of pairs) measure. An advantage of the former measure is that it attempts to infer the minimum number of substitutions and gap events required for a given tree, but it is more difficult to compute than the SP measure. The latter, however, may be unduly influenced by closely related sequences (which include redundant information), because all pairwise alignments are treated equally. This problem can be circumvented by giving less weights to more closely related pairs (Lipman et al. 1989).

The major difficulty in aligning multiple sequences is computational. This difficulty is particularly apparent for methods using dynamic programming, which is the most widely accepted algorithm, because an explicit, realistic measure for the cost of an alignment can be used. For a dynamic-programming algorithm for aligning multiple sequences, see Carrillo and Lipman (1988) and Lipman et al. (1989).

It has long been recognized that alignment of multiple sequences should take into consideration the branching order of the sequences under study, for the order in which the sequences are aligned can influence the final outcome (Sankoff et al. 1972; Sankoff 1975; Hogeweg and Hesper 1984). If the phylogeny of the sequences is known, alignment can be made progressively from more related to less related sequences. If the phylogeny is not known, then sequence alignment and phylogenetic reconstruction should be done simultaneously—this is known as the integrated or unified approach. As this approach may require many cycles of tree reconstruction and sequence alignment, it is computationally intensive, and so most algorithms use heuristic approaches (e.g., Feng and Doolittle 1987, 1990; Konings et al. 1987; Higgins et al. 1992; Hein 1990). For example, the algorithm of Hogeweg and Hesper (1984) and Konings et al. (1987) includes the following steps: (1) calculate all pairwise similarities by the Needleman-Wunsch algorithm modified by Fitch and Smith (1983), (2) construct an initial tree from these similarities, (3) realign the sequences progressively in order of their relatedness according to the inferred tree, (4) construct a new tree from the pairwise similarities calculated from the new alignment, and (5) reiterate the process, if the new tree is not identical to the previous one.

PROBLEMS

1. Note that the proportion of differences between two sequences according to Kimura's two-parameter model is $P + Q$. Show that Equation 4.4 reduces to Equation 4.2 when $P = P/3$ and $Q = 2P/3$.

2. Estimate the number of nucleotide substitutions (K) for the two sequences in Figure 4.3 by using the Jukes-Cantor method, Kimura's two-parameter method, and Formula (4.9).

3. Compute the K values for $p = 0.5$ by using (a) Formula (4.2) and (b) Formula (4.10) for $a = 0.5$, 1.0, and 2.0.

```
                                                                        70
1  GTCTGTTCCAAGGGCCTTTGCGTCAGGTGGGCTCAGGG-----------------CCAGGGTGGCTGGAC
2  GTCTGTTCCAAGGGCCTTCGCGTCAGGTGGGCTCAGGGCTGCCCACTTGGGGGTTCCAGGGTGGCTGGAC


                                                                       140
1  CCCAGGCCCCAGCTCTGCAGCAGGGAGGACGTGGCTGGGCTCGTGAAGCATGTGGGGGTGAGCCCAGGGG
2  CCCAGGCCCCAGCTCTGCAACAGGGAGGACATGGCTGGGCTCTTGAAGCGTTTGAGGGTGAACCCAGGGG


                                                 194
1  CCCCAAGGCAGGGCACCTGGCCTTCAGCCTGCCTCAGCCCTGCCTGTCTCCCAG
2  CCC-AGGGCAG-GCACCTG-GCCTCAGCTGGCCTCAGG-CTGCCTGTCTCTCAG
```

Figure 4.3   Alignment of the sequences of the first introns of human (sequence 1) and green monkey (sequence 2) insulin genes. Notation: – denotes a gap.

4. Calculate (a) the number of synonymous substitutions per synonymous site and (b) the number of nonsynonymous substitutions per nonsynonymous site for the following two sequences:

Ser	Thr	Glu	Met	Cys	Leu	Met	Gly	Gly
TCA	ACT	GAG	ATG	TGT	TTA	ATG	GGG	GGA

TCG	ACA	GGG	ATA	TAT	CTA	ATG	GGT	ATA
Ser	Thr	Gly	Ile	Tyr	Leu	Met	Gly	Ile

5. Use the dot matrix method to align the following two sequences:

$$AATGCTTGCATGGGGCTAGTT$$

$$ATTGCTGCATGAGGCGCGCTAGT$$

Choose two possible alignments and decide which one is better by using a constant gap penalty of 2 per nucleotide. Will the choice be affected by using a much bigger gap penalty, say, 10?

6. Find a pair of a and b values in the linear penalty system $w_k = a + bk$ for which the following alignment.

```
T C A G A C G A G - T G
: :     :       : :   : :
T C - G - - G A G C T G
```

has the same similarity measure as alignments I, II, and III given in the text.

7. In the similarity approach, if the linear penalty system $w_k = a + bk$ is used, show that alignment I is always better than alignments II and III, if $a > 1$; is as good as alignments II and III, if $a = 1$; and is worse than alignments II and III, if $a < 1$.

CHAPTER 5

Molecular Phylogenetics: Methods

M OLECULAR PHYLOGENETICS IS THE STUDY of evolutionary relationships among organisms or genes by a combination of molecular biology and statistical techniques. It is also commonly called **molecular systematics**, if the relationships of organisms are the concern. The study of the evolutionary relationships among genes in a gene family will be discussed in Chapter 10.

The molecular approach to systematics was initiated at the turn of the century by Nuttall (1904), who used serological cross-reactions to study the phylogenetic relationships among various groups of animals (see the Introduction). However, extensive use of molecular data in phylogenetic studies did not begin until the early 1960s, i.e., until after the introduction of protein sequencing, protein electrophoresis, and other molecular techniques into the field. Protein sequence data allowed, for the first time, the investigation of long-term evolution such as the relationships among mammalian orders or even more distantly related taxa (see Fitch and Margoliash 1967). On the other hand, less expensive and more expedient methods, such as protein electrophoresis, DNA–DNA hybridization, and immunological methods, though less accurate than protein sequencing, were extensively used to study the phylogenetic relationships among populations or closely related species (see Goodman 1962; Nei 1975; Ayala 1976; Wilson et al. 1977). The advent of various recombinant DNA techniques since the 1970s has led to a rapid accumulation of DNA sequence data, thus stimulating even greater interest in molecular systematics. Yet, the development of the polymerase chain reaction (PCR) method (Saiki et al. 1985) has made systematic studies even easier, resulting in an unprecedented high level of actitivies in phylogenetic reconstruction.

There are several reasons why molecular data, particularly DNA sequence data, are much more powerful for evolutionary studies than morphological and physiological data. First, DNA and protein sequences generally evolve in a much more regular manner than do morphological and physiological characters and therefore can provide a clearer picture of relationships of organisms. Second, molecular data are often much more amenable to quantitative treatments than are morphological data. In fact, sophisticated mathematical and statistical theories have already been developed for analyzing DNA sequence data (discussed later; Nei 1987; Felsenstein 1988; Swofford and Olsen 1990; Miyamoto and Cracraft 1991). Third, molecular data are much more abundant. This abundancy is especially useful in studying microorganisms, in which only a limited number of morphological or physiological characters are available for phylogenetic studies. Indeed, molecular approaches have completely revolutionized the study of microbial taxonomy (e.g., Woese 1987). Of course, we should not abandon traditional means of evolutionary inquiry such as morphology, anatomy, physiology, and paleontology. Rather, different approaches provide complementary data. Morphological and anatomical data are necessary for taxonomical studies, and paleontological information is required for constructing a time frame for evolutionary studies.

Molecular data have proved useful for studying not only the phylogenetic relationships among closely related populations or species such as the relationships among human populations (e.g., Cann et al. 1987; Vigilant et al. 1991; Hedges et al. 1992; Templeton 1992; Horai et al. 1993; Torroni et al. 1993; Bailliet et al. 1994) and those between humans and apes (Chapter 6) but also very ancient evolutionary occurrences such as the origin of mitochondria and chloroplasts (see Cedergren et al. 1988; Giovannovi et al. 1988; Lockhart et al. 1994) and the divergence of phyla and kingdoms (e.g., Woese 1987; Sogin et al. 1986, 1989; Wainright et al. 1993). In the future, most phylogenetic issues are likely to be resolved by molecular data and we may eventually be able to fulfill Darwin's dream of having "a fairly true genealogical tree of each great kingdom of Nature."

To be able to take advantage of molecular data, however, one must understand the methodology of molecular phylogenetics. The aim of this chapter is to provide basic principles of tree reconstruction; a number of interesting examples of phylogenetic trees based on molecular data will be presented in the next chapter.

PHYLOGENETIC TREES

All life forms are related by descent to one another. Closely related organisms are descended from more recent common ancestors than are distantly related ones. The purposes of phylogenetic studies are (1) to reconstruct the correct genealogical ties between organisms and (2) to estimate the time of divergence between organisms since they last shared a common ancestor.

In phylogenetic studies, the evolutionary relationships among a group of organisms are illustrated by means of a **phylogenetic tree**. A phylogenetic tree is a graph composed of nodes and branches, in which only one branch connects any two adjacent nodes (Figure 5.1). The **nodes** represent the taxonomic units, and the **branches** define the relationships among the units in terms of descent and ancestry. The branching pattern of a tree is called the **topology**. The **branch length** usually represents the number of changes that have occurred in that branch. The taxonomic units represented by the nodes can be species, populations, individuals, or genes.

When referring to the nodes of a tree, we distinguish between **external** (or **terminal**) nodes and **internal** nodes. For example, in Figure 5.1 nodes A, B, C, D, and E are external, whereas all others are internal. External nodes represent the

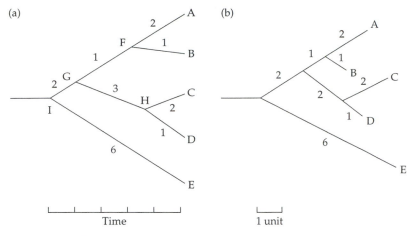

Figure 5.1 Two alternative representations of a phylogenetic tree for five OTUs. (a) Unscaled branches: extant OTUs are lined up and nodes are positioned proportionally to times of divergence. (b) Scaled branches: lengths of branches are proportional to the numbers of molecular changes. From Li and Graur (1991).

extant taxonomic units under comparison and are referred to as **operational taxonomic units (OTUs)**. Internal nodes represent ancestral units.

The branches of a tree are also known as the **edges** and can be classifed as **internal** or **external**. For example, in Figure 5.1 the branches leading to nodes A, B, C, D, and E are external, whereas all others are internal. External branches are also called **peripheral branches**.

Figure 5.1 illustrates two common ways of drawing a phylogenetic tree. In Figure 5.1a, the branches are **unscaled**; their lengths are not proportional to the number of changes, which are indicated on the branches. This presentation allows us to line up the extant OTUs and also to place the nodes representing divergence events on a **time scale** on which the times of divergence are known or have been estimated. In Figure 5.1b, the branches are **scaled**, that is, their lengths are proportional to the numbers of changes.

A tree is said to be **additive** if the distance between any two OTUs is equal to the sum of the lengths of all the branches connecting them. For example, if additivity holds, the distance between OTUs A and C in Figure 5.1a should be equal to $2 + 1 + 3 + 2 = 8$. The distance between two OTUs is calculated directly from molecular data (e.g., DNA sequences), while the branch lengths are estimated from the distances between OTUs according to certain rules (see the section on estimation of branch length). Additivity usually does not hold if multiple substitutions between sequences have occurred at any nucleotide sites.

An internal node is **bifurcating** if it has only two immediate descendant lineages, but **multifurcating** if it has more than two immediate descendant lineages. For simplicity, we shall only consider bifurcating trees like those in Figure 5.1.

Rooted and Unrooted Trees

Phylogenetic trees can be either **rooted** or **unrooted** (Figure 5.2). In a rooted tree there exists a particular node, called the root (R in Figure 5.2a), from which a

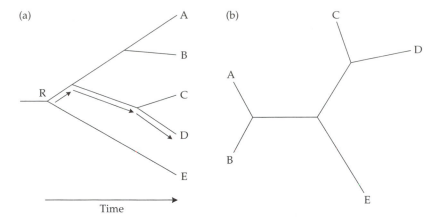

Figure 5.2   (a) Rooted and (b) unrooted phylogenetic trees. Arrows indicate the unique path leading from the root (R) to OTU D. From Li and Graur (1991).

unique path leads to any other node. The direction of each path corresponds to evolutionary time, and the root is the common ancestor of all the OTUs under study. An unrooted tree is a tree that only specifies the relationships among the OTUs but does not define the evolutionary path (Figure 5.2b). Unrooted trees do not make assumptions or require knowledge about common ancestors.

The number of bifurcating unrooted trees (N_U) for n OTUs ($n \geq 3$) is given by

$$N_U = (2n - 5)! / \left[2^{n-3}(n - 3)!\right] \tag{5.1}$$

while the number of bifurcating rooted trees (N_R) for $n \geq 2$ is

$$N_R = (2n - 3)! / \left[2^{n-2}(n - 2)!\right] \tag{5.2}$$

Note that the number of possible rooted trees for n OTUs is equal to the number of possible unrooted trees for $n + 1$ OTUs. Table 5.1 shows that both N_U and N_R increase very rapidly with n—for 10 OTUs there are already more than 2 million bifurcating unrooted trees and close to 35 million rooted ones. Since only one of these trees represents correctly the true evolutionary relationships among the OTUs, it is usually very difficult to infer the true phylogenetic tree when n is large.

The majority of tree-making methods yield unrooted trees. To root an unrooted tree, we usually need an **outgroup** (i.e., an OTU for which external information, such as paleontological evidence, clearly indicates that it has branched off earlier than the taxa under study). The root is then placed between the outgroup and the node connecting it to the other OTUs. For example, if OTU E in Figure 5.2b is known to be the outgroup, then the root can be placed as in Figure 5.2a.

True and Inferred Trees

The sequence of speciation events that has led to the formation of any group of OTUs is historically unique. Thus, only one of all the possible trees that can be built with a given number of OTUs represents the true evolutionary history. A

TABLE 5.1	Possible numbers of rooted and unrooted trees up to 10 OTUs	
Number of OTUs	*Number of rooted trees*	*Number of unrooted trees*
2	1	1
3	3	1
4	15	3
5	105	15
6	954	105
7	10,395	954
8	135,135	10,395
9	2,027,025	135,135
10	34,459,425	2,027,025

From Felsenstein (1978).

tree that is obtained by using a certain set of data and a certain method of tree reconstruction is called an **inferred tree**. An inferred tree may or may not be identical with the true tree.

Gene Trees and Species Trees

A phylogenetic tree that represents the evolutionary pathways of a group of species is called a **species tree**. When a phylogenetic tree is constructed from one gene from each species, the inferred tree is a **gene tree** (Nei 1987). It can differ from the species tree in two respects. First, the divergence of two genes sampled from two different species may predate the divergence of the two species (for example, alleles a and f in Figure 5.3). Such predating will result in an overestimate of the branch length and may present a problem if we are concerned with short-term evolution, in which the component of divergence due to genetic polymorphism within species cannot be ignored.

The second problem with gene trees is that the branching pattern of a gene tree (i.e., its topology) may be different from that of the species. Figure 5.4 shows three different possible relationships between the two trees. The topologies of gene trees in (a) and (b) are identical with those of the corresponding species trees (e.g., X and Y form a cluster). The gene tree in (c), however, is different from the true species tree, since now Y and Z are sister groups when considering this gene. The probability of obtaining the erroneous tree (c) is quite high when the time interval between the first and second species splitting $(t_2 - t_1)$ is short, as is probably true in the case of the phylogenetic relationships among humans, chimpanzees, and gorillas. To avoid this type of error, one needs to use many genes in the reconstruction of the phylogeny. A large amount of data is also required to avoid stochastic errors, which can occur because nucleotide substitutions occur randomly, so that, for instance, lineage Z in Figure 5.4a may have by chance accumulated fewer substitutions than lineages X and Y, despite the fact that it has branched off earlier in time.

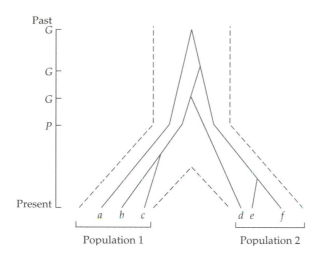

Figure 5.3   Diagram showing that gene splitting (G) usually occurs earlier than population splitting (P) if the population is genetically polymorphic at time P. The evolutionary history of gene splitting resulting in the six alleles denoted *a–f* is shown in solid lines, and population splitting is shown in broken lines. After Nei (1987).

Monophyletic Groups and Clades

A group of taxa is said to be **monophyletic** if they are derived from a single common ancestor but is said to be **polyphyletic** if they are derived from more than one common ancestor. For example, in Figure 5.1 OTUs A and B form a monophyletic group because they are derived from the common ancestor F, whereas OTUs A and C are polyphyletic because they are derived from two different ancestors, F and H, respectively. However, the entire group of A, B, C, and D is monophyletic because of common ancestor G. A group of taxa is said to be **paraphyletic** if the taxa are derived from a common ancestor but the group does not include all decendent taxa of the same common ancestor. For example, the group of OTUs A, B, and E in Figure 5.1 is paraphyletic because it does not include all the descendent OUTs of the common ancestor I, i.e., it does not include OTUs C and D.

In systematic studies one is often interested in the identification of clades. A **clade** is defined as a monophyletic group that includes all the descendent species of the common ancestor (in the literature, a clade and a monophyletic group are often used exchangeably). Figure 5.5 shows the evolutionary relationships among three classes of vertebrates: birds, reptiles, and mammals. The classical taxonomic assignment of reptiles to a separate class does not fit the definition of a clade, for the three groups of reptiles share a common ancestor with another group, the class of birds, which is not included within the definition of the Reptilia. In other words, the class Reptilia is paraphyletic. Birds and crocodiles, on the other hand, do constitute a natural clade, the Archosauria, since they share a common ancestor not shared by any extant organism other than birds and crocodiles. Similarly, birds and reptiles taken together constitute a natural clade, and of course, a monophyletic group.

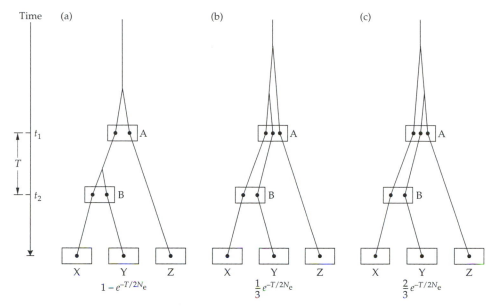

Figure 5.4 Three possible relationships between a species tree (rectangles) and a gene tree (dots). In (a) and (b), the topologies of the species trees are identical to those of the gene trees. Note that in (a) the time of divergence between the genes is roughly equal to the time of divergence between the populations. In (b), on the other hand, the time of divergence between genes X and Y greatly predates the time of divergence between the respective populations. The topology of the gene tree in (c) is different from that of the species tree. For neutral alleles, the probability of occurrence of each tree is given underneath the tree. t_1 is the time at which the first speciation event occurred and t_2 is the time at which the second speciation event occurred. $T = t_2 - t_1$, and N_e is the effective population size. From Nei (1987).

METHODS OF TREE RECONSTRUCTION

Numerous tree-making methods (e.g., see Sneath and Sokal 1973; Nei 1987; Felsenstein 1988) have been proposed, because no single method performs well under all circumstances. A number of methods that have been frequently used or are convenient for illustrating basic principles will be discussed in this section. For simplicity, we consider nucleotide sequence data, though most of the methods described are also applicable to other types of molecular data such as amino acid sequences.

The methods described below can be classified into four types: distance matrix methods, maximum parsimony methods, maximum likelihood methods and methods of invariants. In **distance matrix methods**, evolutionary distances (usually numbers of nucleotide or amino acid substitutions between sequences) are computed for all pairs of taxa, and a phylogenetic tree is constructed by using an algorithm based on some functional relationships among the distance values. The first five methods described below are distance matrix methods. In **maximum parsimony methods**, character states (e.g., the nucleotide or amino acid at a site) are used, and the shortest pathway leading to these character states is chosen as the best tree. In **maximum likelihood methods** one searches for the maxi-

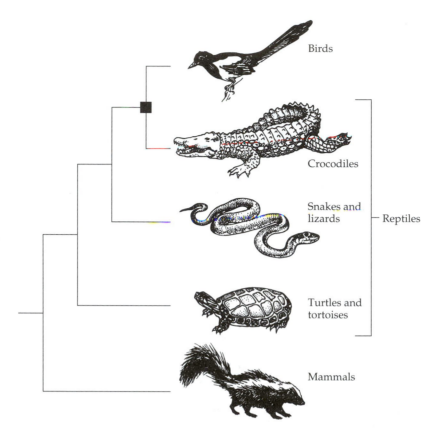

Figure 5.5   Phylogenetic tree of birds, reptiles, and mammals. The reptiles do not constitute a natural clade since they share ancestors with the birds, which are not included in the Reptilia. Birds and crocodiles, on the other hand, constitute a natural clade (Archosauria) since they share a common ancestor (black box) not shared by any other organism. From Li and Graur (1991).

mum likelihood (ML) value for the character state configurations among the sequences under study for each possible tree and chooses the one with the largest ML value as the preferred tree. (A nucleotide configuration at a site means the pattern of nucleotide differences at that site among the sequences involved.) The **methods of invariants** study some particular functions of the character states that have the expected value 0 under certain trees but have nonzero expectations under other trees. The invariant methods can be skipped if a reader is not deeply concerned with tree reconstruction or finds the material difficult. A description of methods is given in the following sections and a comparison of the strengths and weaknesses of methods will be presented in a later section.

Unweighted Pair-Group Method with Arithmetic Mean (UPGMA)

The **unweighted pair-group method with arithmetic mean (UPGMA)**, the simplest method for tree reconstruction, was originally developed for constructing taxonomic phenograms, i.e., trees that reflect the phenotypic similarities between

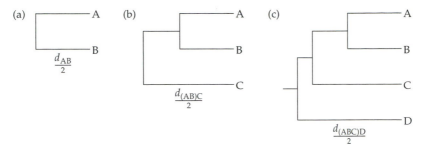

Figure 5.6 Diagram illustrating the stepwise construction of a phylogenetic tree for four OTUs according to the UPGMA method (see text). From Li and Graur (1991).

OTUs (Sokal and Michener 1958). However, it can also be used to construct phylogenetic trees if the rates of evolution are approximately constant among the different lineages so that an approximately linear relation exists between evolutionary distance and divergence time (Nei 1975). For such a relation to hold, linear distance measures such as the number of nucleotide (or amino acid) substitutions should be used.

The UPGMA method employs a sequential clustering algorithm, in which local topological relationships are inferred in order of decreasing similarity and a phylogenetic tree is built in a stepwise manner. That is, we first identify the two OTUs that are most similar to each other (i.e., have the shortest distance) and treat them as a new single OTU. Such an OTU is referred to as a **composite OTU**. Subsequently, from among the new group of OTUs, we identify the pair with the highest similarity, and so on, until only two OTUs are left.

To illustrate the method, let us consider a case of four OTUs. The pairwise evolutionary distances, such as Jukes and Cantor's (1969) estimates (Chapter 4), are given by the following matrix:

	OTU		
OTU	A	B	C
B	d_{AB}		
C	d_{AC}	d_{BC}	
D	d_{AD}	d_{BD}	d_{CD}

In this matrix, d_{ij} stands for the distance between OTUs i and j. The first two OTUs to be clustered are the ones with the smallest distance. Let us assume that d_{AB} has the smallest value. Then, OTUs A and B are the first to be clustered, and the branching point is positioned at a distance of $d_{AB}/2$ substitutions (Figure 5.6a).

After the first clustering, A and B are treated as a single composite OTU, and a new distance matrix is computed.

	OTU	
OTU	(AB)	C
C	$d_{(AB)C}$	
D	$d_{(AB)D}$	d_{CD}

In this matrix, $d_{(AB)C} = (d_{AC} + d_{BC})/2$, and $d_{(AB)D} = (d_{AD} + d_{BD})/2$. In other words, the distance between a simple OTU and a composite OTU is the average of the distances between the simple OTU and the constituent OTUs of the composite OTU. If $d_{(AB)C}$ turns out to be the smallest distance in the new matrix, then OTU C will be joined to the composite OTU (AB) with a branching node at $d_{(AB)C}/2$ (Figure 5.6b).

The final step consists of clustering the last OTU, D, with the composite OTU (ABC). The root of the entire tree is positioned at $d_{(ABC)D}/2 = [(d_{AD} + d_{BD} + d_{CD})/3]/2$. The final tree inferred is shown in Figure 5.6c.

In the UPGMA method, the distance between two composite OTUs is computed as the arithmetic mean of the pairwise distances between the constituent OTUs of the two composite OTUs. For example, the distance between the composite OTUs (ij) and (mn) is

$$d_{(ij)(mn)} = \left(d_{im} + d_{in} + d_{jm} + d_{jn}\right) / 4 \tag{5.3}$$

In general, the distance between two clusters of OTUs X and Y is

$$d_{XY} = \sum_{i,j} d_{ij} / (n_X n_Y) \tag{5.4}$$

where the summation is over every i in cluster X and every j in cluster Y and n_X and n_Y are the numbers of OTUs in X and Y, respectively. For example, for the composite OTUs (ijk) and (mn),

$$d_{(ijk)(mn)} = \left(d_{im} + d_{in} + d_{jm} + d_{jn} + d_{km} + d_{kn}\right) / 6$$

Transformed Distance Method

If the assumption of rate constancy among lineages does not hold, UPGMA may give an erroneous topology. For example, suppose that the phylogenetic tree in Figure 5.7a is the true tree. The pairwise evolutionary distances are given by the following matrix:

		OTU	
OTU	A	B	C
B	8		
C	7	9	
D	12	14	11

By using the UPGMA method, we obtain an inferred tree that differs from the true tree in its branching pattern (Figure 5.7b). For example, OTUs A and C are grouped together, whereas in the true tree, A and B are sister OTUs. (Note that additivity does not hold in this case; the true distance between A and B is 8, whereas the sum of the lengths of the branches connecting A and B is $3.50 + 0.75 + 4.25 = 8.50$.)

The topological errors might be remedied, however, by using a correction called the **transformed distance method** (Farris 1977; Klotz et al. 1979). Briefly, this method uses an outgroup as a reference to make corrections for unequal rates of evolution among the lineages under study and then applies UPGMA to the new distance matrix to infer the topology of the tree.

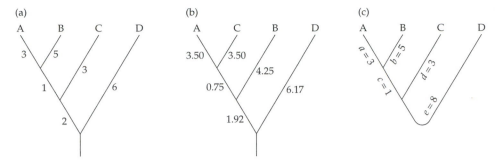

Figure 5.7 (a) The true phylogenetic tree. (b) The erroneous phylogenetic tree reconstructed by using the UPGMA method, which does not take into account the possibility of unequal substitution rates along different branches. (c) The tree inferred by the transformed distance method. The root must be between OTU D and the node of the common ancestor of OTUs A, B, and C, but its exact location cannot be determined. From Li and Graur (1991).

In the present case, let us assume that taxon D is an outgroup to all other taxa. D can then be used as a reference to transform the distances by the following equation:

$$d'_{ij} = \left[\left(d_{ij} - d_{iD} - d_{jD} \right) / 2 \right] + \bar{d}_D \qquad (5.5)$$

where d'_{ij} is the transformed distance, i = A, B, or C, and $\bar{d}_D = (d_{AD} + d_{BD} + d_{CD})/3$. The term $\bar{d}_D$ was introduced to make all d'_{ij} values positive, because in practice a distance can never be negative. For the general case of n OTUs (not including the outgroup), $\bar{d}_D = \sum d_{iD}/n$.

In our example, $\bar{d}_D = 37/3$, and the new distance matrix for taxa A, B, and C is

	OTU	
OTU	A	B
B	10/3	
C	13/3	13/3

Since d'_{AB} has the smallest value, A and B are the first to be clustered together, and subsequently, C is added to the tree. By definition, the outgroup OTU, D, determines the root of the tree and is the last to be added. This gives the correct topology (Figure 5.7c). In the above example, we considered only three taxa with one outgroup, but the method can be easily extended to more taxa and/or more outgroups.

In many instances it is impossible to decide *a priori* which of the taxa under consideration is an outgroup. To overcome this difficulty, a two-stage approach has been proposed (Li 1981). In the first step, one infers the root of the tree by using the UPGMA method; of course, the inferred root may not be the true root. After that, the taxa on one side of the root are used as references (outgroups) for making corrections for the unequal rates of evolution among the lineages on the other side of the root, and vice versa. In our example, this approach also identifies the correct tree.

Neighbors-Relation Methods

In an unrooted bifurcating tree, two OTUs are said to be **neighbors** if they are connected through a single internal node. For example, in Figure 5.8a, A and B are neighbors and so are C and D. In comparison, in Figure 5.8b, neither A and C

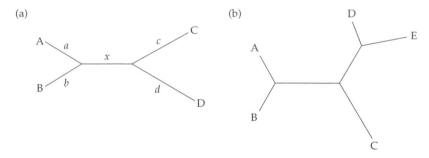

Figure 5.8   Bifurcating unrooted trees with (a) four OTUs and (b) five OTUs. From Li and Graur (1991).

nor B and C are neighbors. However, if we combine OTUs A and B into one composite OTU, then the combined OTU (AB) and OTU C become a new pair of neighbors.

Let us now assume that the tree shown in Figure 5.8a is the true tree. Then, if additivity holds, we should have

$$d_{AC} + d_{BD} = d_{AD} + d_{BC} = a + b + c + d + 2x = d_{AB} + d_{CD} + 2x \qquad (5.6)$$

where x is the length of the internal branch. Therefore, the following two conditions hold:

$$d_{AB} + d_{CD} < d_{AC} + d_{BD} \qquad (5.7a)$$

$$d_{AB} + d_{CD} < d_{AD} + d_{BC} \qquad (5.7b)$$

These two conditions are collectively known as the **four-point condition** (Buneman 1971). They may hold even if additivity holds only approximately.

Conversely, for four OTUs with unknown phylogenetic relationships, the above two conditions can be used to identify the neighbors (A and B; C and D). Once the two pairs of neighbors are determined, so is the topology of the tree.

Sattath and Tversky (1977) proposed the following method for dealing with more than four OTUs. First, compute a distance matrix as in the case of UPGMA. For every possible quadruple, say OTUs i, j, m, and n, compute $d_{ij} + d_{mn}$, $d_{im} + d_{jn}$, and $d_{in} + d_{jm}$. Suppose that the first sum is the smallest. Then, assign a score of 1 to both the pair i and j and the pair m and n; pairs i and m, i and n, j and n, and j and m are assigned a score of 0. If, on the other hand, $d_{im} + d_{jn}$ has the smallest value, then we assign the pair i and m and the pair j and n each a score of 1 and assign the other four possible pairs a score of 0. When all the possible quadruples are considered, the pair with the highest total score is selected as the first pair of neighbors and treated as a single OTU. Next, compute a new distance matrix as in the case of UPGMA and then repeat the same process to select the second pair of neighbors. This process is continued until the number of OTUs is reduced to three, which can form only one unrooted tree. A detailed illustration of this method will be given when we consider the phylogeny of apes and humans (Chapter 6).

Another method (the neighborliness method) based on the neighbors-relation concept has been proposed by Fitch (1981). A similar approach, called the neighbor-joining method (Saitou and Nei 1987), is discussed in the following section.

Neighbor-Joining Method

The principle of the neighbor-joining method is to find neighbors sequentially that may minimize the total length of the tree. This method starts with a starlike tree, as given in Figure 5.9a, in which there is no clustering of OTUs. The first step is to separate a pair of OTUs (e.g., 1 and 2) from all the others (Figure 5.9b). In this tree there is only one interior branch, that is, the branch connecting nodes X and Y, where X is the common node for OTUs 1 and 2 and Y is the common node for the others (3, 4, . . . , N). For this tree the sum of all branch lengths is

$$S_{12} = \frac{1}{2(N-2)} \sum_{K=3}^{N} (d_{1k} + d_{2k}) + \frac{1}{2} d_{12} + \frac{1}{N-2} \sum_{3 \leq i \leq j \leq N} d_{ij} \qquad (5.8)$$

Any pair of OTUs can take the positions of 1 and 2 in the tree, and there are $N(N-1)/2$ ways of choosing them. Among these possible pairs of OTUs, the one that gives the smallest sum of branch lengths is chosen. This pair of OTUs is then regarded as a single OTU, and the arithmetic mean distances between OTUs are computed to form a new distance matrix. The next pair of OTUs that gives the smallest sum of branch lengths is then chosen. This procedure is continued until all $N-3$ interior branches are found. Saitou and Nei (1987) showed that in the case of four OTUs the necessary condition for this method to obtain the correct tree topology is also given by the four-point condition.

Instead of the S criterion defined in Equation 5.8, Studier and Keppler (1988) proposed the Q criterion, defined as follows:

$$Q_{12} = (N-2)d_{12} - \sum_{i=1}^{N} d_{1i} - \sum_{i=1}^{N} d_{2i} \qquad (5.9)$$

Gascuel (1994) showed that S and Q are related by the following equation.

$$S_{12} = \frac{1}{2(N-2)} Q_{12} + \frac{1}{(N-2)} \sum_{1 \leq i \leq j \leq N} d_{ij} \qquad (5.10)$$

Since the last term in the above equation is a constant for all possible trees, minimizing S is equivalent to minimizing Q. However, the Studier-Keppler algorithm

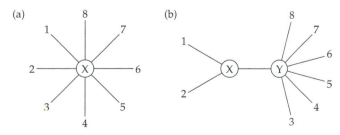

Figure 5.9 (a) A starlike tree with no hierachical structure. (b) A tree in which OTUs 1 and 2 are clustered. From Saitou and Nei (1987).

is computationally faster than the Saitou-Nei algorithm. Gascuel (1994) further provided a theoretical explanation for why the neighbor-joining method and the neighbors-relation method of Sattath and Tversky (1977) give identical or similar tree topologies.

Minimum-Evolution Method

For each possible alternative tree one can estimate the length of each branch from the estimated pairwise distances (d_{ij}) between taxa and then compute the sum (S) of all branch-length estimates. The minimum-evolution criterion is to choose the tree with the smallest S value as the best tree (Cavalli-Sforza and Edwards 1967; Saitou and Imanishi 1989). Rzhetsky and Nei (1993) showed that the expectation of S for the true tree is indeed the smallest among all possible trees, provided that the evolutionary distances (d_{ij}) used are statistically unbiased and that the branch lengths are estimated by the least-squares method, which is explained in the next section. This suggests that the minimum-evolution criterion is in fact a good one for choosing trees. In practice, however, it is too tedious to compute the S values for all possible alternative trees when the number of taxa is large. Since the neighbor-joining (NJ) method is a fast, approximate method for finding the minimum-evolution tree (i.e., the tree with the smallest S), the true tree is likely to be the NJ tree or a tree that differs from the NJ tree by only a few topological differences. Thus, Rzhetsky and Nei (1992) proposed to compare the S values for the NJ tree and all the trees that are very similar to the NJ tree in topology and then to choose the tree with the minimum S value as the best tree. Note that under the minimum-evolution criterion a tree is worse than another tree only if its S value is significantly larger than that of the other tree. Thus, all trees whose S values are not significantly different from the minimum S value should be regarded as candidates for the true tree. The statistical procedure for testing differences in S values will be discussed later.

Maximum Parsimony Methods

The principle of maximum parsimony searches for a tree that requires the smallest number of evolutionary changes to explain the differences observed among the OTUs under study. Such a tree is called a **maximum parsimony tree**. Often more than one tree with the same minimum number of changes are found, so that no unique tree can be inferred. The method discussed below was first developed for amino acid sequence data (Eck and Dayhoff 1966) and was later modified for use on nucleotide sequences (Fitch 1977b).

We start with the definition of **informative sites**. A nucleotide site is phylogenetically informative only if it favors some trees over the others. To illustrate the distinction between informative and noninformative sites, consider the following four hypothetical sequences:

					Site				
Sequence	1	2	3	4	5	6	7	8	9
1	A	A	G	A	G	T	G	C	A
2	A	G	C	C	G	T	G	C	G
3	A	G	A	T	A	T	C	C	A
4	A	G	A	G	A	T	C	C	G
					*		*		*

There are three possible unrooted trees for four OTUs (Figure 5.10). Site 1 is not informative because all sequences at this site have A, so that no change is required in any of the three possible trees. At site 2, sequence 1 has A while all other sequences have G, and so a simple assumption is that the nucleotide has changed from G to A in the lineage leading to sequence 1. Thus, this site is also not informative, because each of the three possible trees requires 1 change. As shown in Figure 5.10a, for site 3 each of the three possible trees requires 2 changes and so it is also not informative. Note that if we assume that the nucleotide at the node connecting OTUs 1 and 2 in tree I in Figure 5.10a is C instead of G, the number of changes required for the tree remains 2. Figure 5.10b shows that for site 4, each of the three trees requires 3 changes and thus site 4 is also noninformative. For site 5, tree I requires only 1 change, whereas trees II and III require 2 changes each (Figure 5.10c). Therefore, this site is informative.

From these examples, we see that a site is informative only when there are at least two different kinds of nucleotides at the site, each of which is represented in at least two of the sequences under study. In the above figure, the informative sites (i.e., sites 5, 7, and 9) are indicated by asterisks. For these sites, tree I requires 1, 1, and 2 changes, respectively; tree II requires 2, 2, and 1 changes; and tree III requires 2, 2, and 2 changes. Thus, tree I is chosen because it requires the smallest number of changes (4) at the informative sites.

In the case of four OTUs, an informative site favors only one of the three possible alternative trees. For example, site 5 favors tree I over trees II and III, and is said to support tree I. It is easy to see that the tree supported by the

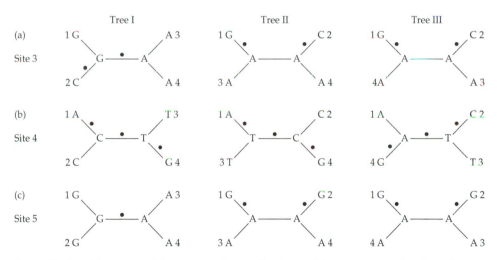

Figure 5.10 Three possible unrooted trees for four DNA sequences that have been used to choose the most parsimonious tree (see text). The terminal nodes indicate the nucleotide type at homologous positions in the extant species. Each dot on a branch means a substitution is inferred on that branch. Note that the nucleotides at the two internal nodes of each tree represent one possible reconstruction from among several alternatives. For example, the nucleotide at both the internal nodes of tree III(c) (bottom right) can be G instead of A. In this case, the two substitutions will be positioned on the branches leading to species 3 and 4. However, the minimum number of required substitutions remains the same. From Li and Graur (1991).

largest number of informative sites is the maximum parsimony tree. For instance, in the above example, tree I is supported by two sites, tree II by one site, and tree III by none.

When the number of OTUs under study is larger than four, the situation becomes more complicated because there are many more possible trees to consider and because inferring the number of substitutions for each alternative tree becomes more tedious. However, the basic principle remains simple—it is to infer the minimum number of substitutions required for a given tree.

This inference can be made by using Fitch's (1971) method. Let us consider the case of six OTUs and assume that at a particular nucleotide site the nucleotides in the six sequences are C, T, G, T, A, and A, as shown in Figure 5.11a. We want to infer the nucleotides at the internal nodes 7, 8, 9, 10, and 11 from the

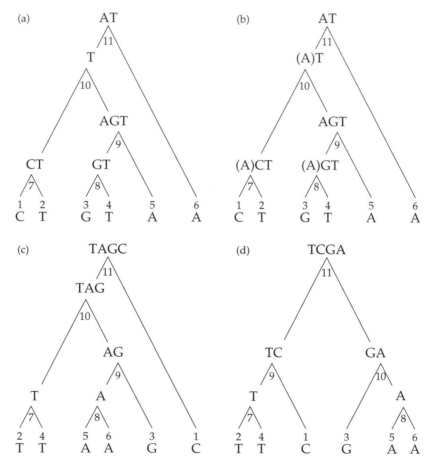

Figure 5.11 Nucleotides in six extant species and the possible nucleotides in five ancestral species. (a), (c), and (d) show three possible trees; (b) shows that one may include A as a possible ancestral nucleotide at nodes 7, 8, and 10 in addition to those indicated in (a), because A is a possible common ancestral nucleotide (node 11) of all the six sequences. Modified from Fitch (1971).

nucleotides at the tips of the tree. The nucleotide at node 7 cannot be determined uniquely, but must be either C or T under the parsimony rule. Therefore, the set of candidate nucleotides at node 7 consists of C and T. Similarly, the set of candidate nucleotides at node 8 consists of G and T and that at node 9 consists of A, G, and T. However, at node 10, T is chosen because it is shared by the sets at the two descendant nodes, 7 and 9. Finally, the nucleotide at node 11 cannot be determined uniquely but parsimony requires it be either A or T.

In mathematical terms, the rule used is as follows: the set at an interior node is the intersection of its two immediately descendant sets if the intersection is not empty (e.g., the nucleotide at node 10 is the intersection of the sets at nodes 7 and 9); otherwise, it is the union of the descendant sets (e.g., the set at node 9 is the union of the sets at nodes 8 and 5). For every occasion that a union is required to form the nodal set, a nucleotide substitution at this position must have occurred at some point during the evolution of this position. Thus, counting the number of unions gives the minimum number of substitutions required to account for descendent nucleotides from a common ancestor, given the phylogeny assumed at the outset. In the present example, this number is 4.

The alternative trees in Figure 5.11c and d each require 3 unions for forming the nodal sets and thus 3 nucleotide substitutions. This is the minimum number of substitutions required to explain the differences among the nucleotides at this position because all of the four types of nucleotides are present at this position. Therefore, there must be at least 3 mutational changes. There are many other alternative trees, each of which requires 3 substitutions. Thus, unlike the case of four OTUs, when more than four OTUs are involved, an informative site may favor many alternative trees.

Although inferring the minimum number of substitutions is straightforward, inferring the evolutionary path (i.e., the true sequence in which the nucleotide had changed in each step) is often difficult; see Fitch (1971) for the procedure.

The procedure for inferring a maximum parsimony tree can be summarized as follows: First, identify all the informative sites. Next, for each possible tree, infer the minimum number of substitutions at each informative site and sum up the numbers over all informative sites. Finally, choose the tree or trees with the smallest number of substitutions; if there is more than one maximum parsimony tree, no unique tree is inferred. Note further that the above procedure neglects all substitutions at noninformative sites. Such substitutions can be easily included because the number of substitutions at a noninformative site is equal to the number of different nucleotides present at that site minus one. For example, if the nucleotides at a site are A, T, T, C, T, and T in lineages 1, 2, . . . , 6, then the number of substitutions is 3 − 1, because regardless of the tree topology we can assume that a T → A substitution has occurred in the first lineage and a T → C substitution has occurred in the fourth lineage.

When the number of OTUs under study is 7 or larger, the number of possible alternative trees is large, so that the calculation requires a computer program.

Note that the minimum-evolution (ME) method described earlier is different from maximum parsimony methods. In parsimony methods, discrete characters are used, and a topology that requires the smallest number of character-state changes (e.g., nucleotide changes) is adopted as the most likely tree. In contrast, the ME method uses distance data.

Maximum Likelihood Methods

The first application of a maximum likelihood (ML) method to tree reconstruction was made by Cavalli-Sforza and Edwards (1967), who used gene frequency data. Later, Felsenstein (1973, 1981) developed ML algorithms for amino acid or nucleotide sequence data. In the past, however, the ML method has not been used frequently, largely because of computational difficulties. Fortunately, computers are now much faster and in recent years many authors have become interested in this method (e.g., Hasegawa et al. 1985; Felsenstein 1988; Saitou 1988; Kishino et al. 1990); see also the review by Li and Gouy (1991).

The ML method requires a probabilistic model for the process of nucleotide substitution. That is, we must specify the transition probability from one nucleotide state to another in a time interval in each branch. For example, consider the one-parameter model with the rate of substitution λ per site per unit time. Assume that the nucleotide at a given site is i at time 0. Then, from Equations 3.11 and 3.12, the probability that the nucleotide at time t is i is given by

$$P_{ii}(t) = \frac{1}{4} + \frac{3}{4}e^{-4\lambda t/3} \tag{5.11}$$

and the probability that the nucleotide at time t is j, $j \neq i$, is given by

$$P_{ij}(t) = \frac{1}{4} - \frac{1}{4}e^{-4\lambda t/3} \tag{5.12}$$

The next step is to set up the likelihood function. Let us use the case of four sequences (taxa) as an example. First, we assume a constant rate of substitution and consider the hypothetical tree given in Figure 5.12a. The likelihood function for a nucleotide site with nucleotides i, j, k, and ℓ in sequences 1, 2, 3 and 4, respectively, can be computed as follows. If the nucleotide at the ancestral node (the root) was x, the probability of having nucleotide ℓ in sequence 4 is $P_{x\ell}(t_1 + t_2 + t_3)$ because $t_1 + t_2 + t_3$ is the total amount of time between the two nodes, the probability of having nucleotide y at the common ancestral node of sequences 1, 2, and 3 is $P_{xy}(t_1)$, and so on. Therefore, given x, y, and z at the ancestral node and the two other internal nodes, the probability of observing i, j, k, and ℓ at the tips of the tree is equal to

$$P_{x\ell}(t_1 + t_2 + t_3)P_{xy}(t_1)P_{yk}(t_2 + t_3)P_{yz}(t_2)P_{zi}(t_3)P_{zj}(t_3)$$

All the above transition probabilities can be computed by using Equations 5.11 and 5.12. Since in practice we do not know the ancestral nucleotide, we can only assign a probability g_x, which is usually assumed to be the frequency of nucleotide x in the sequence. Noting that x, y, and z can be any of the four nucleotides, we sum over all possibilities and obtain the following likelihood function

$$h(i, j, k, \ell) = \sum_x g_x P_{x\ell}(t_1 + t_2 + t_3)$$
$$\times \sum_y P_{xy}(t_1)P_{yk}(t_2 + t_3)\sum_z P_{yz}(t_2)P_{zi}(t_3)P_{zj}(t_3) \tag{5.13}$$

Note that this likelihood function represents the probability of observing the configuration of nucleotides i, j, k, and ℓ in sequences 1, 2, 3, and 4 for the given

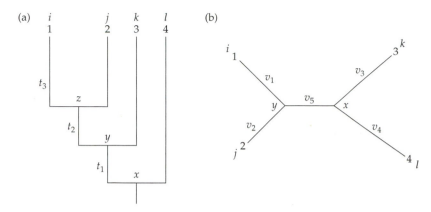

Figure 5.12 Model trees for the derivation of the likelihood function. From Li and Gouy (1991).

hypothetical tree in Figure 5.12a. If the hypothetical tree is different from that in Figure 5.12a, then the likelihood function is different from that of Equation 5.13. That is, the likelihood function depends on the hypothetical tree.

A formulation without the assumption of rate constancy can be done in a similar manner. In this case, it is usually more convenient to consider the transition probability in terms of the branch length; for example, we consider $P_{ij}(v_\alpha)$ instead of $P_{ij}(t_\alpha)$, where $v_\alpha = \lambda_\alpha t_\alpha$ and λ_α and t_α are the values of λ and t for the αth branch. For the unrooted tree given in Figure 5.12b, we obtain the following likelihood function

$$h(i, j, k, \ell) = \sum_x g_x P_{x\ell}(v_4) P_{xk}(v_3) \sum_y P_{xy}(v_5) P_{yi}(v_1) P_{yj}(v_2) \tag{5.14}$$

if we assume that the internal node connecting taxa 3 and 4 is the ancestral node (Felsenstein 1981; Saitou 1988). If the process is time reversible, which is the case for the one-parameter model and many other current models of nucleotide substitution (see Chapter 3), then any node or point in the tree can be taken as the "ancestral" node. This is called the pulley principle (Felsenstein 1981). However, if the process is not time reversible, then we must assume or infer the root of the tree.

In the above we considered a single site. The likelihood for all sites is the product of the likelihoods for individual sites if all the nucleotide sites evolve independently. Suppose that there are s homologous sequences each with N nucleotides. Let $\mathbf{X}_\kappa = (x_{1\kappa}, \ldots, x_{s\kappa})$ be the vector representing the nucleotide configuration at the κth site, i.e., $x_{\zeta\kappa}$ is the nucleotide at the κth site in the ζth sequence. For a given phylogenetic tree T, let $f(\mathbf{X}_\kappa|\theta_1, \ldots, \theta_\eta, T)$ be the likelihood of tree T for the κth site, where $\theta_1, \ldots, \theta_\eta$ are the unknown parameters such as the branching dates and the rates of nucleotide substitution. For example, for the configuration in the tree in Figure 5.12b, we have $\mathbf{X}_\kappa = (i, j, k, \ell)$, $\theta_i = v_i$, and $f(\mathbf{X}_\kappa|\theta_1, \ldots, \theta_\eta, T) = h(i, j, k, \ell)$, which is given by Equation 5.14. For simplicity, let us assume that the sequences are homogeneous so that all sites on the sequences are equivalent (i.e., they evolve at the same rates). Then the likelihood function for the entire sequence for tree T is

$$L\left(\theta_1,\ldots,\theta_\eta\middle|\mathbf{X}_1,\ldots,\mathbf{X}_N,\mathrm{T}\right) = \prod_{\kappa=1}^{N} f\left(\mathbf{X}_\kappa\middle|\theta,\mathrm{T}\right) \tag{5.15}$$

where $L(\theta_1,\ldots,\theta_\eta\,|\,\mathbf{X}_1,\ldots,\mathbf{X}_N,\mathrm{T})$ signifies the fact that given the data (observations) $\mathbf{X}_1,\ldots,\mathbf{X}_N$ and tree T, the likelihood is a function of $\theta_1,\ldots,\theta_\eta$.

After the likelihood function L is derived under a hypothetical tree, one can use a computer to search for the $\theta_1,\ldots,\theta_\eta$ values that maximize L. For example, for the tree in Figure 5.12a, one searches for the t_α values that maximize L. (The maximum likelihood value is usually denoted ML). Finally, one chooses from among all possible trees the one with the largest ML value as the true tree. Note that since the likelihood functions such as Equations 5.13 and 5.14 depend on the model of nucleotide substitution, so do the ML values for different trees. Thus, a tree with the largest ML value under one substitution model may not have the largest ML value under another model. So, one should have an explicit model when deriving the likelihood functions.

Under the assumption that all sites are independent and equivalent, the likelihood function can also be derived by considering the nucleotide configurations among the sequences (Saitou 1988). Let us consider the case of 4 sequences. As shown in Table 5.2 there are 15 configuration patterns. The first pattern (i, i, i, i) means that the nucleotide is the same in all four sequences, the second pattern (i, i, i, j) means that the nucleotide is the same for the first three sequences, but is different for the fourth sequence, and so on. For the unrooted tree given in Figure

TABLE 5.2   Nucleotide configurations for four sequences

No.	Configuration[a]			
	A	B	C	D
1	i	i	i	i
2	i	i	i	j
3	i	i	j	i
4	i	j	i	i
5	j	i	i	i
6	i	i	j	j
7	i	j	i	j
8	i	j	j	i
9	i	i	j	k
10	i	j	i	k
11	j	i	i	k
12	j	k	i	i
13	j	i	k	i
14	i	j	k	i
15	i	j	k	ℓ

From Saitou (1988).

[a]A, B, C, and D are sequences and $i, j, k,$ and ℓ are different nucleotides.

5.12b and under the one-parameter substitution model, we can show that the probability of obtaining the first pattern is equal to

$$U_1 = 4h(i,i,i,i)$$

where $h(i, i, i, i)$ is given by Equation 5.14, and the factor 4 arises because there are four possible nucleotides. The probability of obtaining the second pattern is equal to

$$U_2 = 12\,h(i,i,i,j)$$

in which the factor 12 arises because there are four possibilities for i and for each i there are three possibilities for j. In the same manner, we can obtain the probabilities for the other patterns in Table 5.2 (Saitou 1988). The likelihood function, L_j, for the jth topology is

$$L_j = C\prod_{i=1}^{15} U_{ji}^{N_i} \tag{5.16}$$

where U_{ji} is the probability of obtaining the ith pattern for the jth topology, N_i is the observed number of the ith pattern, $C = N!/(N_1!\,N_2!\,\ldots\,N_{15}!)$, and N is the total number of nucleotide sites examined and is equal to the sum of N_i's. This approach reduces the amount of computation considerably.

Methods of Invariants

Let us start with the definition of invariants. First, consider the evolution of a nucleotide site in two descendant lineages A and B (Figure 5.13). Assume that the probability of changing from nucleotide i to nucleotide j per unit time is a_{ij} in lineage A and b_{ij} in lineage B. Let $P(xy)$ be the probability that at time t the nucleotides in lineages A and B are x and y, respectively. Consider, for example, the function

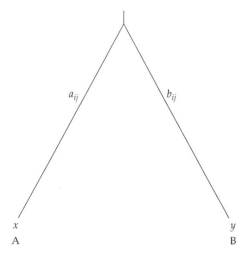

Figure 5.13 Evolution of a nucleotide site in two lineages. a_{ij} is the probability of changing from nucleotide i to j per unit time in lineage A; b_{ij} is similarly defined for lineage B.

$$f = P(\text{AT}) - P(\text{TA}) \tag{5.17}$$

If the nucleotide at the root of the two lineages, i.e., at $t = 0$, is G, then at $t = 1$, $P(\text{AT}) = a_{\text{GA}}b_{\text{GT}}$ because the probability of changing from G to A in lineage A is a_{GA} and the probability of changing from G to T in lineage B is b_{GT}. Similarly, $P(\text{TA}) = a_{\text{GT}}b_{\text{GA}}$. Therefore, at $t = 1$,

$$f = a_{\text{GA}}b_{\text{GT}} - a_{\text{GT}}b_{\text{GA}}$$

If we use the one-parameter model of nucleotide substitution such that $a_{ij} = \alpha_1$ for $i \neq j$ and $b_{ij} = \alpha_2$ for $i \neq j$, then

$$f = \alpha_1\alpha_2 - \alpha_1\alpha_2 = 0$$

for any values of α_1 and α_2. Actually, $f = 0$ for any time t; this can be shown by using Formula 5.12. Therefore, under the above conditions the function defined by (5.17) does not change with time, i.e., $f = 0$ for all t, and it is said to be an invariant. (Note that if $\alpha_1 \neq \alpha_2$, the rates in the two lineages are unequal.) However, if the nucleotide at the root is A instead of G, then at $t = 1$

$$f = a_{\text{AA}}b_{\text{AT}} - a_{\text{AT}}b_{\text{AA}}$$

$$= (1 - 3\alpha_1)\alpha_2 - \alpha_1(1 - 3\alpha_2)$$

because $a_{\text{AA}} = 1 - 3\alpha_1$ and $b_{\text{AA}} = 1 - 3\alpha_2$. So, if $\alpha_1 \neq \alpha_2$, then f is not 0 and thus not an invariant. Therefore, whether the function defined by (5.17) is an invariant depends on the initial nucleotide at the root and on the assumption of equal rates in the two lineages.

In general, it is more useful to have invariants that are independent of the initial nucleotide, which is usually unknown. Further, it is better to remove the assumption of rate constancy, which often does not hold. The following function is such an invariant under the one-parameter model

$$f = P(\text{AC}) - P(\text{AT}) - P(\text{GC}) + P(\text{GT}) \tag{5.18}$$

because it can be shown that $f = 0$ for any initial nucleotide at the root and for any values of α_1 and α_2.

Next, we extend the above idea to a sequence of any length, assuming that all sites in the sequence evolve independently. We consider, instead of the probability $P(xy)$ of observing the configuration xy at a particular site, the number of sites, N_{xy}, in the entire sequence, each of which has nucleotide x in lineage A and nucleotide y in lineage B. Instead of (5.18) we define

$$F = N_{\text{AC}} - N_{\text{AT}} - N_{\text{GC}} + N_{\text{GT}} \tag{5.19}$$

The expectation of F is

$$E(F) = NP(\text{AC}) - NP(\text{AT}) - NP(\text{GC}) + NP(\text{GT})$$

which is 0 under the same conditions for $f = 0$ in (5.18). Thus, an **invariant** under a set of conditions can be defined as a random variable whose expectation is 0 under the given conditions. Since the random variable defined in (5.19) is a linear function of the N_{ij}'s, it is said to be a linear invariant. By contrast, an invariant is nonlinear if it is a nonlinear function of the N_{ij}'s. Here, we consider only linear invariants, an approach initiated by Lake (1987); Lake used the term **evolutionary parsimony**, but it is not based on any parsimony principle. For the theory of nonlinear invariants, which was initiated by Cavender and Felsenstein (1987), readers may refer to Drolet and Sankoff (1990), Fu and Li (1991), and Sankoff (1990).

We now use the case of four DNA sequences A, B, C, and D to illustrate how the theory of linear invariants can be used to infer a phylogeny for four taxa. Since there are four possible types of nucleotides, there are $4^4 = 256$ different configurations (combinations) of nucleotides at each site for four sequences. Lake (1987), however, pools these configurations into 36 configuration groups by the following rule. At a given site, any nucleotide of sequence A is denoted "1" and so are the nucleotides in the other sequences if they are the same as the one in sequence A; those related to the latter by a transitional change are represented by "2", and the two possible transversions are represented by "3" and "4", in the order we come across them, if at all. [Recall that exchanges between A and G (purines) and those between T and C (pyrimidines) are transitions, whereas all other types are transversions.] With this notation, any configuration of four nucleotides can be represented by one of 36 four-dimensional vectors. For example, the four configurations AGCT, GCCA, ACTA, and GCTA are recoded as vectors 1234, 1332, 1341, and 1342, respectively.

Denote the three possible unrooted trees for four species by X, Y, and Z as shown in Figure 5.14. For a set of four nucleotide sequences, let N_{ijmn} be the total score of the configuration $ijmn$'s over the length of the sequences. Lake (1987) defines the following three quantities:

$$X = N_{1133} + N_{1234} - N_{1233} - N_{1134}$$

$$Y = N_{1313} + N_{1324} - N_{1323} - N_{1314}$$

$$Z = N_{1331} + N_{1342} - N_{1332} - N_{1341}$$

Under a particular six-parameter model of nucleotide substitution (Cavender 1989; Jin and Nei 1990), we have the following results: If tree X is the true tree, then Y and Z are invariants whereas X is not. If tree Y is the true tree, then X and Z are invariants whereas Y is not. If tree Z is the true tree, then X and Y are

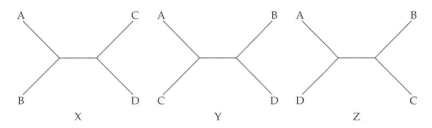

Figure 5.14   Three alternative unrooted trees for four taxa.

invariants whereas Z is not. Therefore, if one of X, Y, and Z, say X, is significantly greater than 0 whereas the other two (i.e., Y and Z) are not, then tree X may be taken as the true tree. (However, if more than one of X, Y, and Z are significantly greater than 0, the problem is unresolved.)

   Lake (1987) proposed to test the significance of X, say, as follows. Let $A = N_{1133} + N_{1234}$ and $B = N_{1233} + N_{1134}$. Then, $(A - B)^2/(A + B)$ is taken as a χ^2 with one degree of freedom; if A and B are small, one should test the equality of A and B by the binomial test (Holmquist et al. 1988; Navidi et al. 1991). A tree is inferred if its associated χ^2 is significant whereas the other two χ^2's are not. Note, however, since three independent χ^2's instead of one are tested, the probability for one of the χ^2's to be significant at the level of α can be up to $1 - (1 - \alpha)^3 \approx 3\alpha$; however, under the assumption that one of the three trees is the true tree, the probability of misidentifying the true tree is about 2α. For further theory of linear invariants, see Cavender (1989, 1991), Fu and Li (1992a,b), and Nguyen and Speed (1992).

Computer Programs

A computer program can usually be obtained from the original authors. In addition, there are software packages available for phylogenetic analysis. PHYLIP (J. Felsenstein) contains computer programs for most of the methods discussed above; information for more than 50 phylogenetics packages can be obtained electronically at:

http://evolution.genetics.washington.edu/phylip/software.html.

PAUP (Swofford 1993) contains an excellent program for the maximum parsimony method and programs for maximum likelihood and distance methods. MEGA (Kumar et al. 1993) contains programs for distance methods.

ESTIMATION OF BRANCH LENGTHS

Estimation of branch lengths has received much less attention than estimation of tree topology. This is understandable because the most important step in tree reconstruction is to infer the correct phylogenetic relationships among the taxa under study. However, branch lengths also contain very useful information because they can provide a rough idea about the degree of separation between nodes or the divergence time between lineages if the rates of evolution are similar among lineages. We shall discuss how to estimate branch lengths, assuming that the tree topology has already been inferred.

   For the UPGMA method, the procedure for estimating branch lengths has already been presented together with the procedure for inferring the branching order. For the maximum parsimony method, we have also described the procedure for inferring the minimum number of substitutions in each branch. This procedure tends to underestimate the branch lengths, for it is intended to minimize the number of substitutions required. The degree of underestimation may not be severe if the tree contains only short branches; otherwise, the underestimation may become serious (see Saitou 1989; Tateno et al. 1994). For the maximum likelihood method, branch lengths and branching order are usually estimated together; since the computation is tedious, one usually uses computer

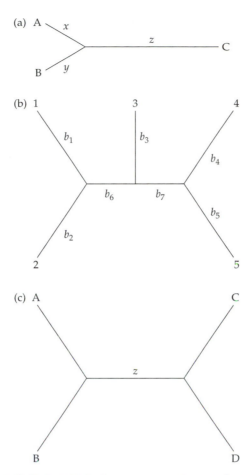

Figure 5.15 Unrooted phylogenetic trees used to illustrat the computation of branch lengths by Fitch and Margoliash's (1967) and the least-squares methods: (a) a tree with three OTUs, (b) a tree with five OTUs, (c) a tree with four composite OTUs.

programs. For distance matrix methods, branch lengths are usually estimated by Fitch and Margoliash's (1967) method or the (ordinary) least-squares method, which will be described below. It should be emphasized that the distances between sequences should be the distances corrected for multiple hits but not the observed distances, for the latter are likely to give underestimates of branch lengths.

We start with Fitch and Margoliash's (1967) method. First, let us consider the simplest case, i.e., an unrooted tree with three OTUs (A, B, and C) and a single node (Figure 5.15a). Let x, y, and z be the lengths of the branches leading to A, B, and C, respectively. The following equations hold:

$$d_{AB} = x + y \tag{5.20a}$$

$$d_{AC} = x + z \tag{5.20b}$$

$$d_{BC} = y + z \tag{5.20c}$$

From these equations, we obtain the following solutions:

$$x = (d_{AB} + d_{AC} - d_{BC})/2 \qquad (5.21a)$$

$$y = (d_{AB} + d_{BC} - d_{AC})/2 \qquad (5.21b)$$

$$z = (d_{AC} + d_{BC} - d_{AB})/2 \qquad (5.21c)$$

Let us now deal with the case of more than three OTUs. For simplicity, let us assume that there are five OTUs (1, 2, 3, 4, and 5) and that the topology and the branch lengths are as shown in Figure 5.15b. Suppose that OTUs 1 and 2 are the first OTUs to be clustered together in the tree reconstruction process. We then use A and B to denote OTUs 1 and 2, respectively, and put all the other OTUs into a composite OTU denoted as C. By this arrangement, we can apply Equations 5.20a–c to estimate the lengths of the branches leading to A, B, and C, except that now $d_{AC} = d_{1(345)} = (d_{13} + d_{14} + d_{15})/3$, and $d_{BC} = d_{2(345)} = (d_{23} + d_{24} + d_{25})/3$. Then we have $b_1 = x$ and $b_2 = y$. OTUs 1 and 2 are subsequently considered as a single composite OTU. In the next step, suppose that OTUs (12) and 3 are the next pair to be joined together. Then we denote OTUs (12) and 3 by A and B, respectively, and put the other OTUs (i.e., 4 and 5) into the composite OTU C. In the same manner as above, we obtain x, y, and z. Note that $b_3 = y$ and $b_6 + (b_1 + b_2)/2 = x$. From the values for b_1 and b_2, which have been obtained previously, we can calculate b_6. The process is continued until all branch lengths are obtained.

Note that sometimes an estimated branch length can be negative. Since the true length can never be negative, it is better to replace such an estimate by 0.

As an example of using the above method, let us compute the branch lengths of the tree in Figure 5.7c. For convenience, we again present the distance matrix that was used to infer the topology of this tree; however, to avoid confusion with the notation in Equations 5.21 we rename OTUs A, B, C, and D as OTUs 1, 2, 3, and 4.

	OTU		
OTU	1	2	3
2	8		
3	7	9	
4	12	14	11

Since OTUs 1 and 2 were clustered first, we first compute the lengths (a and b) of the branches leading to these two OTUs by putting OTUs 3 and 4 into a composite OTU C. We then have $d_{AB} = d_{12} = 8$, $d_{AC} = (d_{13} + d_{14})/2 = (7 + 12)/2 = 9.5$, and $d_{BC} = (d_{23} + d_{24})/2 = 11.5$. From Equations 5.21, we have $a = x = (8 + 9.5 - 11.5)/2 = 3$ and $b = y = (8 + 11.5 - 9.5)/2 = 5$. Next we treat OTUs 1 and 2 as a single OTU (12) and denote it by A. Denote OTU 3 by B and OTU 4 by C. We then have $d_{AB} = d_{(12)3} = (d_{13} + d_{23})/2 = (7 + 9)/2 = 8$, $d_{AC} = d_{(12)4} = (d_{14} + d_{24})/2 = (12 + 14)/2 = 13$, and $d_{BC} = d_{34} = 11$. Therefore, from Equations 5.21, we have $x = (8 + 13 - 11)/2 = 5$, and $d = y = (8 + 11 - 13)/2 = 3$, and $e = z = (13 + 11 - 8)/2 = 8$. We note from Figure 5.7c that $(a + b)/2 + c = x$, and so $c = 1$. This completes the computation. Note, however, that since we do not know the exact location of the root, we cannot estimate the length of the branch connecting the root and OTU D but can only estimate the length from the common ancestral node of OTUs A, B, and C through the root to OTU D, i.e., $e = 8$.

We now discuss the least-squares method. Let us use the tree in Figure 5.15b as an example. For the least-squares method we set up the following equations.

$$d_{12} = b_1 + b_2 + \qquad\qquad\qquad\qquad\qquad\qquad e_{12}$$
$$d_{13} = b_1 + \qquad b_3 + \qquad\qquad\qquad b_6 + \qquad e_{13}$$
$$d_{14} = b_1 + \qquad\qquad b_4 + \qquad\qquad b_6 + b_7 + e_{14}$$
$$d_{15} = b_1 + \qquad\qquad\qquad b_5 + b_6 + b_7 + e_{15}$$
$$d_{23} = \qquad b_2 + b_3 + \qquad\qquad\qquad b_6 + \qquad e_{23}$$
$$d_{24} = \qquad b_2 + \qquad b_4 + \qquad\qquad b_6 + b_7 + e_{24}$$
$$d_{25} = \qquad b_2 + \qquad\qquad\qquad b_5 + b_6 + b_7 + e_{25}$$
$$d_{34} = \qquad\qquad b_3 + b_4 + \qquad\qquad\qquad b_7 + e_{34}$$
$$d_{35} = \qquad\qquad b_3 + \qquad\qquad b_5 + \qquad\qquad b_7 + e_{35}$$
$$d_{45} = \qquad\qquad\qquad b_4 + b_5 + \qquad\qquad\qquad + e_{45} \qquad (5.22)$$

where e_{ij}'s are sampling errors; e_{ij} has mean 0 and variance $V(d_{ij})$. The least-squares estimates of the b_i's are estimates $\hat{b}_i$ such that the following sum of squares is the smallest:

$$\sum_{i,j} \left(d_{ij} - \hat{d}_{ij} \right)^2$$

where $\hat{d}_{12} = \hat{b}_1 + \hat{b}_2$, $\hat{d}_{13} = \hat{b}_1 + \hat{b}_3 + \hat{b}_6$, and so on.

Rzhetsky and Nei (1993) have developed a simple computational procedure for implementing the least-squares estimators. Consider Figure 5.15b as an example. If we choose one particular interior branch of this tree, the tree can be drawn in the form of Figure 5.15c, where A, B, C, and D each represent a cluster of OTUs. For example, for branch b_6 in Figure 5.15b, A = (1), B = (2), C = (3), and D = (4,5). In general, let d_{AB} be the distance between clusters A and B as defined by Equation 5.4; d_{AC}, d_{AD}, d_{BC}, d_{BD}, and d_{CD} are similarly defined. Then the least-squares estimator of the branch length z in Figure 5.15c is given by

$$\hat{z} = \frac{1}{2}\left[\gamma(d_{AC} + d_{BD}) + (1 - \gamma)(d_{BC} + d_{AD}) - d_{AB} - d_{CD} \right] \qquad (5.23)$$

where $\gamma = (n_B n_C + n_A n_D)/[(n_A + n_B)(n_C + n_D)]$ and n_A, n_B, n_C, and n_D are the numbers of OTUs in clusters A, B, C, and D, respectively. By contrast, the least-squares estimator of the length z of an exterior branch of the tree in Figure 5.15a is given by Equation 5.21c, except that A, B, and C now may be composite OTUs and d_{AB}, d_{AC}, and d_{BC} are computed by using Equation 5.4.

Note that the Fitch-Margoliash method gives the least-squares estimates when the number of OTUs is five or smaller, but this may not be the case when the number of OTUs is six or larger.

PHENETICS VERSUS CLADISTICS

A long-standing controversy in taxonomy has been the often acrimonious dispute between **cladists** and **pheneticists**. The term **cladistics** can be defined as the study of the pathways of evolution. In other words, cladists are interested in such questions as: how many branches are there among a group of organisms; which branch connects to which other branch; and what is the branching sequence (Sneath and Sokal 1973). A treelike network that expresses such ancestor–descendant relationships is called a **cladogram**. To put it another way, a cladogram refers to the topology of a rooted phylogenetic tree.

On the other hand, **phenetics** is the study of relationships among a group of organisms on the basis of the degree of similarity between them, be that similarity molecular, phenotypic, or anatomical. A treelike network expressing phenetic relationships is called a **phenogram**. While a phenogram may serve as an indicator of cladistic relationships, it is not necessarily identical with the cladogram. If there is a linear relationship between the time of divergence and the degree of genetic (or morphological) divergence, the two types of trees may become identical to each other.

Among the methods discussed above, the maximum parsimony method is a typical representative of the cladistic approach, whereas the UPGMA method is a typical phenetic method. The other methods, however, cannot be classified easily according to the above criteria. For example, the transformed-distance method and the neighbor-joining method have often been said to be phenetic methods, but this is not an accurate description. Although these methods use similarity (or dissimilarity, i.e., distance) measures, they do not assume a direct connection between similarity and evolutionary relationship, nor are they intended to infer phenetic relationships.

In molecular phylogenetics, a better classification of methods would be to distinguish between **distance** and **character-state approaches**. Methods belonging to the former approach are based on distance measures, such as the number of nucleotide or amino acid substitutions, while methods belonging to the latter approach rely on the state of the character, such as the nucleotide or amino acid at a particular site, or the presence or absence of a deletion or an insertion at a certain DNA location. According to this classification, the first five methods described in this chapter are distance methods, while the maximum parsimony method is a character-state method. The method of linear invariants also uses character-state information and thus is also a character-state method. Likelihood methods use character-state information to form the likelihood function but search for branch lengths (or times) that maximize the likelihood for a given tree. Therefore, they cannot be easily classified into either of the two categories.

It has often been argued that character-state methods are more powerful than distance methods, because the raw data are a string of character states (e.g., the nucleotide sequence) and in transforming character-state data into distance matrices some information is lost. Note, however, that while the maximum parsimony method indeed uses the raw data, it usually uses only a small fraction of the available data. For instance, in the example in the section on maximum parsimony methods, only three of the nine sites are used. For this reason, this method is often less efficient than some distance matrix methods (see below). Of course, if the number of informative sites is large, the maximum parsimony method is generally very effective. The maximum likelihood method uses the character-state information at all sites and thus can be said to use the "full" information; however, it requires assumptions on the model of nucleotide substitution. The next section gives a more detailed comparison of methods.

COMPARISON OF METHODS

Phylogenetic reconstruction has always been controversial and is likely to continue to be so for a long time, because the issue is complex and because there are differences in personal preference or philosophy. Since currently no method works well under all conditions or can be claimed to be better than all others, it is useful to know the strengths and weaknesses of different methods. We therefore consider the assumptions explicitly or implicitly involved in a method, the computational time, the statistical consistency of a method, and the relative performances of different methods under various conditions.

The performance of a method is usually studied by using either computer simulation or empirical tests. Empirical tests use cases where the evolutionary history is known. Fitch and Atchley (1985) and Atchley and Fitch (1991) used genetic data to test whether current methods of phylogenetic reconstruction can indeed recover the known relations among inbred strains of laboratory mice. Gouy and Li (1989) tested Lake's method of linear invariants, the maximum parsimony method, and the neighbor-joining method by using the small subunit rRNA sequences from human, *Drosophila*, rice, and *Physarum*. Hillis et al. (1992, 1994) developed an ingenious experimental approach by taking advantage of the facts that viruses (e.g., bacteriophage T7) can be manipulated in the laboratory through thousands of generations per year and that mutation rates of viruses can be easily elevated through the use of mutagens. However, the number of empirical studies is still too limited to draw any general conclusion about the relative performances of different methods; as molecular evolution is stochastic, a large number of replications is needed for testing methods.

Assumptions and Computational Time

Few methods are completely explicit in assumptions about the pattern of and rate of nucleotide substitution; indeed, some methods (e.g., the maximum parsimony method) make no explicit assumption at all. Knowing the assumptions (explicitly or implicitly) stipulated in a method may help us understand its range of applicability and become cautious about an inferred tree when any of the assumptions appear to be violated; for testing the assumptions of a model or method, see Goldman (1993) and Tamura (1994). It should be emphasized that when a method makes no explicit assumption, it does not mean that the method works under all conditions. It should also be emphasized that a method requiring fewer assumptions may not necessarily perform better than one requiring more assumptions. The performance of a method depends not only on assumptions but also on the computational procedure or the optimality criterion used (see below).

The UPGMA method implicitly assumes a constant rate for all branches. There is now strong evidence that this assumption is often violated (see Chapter 8). Moreover, if the sequences used are not long, some branches may by chance happen to have evolved faster than others, thus in effect violating the rate-constancy assumption. For this reason, an inferred UPGMA tree often involves errors in branching order (i.e., topological errors), even if the assumption of equal rates holds (e.g., see Saitou and Nei 1987).

The transformed distance method, the neighbors-relation method, the neighbor-joining method and many other distance matrix methods do not assume rate-constancy. However, since all these methods assume that the effect of unequal rates among branches can be "corrected" from the distances in the distance matrix, their performance is affected by the accuracy of the distances estimated. When the distances are small, fairly accurate estimates of distances can be obtained, and these methods may perform well even under nonconstant rates of evolution, if the sequences used are long. Indeed, when the distances are small, one may even use the observed (i.e., uncorrected) distances (Saitou and Nei 1987). Note, however, if the sequences are short, then estimates of distances are subject to large statistical errors. Moreover, if some distances are large or if the rate varies greatly among sites, then accurate estimation of distances is difficult (Chapter 4). Under any of these situations, the performance of a distance method can be compromised.

The maximum parsimony (MP) method makes no explicit assumptions, except that the parsimony criterion assumes that a tree that requires fewer substitutions is better than one that requires more. Note that a tree that minimizes the number of substitutions also minimizes the number of "homoplastic events," i.e., parallel, convergent, and back (reversal) substitutions (see definitions in Chapter 3). When the degree of divergence between sequences is small so that homoplasies are rare, the parsimony criterion usually works well. However, when the degree of divergence is large so that homoplasies are common, it may not work well. In particular, if some sequences have evolved much faster than others, then homoplastic events are likely to have occurred more often among these sequences than among others and the parsimony criterion may become misleading. Note also that the chance of homoplasy depends on the substitution pattern. For example, if transitional substitutions occur more often than transversional substitutions, then the chance of homoplasy will be higher than the case of random substitution among the four nucleotides. Therefore, the performance of the MP method is affected by unequal rates of evolution and by the pattern of nucleotide substitution, though it makes no explicit assumptions about either rates of evolution or the substitution model. Of course, some of these effects can be taken into account in a computer algorithm (see Swofford 1993); e.g., transitional bias can be taken into account by assigning different weights to transitional and transversional changes, but this requires an assumption of the transition/transversion ratio.

The maximum likelihood (ML) method usually makes explicit assumptions about the rate of evolution and the pattern of nucleotide substitution. It has been commonly believed that this method is insensitive to violations of assumptions, but the simulation study of Tateno et al. (1994) suggested that the ML method may not be very robust. One advantage of this method is that one may develop a computer program for it with options for choosing rates and patterns of substitution, so that it may be able to handle various situations. However, the computation may become very tedious and current computer packages may not have many user options for choosing substitution models and rates.

The method of linear invariants has become explicit in assumptions, owing to works by Cavender (1989, 1991), Jin and Nei (1990), Fu and Li (1992a,b), and others. For example, these authors showed that the most general conditions for the existence of linear invariants is a six-parameter model of nucleotide substitution. One advantage of linear invariants is that such invariants exist without the assumption of equal rates among lineages.

It has often been said that distance matrix methods such as the neighbor-joining (NJ) method use no optimality criterion. That is not quite true — although these pairwise clustering methods use no global criterion, they do use local criteria. For example, the neighbors-relation method uses the score of neighbors and the NJ method uses the S_{ij} value defined by Equation 5.8 as a criterion for clustering OTUs. Notice also that the minimum-evolution method does use a global criterion, though, like the maximum parsimony method, it may be too tedious to do an exhaustive search for a tree with the minimum total tree length.

We now consider the computational time. In the first four distance matrix methods described in this chapter, the OTUs are sequentially clustered to derive a final tree. For this reason, the computer algorithms for these methods are fast (see Kuhner and Felsenstein 1994). In comparison, the minimum-evolution (ME) and maximum parsimony (MP) methods compare all possible alternative trees. This comparison is feasible only when the number of OTUs is small or moderate and the sequences under study are not long. For example, for 10 OTUs there are more than 2 million possible unrooted trees to be considered (see Table 5.1) and so the computer time required becomes very large if the sequences are long. Thus, when the number of OTUs is large, say >11, one does not do an exhaustive search. We mentioned earlier how this difficulty might be overcome in the case of the ME method. The MP method uses the branch-and-bound technique to reduce the number of potentially optimal trees to be considered (Hendy and Penny 1982) or even uses a heuristic approach such as "stepwise addition" or "branch swapping," which does not guarantee obtaining the optimal tree (see Swofford and Olsen 1990). The maximum likelihood (ML) method is even more time-consuming because it considers all alternative trees and for each tree it searches for the maximum likelihood value (Kuhner and Felsenstein 1994). Thus, when the number of OTUs is large, it uses heuristic approaches to reduce the number of trees to be considered (see Felsenstein 1981; Saitou 1988). The method of linear invariants is computationally much less tedious than the MP and ML methods but more tedious than distance matrix methods.

Consistency

In statistics an estimation method (estimator) is said to be consistent if it approaches the true value of the quantity to be estimated as more and more data are accumulated. Phylogenetic reconstruction can be regarded as a type of statistical estimation—the "quantity" to be estimated is the true phylogeny. Obviously, a desired property of any tree reconstruction method is that it be consistent.

The fact that the maximum parsimony method can become inconsistent under unequal rates of evolution was first discovered by Felsenstein (1978; see also Cavender 1978). Figure 5.16 shows schematically why this can happen. Suppose that branches 1 and 3 are long, whereas the other branches are short. Then, the nucleotides at the tips of lineages 2 and 4 are likely to remain the same as the

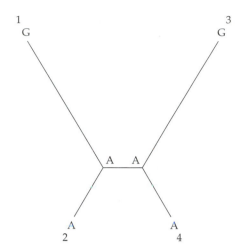

Figure 5.16   A schematic illustration for parallel changes in the two long lineages 1 and 3.

nucleotide, say A, at their common ancestral node. On the other hand, the nucleotides at the tips of lineages 1 and 3 may be different from A and by chance both may become G (C or T). Under the parsimony criterion, such a site supports the erroneous grouping of lineages 1 and 3 in a cluster.

The above example provides an intuitive argument for how the MP method can become inconsistent. To have a better understanding of the conditions for inconsistency to occur, let us consider a more rigorous analysis using the model tree (tree 1) in Figure 5.17, in which OTU 4 is known to be an outgroup (Zharkikh and Li 1992b). Let p_1, p_2, and p_3 be the probabilities that a randomly chosen site will support trees 1, 2, and 3, respectively. The MP method is consistent as long as p_1 is larger than both p_2 and p_3. This condition is violated only when substantial differences occur among the rates in lineages 1, 2, and 3. Let T_1, T_2, and T_3 be the divergence times as shown in Figure 5.17. Assume that the substitution rate is u in all branches except that $u_1 = wu$ in branch 1. Further, assume that $T_1 = 100$ million years (Myr), $T_2 = 50$ Myr, and $T_3 = 45$ Myr. For $u = 10^{-8}$ substitutions per site

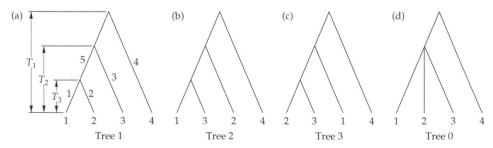

Figure 5.17   Tree 1 is the model tree for the evolution of DNA sequences in four lineages. Trees 2 and 3 are the two erroneous alternative trees and tree 0 is a trifurcating tree. From Zharkikh and Li (1992a).

per year, if $u_1 > 1.3u$ (i.e., $w > 1.3$), then $p_3 > p_1 > p_2$ and the MP method becomes inconsistent. In contrast, for $u = 10^{-9}$ the MP method becomes inconsistent only when $w > 4$. This example shows that the conditions for inconsistency to occur depend not only on rate differences but also on the degree of sequence divergence and that one way to avoid the inconsistency of the MP method is to use more slowly evolving sequences (see Zharkikh and Li 1992b; Huelsenbeck and Hillis 1993). Note that when w exceeds a critical value so that $p_3 > p_1 > p_2$, the data will tend to support tree 3 or, in other words, lineages 1 and 4, which are the longer edges, will tend to be clustered together in an unrooted tree. This tendency has been dubbed "long edges attract" (Penny et al. 1987).

In Figure 5.17a if the rate is the same in all branches except that $u_2 = wu$ for lineage 2, then the effect is the same as that in the above case except now that tree 2 instead of tree 3 is favored when inconsistency occurs. Inconsistency of the MP method can also occur when the rate in lineage 3 evolves more slowly than lineages 1 and 2. However, it cannot occur if lineage 3 evolves faster than lineages 1 and 2, because lineages 3 and 4 will then be the longer edges.

Surprisingly, in contrast to the case of four taxa, in the case of five or more taxa the MP method can become inconsistent even under equal rates of evolution (Hendy and Penny 1989; Zharkikh and Li 1993; Takezaki and Nei 1994). Under some models of nucleotide substitution, the inconsistency problem of the MP method may be avoided by making certain nonlinear transformations of the nucleotide configuration patterns among sites (Steel et al. 1993), but it may be difficult to find such transformations in general.

Distance matrix methods are consistent under equal rates of evolution. When unequal rates occur, the UPGMA method often becomes inconsistent, whereas the transformed distance method, the neighbors-relation method, and the neighbor-joining method remain consistent if the distances are estimated accurately. The latter methods, however, may become inconsistent if distances are not estimated accurately (see Felsenstein 1988; Jin and Nei 1990; DeBry 1992; Huelsenbeck and Hillis 1993). For example, in the case of four taxa if the observed distances instead of the distances corrected for multiple hits are used, then the conditions for the transformed distance method, the neighbors-relation method and the neighbor-joining method to become inconsistent are similar to those for the MP method (Penny et al. 1992; Zharkikh and Li 1992b). As discussed in Chapter 4, accurate estimation of distances is difficult when the rate varies among sites or when the substitution pattern is complicated.

Maximum likelihood methods are commonly believed to be consistent (see Felsenstein 1988). However, if the assumptions used to construct the likelihood function are unrealistic, ML methods may become inconsistent. For example, inconsistency may occur if the substitution model used is not realistic for sequences under study or if the rate of evolution is assumed to be uniform (either among sites or among lineages) but this is in fact not the case.

The method of linear invariants can become inconsistent if some of the necessary conditions for the existence of linear invariants are violated. For example, Lake's method can become inconsistent if the condition of balanced transversion is violated (Jin and Nei 1990; Huelsenbeck and Hillis 1993); "balanced transversion" means that the rates for the two types of transversion for a nucleotide are equal (e.g., $f_{AT} = f_{AC}$).

Simulation Studies

Computer simulation provides a convenient means for comparing the performances of different methods because many different conditions can be investigated and because the true tree (i.e., the model tree) is known, so that the probability for a method to recover the true tree can be evaluated. In contrast, in empirical studies the true tree is often unknown and there is usually little information about the conditions under which the molecules used have evolved. Moreover, in computer simulation a large number of replications can be conducted, a situation that is usually not feasible in empirical or experimental studies. Of course, one should be cautious about the direct applicability of simulation results to real-world situations because simulation studies usually make simplfying assumptions. However, if a method does not perform well under simple conditions, it is unlikely to do well in complex situations. Thus, simulation studies can help us understand the limitations of a method.

There is a wealth of literature on the simulation study of tree-making methods; see the reviews by Felsenstein (1988) and Nei (1991). Here we review some simulation studies that show the relative performances of different methods under certain conditions. The criterion for the performance of a method is the probability of recovering the correct tree. Since the simulation study of Saitou and Nei (1987) showed that the UPGMA method is inferior to the transformed distance method, the neighbors-relation method, and the neighbor-joining method even under the assumption of rate-constancy and that the neighbors-relation method and the neighbor-joining (NJ) method perform approximately equally well, the UPGMA and the neighbors-relation methods will not be considered in subsequent comparisons. For Lake's (1987) method of linear invariants, we note that this method was intended to deal with unequal rates of evolution. Simulation studies (Li et al. 1987b; Jin and Nei 1990; Huelsenbeck and Hillis 1993) showed that when rates are highly unequal among lineages, the method of linear invariants is indeed superior to the MP method, provided that the necessary conditions for the existence of linear invariants hold. However, because it relies mainly on transversional substitutions and thus uses only a small fraction of the sequence data, it is inferior to the NJ method, if a proper distance measure is used for the latter. Indeed, it is even inferior to the MP method when the rates of substitution are nearly equal among lineages (Li et al. 1987b; Jin and Nei 1990; Huelsenbeck and Hillis 1993).

Let us first consider the case of a constant rate of evolution. Using tree a in Figure 5.18 as a model tree, Sourdis and Nei (1988) simulated the evolution of a nucleotide sequence along each branch starting from a random sequence of nucleotides at the root of the tree. They used the one-parameter or Kimura's two-parameter model of nucleotide substitution to simulate the change of nucleotide at each site. For each simulation replicate they used the simulated nucleotide sequences at the tips of the model tree to reconstruct a tree by the the maximum parsimony (MP) method, the neighbor-joining (NJ) method and the transformed distance (TD) method. For model tree a (i.e., a constant rate), the MP method is as good as or slightly better than the TD and NJ methods only when the number of nucleotide substitutions is small ($U = 0.05$) and when the number of nucleotides in a sequence is $L \geq 1200$ (Table 5.3). In all other cases, the TD and NJ methods perform better than the MP method.

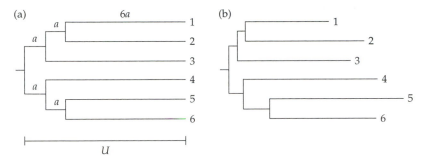

Figure 5.18   Model trees. (a) Constant rate of nucleotide substitution. U is the expected number of substitutions per site from the common ancestral sequences to the extant sequences and a defines a unit branch length such that $8a = U$. (b) Varying rates among lineages. From Nei (1991).

Saitou and Imanishi (1989) and Hasegawa et al. (1991) compared the MP method, the NJ method, the minimum-evolution (ME) method and the maximum likelihood (ML) method, using the model trees in Figure 5.19. Again, the MP method is less efficient than the NJ method (Table 5.4). The performances of the ME and NJ methods are about the same. The performance of the ML method depends to some extent on the version of the DNAML algorithm of Felsenstein's PHYLIP package; version 2.3 uses the one-parameter model of nucleotide substitution [i.e., equal transition (α) and transversion (β) rates], which was the model used in the computer simulation, while version 3.1 assumes $\alpha = 2\beta$. In the present case (Table 5.4), version 2.3 performs better than version 3.1 because the assumption of $\alpha = \beta$ agrees with the simulation model. Under trees a and b (constant rate)

TABLE 5.3   **Probabilities (%) of obtaining the correct tree for three tree-making methods for model trees a and b in Figure 5.18**[a]

Method	Constant rate (Tree a)			Varying rate (Tree b)		
	300[b]	600	1,200	300	600	1,200
		$U = 0.05$				
MP	46.3	79.0	97.0	16.7	39.3	58.0
NJ	58.7	82.0	95.3	30.0	46.7	63.3
TD	54.7	82.0	95.0	25.3	42.3	62.3
		$U = 0.5$				
MP	44.7	69.3	84.0	16.7	22.3	31.0
NJ	62.7	84.3	97.0	25.0	34.7	49.3
TD	64.3	84.3	96.0	25.0	33.3	48.0

From Sourdis and Nei (1988).

[a]The methods used are: MP, maximum-parsimony method; NJ, neighbor-joining method; and TD, transformed-distance method with Li's (1981) algorithm. The number of replications is 300.

[b]Number of nucleotides in a sequence.

TABLE 5.4 Probabilities (%) of obtaining the correct tree with the maximum-parsimony (MP), the neighbor-joining (NJ), the minimum-evolution (ME), and the maximum-likelihood (ML) method for the model trees in Figure 5.19[a]

| | | | | ML[b] | |
| | | | | DNAML version 2.3 | DNAML version 3.1 |
Model tree	MP	NJ ($\alpha = \beta$)	ME ($\alpha = \beta$)	($\alpha = \beta$)	($\alpha = 2\beta$)
Tree a (constant rate)					
$8a = 0.05^c$					
300 bp	34	40	40	52	38
600 bp	76	82	80	88	80
$8a = 0.50$					
300 bp	60	46	42	48	48
600 bp	84	82	82	78	70
Tree b (constant rate)					
$8a = 0.05$					
300 bp	54	70	68	80	62
600 bp	84	86	80	90	88
$8a = 0.50$					
300 bp	48	60	60	66	56
600 bp	58	70	70	80	76
Tree c (nonconstant rate)					
$a = 0.01$					
300 bp	64	72	72	84	78
600 bp	90	92	92	96	98
$a = 0.05$					
300 bp	24	68	68	92	92
600 bp	20	96	96	100	100
Tree d (nonconstant rate)					
$a = 0.01$					
300 bp	68	74	74	86	80
600 bp	94	92	92	96	96
$a = 0.05$					
300 bp	26	78	78	96	96
600 bp	46	100	100	100	100

From Saitou and Imanishi (1989) and Hasegawa et al. (1991).
[a]Simulations were replicated 50 times based on the one-parameter model of nucleotide substitution.
[b]DNAML version 2.3 and version 3.1 of Felsenstein's PHYLIP package.
[c]a is a unit branch length in the model trees.

in Figure 5.19, the ML method performs better than the NJ method if version 2.3 is used, but slightly worse if version 3.1 is used.

In an extensive simulation study with Kimura's (1980) two-parameter model and with 10 OTUs, Kuhner and Felsenstein (1994) found that under a constant rate of evolution the MP method was slightly superior to the NJ method for the cases with short sequences and low rates or with long sequences and high rates, but inferior to the NJ method for the intermediate cases. The ML method was superior to both the NJ and MP methods in all cases considered.

We now consider nonconstant rates of substitution. For the model tree in Figure 5.18b (Table 5.3), the MP method is less efficient than the TD and NJ methods. Moreover, Table 5.4 shows that for model trees in Figure 5.19c,d the MP method is also less efficient than the NJ method, which is in turn less efficient than the ML method. Using a four-taxa tree, Hasegawa et al. (1991) showed that the ML method performed well under unequal rates. Therefore, the ML method seems to be quite robust against the effects of unequal rates. The same conclusions were reached in the above-mentioned study by Kuhner and Felsenstein (1994). However, if transitional substitution occurs several times more frequently than transversional substitution, then the NJ method performs somewhat better than the ML method (Tateno et al. 1994).

Only a few studies of the effect of rate variation among sites on the performance of a method have been conducted (Olsen 1987; Jin and Nei 1990; Kuhner and Felsenstein 1994; Tateno et al. 1994), though this is an important situation. Kuhner and Felsenstein's (1994) study showed that the MP method performed less well than the NJ method, which was in turn inferior to the ML method. In this study, however, unweighted parsimony (all changes were given equal weights) was used. Tateno et al. (1994) showed that weighting (i.e., transversions were given more weight than transitions and more conservative sites were given more weight than less conservative sites) improved the performance of the MP method, but it remained inferior to the NJ and ML methods. In Kuhner and Felsenstein (1994), the distances used in the NJ method were estimated without correction of rate heterogeneity. Jin and Nei (1990) and Tateno et al. (1994) showed that corrections for rate heterogeneity with the assumption of a gamma distribution of substitution rates can improve considerably the performance of the NJ method, even when the underlying distribution of substitution rates is not

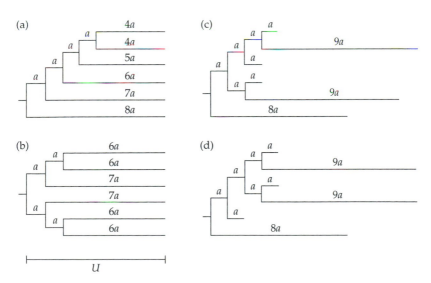

Figure 5.19 Model trees used by Saitou and Imanishi's (1989) computer simulation.

close to a gamma distribution; indeed, under certain conditions the performance of the NJ method may become better than the ML method. To date, no simulation study seems to have been conducted on the performance of the ML method with rate variation among sites; Yang (1993) has developed such a method for four taxa. Such a study would require a large amount of computer time, but should be of great interest.

In summary, simulation studies have shed much light on the relative performances of tree reconstruction methods. In general, under nonuniform rates either among lineages or among sites, the MP method tends to perform less well than the NJ method, while the ML method seems to be more robust than the MP and NJ methods. However, because performance is a complex issue, more research remains to be done, especially in the case of rate variation among sites.

STATISTICAL TESTS: ANALYTICAL METHODS

There exist only a few methods for evaluating the statistical confidence of an inferred phylogeny or for testing whether one phylogeny is significantly better than another. The statistical methodology for testing phylogenies is not well developed for two reasons. First, although phylogenetic reconstruction has long been recognized as a problem in statistical inference (Edwards and Cavalli-Sforza 1964), few authors have formulated the problem in a statistical framework. Indeed, most current methods yield one or a few trees and do not provide information concerning the confidence level of estimated phylogenies. Second, the problem is extremely complex, largely because the number of possible alternative trees is large even when only a moderate number of taxa are involved. For this reason, most current statistical tests are heuristic when more than five taxa are involved. Fortunately, there has recently been a strong trend to make phylogenetic reconstruction more statistical.

Statistical tests can be classified as analytical or resampling. Resampling methods (e.g., the bootstrap and the jackknife) resample the data to infer empirically the variability of the estimate obtained by a tree-making method. In phylogenetic study, bootstrap (Felsenstein 1985a) is the most popular resampling method and will be discussed in the next section. Analytical tests can be based on parsimony methods, distance methods, likelihood methods, or invariant methods. The likelihood approach is more involved and readers are referred to the review by Li and Gouy (1991) and the studies by Kishino et al. (1990) and Tateno et al. (1994). Invariant methods have been discussed briefly earlier. We discuss below the other approaches.

Parsimony Tests

The analytic approach using parsimony tests was initiated by Cavender (1978). He studied the confidence limits that can be placed on a phylogeny inferred from a four-species data set. The results were extended by Felsenstein (1985b), who obtained narrower confidence limits by adding the assumption of a constant rate of evolution, i.e., an evolutionary clock. Felsenstein's results will be discussed below. We shall also discuss the case of more than four species.

FOUR SPECIES WITH A MOLECULAR CLOCK. We assume that species 4 is a known outgroup, so that it can be used as a reference to infer the branching order of the other three species. Trees 1, 2, and 3 in Figure 5.17 are the three possible rooted bifurcating trees. We assume that each randomly chosen informative site supports tree i with probability p_i, $i = 1, 2,$ or 3. Then the probability that among n random informative sites there are i sites supporting tree 1, j sites supporting tree 2, and k sites supporting tree 3 is trinomial, that is,

$$\text{Prob } (i, j, k) = \frac{n!}{i!\, j!\, k!}\, p_1^i p_2^j p_3^k \tag{5.24}$$

For the moment let us assume that tree 1 is the true tree. Under the assumption of rate constancy, an informative site has a higher probability of supporting the true tree than supporting an incorrect tree, i.e., $p_1 \geq p_2 = p_3$. The probability that among n random informative sites the number (C) of sites supporting a particular incorrect tree is m or larger is given by

$$\text{Prob } (C \geq m) = \sum_{i=m}^{n} \frac{n!}{i!\,(n-i)} p_2^i (p_1 + p_2)^{n-i} \tag{5.25}$$

which is obtained by expanding $[p_2 + (p_1 + p_3)]^n$ with the condition $p_3 = p_2$. Since $p_1 + 2p_2 = 1$ and since $p_2 \leq p_1$, Formula 5.25 assumes the maximum value when $p_1 = p_2 = 1/3$, i.e., if the three species represent a true trichotomy (Figure 5.17d).

Now suppose that we do not know which tree is the true tree. Suppose further that in a sequence data set with n informative sites the best supported tree is favored by m sites. Is this tree the true tree? Let us assume that this tree is incorrect and has by chance been supported by so many sites. The probability for a particular incorrect tree to be supported by m or more sites is given by Equation 5.25. Therefore, the probability for one of the two incorrect trees to be supported by m or more sites is

$$P = 2\,\text{Prob } (C \geq m) \tag{5.26}$$

If P is smaller than α, then the support for the best tree is significant at the level of α. This is Felsenstein's (1985b) test.

In practice, p_1 is unknown and it is necessary to use the least favorable value, 1/3. A table for the test based on Equation 5.25 has been given by Li and Gouy (1991). If n is greater than 50, Equation 5.25 can be computed by assuming that m/n follows a normal distribution with mean equal to 1/3 and variance equal to $(1/3 \times 2/3)/n$.

In addition to the above test, Felsenstein (1985b) has also developed a statistic for testing whether the best supported tree is significantly better than the next best tree. Suppose that these two trees are supported by n_1 and n_2 sites, respectively. Let $s = n_1 - n_2$. Then the probability that the best supported tree is at least s steps better is given by

$$\text{Prob } (S \geq s) = \sum_{(i,j,k)} \text{Prob } (i, j, k) \tag{5.27}$$

where

$$j - \max(i, k) \geq s$$
$$\text{or } k - \max(i, j) \geq s$$

If Prob $(S \geq s)$ is smaller than α, then the best supported tree is significantly better than the next best tree at the α level. Again we do not know p_1 and so we do the worst-case analysis assuming $p_1 = p_2 = p_3 = 1/3$. Felsenstein (1988) gave a table of results computed from this analysis.

The test given by Formula 5.27 is computationally somewhat more complicated than that given by Formula 5.26, but the powers of the two tests are about the same (Felsenstein 1985b). This result is not surprising because if the best supported tree is significant, it is likely to be significantly better than the next best tree (and also the worst tree).

Equation 5.26 is based on the assumption that the chance to have a truly trifurcating tree (Figure 5.17d) is negligible, so that one of the three bifurcating trees (Figure 5.17a–c) is the true tree. It is interesting to see if one can indeed reject the trifurcating tree. Under this null model we have $p_1 = p_2 = p_3 = 1/3$. So, testing the null hypothesis is equivalent to testing whether the trinomial distribution with $p_i = 1/3$ can fit the observed data. If the expected number of sites supporting a bifurcating tree, which is equal to $n/3$ under the null model, is not too small, say ≥ 5, the following statistic

$$\chi^2 = \sum_{i=1}^{3} \frac{(n_i - n/3)^2}{n/3} \tag{5.28}$$

has a χ^2 distribution with 2 degrees of freedom, where n_i is the observed number of sites supporting tree i, $i = 1, 2, 3$. (The degree of freedom of the test is 2 because of the constraint $n_1 + n_2 + n_3 = n$.) It should be noted that rejection of the null hypothesis does not necessarily imply significant support for one of the bifurcating trees; to draw this conclusion, one needs to conduct test (5.26) or (5.27). For example, if $n_1 = 12$, $n_2 = 10$, and $n_3 = 2$, then $\chi^2 = 15$ so that the null model can be rejected, but tree 1 is obviously not significantly better than tree 2.

The above formulation did not take into account the effect of unequal base compositions among sequences, but parsimony tends to group sequences together according to their base compositions. This problem has recently been treated by Steel et al. (1993), who proposed a frequency-dependent significance test (the FD test); see also Li and Zharkikh (1995).

Although the above results were derived under the assumption that one of the four species is known to be an outgroup, they apply equally well in the absence of this knowledge if we consider unrooted trees and if the rate-constancy assumption holds.

MORE THAN FOUR SPECIES.   In theory the above analysis can be extended to the case of more than four species. However, the situation now becomes very complicated because the number of possible alternative trees increases rapidly with the number of species; for example, for five species there are 15 possible unrooted trees. Moreover, the above formulation is based on the worst-case analysis, so when the number of possible alternative trees is large, the test developed is likely to have a very low power.

In view of the complexity of the problem Kishino and Hasegawa (1989) developed a heuristic test in a manner similar to that of Templeton (1983). They considered whether two trees X and Y differ significantly in terms of the number of substitutions. Assume that all nucleotide sites are independent and equivalent. Let the minimum number of base substitutions at the ith informative site for tree X be X_i and let

$$X = X_1 + X_2 + \cdots + X_n \tag{5.29}$$

We define Y in the same manner. Let $Z = Y - X$. The mean of Z can be estimated by the inferred values of X and Y. The variance of Z can be estimated by the sample variance over sites, that is,

$$V(Z) = \frac{n}{n-1} \sum_{i=1}^{n} \left(Z_i - \frac{1}{n} \sum_{k=1}^{n} Z_k \right)^2 \tag{5.30}$$

where $Z_i = X_i - Y_i$. When n is large, $[Z - E(Z)]/V(Z)^{1/2}$ approaches the standard normal distribution, and the standard normal test can be used to test whether Z is significantly different from 0. Note that a two-sided test should be used because by chance Y can be larger or smaller than X. If $|Z|$ is significantly greater than 0, then $Z > 0$ implies that tree X is better than tree Y whereas $Z < 0$ implies that tree Y is better than tree X. To take account of variation in base composition among sequences, Steel et al. (1993) proposed a modified test.

It should be noted that when one tree is significantly more parsimonious than another tree, it does not mean that it is the most parsimonious tree. To draw this conclusion one must show that the tree is more parsimonious than all of the other possible trees. In practice, however, when the number of possible trees is large, we cannot compare all of them and so instead we usually compare only those that seem to be most plausible. In this respect, one should note that two trees usually differ only in a small part of the whole topology, and in comparing two trees we implicitly assume that the other part of the tree topology is correct. Since this assumption may not be true, it can complicate the test.

Distance Tests

Two types of test are available for testing the significance of a phylogeny inferred from a distance matrix method. One approach is to test the significance of estimated internodal distances; if all internodal distances are significantly larger than 0, the inferred tree may be considered significant. Li (1989) and Tajima (1992) have developed methods for this test. The other approach is based on the minimum-evolution criterion: if the total length of a tree is significantly shorter than every other alternative tree, it is considered significant. This approach appears to be superior to the first test and will be discussed below.

Under the minimum-evolution (ME) criterion, tree A is significantly better than tree B if the sum (S_A) of branch-length estimates for tree A is significantly smaller than that (S_B) for tree B. The test proposed by Rzhetsky and Nei (1992) is as follows. The sum (S) of branch lengths for a tree is estimated from the pairwise distances between taxa. Let us use tree A in Figure 5.20 as an example. Let $\hat{b}_i$ be the least-squares estimate of b_i. Then, by definition

$$S_A = \hat{b}_1 + \cdots + \hat{b}_7$$

In terms of d_{ij}, we have

$$
\begin{aligned}
S_A = {} & d_{12}/2 + d_{13}/4 + d_{14}/8 + d_{15}/8 + d_{23}/4 \\
& + d_{24}/8 + d_{25}/8 + d_{34}/4 + d_{35}/4 + d_{45}/2
\end{aligned} \tag{5.31}
$$

Similarly, for tree B we have

$$S_B = d_{12}/4 + d_{13}/2 + d_{14}/8 + d_{15}/8 + d_{23}/4$$
$$+ d_{24}/4 + d_{25}/4 + d_{34}/8 + d_{35}/8 + d_{45}/2 \tag{5.32}$$

Therefore,

$$D = S_B - S_A$$
$$= -d_{12}/4 + d_{13}/4 + d_{24}/8 + d_{25}/8 - d_{34}/8 - d_{35}/8 \tag{5.33}$$

The null hypothesis is $D = 0$ and a two-sided test should be used because D can be greater or smaller than 0. If $|D|$ is significantly greater than 0 and if S_A is smaller than S_B, then tree A is better than tree B. The variance of D is given in Rzhetsky and Nei (1992) and can be used to test the significance of D.

D can also be written as

$$D = \frac{1}{2}b_6 - \frac{1}{8}(2e_{12} - 2e_{13} - e_{24} - e_{25} + e_{34} + e_{35}) \tag{5.34}$$

where e_{ij} is the sampling error of d_{ij} as defined in Equation 5.22. Therefore, testing $D = 0$ is equivalent to testing $b_6 = 0$ and if the branch length b_6 is significantly greater than 0, then tree A is better than tree B. Similarly, for the comparison of trees A and C in Figure 5.20, we have

$$S_C - S_A = \frac{3}{4}(b_6 + b_7) - \frac{3}{8}(e_{12} - e_{14} - e_{25} + e_{45}) \tag{5.35}$$

Therefore, we are testing the null hypothesis that both b_6 and b_7 in tree A are equal to 0.

The above principle applies to any pair of bifurcating trees, and if a tree is found to have an S value significantly smaller than that for any other tree, it can be taken as the best tree. However, when the number of taxa is not small, it is impractical to compare all the trees. As mentioned earlier, Rzhetsky and Nei (1992) proposed a fast approximate search for the ME tree. Their search scheme is based on the observation that the neighbor-joining (NJ) tree is almost always identical with the ME tree. Briefly, their procedure is as follows: (1) Construct an NJ tree by using Saitou and Nei's (1987) method. (2) Obtain all trees that differ from the NJ tree by only one or two topological differences. (3) Estimate S for each of these topologies and compute $D = S - S_{NJ}$ and the standard error of D for each tree, where S_{NJ} is the S value for the NJ tree. (4) If D is significantly greater than 0 for each tree, adopt

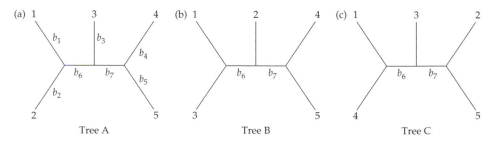

Figure 5.20 Model trees illustrating computation of branch lengths and S values. From Rzhetsky and Nei (1992).

the NJ tree as the most probable tree. However, if there are alternative trees whose S values are not significantly different from S_{NJ}, they should be regarded as good as the NJ tree. Furthermore, if there is any tree whose S is significantly smaller than S_{NJ}, it should be considered better than the NJ tree. A computer program for the above computations is available from Rzhetsky and Nei.

STATISTICAL TESTS: THE BOOTSTRAP

In statistics, the bootstrap is a computational technique for estimating statistics or parameters when the distribution is difficult to derive analytically (Efron 1982). Since its introduction into phylogenetic study by Felsenstein (1985a), the bootstrap technique has been frequently used as a means to estimate the confidence level of phylogenetic hypotheses. The statistical properties of this technique in the context of phylogenetic study appear to be complex, but recent studies (e.g., Zharkikh and Li 1992a,b, 1995; Felsenstein and Kishino 1993; Hillis and Bull 1993) have led to a better understanding of the technique. For an explanation of the technique under simple models, see Felsenstein and Kishino (1993) and Li and Zharkikh (1994), and for a review of recent progress, see Li and Zharkikh (1995).

The Procedure

To explain the technique, let us use the simple example shown in Figure 5.21; OTU 5 is assumed to be a known outgroup. In principle, one should specify the null hypothesis before taking or analyzing data, and we assume this has been done. Note that the null hypothesis as depicted by the tree in Figure 5.21 consists of two subhypotheses: (1) OTUs 1 and 2 belong to one clade and (2) OTUs 3 and 4 belong to another clade. Now suppose that one DNA sequence of length N is obtained from each OTU and that the sequences are as follows:

Taxa	Sequence
1	GCAGTACT...
2	GTAGTACT...
3	ACAATACC...
4	ACAACACT...
5	GCGGCATT...

To estimate the confidence levels of subhypotheses 1 and 2, we do bootstrapping on the sampled sequences to generate N_b pseudosamples. Note that sampling is with replacement, i.e., a site sampled can be sampled again with the same probability as can each of all other sites. Each pseudosample is generated as follows. First, a site is randomly chosen from the original sample; say it is the sixth site, and this site is placed as the first site in the following figure.

A pseudosample:

1	AGATACTC...
2	AGATGCTT...
3	AAACACTC...
4	AAACACCC...
5	AGACATCC...

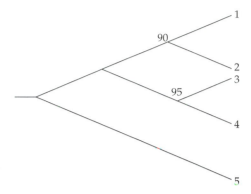

Figure 5.21 Hypothetical tree for illustrating the bootstrap technique. The numbers on the interior nodes are the bootstrap proportions supporting the subsets (1,2) and (3,4).

Next, another site is randomly chosen from the *original* sample; say, it is the first site, and this site is placed in the second position in the pseudosample. Third, another site is randomly chosen from the *original* sample; say, it is the sixth site and placed in the third position in the pseudosample. (Note that the sixth site has now been chosen twice.) This sampling procedure is continued until N sites are chosen. This pseudosample is then used to construct a tree and the subset (1,2) is given a score of 1 if OTUs 1 and 2 appear in one group, but a score of 0 otherwise; the score for the subset (3,4) is similarly decided. This completes one bootstrap replicate. The above procedure is repeated N_b times, and the proportion of bootstrap replicates in which the subset (1,2) appeared is taken as the bootstrap value supporting this subset; the bootstrap supporting the subset (3,4) is similarly obtained. These bootstrap proportions are indicated at the nodes of the tree (Figure 5.21) and are usually taken as the confidence levels for the subsets.

In essence, the bootstrap technique assumes that the original sample can represent the original sample space, so that each pseudosample can be treated as a real sample, i.e., a sample from the original sample space. Obviously, for this assumption to hold well, the original sample has to be large enough. Note also that resampling is with replacement; if resampling is without replacement, then all pseudosamples are the same except that sites are arranged in different orders, and this will not produce any variability among pseudosamples.

Effective Number of Competing Trees

Before going further, let us introduce the concept of the effective number of competing alternative trees (K_e). For this purpose, let us consider the maximum parsimony method and the two model trees in Figure 5.22a. In each tree there are four taxa with one known outgroup, so that there are in total 15 possible alternative trees as shown in Figure 5.22b. Intuitively, it seems that under model tree I (a tetrafurcating tree) and under the assumption of a molecular clock, the 15 possible alternative trees should be, on average, equally parsimonious, that is, should occur with equal probability, but this is not the case (Hendy and Penny 1989; Zharkikh and Li 1993). In fact, each of the 12 asymmetrical bifurcating trees (trees a to l) is more likely to be inferred (i.e., more parsimonious) than each of the three symmetrical bifurcating trees (trees m to o) (Zharkikh and Li 1993). Indeed, when the sequence length, L, is large, the probability of inferring a symmetrical tree becomes very small, and hence the probability of each of the 12 asymmetrical trees becomes close to 1/12 instead of 1/15. In other words, the effective number of competing alternative topologies (K_e) is close to 12 rather than 15 (it may not be an integer). Under model tree II, taxa 4 branched off much earlier than the other three taxa, which form a trichotomy. In this case the major competing trees are trees a, b, and c, while the other 12 trees are much less likely. In other words, K_e is close to 3 instead of 15. These two examples show that K_e depends on the model tree or the true tree. Obviously, it is easier to obtain tree a under model tree II than under model tree I. Since in practice we do not know the true tree, we do not know K_e, although the total number of possible alternative trees can be computed from the number of taxa under study.

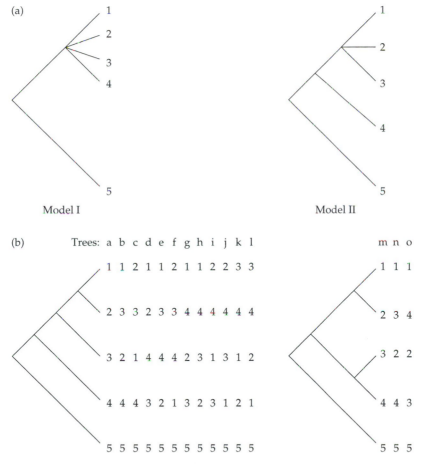

Figure 5.22 (a) Two model trees for four taxa with one known outgroup. Model tree I represents a tetrafurcating tree. In model tree II, species 4 branched off much earlier than species 1, 2, and 3, which form a trichotomy. (b) Fifteen possible alternative trees. The first 12 alternatives (a to l) are asymmetrical trees while the last three alternatives are symmetrical trees. From Li and Zharkikh (1995).

In the above we considered the entire tree topology. We may also consider only subsets of taxa. For example, let us consider the possible subsets with two taxa. Under model tree I, the subsets (1,2), (1,3), (1,4), (2,3), (2,4), and (3,4) are equally likely to occur, so that the effective number of subsets with two taxa is 6. On the other hand, under model tree II, the subsets (1,2), (1,3), and (2,3) are more likely to occur than the other subsets, so that the effective number of subsets is between 3 and 6 and becomes close to 3 as the sequence length (the number of informative sites) increases.

Estimation Bias

Theoretical and simulation studies (Zharkikh and Li 1992a,b, 1995; Felsenstein and Kishino 1993; Hillis and Bull 1993) showed that the bootstrap technique usually gives underestimates of the confidence level of a phylogenetic hypothesis. This result can be briefly explained as follows. Let P_1 be the (true) probability of obtaining tree 1 (the true tree) for a given sample size, P_1^* be the theoretical estimate of P_1 for a given sample, and P_1^{**} be a bootstrap estimate of P_1^*. Given P_1^*, the expectation of P_1^{**} is P_1^*, as indicated by the following conditional expection:

$$E\left(P_1^{**}\big|P_1^*\right) = P_1^* \tag{5.36}$$

In other words, P_1^{**} is an unbiased estimator of P_1^*. However, it is a biased estimator of P_1 because P_1^* is a biased estimator of P_1 (Zharkikh and Li 1992a, 1995; Felsenstein and Kishino 1993); the bias arises because P_1^* is a nonlinear function of the proportions of informative sites. Note also that bootstrapping introduces an additional variance, i.e.,

$$V\left(P_1^{**}\big|P_1^*\right) = P_1^*\left(1 - P_1^*\right) / N_b \tag{5.37}$$

(Hedges 1992; Zharkikh and Li 1992a). The error in the estimation of P_1^* by bootstrapping can be reduced by increasing the number of bootstrap replications N_b. For example, if $P_1^* = 0.95$ and $N_b = 400$, then the standard error introduced by bootstrapping is 0.02, or the 95% confidence interval of P_1^* is approximately (0.93, 0.97). Thus, $N_b \geq 400$ is recommended in general for bootstrap estimation. Note, however, that the total variance of the estimate of P_1 by P_1^{**} contains an additional term $V(P_1^*)(1 - 1/N_b)$, which is almost independent of N_b and cannot be reduced to 0 by increasing N_b (Zharkikh and Li 1992a; Li and Zharkikh 1995).

The degree of underestimation of P_1 by P_1^{**} increases with the effective number of competing trees K_e (Zharkikh and Li 1995). To see this, let us consider a simple artificial model, called the multinomial model of K alternatives. It assumes that there are K types of informative site, each of which supports one and only one topology. Let p_i be the probability of occurrence of the ith type site ($i = 1, 2, \ldots, K$). Then the distribution of the K types of informative site in a sample of N sites follows the multinomial distribution. For simplicity, let us assume that N is large, so that the chance of having more than one parsimonious tree is negligible. We assume further that $p_1 \geq p_2 = \cdots = p_K$. Then, P_1 increases from $1/K$ to 1 as the difference $p_1 - p_2$ increases from 0 to 1. Table 5.5 shows that the expected bootstrap value $\bar{P}_1^{**}$ is smaller than P_1 and the difference increases with K. This means that the underestimation of P_1 by bootstrapping tends to be more serious as K increases. Thus, if we are considering the probability (P_1) of obtaining the subset

---

TABLE 5.5 Relationship between P_1 and $\bar{P}_1^*$ ($\bar{P}_1^{**}$) as a function of the number of alternatives (K) in the multinomial model

| | $\bar{P}_1^*$ | | | |
P_1	$K = 2$	$K = 3$	$K = 5$	$K = 10$
0.50	0.50	0.45	0.40	0.35
0.60	0.57	0.48	0.47	0.42
0.70	0.65	0.60	0.55	0.50
0.80	0.72	0.68	0.64	0.59
0.90	0.82	0.78	0.75	0.70
0.95	0.88	0.85	0.82	0.79
0.99	0.95	0.94	0.92	0.90

From Li and Zharkikh (1995).

(1,2), then bootstrapping tends to underestimate P_1 more seriously under model tree I than under model tree II in Figure 5.22, because K_e is 6 for the former model but only between 3 and 6 for the latter (see above).

The Complete-and-Partial Bootstrap Technique

To correct the bias in bootstrap estimation, Zharkikh and Li (1995) developed the **complete-and-partial (CP) bootstrap technique**. For simplicity, let us consider the multinomial model described above. Suppose that we want to estimate P_1. Then, we first obtain an estimate (P_1^{**}) by the standard bootstrap technique, which is called complete bootstrapping because the size of each pseudosample is the same as the original sample size N. Next, we obtain a second estimate (P_{1r}^{**}) by taking pseudosamples of smaller size, say $N_r = N/r$, where $r > 1$; this second procedure is called partial bootstrapping. From these two estimates, an unbiased estimate of P_1 can be obtained; for details, see Zharkikh and Li (1995). The reason for requiring two bootstrap estimates is because P_1 depends on two unknowns, K and the difference $p_1 - \max(p_2, \ldots, p_K)$, and a single bootstrap value is not sufficient for solving the two unknowns. Bias corrections can also be made by the iterated bootstrap technique (Hall and Martin 1988; Rodrigo 1993), which is, however, more tedious than the CP bootstrap technique (Zharkikh and Li 1995).

In the above we assumed that the null hypothesis is specified in advance. In phylogenetic study, however, the null hypothesis is usually not prespecified. The lack of prespecification can lead to overestimation of the confidence level (see Zharkikh and Li 1992a; Li and Zharkikh 1995). This effect can also be taken into account by the CP bootstrap technique. A computer program is available for the CP bootstrap technique with either the maximum parsimony method or the neighbor-joining method (send request to Zharkikh or Li).

It should be emphasized that the bootstrap technique assumes that the sample is representative of the sample space. Thus, the technique is not recommended when the sample size is small. Note also that the current CP bootstrap technique has been developed under the assumption that the method is statistically consistent and that the sites on the DNA sequences are independently and identically distributed, i.e., all sites are independent of one another and evolve at the same rate. Since violations of these conditions are likely to occur, the results obtained by the current CP bootstrap technique should be taken with caution. How to remove the assumptions of independence and equivalence of sites is a subject of great interest for future research.

PROBLEMS

1. In computing the average distance between two composite OTUs or two OTU clusters, Formula 5.7 gives the same weight for all pairwise distances, i.e., it takes the arithmetic average. What are the advantages and disadvantages of this method? Can you think of an alternative method and compare it with Formula 5.7?

2. Show that in the case of four OTUs, the S_{ij} criterion given by Equation 5.8 and the minimum-evolution criterion are the same as the four-point condition.

3. Identify all the informative sites in the following four hypothetical sequences and infer the maximum parsimony (MP) tree. Is the MP tree statistically significant? [A conclusion can be drawn without a rigorous test. For a rigorous test, consult Table 1 in Li and Gouy (1991).]

50

1 GTCTGTTCCAAGGGCCTTTGCGTCAGGCTGGGCCTCAGGGTTGCCCCACT

2 GTCTGTTACAAGGGCCTTCGCGCCAGGCTAGGCCTCAGGGTTGCCTCACT

3 GTGTGTTTCAAGGGCCTTTGCGCCAGTCTGGGCCCCAGGGCTGCCCCACT

4 GTGCGTTACATGGGCCTTCGCGTCAGTCTGGGCCCCAGGGCTGCCTCACT

100

1 CGGGGTTCCAGGGCAGCTGGACCCCAGGCCCCAGCTCTGCATCAGGGAGG

2 CGAGGTGCCAGGGCGGCTGGTCCCCAGGCCTCAGCTCTGCAGCAAGGAGG

3 CGTGGTACCAGAGCAGTTGGACCCCAGGTCCCAGCTCTACAGCATTGAGG

4 TGCGGTTCCAGAGCGGTTGGACCCTAGGTCTCAACTCTACAGCAGGGAGG

140

1 ACGTGGCTGGGCCCGTGAAACATGTGTGGGTGAGTCCAGG

2 GCGTGGTTGGGCTCGTGAAGCATGTGGGGGTGAGTCCGGG

3 ATGTGGCTGGACTCGTGAAGCATTTGTGGGTGAGCCCGGG

4 GTGTGGTTGGGCCCGTGAAGCATTTGGGGGTAAGCCCAGG

4. Given that the pairwise distances for the four OTUs A, B, C, and D in Figure 5.8a are d_{AB}, d_{AC}, d_{AD}, d_{BC}, d_{BD}, and d_{CD}. Derive a formula for the length of the central branch (x).

5. Compute the pairwise distances for the four sequences in Problem 3 by the Jukes-Cantor method. Infer the tree by the four-point condition and compute the central branch length.

6. Show that in Figure 5.11b it is equally parsimonious to assume that the nucleotides at nodes 11, 10, 9, 8, and 7 were A instead of T, though A was not included at nodes 10, 8, and 7 in Figure 5.11a.

7. Explain the differences between the minimum-evolution method and the maximum parsimony method.

8. In Figure 5.12a, assume that $x = A$, $y = G$, and $z = T$. Use Formulas 5.11 and 5.12 (i.e., the one-parameter model) to derive the probability of having the configuration $i = j = T$, $k = C$ and $\ell = A$ at the tips of the tree.

9. Assume that $g_x = 1/4$ for all possible x's (A, T, C, G) in Formula 5.13. Derive the likelihood function $h(A,A,G,T)$ for Figure 5.12a under the one-parameter model of nucleotide substitution.

10. For the sequences in Problem 3, use Formula 5.16 to construct the likelihood function for the tree in Figure 5.12a.

11. Show that under the one-parameter model the function defined by Equation 5.18 is an invariant for any values of α_1 and α_2.

12. Explain why in the bootstrap method the sequence length, or, more precisely, the number of variable or phylogenetically informative sites should not be small.

CHAPTER 6

Molecular Phylogenetics: Examples

APPLICATION OF MOLECULAR BIOLOGY TECHNIQUES and advances in tree reconstruction methodology (see the preceding chapter) have led to tremendous progress in phylogenetic studies in the last two decades, resulting in a better understanding of the evolutionary history of almost every taxonomic group. This chapter presents several examples where molecular studies have resolved a long-standing issue, led to a drastic revision of a traditional view, or pointed to a new direction of research. As the field of molecular phylogenetics is progressing rapidly, some of the views presented here may soon be revised.

MAN'S CLOSEST RELATIVES

The question "Who are man's closest evolutionary relatives?" has always intrigued biologists. Darwin (1871), for instance, claimed that the African apes, the chimpanzee (*Pan*) and the gorilla (*Gorilla*), are man's closest relatives, and hence, he suggested that man's evolutionary origins were to be found in Africa. Darwin's view fell into disfavor for various reasons, and for a long time taxonomists believed that the genus *Homo* was only distantly related to the apes, and thus *Homo* was given a family of its own, Hominidae. Chimpanzees, gorillas, and orangutans (*Pongo*) were usually placed in a separate family, the Pongidae (Figure 6.1a). The gibbons (*Hylobates*) were either classified separately or with the Pongidae (Figure 6.1b; see Simpson 1961). Goodman (1963) correctly recognized that this systematic arrangement is anthropocentric, because humans represent "a new grade of phylogenetic development, one which is 'higher' than the pongids and all other preceding grades." Indeed, placing the various apes in one family and humans in another implies that the apes share a more recent common ancestry with each other than with humans. When *Homo* was put in the same clade with a living ape, it was usually with the Asian ape, the orangutan (Figure 6.1c; Schultz 1963).

By using a serological precipitation method, Goodman (1962) was able to demonstrate that humans, chimpanzees, and gorillas constitute a natural clade (Figure 6.1d), with orangutans and gibbons having diverged from the other apes at much earlier dates. From microcomplement fixation data, Sarich and Wilson (1967) estimated the divergence time between humans and chimpanzees or gorillas to be

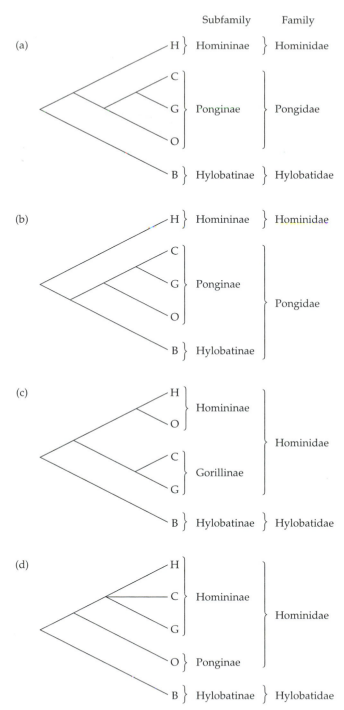

Figure 6.1 Four alternative phylogenies and classifications of modern apes and humans (Hominoidea). Traditional classifications setting humans apart are shown in (a) and (b). The clustering of humans with the orangutan is shown in (c). Cumulative molecular as well as morphological evidence favors the classification in (d). H, human; C, chimpanzee; G, gorilla; O, orangutan; and B, gibbon. From Li and Graur (1991).

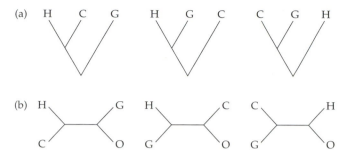

Figure 6.2   (a) Three possible rooted tree for gorillas, chimpanzees, and humans. (b) Comparable unrooted trees with the orangutan as an outgroup. C, chimpanzee (*Pan troglodytes*); H, human (*Homo sapiens*); G, gorilla (*Gorilla gorilla*); O, orangutan (*Pongo pygmaeus*). From Li and Graur (1991).

as recent as 5 million years ago, rather than a minimum date of 15 million years ago, as was commonly accepted by paleontologists at that time.

However, serological studies, electrophoretic studies, and amino acid sequences could not resolve the evolutionary relationships among humans and the African apes, and the so-called human-gorilla-chimpanzee trichotomy remained unsolved and continued to be an extremely controversial issue (Figure 6.2). Early studies of mitochondrial DNA by either restriction or sequence analysis (Ferris et al. 1981; Brown et al. 1982; Hixson and Brown 1986) slightly favored a closer relation of chimpanzee and gorilla, whereas the DNA-DNA hybridization data of Sibley and Ahlquist (1984) supported the human-chimpanzee clade. The first large set of DNA sequence data for resolving the controversy was obtained by M. Goodman and colleagues (see Miyamoto et al. 1987). In the following discussion, this data set and that of Maeda et al. (1988) will be used to illustrate some tree-making methods discussed in the previous chapter. Further data concerning the issue will be discussed later.

Table 6.1 shows the number of nucleotide substitutions per 100 sites between each pair of the following OTUs: humans (H), chimpanzees (C), gorillas (G), orangutans (O), and rhesus monkeys (R). Let us first apply the UPGMA method

TABLE 6.1   **Mean (below diagonal) and standard error (above diagonal) of the number of nucleotide substitutions per 100 sites between OTUs[a]**

OTU	OTU				
	Human	Chimpanzee	Gorilla	Orangutan	Rhesus monkey
Human		0.17	0.18	0.25	0.41
Chimpanzee	1.45		0.18	0.25	0.42
Gorilla	1.51	1.57		0.26	0.41
Orangutan	2.98	2.94	3.04		0.40
Rhesus monkey	7.51	7.55	7.39	7.10	

From Li et al. (1987b).

[a]The sequence data used are 5.3 kb of noncoding DNA, which is made up of two separate regions: (1) the η–globin locus (2.2 kb) described by Koop et al. (1986), and (2) 3.1 kb of the η–δ–globin intergenic region sequenced by Maeda et al. (1983, 1988).

to these distances. The distance between humans and chimpanzees is the shortest ($d_{HC} = 1.45$). Therefore, we join these two OTUs first and place the node at $1.45/2 = 0.73$ (Figure 6.3a). Following the procedure in Chapter 5 ("Methods of Tree Reconstruction"), we compute the distances between the composite OTU (HC) and each of the other species, and obtain a new distance matrix:

		OTU	
OTU	(HC)	G	O
G	1.54		
O	2.96	3.04	
R	7.53	7.39	7.10

Since (HC) and G are separated by the shortest distance, they are the next to be joined together, and the connecting node is placed at $1.54/2 = 0.77$. Continuing the process, we obtain the tree in Figure 6.3a. We note that the estimated branching node for H and C is very close to that for (HC) and G. In fact, the distance between the two nodes is smaller than all the standard errors for the estimates of the pairwise distances among H, C, and G (Table 6.1). Thus, although the data suggest that man's closest living relatives are the chimpanzees, the data do not provide a conclusive resolution of the branching order. On the other hand, it is unequivocal that the orangutan is an outgroup to the human-chimpanzee-gorilla clade.

Next, we use Sattath and Tversky's neighbors relation method (page 109). We consider four OTUs at one time. Since there are five OTUs, there are $5!/[4!(5-4)!] = 5$ possible quadruples. We start with OTUs H, C, G, and O and compute the following sums of distances (data from Table 6.1): $d_{HC} + d_{GO} = 1.45 + 3.04 = 4.49$, $d_{HG} + d_{CO} = 4.45$, and $d_{HO} + d_{CG} = 4.55$. Since the second sum is the smallest, we choose H and G as one pair of neighbors and C and O as the other (Table 6.2). Similarly, we consider the four other possible quadruples. The results are shown in Table 6.2. Noting from the bottom of that table that (OR) has the highest neighbors-relation score among all neighbor pairs, we choose (OR) as the first pair of neighbors. Treating this pair as a single OTU, we obtain the following new distance matrix:

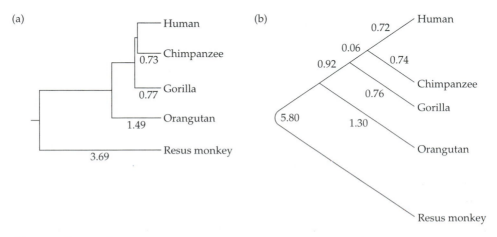

Figure 6.3 Phylogenetic tree for humans, chimpanzees, gorillas, orangutans, and rhesus monkeys inferred (a) from the UPGMA method and (b) from Sattath and Tversky's method. From Li and Graur (1991).

TABLE 6.2 **Neighbors-relation scores obtained from the distance matrix in Table 6.1**

OTUs compared[a]	Sum of pairwise distances	Neighbor pairs chosen
H, C, G, O	$d_{HC} + d_{GO} = 4.49$	
	$d_{HG} + d_{CO} = 4.45$	(HG), (CO)
	$d_{HO} + d_{CG} = 4.55$	
H, C, G, R	$d_{HC} + d_{GR} = 8.84$	(HC), (GR)
	$d_{HG} + d_{CR} = 9.06$	
	$d_{HR} + d_{CG} = 9.08$	
H, C, O, R	$d_{HC} + d_{OR} = 8.55$	(HC), (OR)
	$d_{HO} + d_{CR} = 10.53$	
	$d_{HR} + d_{CO} = 10.45$	
H, G, O, R	$d_{HG} + d_{OR} = 8.61$	(HG), (OR)
	$d_{HO} + d_{GR} = 10.37$	
	$d_{HR} + d_{GO} = 10.55$	
C, G, O, R	$d_{CG} + d_{OR} = 8.67$	(CG), (OR)
	$d_{CO} + d_{GR} = 10.33$	
	$d_{CR} + d_{GO} = 11.59$	

Total scores: (HC) = 2, (HG) = 2, (HO) = 0, (HR) = 0, (CG)=1,
(CO) = 1, (CR) = 0, (GO) = 0, (GR) = 1, (OR) = 3

From Li and Graur (1991).
[a]Notation: H, human; C, chimpanzee; G, gorilla; O, orangutan; and R, rhesus monkey.

	OTU		
OTU	H	C	G
C	1.45		
G	1.51	1.57	
(OR)	5.25	5.25	5.22

As only four OTUs are left, it is easy to see that $d_{HC} + d_{G(OR)} = 6.67 < d_{HG} + d_{C(OR)} = 6.76 < d_{H(OR)} + d_{CG} = 6.82$. Therefore, we choose H and C as one pair of neighbors and choose G and (OR) as the other. The final tree obtained by this method is shown in Figure 6.3b. The topology of this tree is identical to that in Figure 6.3a. Note, however, that in this method O and R rather than H and C were the first pair to be joined to each other. The reason is that, in an unrooted tree, O and R are in fact neighbors. The branch lengths in Figure 6.3b were obtained by the Fitch-Margoliash method (see page 123).

Third, we use the neighbor-joining method. It is simpler to use Studier and Keppler's (1988) algorithm (i.e., Equation 5.9) than Saitou and Nei's (1987) algorithm (i.e., Equation 5.8). From Equation (5.9), the Q value for the pair of OTUs i and j is

$$Q_{ij} = (N - 2)d_{ij} - \sum_{k=1}^{N} d_{1k} - \sum_{k=1}^{N} d_{2k} \qquad (6.1)$$

In the present case, $N = 5$, and for OTUs 1 and 2 we have

$$Q_{12} = 3d_{12} - (d_{12} + d_{13} + d_{14} + d_{15}) - (d_{12} + d_{23} + d_{24} + d_{25})$$

With the d_{ij} values in Table 6.1, we have $Q_{12} = -2261$. In a similar manner, we obtain the other Q values under step 1 in Table 6.3. Since Q_{45} is the smallest, we take OTUs 4 and 5 (i.e., orangutan and rhesus monkey) as the first pair of neighbors. This is the same as in the case of the neighbors-relation method (Table 6.2), and when OTUs 4 and 5 are regarded as a composite OTU, the new distance matrix is the same as the preceding matrix. Using the new distance matrix and the same procedure as above, we obtain the Q values under step 2 in Table 6.3. Since Q_{12} and $Q_{3(45)}$ are the smallest, we choose OTUs 1 and 2 as one pair of neighbors and OTUs 3 and (45) as another pair of neighbors. Thus, the final tree obtained by this method is exactly the same as that by the neighbors-relation method (Figure 6.3b).

Finally, let us consider the maximum parsimony method. For simplicity, let us consider only humans, chimpanzees, gorillas, and orangutans (Figure 6.2). Table 6.4 shows the informative sites for the 10.2-kb region including the η–globin pseudogene and its surrounding regions (Koop et al. 1986; Miyamoto et al. 1987; Maeda et al. 1988). For each site, the hypothesis supported is given in the last column. If we consider base changes only, then there are 15 informative sites, of which eight support the human–chimpanzee clade (hypothesis I), four support the chimpanzee–gorilla clade (hypothesis II), and three support the human–gorilla clade (hypothesis III). In addition, all four informative sites involving a gap support the human–chimpanzee clade. Therefore, the human–chimpanzee clade is chosen as the best representation of the true phylogeny. In another analysis with more sequence data, Williams and Goodman (1989) showed that support for the human–chimpanzee clade is statistically significant at the 1% level.

The clustering of humans and chimpanzees in one clade is, however, not supported by the repeat pattern in the involucrin gene, which favors instead the chimpanzee–gorilla clade (Djian and Green 1989). Nevertheless, the overall molecular evidence is now strongly in favor of the human–chimpanzee clade. In addition to the 10.2-kb sequence data discussed above, this clade is supported by extensive

---

TABLE 6.3 The Q values for the distance matrix in Table 6.1

Step 1	Step 2
$Q_{12} = -2261$	$Q_{12} = -1357$
$Q_{13} = -2243$	$Q_{13} = -1348$
$Q_{14} = -2057$	$Q_{1(45)} = -342$
$Q_{15} = -2047$	$Q_{23} = -1342$
$Q_{23} = -2231$	$Q_{2(45)} = -1348$
$Q_{24} = -2075$	$Q_{3(45)} = -1357$
$Q_{25} = -2041$	
$Q_{34} = -2045$	
$Q_{35} = -2089$	
$Q_{45} = -2431$	

Note: OTUs 1, 2, 3, 4, and 5 are human, chimpanzee, gorilla, orangutan, and rhesus monkey, respectively.

TABLE 6.4 Informative sites among human, chimpanzee, gorilla, and orangutan sequences

Site[a]	Human	Chimpanzee	Gorilla	Orangutan	Hypothesis supported[b]
Data from Miyamoto et al. (1987)					
34	A	G	A	G	III
560	C	C	A	A	I
1287	*[c]	*	T	T	I
1338	G	G	A	A	I
3057–3060	*****	*****	TAAT	TAAT	I
3272	T	T	*	*	I
4473	C	C	T	T	I
5153	A	C	C	A	II
5156	A	G	G	A	II
5480	G	G	T	T	I
6368	C	T	C	T	III
6808	C	T	T	C	II
6971	G	G	T	T	I
Data from Maeda et al. (1988)					
127–132	*****	*****	AATATA	AATATA	I
1472	G	G	A	A	I
Δ2131	A	A	G	G	I
Δ2224	A	G	A	G	III
2341	G	C	G	C	III
2635	G	G	A	A	I

From Williams and Goodman (1989) and Li and Graur (1991).

[a]Site numbers correspond to those given in the original sources. The total length of the sequence used is 10.2 kb, about twice that used in Table 6.1.

[b]Hypotheses: I, human and chimpanzee in one clade; II, chimpanzee and gorilla in one clade; and III, human and gorilla in one clade.

[c]Each asterisk denotes the deletion of a nucleotide at the site.

DNA–DNA hybridization data (Sibley and Ahlquist 1987; Caccone and Powell 1989), by two-dimensional protein electrophoresis data (Goldman et al. 1987), and especially by extensive mitochondrial DNA sequence data (Ruvolo et al. 1991; Horai et al. 1992), including the complete mtDNA sequences from three humans, one common and one pigmy chimpanzee, one gorilla and one orangutan (Horai et al. 1995). Thus, the closest extant relatives of humans are the two chimpanzee species, followed by gorillas, orangutans, and the nine gibbon species. This is contrary to the conclusion of a closer relatedness of chimpanzee and gorilla derived from morphological and physiological comparisons (Ciochon 1985; Andrews 1987).

WHALE STORIES

Whales, dolphins, and porpoises, which form the order Cetacea, are among the most fascinating and spectacular of all vertebrates. They are intelligent and can be trained to play various tricks, they have an elaborate communication system (particularly, the sonar system in toothed whales), and large whales far exceed the bulk of the largest dinosaurs.

The origin of Cetacea has been an enduring evolutionary mystery since Aristotle, for the transition from terrestrial life to an aquatic way of life that is rivaled by no other vertebrates required many drastic yet coordinated changes in biological systems. For example, living cetaceans are unique among mammals in that they lack external hind limbs and that all swim by dorsoventral oscillation of a heavily muscled tail.

A link between cetaceans and ungulates was suggested more than a century ago by Flower (1883) on the basis of comparative anatomical information. This view was accepted by Gregory (1910), but Simpson (1945) and Romer (1966) suggested that the cetacean lineage goes back to the very root of the eutherian radiation. However, Flower's view was later endorsed by Van Valen (1966) and Szalay (1969), who argued on the basis of dental and other grounds for a connection between cetaceans and mesonychid condylarths, an early Tertiary assemblage of ungulates. The first paleontological evidence for a connection between cetaceans and artiodactyls was provided by remains of a middle-Eocene (~45 Myr old) whale exhibiting an artiodactyl-like paraxonic arrangement of the digits on its vestigial hind limbs (Gingerich et al. 1990; Wyss 1990). Recent fossil findings by Gingerich et al. (1994) and Thewissen et al. (1994) have provided some insight into the terrestrial-aquatic transition (Novacek 1994).

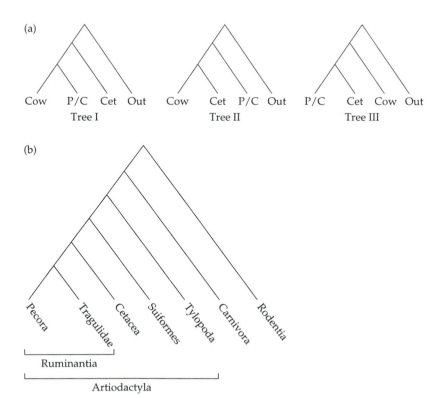

Figure 6.4   (a) Three possible phylogenetic trees for cetacean (Cet), cow (Cow), pig or camel (P/C), and an outgroup (Out). Tree I represents the traditional view that the artiodactyls form a natural monophyletic clade. (b) Schematic molecular phylogenetic tree for the taxa under study. From Graur and Higgins (1994).

TABLE 6.5   **Analysis of mitochondrial genes for three alternative phylogenies among pigs, cows, and cetaceans, with mouse as outgroup**[a]

Methods	Cow-Pig	Cow-Cetacean	Pig-Cetacean
Parsimony (131 informative sites)[b]	38 (641)	55 (624)	38 (641)
ML bootstrap (100 replicates)[c]	0	89	11
NJ bootstrap (1000 replicates)[d]	45	908	47
ML test[e]	-62.1 ± 24.4	ML	-39.8 ± 26.8
Internal branch test[f]		0.0058 ± 0.0021	

From Graur and Higgins (1994).

[a]The genes used: cytochrome *b*, ATPase subunit 6, NADH dehydrogenase subunit 1, 12S rRNA, and tRNAs.

[b]Number of informative transversions; the figures in parentheses are the minimum numbers of transversional substitutions for an entire tree.

[c]Number of occurrences in 100 maximum-likelihood bootstrap trees using a transition/transversion ratio of 10.0/1.

[d]Number of occurrences in 1000 neighbor-joining trees based on transversion distances.

[e]Kishino and Hasegawa's (1989) ML test. The differences and estimated standard errors of the differences between the maximum-likelihood tree (ML) and the two other trees are given.

[f]Li's (1989) test. The least-squares estimate of the length of the internal branch supporting the cow–cetacean tree is given with its standard error. The estimates of the internal branch lengths in the other two trees are negative.

The molecular evidence for a close relationship between Cetacea and Artiodactyla has been increasing since the 1980s. Goodman et al. (1982) analyzed seven protein sequences and concluded that Cetacea belongs with Artiodactyla. This conclusion received further support from studies on mtDNA sequences (e.g., Irwin et al. 1991; Milinkovitch et al. 1993; Cao et al. 1994). A recent phylogenetic analysis of mtDNA and protein sequence data suggested that the cetaceans are in fact not just related to but within the artiodactyls (see below; Graur and Higgins 1994).

The order Artiodactyla is traditionally divided into three suborders: Suiformes (e.g., pigs and hippopotamuses), Tylopoda (camels and llamas), and Ruminantia (cows, goats, giraffes, deer, and elk). Graur and Higgins (1994) inferred the phylogenetic position of Cetecea in relation to the three artiodactyl suborders by using protein and DNA sequence data from pig, cow, camel, several cetacean species, and an outgroup (e.g., mouse). In Figure 6.4a, tree I represents the traditional view that the cetacean lineage branched off prior to the divergence between the two artiodactyls, cow and pig (or camel). Trees II and III represent the alternatives, i.e., cetaceans are clustered with one of the two artiodactyl species.

Table 6.5 shows the results of the cow-pig-cetacean analysis of mtDNA data with mouse as the outgroup. Among the 131 informative transversions, tree II was supported by 55 informative sites, whereas trees I and III were each supported by only 38 sites, so that tree II was significantly more parsimonious than trees I and III ($P < 5\%$). When Kishino and Hasegawa's (1989) test of difference between DNA maximum-likelihood trees was used, tree II was significantly better than tree I ($P < 5\%$), though tree III could not be rejected. Moreover, none of the 100 maximum-likelihood bootstrap replicates supported tree I and only 45 of the 1000 neighbor-joining bootstrap replicates supported tree I. By using Li's (1989) test, the length of the internal branch in tree II was found to be significantly greater than 0 ($P < 1\%$). Thus, the mtDNA sequence data strongly supported tree II. Graur and Higgins also used the sequences of 11 proteins, and their analyses using the parsimony and the neighbor-joining methods again supported tree II, though not with statistical significance. Five of the 11 proteins used were also available for camel, and a parsimony analysis suggested that cetaceans are more closely related to pig than to camel.

Based on the above analyses, Graur and Higgins proposed the phylogeny shown in Figure 6.4b, in which Ruminantia and Cetacea are more closely related to each other, followed by the other two suborders Suiformes and Tylopoda. In Figure 6.4b Pecora (e.g., cow) and Tragulidae (chevrotain) represent two of the most distantly related infraorders of Ruminantia and the placement of Cetecea outside the Pecora-Tragulidae clade was based on a parsimony analysis of mitochondrial 12S and 16S rRNA genes. On morphological and paleontological grounds, pig and cow are estimated to have diverged ~55–60 Myr ago while the divergence between Tragulus and cow is put at ~45 Myr ago. Using these dates and the transversion divergences of mtDNA, Graur and Higgins estimated that the cetacean and ruminant lineages diverged 45–49 Myr ago. This estimate is compatible with the early whale fossils with dated ages of 52 and 46–47 Myr (Thewissen et al. 1994; Gingerich et al. 1994).

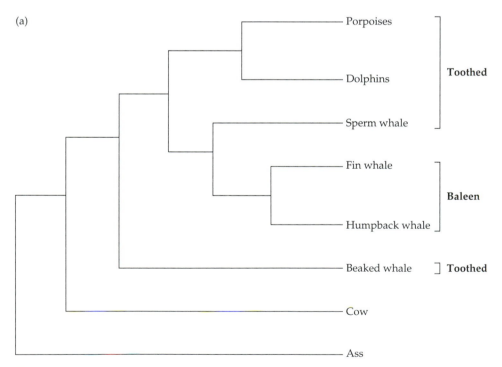

Figure 6.5 (a) A simplified schematic presentation of the revised phylogeny of whales proposed by Milinkovitch et al. (1993). (b) Phylogenetic tree based on the cytochrome *b* gene sequences of the sperm whale, 2 dolphins, and 11 mysticetes, with the cow as an outgroup (from Arnason and Gullberg 1994). The tree was

Living cetaceans are subdivided into two highly distinct suborders, Odontoceti (the echolocating toothed whales, about 67 species) and Mysticeti (the filter-feeding baleen whales, 11 species). The two suborders are thought to have had long independent histories, having separated from the extinct suborder Archeoceti more than 35–40 Myr ago (Barnes et al. 1985). This view was challenged by Milinkovitch et al. (1993), who proposed that the sperm whale (a toothed whale) is more closely related to the baleen whales (represented by the fin whale and the humpback whale) than to other toothed whales (Figure 6.5a). This proposal was based on the parsimony, neighbor-joining, and maximum likelihood analyses of sequenced portions (930 bp) of the 12S and 16S mitochondrial rRNA genes from 16 species of cetaceans, the cow, the wild ass (a perissodactyl), a human, and a sloth. This revised phylogeny, which was also supported by the available myoglobin sequence data, suggests that baleen whales lost the capacity for echolocation rather than never evolved it; the other alternative would be less likely because it requires that the elaborate echolocating system evolved at least twice, once in the sperm whale and once in other toothed whales. The proposed revision, however, was claimed to be at variance with the cytochrome *b* gene sequence data (Arnason and Gullberg 1994), which suggested that the baleen whales are separated from the sperm whale and dolphins (toothed whales, Figure 6.5b). However, Arnason and Gullenberg's study used only one outgroup (cow), but the inference was shown to be sensitive to

(b)

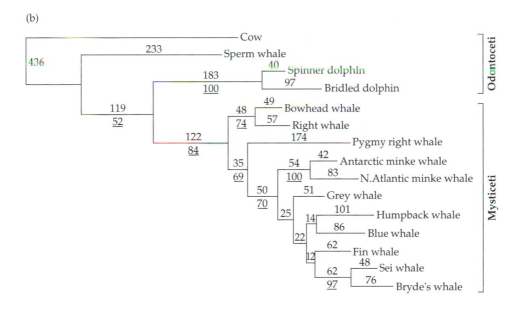

inferred from a heuristic search of the most parsimonious tree by using PAUP 3.1.1 (Swofford 1993). Differences in codon positions 1, 2, and 3 were given the weights 4, 17, and 1, respectively. Bootstrap values (50% majority rule, 1000 replicates) are underlined. Values not underlined show branch lengths (number of substitutions).

the choice of an outgroup (Adachi and Hasegawa 1995). Moreover, their analysis used equal weights for transitional and transversional changes, but when transitional changes were excluded or given less weight, the cytochrome *b* gene sequence data actually supported the sperm whale-baleen whale clade (Milinkovitch et al. 1995). The sister group relationship between sperm whales and baleen whales was also supported by an extended analysis that incorporated two mitochondrial rRNA segments and the cytochrome *b* gene from 21 cetacean species with three artiodactyl outgroups (Milinkovitch et al. 1994). Therefore, there is considerable support for the proposed major revision of the whale phylogeny (Milinkovitch et al. 1993).

An interesting point in Figure 6.5b is that the grey whale is positioned between the minke whales and the remaining rorquals; in an equally parsimonious tree (not shown) the grey whale is placed between the blue/humpback whales and the fin whale (Arnason and Gullberg 1994). In either tree, the placement of the grey whale among the rorquals is of great interest and is supported by other molecular data: sequence analyses of the mtDNA control region (Arnason et al. 1993; Baker and Palumbi 1996), the common cetacean satellite DNA and the heavy mysticete satellite, and the taxonomic distribution of the light mysticete satellite (see references in Arnason and Gullberg 1994). Traditionally, the grey whale has been placed in a distinct, monotypic family, the Eschrichtiidae, on the basis of its similarity to skeletal features of the extinct family Cetotheriidae and its unique adaptation to a benthic feeding ecology. Whereas the right whales and bowhead whales use their long baleen to skim the water surface for small planktonic prey and the rorquals use their vast pleated throats to engulf dense concentrations of plankton or school fish, the gray whale uses its lower jaw like a shovel to scoop up benthic amphipods and mysids, forcing the mud and water back through its coarse baleen plates. The traditional placement of the grey whale in a separate family would imply an ancestral origin of the grey whale's unique benthic ecology, whereas the new finding of the close relationship of the grey whale to the rorquals implies a strikingly rapid morphological adapatation of the grey whale to a new ecological niche (Baker and Palumbi 1996).

ARE THE YEWS CONIFERS OR NOT?

The yews, which form the genus *Taxus*, are the best known members of the family Taxaceae, which is therefore commonly known as the yew family. (The anti-cancer drug Taxol is obtained from the bark of Pacific yew trees.) This family includes about 20 species and has been suggested to consist of either five genera, *Taxus*, *Austrotaxus*, *Pseudotaxus*, *Torreya*, and *Amentotaxus* (e.g., Stewart 1983), or only the first four of the five genera (e.g., Gifford and Foster 1988). The phylogenetic position of Taxaceae is one of the oldest unsolved problems in gymnosperm systematics. The main problem is that the Taxaceae bear uniovulate seeds with the ovule at the terminal position, whereas most other conifers bear multiovulate cones with ovules in the axillary position (Figure 6.6). For this reason there has long been a strong dispute on whether the Taxaceae are conifers.

In the basic taxonomic scheme of extant conifers proposed by Pilger (1926), the order Coniferales included Taxaceae and six other families: Podocarpaceae, Araucariaceae, Cephalotaxaceae, Pinaceae, Taxodiaceae, and Cupressaceae. This

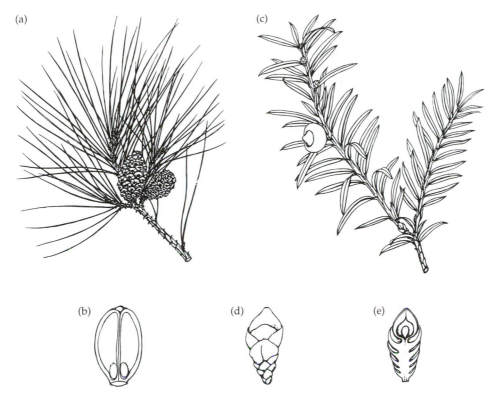

Figure 6.6   (a) and (b) The pine *Pinus luchuensis*: (a) branch with mature female cones and (b) the adaxial view of ovuliferous scale dissected from cone, showing two winged seeds at each side. (c), (d), and (e) The yew *Taxus mairei*: (c) branch with mature fruit and ovuliferous branchlet, (d) the lateral view of ovuliferous branchlet, and (e) the longitudinal section of (d), showing the erect, single ovule.

scheme is different from that of Sahni (1920). Arguing that the seed structures of *Taxus*, *Torreya*, and *Cephalotaxus* were derived from the Paleozoic seeds of the extinct Cordaitales and that these three genera, like *Ginkgo*, are direct descendants of the Cordaitales, Sahni proposed to place them in a distinct phylum, Taxales, equivalent in systematic rank with the Coniferae. Florin (1948, 1954), largely following Sahni, strongly advocated that the terminal uniovulate seed of Taxaceae is a primitive character, its origin being entirely different from that of the conifers. He suggested that Taxaceae represents a distinct class, Taxales, that has evolved in parallel with Coniferales and Ginkgoales, probably since the upper Devonian. Florin's view has been well accepted (e.g., Sporne 1965; Stewart 1983). In contrast, Pulle (1937), Takhtajan (1953), and others thought that the Taxaceae are not fundamentally different from the rest of conifers but possess the most reduced solitary-ovule cones. Takhtajan (1953) suggested that the Taxaceae are descendants of the Podocarpaceae, as evidenced by their pseudoterminal development of the arillate ovules and the structure of the wood. Harris (1976) proposed that the single, terminal ovule of Taxaceae was derived from the fertile, lateral short shoot of Paleozoic conifers by reduction and followed by a shift of the single ovule onto the apex. For more details of the controversy, see Chaw et al. (1993, 1995).

To resolve the above controversy, Raubeson and Jansen (1992) studied the phylogenetic distribution of a rare chloroplast-DNA structural mutation among conifers, nonconiferous seed plants, and pteridophytes. The basic structure of the chloroplast genome, a circular DNA with a large (10–25 kb) inverted repeat, was shown to be conserved across all major plant lineages (Palmer 1985a). However, Strauss et al. (1988) found that one copy of the two repeats was lost in the genomes of Monterey pine and Douglas fir, two members of the family Pinaceae. Since this loss appeared to be a very rare event, Raubeson and Jansen (1992) used it as a cladistic character to examine the question of monophyly of conifers. They examined a broad taxonomic representation of conifers, including at least one member of each of the seven conifer families, plus representatives of each of the other major extant vascular plant lineages. Their results indicated that only one copy of the repeat is present in all conifers, whereas two copies were present in those pteridophytes and nonconiferous seed plants surveyed (Table 6.6). The

TABLE 6.6 **Presence or absence of the inverted repeat in the chloroplast genome**

Present	Absent
Lycopsida	Taxaceae
Lycopodium obscurum	*Taxus X media*
Selaginella sp.	*Torreya nucifera*
Isoetes melanopoda	Araucariaceae
Psilotales	*Araucaria spp.*
Psilotum nudum	*Agathis australis*
Ophioglossales	Cephalotaxaceae
Botrychium virginianum	*Cephalotaxus fortunii*
Marattiales	Cupressaceae
Marattia sp.	*Calocedrus decurrens*
Filicales	*Juniperus communis*
Osmunda cinnamomea	*Cupressus macrocarpa*
Lygodium palmatum	*Callitris presii*
Polypodium vulgare	Pinaceae
Gingkoales	*Larix laricina*
Gingko biloba	*Pinus strobus*
Cycadales	*Tsuga canadensis*
Cycas revoluta	Podocarpaceae
Encephalartos sp.	*Podocarpus spp.*
Gnetales	*Dacrydium cupressinum*
Ephedra tweediana	*Phyllocladus trichomanoides*
Gnetum gnemon	Taxodiaceae
Welwitschia mirabilis	*Metasequoia glyptostroboides*
	Taxodium distichum
	Sequoia sempervirens
	Cryptomeria japonica

From Raubeson and Jansen (1992).

finding that the loss was shared by all conifers but by none of the outgroups studied supported conifer monophyly.

Although the above study provided evidence that Taxaceae and the other six conifer families form a monophyletic group, it provided no information on whether Taxaceae has branched off earlier than the other six families, that is, it remained possible that Taxaceae is actually an outgroup to the other conifers. The studies of Chaw et al. (1993, 1995) rejected this possibility and provided additional support for the monophyly of conifers. They obtained the 18S rRNA sequences from a ginkgo and two members of Gnetales (*Ephedra* and *Genetum*) and from representatives of the following conifer families: Taxaceae, Cephalotaxaceae, Cupressaceae, Taxodiaceae, Araucariaceae, Phyllocladaceae, Podocarpaceae, and Pinaceae. An analysis by the neighbor-joining method with a chlamydomonas as an outgroup suggested that Taxaceae, Cephalotaxaceae, Cupressaceae, Taxodiaceae, and Araucariaceae have very short distances to one another (at most only a few percent divergences) and form a monophyletic group supported by 98% of the bootstrap replicates (Figure 6.7); however, whether Cephalotaxaceae is the closest relative of Taxaceae as suggested by Hart (1987) remains to be seen. The analysis also suggested that all the conifers form a monophyletic group supported by 89% of the bootstrap replicates and that Pinaceae is an outgroup to the other conifers.

Figure 6.7 also sheds some light on the evolutionary position of the genus *Phylocladus*, which was so named because its members possess rhomboidal, flattened, celery-leaf-like photosynthetic organs, i.e., the phylloclades. Robertson (1906) considered the phyloclades to be the expansion of certain stem branches into flattened leaflike structures, implying the phylloclade as a specialized and advanced feature. However, interpreting the phylloclade as a very ancient structure, Keng (1974, 1978) claimed that the genus is the most primitive member of the conifers and proposed to remove the genus from Podocarpaceae to form a monogeneric family, Phyllocladaceae. This proposition was supported by Clifford and Constantine (1980) and others, but opposed by Quinn (1987). From a claddistic analysis of a large set of morphological data, Hart (1987) concluded that the genus is a terminal taxon within the Podocarpaceae. This conclusion is supported by the tree in Figure 6.7—*Phyllocladus* is an outgroup to the other genera of Podocarpaceae, which form a monophyletic group. Since *Phyllocladus* is not a natural clade, it should be included in the family Podocarpaceae. At any rate, it is clear that *Phyllocladus* is not a primitive member of the conifers and that the phylloclade is probably a derived feature.

The above inferences were supported by some insertion and deletion events and by a parsimony analysis (Chaw et al. 1993, 1995). Further support for the monophyly of conifers was provided by the sequence data of the chloroplast *rbc*L gene (Chase et al. 1993). The *rbc*L data also placed Pinaceae as an outgroup to the other conifers.

In conclusion, all three sets of molecular data are not compatible with the proposal that the Taxaceae form a clade distinct from the other conifers; rather, they provide strong evidence for the monophyly of conifers and against the view that the terminal, uniovulate seed of Taxaceae is a primitive character. Moreover, the 18S rRNA sequence data do not support the view that the Taxaceae are direct descendants of the Podocarpaceae. Thus, it seems that the uniovulate seed in Taxaceae and that in some species of *Podocarpus* have different origins, probably all reduced from multiovulate cones.

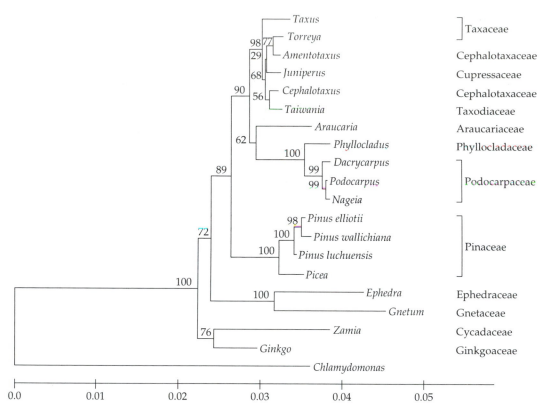

Figure 6.7 A phyloegentic tree of conifers inferred from 18S rRNA or rDNA sequences. Taxaceae, Cephalotaxaceae, Cupressacea, Taxodiaceae, Araucariaceae, Podocarpaceae, and Pinaceae are the seven families of conifers in the classification system of Pilger (1926). Whether Phyllocladaceae should be given a family status or should be a genus of Podocarpaceae is subject to dispute (see text). The other four gymnosperm species included in the tree are not conifers. The chlamydomonas was used as an outgroup. The tree was inferred by the neighbor-joining method. The number of transitional substitutions per site (d_S) and that of transversional substitutions per site (d_V) were computed using Kimura's (1980) two-parameter model and the weighted distance was computed by $\bar{d} = 0.4d_S + 0.6d_V$. The scale is given in $\bar{d}$. The numbers on the branches are bootstrap proportions. From Chaw et al. (1995).

THE ORIGIN OF ANGIOSPERMS

The origin of angiosperms was dubbed by Charles Darwin as "an abominable mystery" and remains a highly controversial issue in plant evolution. Angiosperms began to radiate rapidly about 115 Myr ago (the middle Cretaceous) and became the dominant group of earth's flora by about 90 Myr ago (e.g., Lidgard and Crane 1988). However, no reliable angiosperm fossils older than 120 Myr have been found (Hickey and Doyle 1977; Doyle 1978), and there is little agreement as to when the angiosperms arose and from which branch of the gymnosperms they stem. Since the progymnosperm lineage extends back to at least 370 Myr ago, there is an enormous range of time during which angiosperms might have had their beginnings. Theories as to why there are no fossils of progenitor angiosperms fall into two basic types: either (1) angiosperms did not exist until the early Cretaceous and then radiated explosively (e.g., Hickey and Doyle 1977; Doyle 1978; Thomas and Spicer 1987), or (2) pre-Cretaceous angiosperms lived in habitats so refractory to fossilization that they left no record (e.g., Axelrod 1952, 1970; Takhtajan 1969).

One way to decide between these two views is to estimate the date of divergence between monocotyledons (monocots) and dicotyledons (dicots), the two major classes of angiosperms, because this provides a miminal estimate for the age of angiosperms. The first application of DNA sequence data to estimate this date was made by Martin et al. (1989). Their analysis was based on the comparison of the sequences of the nuclear gene encoding cytosolic glyceraldehyde-3-phosphate dehydrogenase (GAPDH or gapC) from plants, animals, and fungi. By using several divergence dates between animal taxa and between the plant, animal, and fungal kingdoms they estimated the rate of evolution of this gene. From this, they estimated that monocots and dicots diverged about 320 Myr ago. This date appears to be too ancient, because the earliest land plant fossils are only about 420 Myr old (Gensel and Andrews 1984) and so the estimate of Martin et al. would imply that the lineages leading to bryophytes, pteridophytes, gymnosperms, monocots, and dicots all appeared within the first 100 Myr of land plant evolution. Nevertheless, their data provided evidence for pre-Cretaceous origins of angiosperms.

Wolfe et al. (1989b) obtained a different estimate by using three approaches. The first one was based on a calibration of the rate of synonymous substitution in chloroplast genes with the maize–wheat divergence as a reference. Using DNA sequence data, they first demonstrated that maize, wheat, and rice all originated at approximately the same time, i.e., represent approximately a trichotomy (Figure 6.8). Since fossils of rice leaf epidermis have been described from the upper Eocene (about 40 Myr ago), they took 50 Myr as a lower bound for the origin of maize, wheat, and rice. Further, they used 70 Myr as an upper bound for the maize–wheat divergence, as suggested by Stebbins (cited in Chao et al. 1984). From the average number of synonymous substitutions per site between maize and wheat chloroplast genes (Table 6.7), one can readily estimate the rate of synonymous substitution to be $a =$

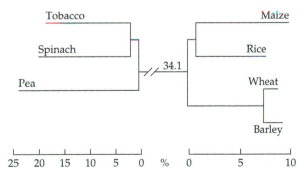

Figure 6.8 Phylogenetic tree for three dicot and four monocot species. The tree was inferred by the neighbor-joining method using the synonymous distances for three chloroplast genes: *rbc*L, *atp*B, and *atp*E. The length (0.7%) of the internal branch leading to the maize–rice pair is less than the standard error (~1.7%), and thus the maize, rice, and wheat/barley lineages are probably close to a trichotomy. From Wolfe et al. (1989b).

TABLE 6.7   Numbers of synonymous substitutions per 100 sites (K_s, %) between maize, wheat (or its siblings), and tobacco for 12 chloroplast genes

Gene	Maize vs. wheat	Maize vs. tobacco	Wheat vs. tobacco	$L_s{}^a$
atpA	14.6 ± 2.2	53.5 ± 5.6	52.3 ± 5.4	344
atpB	17.5 ± 2.5	65.4 ± 6.4	65.3 ± 6.3	347
atpE	19.3 ± 5.2	50.8 ± 10.1	67.3 ± 12.8	88
atpH	8.4 ± 3.9	38.6 ± 9.7	40.9 ± 10.0	64
orf43	7.2 ± 5.2	43.3 ± 15.8	44.4 ± 16.8	29
orf62	14.5 ± 6.2	46.4 ± 13.3	38.7 ± 11.6	45
psaC	29.6 ± 9.0	79.5 ± 18.8	94.8 ± 23.7	52
psbB	14.6 ± 2.2	56.3 ± 5.6	59.1 ± 6.2	343
psbC	18.9 ± 4.1	52.8 ± 8.5	47.8 ± 7.4	153
psbD	20.0 ± 3.2	48.4 ± 5.9	47.5 ± 5.9	239
psbH	20.5 ± 7.4	60.3 ± 14.7	66.9 ± 16.0	51
rbcL	19.9 ± 2.8	70.2 ± 6.9	62.7 ± 6.2	321
Total[b]	17.3 ± 1.0	57.8 ± 2.4	57.6 ± 2.4	2077

From Wolfe et al. (1989a).

[a]Number of synonymous sites compared.

[b]The "total" K_s value for each species pair is a mean over all genes, weighted by the number of sites in each gene.

$0.173/(2T) = 1.73 \times 10^{-9}$ or 1.24×10^{-9} substitutions per site per year, depending on whether the lower bound of $T = 50$ Myr or the upper bound of $T = 70$ Myr is used. The average number of synonymous substitutions per site per year between monocot (maize and wheat) and dicot (tobacco) genes is $(0.578 + 0.576)/2 = 0.577$ (Table 6.7). Therefore, the date of the monocot–dicot divergence (Figure 6.8) is estimated to be $0.577/(2a)$ or between 170 and 230 Myr. The second approach was based on a calibration of the rate of nonsynonymous substitution with the bryophyte–angiosperm divergence as a reference point, which was taken to be between 350 and 450 Myr ago; the nonsynonymous rate was used because synonymous substitutions have been saturated for this degree of divergence and therefore cannot be reliably estimated. This approach gave an estimate between 150 and 260 Myr. The third approach was based on a calibration of the substitution rate in nuclear ribosomal RNA genes with the plant–animal divergence as a reference point, which was assumed to be about 1000 Myr ago. The estimate obtained from the 26S rRNA sequence data was between 200 and 250 Myr and that obtained from the 18S rRNA sequence data was between 200 and 210 Myr. From the above estimates, Wolfe et al. suggested that monocts and dicots diverged about 200 Myr ago with an uncertainty of about 40 Myr. These results are supported by recent estimates from mitochondrial gene sequences (Laroche et al. 1995) and suggest that monocots and dicots could have diverged in the early Jurassic. Therefore, the molecular data strongly support the hypothesis that angiosperms existed long before they came to prominence.

There are other major issues concerning the origin and evolution of the angiosperms. First, what is the common ancestor of the angiosperms? One view is that the angiosperms are derived from a gymnosperm lineage, i.e, from the ancestor of Gnetales (see the review by Crane et al. 1995). This implies that Gnetales and angiosperms share a common ancestor not shared by other gymnopserms, so that the gymnosperms are paraphyletic. Since Haeckel (1894) this view has been espoused by many plant taxonomists (e.g., Friedmann 1990, 1994) because of the angiosperm-like features of the Gnetales, such as the dicotyledonous seeds, vessels in the secondary wood, and reduced female gametophytes. It has gained support from cladistic analysis of morphological data (e.g., Doyle and Donoghue 1986; Loconte and Stevenson 1990; Nixon et al. 1994); parsimony analyses of *rbc*L (Chase et al. 1993) and rRNA genes (Hamby and Zimmer 1992); and parsimony analysis of combined morphological and molecular data (Albert et al. 1994; Doyle et al. 1994). The opposing view is that no extant gymnosperm is a sister group of the angiosperms and the angiosperms are derived from the progymnosperm. This view is supported by the 5S RNA sequence data of Hori et al. (1985), the partial 18S rRNA sequence data of Troistky et al. (1991), the *rbc*L sequence data of Hasebe et al. (1992), and the chloroplast sequence data of Goremykin et al. (1996). This issue remains to be resolved. Second, do the dicots have multiple origins? The traditional view that the dicots are paraphyletic is supported by the partial rRNA sequence data of Hamby and Zimmer (1992) and the *rbc*L tree inferred by Chase et al. (1993), but it requires more extensive data (from different genes) to completely settle the issue. Third, are the monocots monophyletic? This may not be the case according to Hamby and Zimmer (1992), but further research is needed to resolve the question.

THE ORIGIN OF EUKARYOTES

The living world has historically been dichotomously divided into eukaryotes and prokaryotes. Those organisms with a well-formed nucleus are eukaryotes whereas those without such a nucleus are prokaryotes. The eukaryotes include all protists such as ciliates and flagellates, as well as fungi, plants, and animals. The only prokaryotes are the bacteria, including the cyanobacteria, which were formerly known as blue-green algae.

C. R. Woese and coworkers have challenged this dichotomous view (Woese and Fox 1977; Fox et al. 1980). Since the late 1960s they have been studying bacterial relations by comparing the sequences of the ribosomal RNAs from different species. In their earlier studies they used T1 ribonuclease to digest the 16S rRNA sequence into oligonucleotides (words), each of which ends with a single G, e.g., CUAAG and UCAUCG. The oligonucleotides thus produced were short enough to be sequenced by the techniques then available. To reduce the chance of recurrence of the same word in a 16S rRNA molecule, they used only words with six letters or more to form a dictionary characteristic of an organism. They then analyzed the data in terms of an association coefficient, S_{AB}, which is defined as twice the number of nucleotides in the words common to both of dictionaries A and B divided by the number of nucleotides in all words in the two dictionaries. They showed that the S_{AB} values can be used to study the evolutionary relationships among organisms.

Woese and coworkers came across a totally unexpected finding when examining the ribosomal RNA of the methanogenic bacteria. These unusual organisms live only in oxygen-free environments such as the intestinal tract of animals and sewage-treatment plants, and generate methane (CH_4) by the reduction of carbon dioxide. Methanogens are no doubt bacteria because they are the size of bacteria, have no nuclear membrane, and have a low DNA content. Thus, they would be expected to be more closely related to other bacteria than to eukaryotes. However, the S_{AB} values showed that methanogens are related as closely to the latter as to the former and that the group formed by methanogens is phylogenetically as deep as any other bacterial group. On the basis of these results and the argument that methanogenic metabolism (the reduction of carbon dioxide to methane) is ideally suited to the kind of atmosphere thought to have existed on the primitive earth (rich in carbon dioxide but virtually devoid of oxygen), Woese and Fox (1977) proposed to name methanogens and their relatives the archaebacteria, implying that this group of bacteria is at least as ancient as the true bacteria (eubacteria).

The archaebacterial group includes also extreme halophiles (halobacteria) and extreme thermophiles (Woese 1987). Halobacteria require high concentration of salt and grow in salty habitats along the ocean borders and in inland waters such as the Great Salt Lake and the Dead Sea. They are the only photosynthetic archaebacteria; they transduce light into chemical energy by means of a proton pump based on bacteriorhodopsin. The extreme thermophiles grow at high temperatures, some near the boiling point of water. All species grow anaerobically, though a minority of species can also grow aerobically. Most species require sulfur as an engery source and live in hot sulfur springs.

Woese and Fox (1977a) and Fox et al. (1980) proposed that archaebacteria, eubacteria, and eukaryotes were derived from a common ancestor and represent the three primary lines of life because the S_{AB} values suggested that these three groups are about equally distant from one another. Under this hypothesis, the prokaryotes are not monophyletic as traditionally believed but are divided into two groups: the eubacteria and the archaebacteria. Moreover, the eukaryotic lineage is as deep as the two other lineages, contrary to the traditional view that the eukaryotic lineage arose long after the diversification of prokaryotes. The view of Woese and Fox can be represented by Figure 6.9a ("the archaebacterial tree") if one neglects the root of the tree; for simplicity methanogens are not included.

Woese and Fox's proposal has stimulated a great deal of controversy. Van Valen and Maiorana (1980) argued that archaebacteria were derived from other bacteria and gave rise to the eukaryotes. They pointed out that archaebacteria and eukaryotes shared several specific and derived similarities. For example, the archaebaterium *Thermoplasma* has a histone-like protein that produces nucleosome-like condensations of DNA, has actin- and myosin-like proteins, and can invaginate its plasma membrane, an ability that is necessary for a cell that engulfs others. Moreover, *Halbacterium*, another archaebacterium, has a light-inducing pigment almost identical to rhodopsin, and its initiator tRNA produces nonformylated methione, as in eukaryotes.

The proposal that the archaebacteria form a monophyletic group was challenged by Lake et al. (1984), who claimed that archaebacteria are paraphyletic; he grouped extremely thermophilic, sulfur-metabolizing bacteria (which he called eocytes) with eukaryotes, and halobacteria with eubacteria ("the eocyte tree," Figure 6.9b). Lake's view was initially based on a crude analysis of ribosomal morphology but later gained considerable support from an analysis of 16S rRNA sequence data by the method of linear invariants (Lake 1987, 1988). His analysis

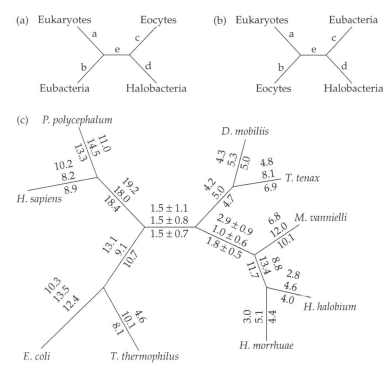

Figure 6.9 (a) The archaebacterial tree; for simplicity, methanogens are not included. (b) The eocyte tree. (c) The unrooted tree inferred by the neighbor-joining and maximum parsimony methods. The branch lengths were estimated by the least-squares method after the topology was inferred. The first, second, and third (below the branch) values on each branch represent the estimated number of substitutions per 100 sites for the SSU data, the LSU data, and the combined data respectively. From Gouy and Li (1989).

showed that the eocyte tree is highly significant ($P < 2 \times 10^{-5}$). However, other studies based on the least-squares method (Olsen, 1987), Fitch and Margoliash's (1967) method (Leffers et al. 1987), or the maximum parsimony method (Cedergren et al. 1988) supported the archaebacterial tree.

This divergence in opinion prompted Gouy and Li (1989) to conduct an examination of the problem. They used Lake's method of linear invariants (LI), the maximum parsimony (MP) method, and the neighbor-joining (NJ) method and considered species for which both the small subunit (SSU, i.e., 16S) and large subunit (LSU, i.e., 28S) rRNA sequences were available so that the congruency of the SSU and LSU trees inferred by a method could be checked. Further, they used only highly conserved regions so that a reliable alignment can be obtained. While application of the LI method to the SSU data did indeed support the eocyte tree, application of this method to the LSU data strongly supported the archaebacterial tree ($P < 5 \times 10^{-6}$). When the SSU and LSU data were combined, the LI method also supported the archaebacterial tree. Application of the NJ method to the SSU and LSU data either separately or jointly all gave the archaebacterial tree (Figure 6.9c). The same was true for the MP method; for the SSU data, the archaebacterial, eocyte, and halobacterial trees were supported by 20, 10, and 9 informative sites, respectively, and for the LSU data, the corresponding numbers were 34, 27, and 14. In conclusion, their study supported the monophyly of archaebacteria.

The study of Rivera and Lake (1992) on elongation factor EF-1α sequences has reopened the controversy on the archaebacterial tree vs. the eocyte tree. Elongation factor EF-1α (termed EF-Tu in bacteria) is an ubiquitous protein that transports aminoacyl-tRNAs to the ribosome and participates in their selection by the ribosome. While the sequences of eubacteria and the archaebacteria *T. celer* (a methanogens species) and *Halobacteria volcanii* contain a 4-amino acid segment (GPMP or its variants) near the guanosine diphosphate binding domain, Matheson et al. (1990) found that the sequence of the eocyte *Sulfolobus acidocaldarius* contains an 11-amino acid segment similar to the eukaryotic one. Later, Rivera and Lake (1992) found that other eocyte species also contain the 11-amino acid segment (Table 6.8). This difference in segment length can be explained by one change under the eocyte tree but by at least two changes under the archaebacterial tree. Rivera and Lake therefore concluded that the eocyte tree is the true tree. This conclusion assumes that gap events occur extremely rarely, so that one difference is sufficient to draw a conclusion on the tree topology. This assumption is, however, not supported by the fact that the sequence of *Methanogens vannielii* contains a 7-amino acid segment instead of the 4-amino acid segment in other methanogens (Table 6.8). At any rate, the issue deserves further investigation, for Rivera and Lake's conclusion is contrary not only to that of Gouy and Li (1989) but also to that of Gogarten et al. (1989), who showed that the ATPase data support the monophyly of the archaebacteria. The monophyly was further supported by a recent study using aminoacyl-tRNA synthetase genes (Brown and Doolittle 1995). However, the expanded database of elongation factor sequences tends to support the paraphyly of archaeons, with the *Sulfolobus/Desulfurococcus* clade as a sister group to the eukaryotes (Baldauf et al. 1996; Hashimoto and Hasegawa 1996). See Baldauf et al. for a discussion of other data.

The phylogenetic analyses discussed above actually provided only unrooted trees—the placement of the root between the eubacterial lineages and the other two lineages was based on other molecular and cellular data (as discussed above). Gogarten et al. (1989) and Iwabe et al. (1989) have come up with an ingenious idea to infer the root of the tree. They used duplicate genes that were derived before the separation of the three lineages (domains), which are called Bacteria, Archaea, and Eucarya by Woese et al. (1990). The idea is illustrated in Figure 6.10. Suppose that gene A was duplicated into A_1 and A_2 before the divergence of the three lineages. Subsequently, as the three lineages diverged, A_1 (and A_2) should also diverge in the same order. Therefore, A_2 sequences are outgroups to A_1 sequences and can be used to root the A_1 sequences, i.e., to root the three lineages. Similarly, A_1 sequences can be used to root the A_2 sequences and the three lineages.

Iwabe et al. (1989) applied this concept to the EF-Tu (EF-1α) and EF-G genes. Since these two elongator genes are duplicate genes present in all prokaryotes and eukaryotes, the duplication event should have occurred before the separation between eukaryotes and prokaryotes. For this reason, the EF-Tu sequences can be used as outgroups to infer the root of the tree for the EF-G sequences, and the tree in the lower part of Figure 6.11 suggests that the eukaryote lineage (as represented by *Dictyostelium* and hamster) is derived from the archaebacterial lineage (as represented by *Methanococcus*) and that the eubacterial lineage is the most ancient. The same conclusion was reached when the EF-G sequences were used as outgroups to

TABLE 6.8 A comparison of elongation factor EF-1α and EF-Tu sequences in, and EF-Tu, EF-G, and IF-2 sequences near, the KNMIT$_{94}$ to QTREH$_{118}$ region

Taxon	Organism	Left primer		11-amino acid segment	4-amino acid segment	Right primer
Sequences of EF-1α and EF-Tu in the KNMITG-QTREH region						
Eukary.	Human	KNMITG	TSQADCAVLI VAAGV	GEFEAGISKNG		QTREH
Eukary.	Tomato	KNMITG	TSQADCAVLI IDSTT	GGFEAGISKDG		QTREH
Eukary.	Yeast	KNMITG	TSQADCAILI IAGGV	GEFEAGISKDG		QTREH
Eocytes	*P. occu.*	KNMITG	ASQADAAILV VSARK	GEFEAGMSAEG		QTREH
Eocytes	*D. muco.*	KNMITG	ASQADAAILV VSARK	GEFEAGMSAEG		QTREH
Eocytes	*A. infe.*	KNMITG	ASQADAAI IAVSAKK	GEFEAGMSEEG		QTREH
Eocytes	*Su. acid.*	KNMITG	ASQADAAILV VSAKK	GEYEAGMSAEG		QTREH
Methan-ogens, relatives	*T. celer*	KNMITG	ASQADAAVLV VAVTD		---GVMP	QTKEH
	Mc.van.	KNMITG	ASQADAAVLV VNVDD		AKSGIQP	QTREH
Halobact.	*H. maris.*	KNMITG	ASQADNAVLV VAADD		---GVQP	QTQEH
Eubact.	*Th. mar.*	KNMITG	AAQMDGAILV VAATD		---GPMP	QTREH
Eubact.	*S. plat.*	KNMITG	AAQMDGAILV VSAAD		---GPMP	QTREH
Eubact.	Mitoch.	KNMITG	AAQMDGAI IVVAATD		---GQMP	QTREH

Molecule	Conserved						Right
Sequences of EF-Tu, EF-G, and IF-2 near the KNMITG-QTREH region							
EF-Tu	*T. celer*	DAPGH	RDFV	KNMITG	A SQADAAVLV VAVTD	---GVMP	QT
EF-Tu	*Mc. van.*	DCPGH	RDFI	KNMITG	A SQADAAVLV VNVDD	AKSGIQP	QT
EF-Tu	*H. maris*	DCPGH	RDFV	KNMITG	A SQADNAVLV VAADD	---GVQP	QT
EF-Tu	*Th. mar.*	DCPGH	ADYI	KNMITG	A AQMDGAILV VAATD	---GPMP	QT
EF-Tu	*S. plat.*	DCPGH	ADYV	KNMITG	A AQMDGAILV VSAAD	---GPMP	QT
EF-2	Hamster	DSPGH	VDFS	SEVTAA	L RVTDGALVVVDCVS	---GVCV	QT
EF-2	*Su. acid.*	DTPGH	VDFS	GRVTRS	L RVLDGSIVVIDAVE	---GIMT	QT
EF-2	*Mc.van.*	DTPGH	VDFG	GDVTRA	MRAIDGAVVVCCAVE	---GVMP	QT
EF-2	*S. plat.*	DTPGH	VDFT	IEVERS	MRVLDGVIAVFCSVG	---GVQP	QS
IF-2	*E. coli*	DTPGH	AAFT	SMRARG	AQATDIVVLVVAADD	---GVMP	QT

From Rivera and Lake (1992).

Note: 11-amino acid and 4-amino acid segments are delimited by solid boxes. The dashed boxes enclose the sequences used as PCR primers. The following designations are used to indicate taxa: *D. muco., Desulfurococcus mucosus; P. occu., Pyrodictium occultum; A. infe., Acidianus infernus; Su. acid., Sulfolobus acidocaldarius; T. celer, Thermococcus celer; Mc.van., Methanococcus vannielii; H. maris., Halobacterium marismortui; Th. mar., Thermotoga maritima; S. plat., Spirulina platensis; E. coli, Escherichia coli;* and Mitoch., yeast mitochondrion. The EF-1α sequences from *Mc. van.*, and *Thermoplasma acidophilum* (not shown) contain three and two, respectively, additional amino acids at the beginning of their 4-amino acid segments.

infer the tree for the EF-Tu sequences. The tree in Figure 6.10 was further supported not only by a more extensive analysis of elongation factor sequences but also by other sets of duplicate genes, such as aminoacyl-tRNA sythetase genes (Brown and Doolittle 1996; Lawson et al. 1996), which are also derived from ancient gene duplications that predated the last common ancestor of all living organisms. Therefore,

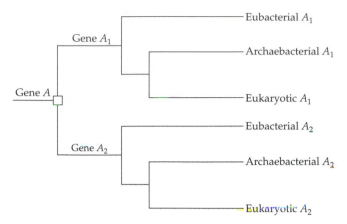

Figure 6.10   Duplication of gene A into A_1 and A_2 prior to the divergence of the eubacterial, archaebacterial, and eukaryotic lineages.

it seems fairly certain that the eukaryotes originated from an archaebacterial ancestor and that the root of the tree of life is between the eubacterial lineage and the ancestral lineage of the archaebacterial and eukaryotic lineages.

There is, however, the proposal that the eukaryotic genome is a chimera derived from a fusion of a (gram-negative) eubacterium and an archaebacterium (Zillig 1991; Gupta and Golding 1993; Golding and Gupta 1995, and references therein). In a maximum likelihood and parsimony analysis of 24 protein sequences from gram

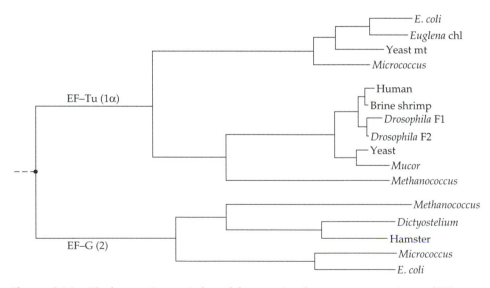

Figure 6.11   Phylogenetic tree inferred from a simultaneous comaprison of EF-Tu(1α) and EF-G(2) sequences from archaebacteria, eubacteria, and eukaryotes. The deepest root was arbitrarily chosen at a point between the two clusters corresponding to the different proteins. chl, chloroplast DNA-coded gene product; mt, nuclear gene-coded mitochondrial isozyme. From Iwabe et al. (1989).

negative and gram positive bacteria, archaebacteria and eukaryotes, Golding and Gupta (1995) found that although nine proteins significantly supported the archaebacteria–eukaryote clade, seven significantly supported the gram negatives–eukaryote clade; the rest gave no significant support to any of the three alternative topologies. They argued that such a large proportion of cases supporting the gram negatives–eukaryote clade is unlikely to be due to horizontal gene transfer, convergent evolution, or methodological errors, especially in view of the fact that none of the proteins significantly supported the gram positives–eukaryote clade. This chimeric origin hypothesis needs further investigation.

ANCIENT DNA

A novel approach to archeology and paleontology was initiated in 1984 when Higuchi et al. (1984) succeeded in extracting DNA from a 140-year old salt-preserved skin of a quagga, an extinct animal that looks like a cross between a zebra and a horse. They cloned DNA fragments isolated from dried skin and connective tissue and obtained mitochondrial DNA sequences totaling 229 bp. These sequences differed by 12 bp from the corresponding sequences of a mountain zebra. These and several sequences obtained in subsequent studies allowed the quagga to be placed into a phylogeny of the horse and its relatives (Figure 6.12; see references in Pääbo et al. 1989). The results argue for a close relationship between the quagga and Burchell's zebra and against the view that the quagga is more closely related to the domestic horse.

The study of Higuchi et al. was soon followed by the cloning of DNA from the skin of an ancient Egyptian mummy dated ~2400 years before present (BP) (Pääbo 1985). However, application of cloning techniques to DNA extracted from ancient material is difficult because of degradation and chemical modification and because of the small amounts that can be extracted. The situation changed when the polymerase chain reaction (PCR) technique was introduced into the study of ancient DNA, because PCR requires minute amounts of DNA, theoretically just a single DNA molecule, and can function efficiently even if the DNA is damaged and degraded to some extent (see the review by Brown and Brown 1994). The power of PCR to ancient DNA was first demonstrated in an application to a 7000-year-old

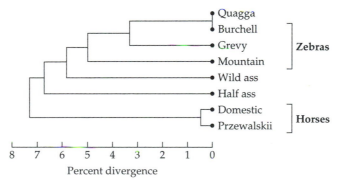

Figure 6.12 Phylogenetic tree relating mitochondrial DNA from the extinct quagga to mtDNAs from other members of the horse genus. From Pääbo et al. (1989).

brain from excavations in Florida (Pääbo et al. 1988) and later in an application to human bone dated to ~5450 years BP (Hagelberg et al. 1989). The latter was a significant step because it was thought that ancient DNA can be found only in artificially preserved material and because bones form the great bulk of ancient organic material. An even more remarkable achievement was the PCR amplification of an *rbc*L DNA segment from a fossilized magnolia leaf from the Miocene, between ~17 and 20 Myr BP (Golenberg et al. 1990); later, Soltis et al. (1992) succeeded in the amplification of an *rbc*L sequence from a Miocene *Taxodium* (bald cypress) from the same site. Because among all modes of fossilization preservation in amber may more consistently yield better quality DNA, even older fossilized DNA has now been amplified (DeSalle et al. 1992; Cano et al. 1993). The list of specimen types shown to contain ancient DNA continued to grow and now includes virtually all kinds of preserved biological material (Table 6.9). Ancient DNA has now become

TABLE 6.9 Types of biological material reported to contain ancient DNA

Type of material	Maximum age reported (years)	Key references
Human material		
Bones and teeth	10,000	Hagelberg et al. (1989), Horai et al. (1989), Hänni et al. (1990)
Mummies	5,000	Pääbo (1985, 1989)
Bog bodies	7,500	Pääbo et al. (1988), Lawlor et al. (1991)
Hair	0.3	Higuchi et al. (1988)
Animal material		
Bones	25,000	Höss and Pääbo (1993)
Naturally preserved skins	13,000	Pääbo (1989)
Museum skins	140	Higuchi et al. (1984) Thomas et al. (1989)
Feathers	130	Ellegren (1991)
Amber specimens	135,000,000	DeSalle et al. (1992, 1993) Cano et al. (1993)
Plant material		
Herbaria specimens	118	Rogers and Bendich (1985) Bruns et al. (1990)
Charred seeds and cobs	4,500	Goloubinoff et al. (1993) Rollo et al. (1991) Allaby et al. (1994)
Mummified seeds and embryos	44,600	Rogers and Bendich (1985)
Anoxic fossils	16,000,000	Golenberg et al. (1990) Soltis et al. (1992)
Amber specimens	40,000,000	Poinar et al. (1993)

From Brown and Brown (1994).

TABLE 6.10   **Examples of research projects that have utilized ancient DNA**

Project	Key references
Archaeology	
Human migrations in the Pacific	Hagelberg and Clegg (1993)
	Hagelberg et al. (1994)
Origins of domesticated maize	Goloubinoff et al. (1993)
Tuberculosis in pre-Columbian native Americans	Salo et al. (1994)
Conservation and population biology	
Bear populations in Europe	Taberlet and Bouvet (1994)
Hybrid status of the North American red wolf	Wayne and Jenks (1991)
Forensic science	
Identifiation of murder victims	Hagelberg et al. (1991)
Identification of war victims	Holland et al. (1993)
Palaeontology	
Taxonomy of extinct insects and plants from fossil material	Golenberg et al. (1990)
	Soltis et al. (1992)
	DeSalle et al. (1992, 1993)
	Cano et al. (1993)
	Poinar et al. (1993)
Zoology	
Taxonomy of recently extinct animals from skins and bones	Higuchi et al. (1984)
	Thomas et al. (1989)
	Janczewski et al. (1992)
	Krajewski et al. (1992)

From Brown and Brown (1994).

an important research tool in disciplines as diverse as archeology, conservation biology, and forensic science (Table 6.10).

However, one should note that contamination is a serious possibility in ancient DNA studies. For example, there is strong evidence that the reported cytochrome *b*–like sequences amplified from dinosaur bone DNA (Woodward et al. 1994) are due to human DNA contamination (Hedges and Schweitzer 1995; Zischler et al. 1995; Collura and Stewart 1995). Recently, Poinar et al. (1996) have developed a technique that uses the extent of racemization of amino acids as a criterion for assessing whether an ancient tissue sample contains endogenous DNA. Their study suggested that many paleontological finds from which DNA sequences purportedly millions of years old have been reported probably did not contain intact DNA because they showed extensive racemization. Fortunately, specimens bound in amber have a high probability to yield ancient DNA.

Another sensational event is the report of revival of bacterial spores in a 25- to 40-million-year-old amber (Cano and Borucki 1995). This bacterium is identified to be most closely related to *Bacillus sphaericus*, based on the sequencing of 530 bp of the 16S rRNA. Reviving an ancient organism enables scientists to study evolution through a true living fossil. Good quality DNA can be obtained in large amounts, and a reliable rate of nucleotide substitution can be estimated, if the ancient organism can be reliably dated. However, there remains considerable reservation about the report because the experiment has not been repeated.

CHAPTER 7

Rates and Patterns of Nucleotide Substitution

T HE MATHEMATICAL THEORY DEVELOPED IN CHAPTER 4 can be used to study the rate of nucleotide substitution, which is a basic quantity in the study of molecular evolution. In order to characterize the evolution of a DNA sequence, we need to know how fast it evolves. It is also interesting to compare the substitution rates among genes and among different DNA regions of a gene, because this can help us understand the mechanism of nucleotide substitution in evolution.

Knowing the rate of nucleotide substitution may also enable us to date evolutionary events such as the divergence between species. However, to do this we need to know whether the rate estimated from one group of species is applicable to another group. This raises the issue of how variable the rate is among different evolutionary lineages or, in other words, whether there exists a molecular clock. This issue will be dealt with in the next chapter.

In this chapter we present data on the rates of nucleotide substitution and discuss the factors that may cause differences in rates and patterns of nucleotide substitution.

ESTIMATION OF SUBSTITUTION RATES

The **rate of nucleotide substitution** is commonly defined as the number of substitutions per site per year and can be calculated by dividing the number of substitutions per site between two homologous sequences (K) by $2T$, where T is the time of divergence between the two sequences (Figure 7.1). That is,

$$r = K/(2T) \tag{7.1}$$

The divergence time, T, is usually assumed to be the same as the time of divergence between the two species from which the two sequences were taken and is usually inferred from paleontological data.

From Equation 7.1 it is easy to see that the accuracy of r depends on the accuracy of the estimates of K and T. When K is small, estimates of K have large standard errors relative to its expected value (i.e., large coefficients of variation), unless the sequence is long. Moreover, if T is small, then even a small error in the estimation of T can lead to a substantial error in r, because T is in the denomina-

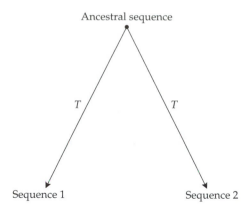

Figure 7.1   Divergence of two homologous sequences from a common ancestral sequence *T* years ago. From Li and Graur (1991).

tor. For these reasons, DNA sequences from closely related species are not suitable for estimating *r*. On the other hand, if *K* is large, it is difficult to obtain a reliable estimate of *K* owing to difficulties in correcting for multiple substitutions at the same site (see Chapter 4). Thus, highly divergent sequences are also not suitable for estimating *r*.

In this chapter we will deal with the issue of rate variation among genes and among different regions in a gene. For this purpose, it is advisable to use the same species pair for all the genes under consideration. The reason is twofold. First, there are usually considerable uncertainties about paleontological estimates of divergence times. By using the same pair of species we can compare rates of substitution among genes without knowledge of the divergence time. Second, the rate of substitution may vary considerably among lineages (see Chapter 8). In this case, differences in rates between two genes may be due to differences between lineages (i.e., lineage effects) rather than due to differences that are attributable to the genes themselves.

MAMMALIAN NUCLEAR GENES

At the present time the best data for studying the rate of nucleotide substitution are from mammals, for several reasons. First, DNA sequence data from mammals are more abundant than from most other groups of organisms. Second, genes from different mammalian orders generally exhibit an intermediate level of divergence and so are suitable for estimating *r*. Third, the fossil record for mammals is relatively good, and fairly reliable times of divergence between mammalian species are available.

In the following we shall compare the rates of nucleotide substitution in different genes using DNA sequence data from humans and mice or rats; if both the mouse and rat genes have been sequenced, the mouse gene is used. The human and rodent lineages are chosen because many gene pairs are available for comparison. As will be seen in Chapter 8, the rodent lineage appears to have evolved considerably faster than other mammalian lineages, so the average rate obtained below may be higher than the average for mammalian lineages. For noncoding regions, it is not easy to obtain reliable alignments between human and rodent sequences, and other pairs of species will be used.

As different regions of a gene serve different functions (Chapter 1) and are apparently subject to different stringencies of functional constraints, it has been customary to treat different regions separately (e.g., Miyata et al. 1980; Li et. al. 1985a).

Coding Regions

The protein-coding regions of genes have attracted the most attention from both molecular and evolutionary biologists because of their functional importance. As a consequence, a large amount of sequence data has become available for these regions and many comparative studies of nucleotide sequences in these regions have been published (e.g., Kafatos et al. 1977; Miyata et al. 1980; Li et al. 1985b). These studies have shown that synonymous substitution usually occurs much faster than nonsynonymous substitution.

Table 7.1 lists the rates of synonymous and nonsynonymous substitutions for 47 genes. The rates were obtained by using the method of Li (1993) and Pamilo and Bianchi (1993) (Chapter 4) and by assuming that the divergence time between the human and rodent lineages is 80 million years.

It is clear from Table 7.1 that the rate of nonsynonymous substitution is extremely variable among different kinds of genes, ranging from effectively zero for the histones to about 3×10^{-9} substitutions per nonsynonymous site per year for interferon γ. Nonsynonymous substitution rates are, of course, reflected in the rates of protein evolution, which may vary by over two orders of magnitude (Dayhoff 1972). As is well known, certain structural proteins (e.g., histones, actins and ribosomal proteins) are extremely conservative. Some peptide hormones, e.g., somatostatin-28 and glucagon, are also extremely conservative, but others have evolved at intermediate rates (e.g., erythropoietin and parathyroid hormone) or at high rates (e.g., relaxin). The insulin C-peptide has often been used as an example of rapid evolution, but it actually has evolved considerably more slowly than relaxin. Most enzymes have evolved at low rates. Hemoglobins, myoglobin, and some carbonic anhydrases have evolved at intermediate rates, while apolipoproteins, immunoglobulins, interferons, and interleukins have evolved rapidly. Apolipoprotein B is a huge protein (4536 amino acids) and the relatively high rate in the region included in Table 7.1, which is actually the most conservative part of the protein (the low density lipoprotein receptor-binding domain), implies that the whole protein has evolved at a high rate. In contrast, the entire sequence of myosin β heavy chain has evolved at a low rate, though it is also a large protein (1933 amino acids). The average rate of nonsynonymous substitution for all the genes in Table 7.1 is 0.74×10^{-9} substitutions per nonsynonymous site per year; the method of Li et al. (1985b) gives a slightly higher average rate, i.e., 0.78×10^{-9}. It is worth noting that the rate in the β globin is close to the average rate.

The rate of synonymous substitution also varies considerably from gene to gene, although not so much as the rate of nonsynonymous substitution. Part of the variation is probably due to statistical fluctuations, because the number of synonymous sites compared is small for many genes. For example, the estimate for the synonymous rate for relaxin is very high, but it has a large standard error. The average synonymous rate over all genes is 3.51×10^{-9} substitutions per synonymous site per year. This value is considerably smaller than that given in Li

TABLE 7.1 **Rates of synonymous and nonsynonymous substitution in various mammalian protein-coding genes**[a]

Gene	L[b]	Nonsynonymous rate	Synonymous rate
Histones			
Histone 4	102	0.00 ± 0.00	3.94 ± 0.81
Histone 3	135	0.00 ± 0.00	4.52 ± 0.87
Ribosomal proteins			
S17	134	0.06 ± 0.04	2.69 ± 0.53
S14	150	0.02 ± 0.02	2.16 ± 0.42
Contractile system proteins			
Actin α	376	0.01 ± 0.01	2.92 ± 0.34
Myosin β heavy chain	1933	0.10 ± 0.01	2.15 ± 0.13
Hormones and other active peptides			
Somatostatin-28	28	0.00 ± 0.00	3.10 ± 1.98
Insulin	51	0.20 ± 0.10	3.03 ± 1.02
Insulin C-peptide	31	1.07 ± 0.37	4.78 ± 2.14
Insulin-like growth factor II	179	0.57 ± 0.11	2.01 ± 0.37
Erythropoietin	191	0.77 ± 0.12	3.56 ± 0.53
Parathyroid hormone	90	1.00 ± 0.20	3.47 ± 0.87
Luteinizing hormone	140	1.05 ± 0.17	2.90 ± 0.54
Growth hormone	189	1.34 ± 0.17	3.79 ± 0.63
Urokinase-plasminogen activator	430	1.34 ± 0.11	3.11 ± 0.35
Interleukin-1	265	1.50 ± 0.15	3.27 ± 0.46
Relaxin	53	2.59 ± 0.51	6.39 ± 3.75
Hemoglobins and myoglobin			
α-globin	141	0.56 ± 0.11	4.38 ± 0.77
β-globin	146	0.78 ± 0.14	2.58 ± 0.49
Myoglobin	153	0.57 ± 0.11	4.10 ± 0.85
Apolipoproteins			
β (partial)	273	1.32 ± 0.14	3.64 ± 0.50
E	291	1.10 ± 0.12	3.72 ± 0.51
A-I	235	1.64 ± 0.17	3.97 ± 0.63
Imunoglobulins			
Ig k	106	2.03 ± 0.30	5.56 ± 1.18
Ig V_H	100	1.10 ± 0.20	4.76 ± 1.12
Interferons			
γ	136	3.06 ± 0.37	5.50 ± 1.45
$\alpha 1$	166	1.47 ± 0.19	3.24 ± 0.66
$\beta 1$	159	2.38 ± 0.27	5.33 ± 1.24
Other proteins			
Aldolase A	363	0.09 ± 0.03	2.78 ± 0.33
APRT	179	0.68 ± 0.11	3.56 ± 0.59

(*continued*)

TABLE 7.1 **(continued)**

Gene	L^b	Nonsynonymous rate	Synonymous rate
Albumin	590	0.92 ± 0.07	5.16 ± 0.48
Hydroxanthine phospho-ribosyltransferase	217	0.12 ± 0.04	1.57 ± 0.31
Creatine kinase M	380	0.15 ± 0.03	2.72 ± 0.34
Glyceraldehyde-3-phos-phate dehydrogenase	332	0.20 ± 0.04	2.30 ± 0.30
Lactate dehydrogenase A	331	0.19 ± 0.04	4.06 ± 0.49
Fibrinogen γ	411	0.58 ± 0.07	4.13 ± 0.46
Amylase	506	0.63 ± 0.06	3.42 ± 0.38
ANF	149	0.72 ± 0.13	3.38 ± 0.60
BADR	412	0.45 ± 0.06	3.07 ± 0.35
CA I	260	0.84 ± 0.11	3.22 ± 0.47
EF 2	857	0.02 ± 0.01	4.37 ± 0.36
Glucagon	29	0.00 ± 0.00	2.36 ± 1.08
Glutamine synthetase	371	0.23 ± 0.04	2.44 ± 0.30
Lipoprotein lipase	437	0.19 ± 0.04	2.95 ± 0.33
Prion protein	224	0.29 ± 0.06	3.89 ± 0.63
TNF	231	0.76 ± 0.11	2.91 ± 0.45
Thymidine kinase	232	0.43 ± 0.08	3.93 ± 0.59
Average[c]		0.74(0.67)	3.51(1.01)

[a]All rates are based on comparisons between human and rodent genes and the time of divergence was set at 80 million years ago. Rates are in units of substitutions per site per 10^9years.

[b]L = number of codons compared.

[c]Average is the arithmetic mean, and values in parentheses are the standard deviations, computed over all genes. Histones 3 and 4 were excluded because they are members of two multigene families and the human and rodent genes used may not be orthologous.

and Graur (1991) because they used the earlier method of Li et al. (1985b); when the latter is applied to the data in Table 7.1, it gives a value of 4.31×10^{-9}, which is close to that of Li and Graur (1991).

The sites in a coding region can also be divided into nondegenerate, twofold degenerate, and fourfold degenerate sites. Table 7.2 shows that at fourfold degenerate sites the rate of transitional substitution (2.2×10^{-9}) is higher than the rate of transversional substitution (1.5×10^{-9}), although at each site two types of transversional change and only one type of transitional change can occur. This observation can be largely explained by the fact that transitional mutation occurs more frequently than transversional mutation (Chapter 1). At twofold degenerate sites, the rate of transitional substitution (1.9×10^{-9}) is only slightly lower than that at fourfold degenerate sites, but the rate of transversional substitution (0.4×10^{-9}) is considerably lower than the corresponding rate at fourfold degenerate sites, owing to the fact that all transversional changes at twofold degenerate sites are nonsynonymous. At nondegenerate sites, the rates of transitional and transversional substitution are about the same ($\sim 0.4 \times 10^{-9}$) and are considerably lower than the corresponding values at fourfold degenerate sites because all changes at nondegenerate sites are nonsynonymous. Thus, the (total)

TABLE 7.2	Rates of transitional and transversional substitution at nondegenerate, twofold degenerate, and four-fold degenerate sites of codons[a]		
Type of substitution	Nondegenerate	Twofold degenerate	Fourfold degenerate
Transition	0.40	1.86	2.24
Transversion	0.38	0.38	1.47
Total	0.78	2.24	3.71

[a]The rates are averages over the genes used in Table 7.1, excluding histones 3 and 4. All rates are in units of substitutions per site per 10^9 years.

rate of nucleotide substitution is lowest at nondegenerate sites, intermediate at twofold degenerate sites, and highest at fourfold degenerate sites (Table 7.2).

Noncoding Regions

Data from noncoding regions are much less abundant than data from coding regions, and so only a limited comparative analysis can be done at the present time. (Note that in order to estimate the rate of substitution in a sequence we must have data from at least two species.) Since most published sequences are mRNAs, which do not include introns and flanking regions, the 5' and 3' untranslated regions are the only noncoding regions that can be studied in detail. Table 7.3 shows the substitution rates in these two regions for 16 genes from humans and rodents. In both regions the rates vary greatly among genes, but this variation may largely represent sampling effects due to the fact that both of these regions are usually very short. In almost all genes, the rates in the 5' and 3' untranslated regions are lower than the rate of substitution at fourfold degenerate sites (i.e., sites at which all possible nucleotide substitutions are synonymous). The average rates for the 5' and 3' untranslated regions are 1.96×10^{-9} and 2.10×10^{-9} substitutions per year, respectively, which are both about 55% of the average rate of 3.71×10^{-9} substitutions per year at fourfold degenerate sites.

Pseudogenes are DNA sequences that were derived from functional genes but have been rendered nonfunctional by mutations that prevent their proper expression (Chapter 1). Since they are subject to no functional constraints, they are expected to evolve at a high rate. Table 7.4 shows a comparison between the rate of substitution in cow and goat $\psi\beta^X$ and $\psi\beta^Z$ globin pseudogenes and the rates in the noncoding regions and fourfold degenerate sites in the β- and γ-globin genes. The rate in these pseudogenes is indeed slightly higher than that in the other regions. This result seems to be generally true for pseudogenes, though currently pseudogene data are limited.

Figure 7.2 presents a comparison of the rates of substitution in different regions of the gene, as well as in pseudogenes. The rates for the 5' and 3' untranslated regions, nondegenerate sites, twofold degenerate sites, and fourfold degenerate sites are the average rates for the genes in Tables 7.2 and 7.3. The rate for the 5' flanking region was computed by assuming that the ratio of this rate to that at fourfold degenerate sites is 5.3/8.6 (as suggested by the values in Table 7.4) and that the average rate at fourfold degenerate sites is 3.71×10^{-9}

TABLE 7.3 Rates of nucleotide substitution in 5' and 3' untranslated regions and at fourfold degenerate sites of protein-coding genes, based on comparisons between human and mouse or rat genes[a]

Gene	5' untranslated		3' untranslated		Fourfold degenerate	
	L^b	Rate	L	Rate	L	Rate
ACTH	99	1.87 ± 0.41	97	2.32 ± 0.49	275	2.78 ± 0.34
Aldolase A	124	1.08 ± 0.26	154	1.73 ± 0.32	195	3.16 ± 0.48
Apolipoprotein A-IV	83	3.06 ± 0.68	134	1.73 ± 0.33	160	3.38 ± 0.50
Apolipoprotein E	23	1.27 ± 0.69	84	1.70 ± 0.42	153	4.00 ± 0.60
Na,K-ATPase β	118	2.45 ± 0.45	1,117	0.57 ± 0.06	118	2.87 ± 0.54
Creatine kinase M	70	1.71 ± 0.46	168	1.79 ± 0.30	178	2.81 ± 0.41
α-fetoprotein	47	3.64 ± 1.13	144	2.79 ± 0.49	225	4.14 ± 0.54
α-globin	34	1.56 ± 0.65	90	2.21 ± 0.50	81	4.47 ± 0.98
β-globin	50	1.30 ± 0.46	126	2.85 ± 0.49	78	2.42 ± 0.56
Glyceraldehyde-3-phosphate dehydrogenase	70	1.34 ± 0.38	121	1.74 ± 0.36	170	2.43 ± 0.39
Growth hormone	21	1.79 ± 0.85	91	1.83 ± 0.41	83	3.82 ± 0.78
Insulin	56	2.92 ± 0.80	53	3.09 ± 0.81	62	4.19 ± 1.00
Interleukin-1	59	1.09 ± 0.38	1,046	2.02 ± 0.14	105	2.97 ± 0.60
Lactate dehydrogenase A	95	2.79 ± 0.55	470	2.48 ± 0.23	152	3.64 ± 0.60
Metallothionin-II	61	1.88 ± 0.52	111	2.57 ± 0.48	23	2.37 ± 1.00
Parathyroid hormone	84	1.79 ± 0.43	228	2.21 ± 0.30	38	3.85 ± 1.21
Average[c]		1.96 (0.78)		2.10 (0.61)		3.33 (0.69)

From Li and Graur (1991).

[a]Rates are in units of substitutions per site per 10^9years.

[b]L = number of sites.

[c]Average is the arithmetic mean, and values in parentheses are the standard deviations, computed over all genes.

TABLE 7.4   Divergence (± standard error) in 5′ and 3′ flanking (FL) regions between cow and goat β- and γ-globin genes and between cow and goat β-globin pseudogenes

Statistic	5′FL	5′UT	Fourfold	Introns	3′UT	3′FL	Pseudogenes
Percent divergence	5.3	4.0	8.6	8.1	8.8	8.0	9.1
Standard error	1.2	2.0	2.5	0.7	2.2	1.5	0.9

From Li and Graur (1991).
FL = flanking region; UT = untranslated region; fourfold = fourfold degenerate sites.

substitutions per year (Table 7.2). The rates for introns, the 3′ flanking region, and pseudogenes were computed in the same manner. Since these are rough estimates based on limited data and since the rate in a region varies from gene to gene, the rates shown in Figure 7.2 may not be applicable to any particular gene but are meant to provide a rough general pattern of the relative substitution rates in different DNA regions. With this precaution, we note that the substitution rate in a gene is highest at fourfold degenerate sites; slightly lower in introns and the 3′ flanking region; intermediate in the 3′ untranslated region, the 5′ flanking and untranslated regions, and twofold degenerate sites; and lowest at nondegenerate sites. Pseudogenes have the highest rate of substitution, although only slightly higher than that at the fourfold degenerate sites of a functional gene.

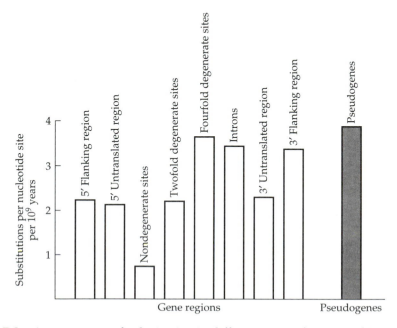

Figure 7.2   Average rates of substitution in different parts of genes and in pseudogenes. After Li and Graur (1991).

CAUSES OF VARIATION IN SUBSTITUTION RATES

We have seen in the preceding section that the rate of nucleotide substitution varies greatly among genes and among different regions of a gene. We now discuss possible causes of these variations. For this purpose, we note that the rate of substitution is determined by two factors: (1) the rate of mutation and (2) the probability of fixation of a mutation (Chapter 2). The probability of fixation depends on whether the mutation is advantageous, neutral, or deleterious and may vary enormously from gene to gene. Since the rate of mutation is unlikely to vary much within a gene but may vary among genes, it is simpler to discuss the rate variation among different regions of a gene and the variation among genes separately.

Variation Among Different Gene Regions

Let us first consider the large difference between the synonymous and nonsynonymous rates in a gene. Since the rate of mutation at synonymous and nonsynonymous sites within a gene should be the same, or at least very similar, the difference in substitution rates is usually attributed to differences in the intensity of purifying selection between the two types of sites. This is understandable in light of the neutral theory of molecular evolution (Chapter 2). Mutations that result in an amino acid substitution have a higher chance of causing deleterious effects on the function of the protein than do synonymous changes. Consequently, the majority of nonsynonymous mutations will be eliminated from the population by purifying selection. The result will be a reduction in the rate of substitution at nonsynonymous sites. In contrast, synonymous changes have a better chance of being neutral or nearly neutral, and more of them can become fixed in a population.

Of course, nonsynonymous substitutions may have a better chance of improving the function of a protein. However, if advantageous selection plays a major role in the evolution of proteins, the rate of nonsynonymous substitution should exceed that of synonymous substitution. Indeed, in some immunoglobulin genes, the nonsynonymous rate in the complementarity-determining regions (CDRs; also known as hypervariable regions) is higher than the synonymous rate. The higher rate has been attributed to overdominant selection for antibody diversity (Tanaka and Nei 1989). However, when the entire immunoglobulin gene is considered, the nonsynonymous rate is still considerably lower than the synonymous rate (Table 7.1). This result indicates that, even in immunoglobulins, most nonsynonymous mutations are disadvantageous and are eliminated from the population. In a case similar to that of immunoglobulins, Hughes and Nei (1989) reported that in certain regions of the major histocompatibility complex genes, the nonsynonymous rate of substitution exceeds the rate of synonymous substitution. They attributed the higher rates of nonsynonymous substitution to overdominant selection. This example will be discussed in detail in Chapter 9.

The contrast between synonymous and nonsynonymous rates in a gene demonstrates a well-known principle in molecular evolution, namely that the stronger the functional constraints on a macromolecule, the slower the rate of evolution. This may be explained under the selectionist view by using Fisher's (1930) argument that the larger the effect a mutation has, the smaller the chance of being advantageous it will have. However, neutralists have provided a more systematic explanation. In particular, Kimura (1977) has formulated the following model. Suppose that a certain fraction, f_0, of all mutations in a certain molecule are selectively neutral or nearly neutral and that the rest are deleterious. (Advantageous mutations are assumed to occur only rarely, such that their relative frequency is effectively zero, and they do not contribute significantly to the overall rate of molecular evolution.) If we denote by v_T the total mutation rate per site per unit time, then the rate of neutral mutation is $v_0 = v_T f_0$. According to the neutral theory of molecular evolution, the rate of substitution is $k = v_0$ (Chapter 2). Therefore,

$$k = v_T f_0 \tag{7.2}$$

Within any given gene, the v_T value can be assumed to be the same for synonymous sites and nonsynonymous sites. However, the f_0 value is higher for synonymous sites than for nonsynonymous sites and so the former should evolve faster than the latter. Thus, although the model is oversimplified, it is helpful for explaining the rate differences among different DNA regions.

According to the above model, the highest rate is expected to occur in a sequence that does not have any function, so that all mutations in it are neutral (i.e., $f_0 = 1$). In fact, pseudogenes do seem to have the highest rate of nucleotide substitution (Table 7.4 and Figure 7.2). The observation that 5' and 3' untranslated regions have lower substitution rates than the rate of synonymous substitutions in coding regions further supports the neutral line of reasoning, because these untranslated regions contain important signals for transcription initiation, termination, and regulation.

Within a protein, different structural or functional domains are likely to be subject to differential functional constraints and to evolve at different rates. A good example is provided by proinsulin, which consists of three segments: A, B, and C (Figure 7.3). Segment C resides in the middle of the proinsulin molecule and is removed during the formation of the active hormone (insulin), which consists of the two remaining segments, A and B. Segment C does not take part in the hormonal activity of insulin and is thought only to facilitate the creation of the proper tertiary structure of the hormone. Consequently, the nonsynonymous substitution rate for the region coding for the C segment is five to six times higher than the average nonsynonymous rate for the regions coding for the A and B chains (Figure 7.3). Nevertheless, considerable constraints must still operate on the C segment, because the nonsynonymous substitution rate in this region is rather low, comparable to that in β globin (Table 7.1).

There are many proteins that contain clearly defined functional or structural domains that are highly conserved in evolution. The steroid and thyroid hormone receptors are such examples (see Green and Chambon 1986). These receptors are factors that regulate the transcription of eukaryotic protein-coding genes in target tissues. The basic structural requirements for the receptors to control gene expression are the ability to bind the hormone with high affinity and specificity and the ability to bind to specific regions of DNA or chromatin. The DNA binding domain is extremely well conserved; e.g., 100% identity at the amino acid level between human and chicken estrogen receptors (Figure 7.4). The hormone binding domain is also well conserved, though not so strongly as the DNA

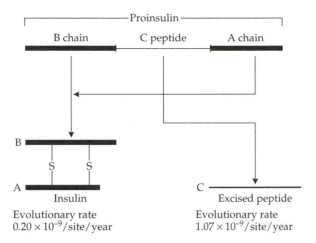

Figure 7.3 Comparison between the rates of nucleotide substitution for DNA regions coding for functional insulin (A and B chains) and the C peptide. A mature insulin molecule consists of one A and B chain, linked by disulfide (S) bonds. After Kimura (1983).

binding domain (Figure 7.4). Domain D is a hydrophilic region that may serve as a hinge between the DNA and hormone binding domains. This domain and the other domains, whose function is unknown, are considerably less well conserved than the DNA and hormone binding domains. Thus, there is an inverse relationship between the stringency of functional requirement and the rate of evolution in different functional domains.

Different regions of the tertiary structure of a protein also usually show different rates of evolution. The best studied case is hemoglobin in vertebrates, which is a tetramer composed of two α and two β chains; the structure and function of hemoglobin will be discussed in detail in a later section. As with other globular

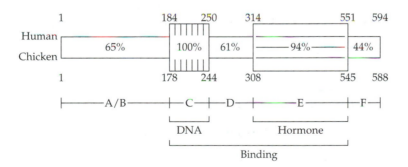

Figure 7.4 Schematic comparison of the human and chicken estrogen receptor sequences. The numbers refer to the position of amino acid residues, excluding the first methionine. The division of the steroid receptors into six domains (A–F) is shown at the bottom of the figure; the first domain was divided into A and B because of different degrees of sequence conservation (data not shown). Domains C and E are the DNA- and hormone-binding domains, respectively. After Green and Chambon (1986).

proteins, the interior of the molecule is filled mostly with hydrophobic amino acids and is well conserved in evolution. In particular, the heme pocket, which surrounds the heme, appears to be structurally very constrained. There are 19 amino acids in the α chain and 21 amino acids in the β chain that are known to form the heme pocket (Perutz and Lehman 1968). Using hemoglobin sequences from mammals, Kimura and Ohta (1973) estimated that the rates of amino acid substitution in the heme pocket are only 0.17×10^{-9} and 0.24×10^{-9} substitutions per amino acid site per year for the α and β chains, respectively (Table 7.5). In contrast, for those sites on the surface of hemoglobin, the rates of amino acid substitution in the α- and β-globin are 1.4×10^{-9} and 2.7×10^{-9}, which are about ten times higher than the corresponding rates in the heme pocket.

Variation Among Genes

To explain the large variation in the rates of nonsynonymous substitution among genes, we again consider the two possible factors: the rate of mutation and the intensity of selection. The assumption of equal rates of mutation for different genes may not hold in this case, since different regions of the genome may have different propensities to mutate. Wolfe et al. (1989c) suggested that different regions of the mammalian nuclear genome may differ from each other by a factor of two in their rates of mutation. However, a twofold difference in mutation rates is too small to account for the thousandfold range in nonsynonymous substitution rates. Thus, the most important factor in determining the rates of nonsynonymous substitution seems to be the selection intensity, which in turn is determined by selective (functional and structural) constraints.

To illustrate the effects of selective constraints, let us consider apolipoproteins and histone 3, which exhibit markedly different rates of nonsynonymous substitution. Apolipoproteins are the major carriers of various lipids in human blood, and their lipid-binding domains consist mostly of hydrophobic residues. Comparative analysis of apolipoprotein sequences from various mammalian orders suggests that in these domains the substitution of a hydrophobic amino acid (e.g., valine, leucine; Chapter 1) for another hydrophobic amino acid is acceptable at many sites (Luo et al. 1989). This lax structural requirement may explain why the nonsynonymous rates in these genes are fairly high (Table 7.1).

TABLE 7.5 **Comparison of evolutionary rates between the heme pocket and the surface in hemoglobin chains[a]**

Region	α-globin	β-globin
Heme pocket	0.17	0.24
Surface	1.35	2.73

From Kimura and Ohta (1973).
[a]All rates are in units of substitutions per amino acid site per 10^9 years.

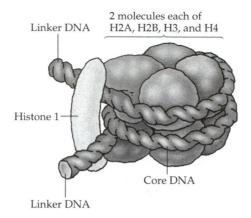

2 molecules each of
H2A, H2B, H3, and H4

Linker DNA

Histone 1

Core DNA

Linker DNA

Figure 7.5 Schematic diagram of a nucleosome. The DNA double helix is wound around the core histones (two each of histones 2A, 2B, 3, and 4). Histone 1 binds to the outside of this core particle and to the linker DNA. From Cooper (1997).

At the other extreme, we have histone 3. Since most amino acids in histone 3 interact directly with either the DNA or other core histones in the formation of the nucleosome (Figure 7.5), it is reasonable to assume that there are very few possible substitutions that can occur without impeding the function of this protein. In addition, histone 3 must retain its strict compactness and its high alkalinity, which are necessary for interaction with an acidic molecule, the DNA. As a consequence, histone 3 is very intolerant of most molecular changes. Indeed, this protein is one of the slowest evolving proteins known, more than one thousand times slower than the apolipoproteins.

Why the rate of synonymous substitution also varies from gene to gene is less clear. There may be two reasons for this variation. First, the rate of mutation may differ among different regions of the genome and the variation in rates of synonymous substitution may simply reflect the chromosomal position of the gene. Since there is currently little data on the variation of mutation rate among different regions of the mammalian genome, the importance of this factor cannot be evaluated. Wolfe et al. (1989c) hypothesized that the mutation rate of a sequence may depend on its GC content. However, analysis of DNA sequences of mammalian genes revealed only minor effects of GC content on the rate of synonymous substitution (Bulmer et al. 1991; Bernardi et al. 1993; Wolfe and Sharp 1993).

The second reason may be that not all synonymous codons are equivalent in fitness, so that some synonymous changes are favored whereas others are selected against. Such selection will create codon usage bias, a subject that will be discussed later. If the intensity of selective contraints on codon usage varies among genes, so will the rate of synonymous substitution. In this connection, it is interesting to note the existence of a positive correlation between the synonymous and nonsynonymous substitution rates (Graur 1985; Li et al. 1985b). In fact, for the genes used in Table 7.1 (excluding the histones), the correlation coefficient between the synonymous rate and the nonsynonymous rate is about 70%. There are several possible causes for this correlation. First, if the rate of mutation varies among genes, then there will be a correlation between synonymous and nonsynonymous rates of substitution. Since little data exist on the variation of mutation rate, the importance of this factor is difficult to evaluate. Second, there might be selection for translational accuracy in mammalian genes, as in the case of *Drosophila* genes (Akashi 1994). Because highly constrained proteins are less tolerant to translational errors, their genes may be more constrained to the use of optimal codons and so will have lower rates of synonymous substitution as well as lower rates of nonsynonymous substitution. Akashi (1994) showed that this is true for *Drosophila* genes and proposed that this might also be the case for mammalian genes. Third, in the presence of codon usage bias, a nonsynonymous change may cause a shift from an optimal codon to a nonoptimal codon. As a hypothetic example, suppose that GUG and GCC are the optimal codons for valine and alanine, respectively. Then, even if the change from valine to alanine is neutral at the protein level, the change from GUG to GCG is not neutral, because GCG is not the optimal codon for alanine. Thus, codon preference can slow down the rate of nonsynonymous substitution, depending on the degree of usage preference. On the other hand, if the nonsynonymous substitution of GCG for GUG has already occurred, then the change from GCG to GCC is advantageous and will be accelerated. Thus, a higher nonsynonymous rate can result in a faster synonymous rate (Lipman and Wilbur 1985). For these reasons, a positive correlation may exist between synonymous and nonsynonymous rates. However, Akashi's (1994) analysis of *Drosophila* genes suggested that this factor is less important than the selection for translational accuracy, because the majority of nonsynonymous changes do not cause a shift in optimal codons. Fourth, the extent of selection at synonymous positions might be affected by the nucleotide composition at adjacent nonsynonymous positions (Ticher and Graur 1989). Fifth, there might be an excess of tandem (doublet) substitutions between adjacent bases, because doublets can cause both synonymous and nonsynonymous substitutions at the same time and thus can create a positive correlation between the two (Wolfe and Sharp 1993).

DROSOPHILA NUCLEAR GENES

Some *Drosophila* species are also well sequenced, though not as extensively as humans and rodents. Table 7.6 provides a list of substitution rates between the *D. melanogaster* and *D. obscura* groups, which are assumed to have diverged about 30 million years ago (Throckmorton 1975; Moriyama and Gojobori 1992). The number of genes compared is smaller than that in Table 7.1, and so is the range of nonsynonymous rates. Nevertheless, a 25-fold difference is seen between the lowest and the highest rates (0.23×10^{-9} vs. 5.68×10^{-9}). The range of synonymous rates in Table 7.6 is from 3.7×10^{-9} to 29.7×10^{-9}, an 8-fold difference, larger than the mammalian range (Table 7.1). As will be discussed later, *Drosophila* genes appear to be subject to codon usage constraints, which are likely to vary among genes, thus causing variation in synonymous rates. This might explain why the range of synonymous rates is larger in *Drosophila* genes than in mammalian genes. The estimated average synonymous rate, 15.4×10^{-9}, is about four to five times higher than that in mammalian genes. Of course, the exact magnitude of difference is not certain, because both estimates involved assumptions about the divergence times.

TABLE 7.6 Rates of synonymous and nonsynonymous substitution in various *Drosophila* genes[a]

Gene	L[b]	Nonsynonymous	Synonymous
Arrestin B	401	0.23 ± 0.07	11.62 ± 1.48
Heat shock protein 83	374	0.30 ± 0.08	9.65 ± 1.27
Myosin alkali light chain 1	155	0.32 ± 0.12	3.67 ± 0.93
Neither-inactivation- nor-after-potential E	353	0.33 ± 0.08	9.48 ± 1.15
Antennapedia	378	0.48 ± 0.10	12.23 ± 1.53
Rhodopsin 3	382	0.78 ± 0.13	15.38 ± 1.90
Adh related gene (Adhr-Dup)	272	0.77 ± 0.15	19.22 ± 2.77
Rhodopsin 2	381	0.87 ± 0.13	17.18 ± 2.10
Alcohol dehydrogenase	254	0.90 ± 0.17	9.50 ± 1.30
Glucose dehydrogenase	612	0.88 ± 0.10	16.90 ± 1.55
Rhodopsin 4	378	0.93 ± 0.13	15.58 ± 1.82
Amylase, distal	494	1.13 ± 0.13	6.80 ± 0.78
Rosy	1334	1.18 ± 0.08	18.78 ± 1.22
Ultrabithorax	249	1.28 ± 0.20	11.25 ± 1.58
Adenine phosphoribosyltransferase	181	1.47 ± 0.27	16.63 ± 2.62
Adenosine 3 (Gart)	1351	1.43 ± 0.10	19.48 ± 1.23
Minute-(3)-99D (rp49)	131	1.52 ± 0.30	11.02 ± 2.15
Urate oxidase	345	1.57 ± 0.20	13.43 ± 1.72
Bicoid	474	1.80 ± 0.18	16.22 ± 1.68
Lethal (2) giant larvae	1161	2.05 ± 0.12	24.82 ± 1.87
Serendipity delta	419	2.23 ± 0.22	29.78 ± 5.13
Pupal cuticle protein	184	2.47 ± 0.35	18.50 ± 3.65
Serendipity beta	342	2.60 ± 0.25	21.77 ± 3.25
Larval cuticle protein 4	111	2.57 ± 0.45	10.30 ± 2.13
Period	1065	2.82 ± 0.15	17.92 ± 1.52
Esterase 6	542	3.03 ± 0.23	21.52 ± 2.37
Chorion protein 16	137	3.05 ± 0.47	14.57 ± 2.58
Toll	1095	3.28 ± 0.17	16.08 ± 1.08
Chorion protein 19	162	3.53 ± 0.45	14.35 ± 2.88
Serendipity alpha	523	4.10 ± 0.28	24.13 ± 2.78
Chorion protein 15	111	5.57 ± 0.75	13.18 ± 2.55
Cp18 Chorion protein 4	169	5.63 ± 0.60	18.38 ± 3.40
Average[c]		1.91 (± 1.42)	15.60 (± 5.50)

From E.N. Moriyama (unpublished).

[a]All rates are based on comparisons between the *melanogaster* and *obscura* groups and the time of divergence was set at 30 million years ago. Rates are in units of substitutions per site per 10^9 years.

[b]L= number of codons compared.

[c]Average = the arithmetic mean. Values in parentheses are the standard deviations, computed over all genes.

RATES OF SUBSTITUTION IN ORGANELLE DNA

Organelle genomes are much smaller and easier to investigate experimentally than nuclear genomes. Moreover, the discovery of an exceptionally high rate of substitution in the mammalian mitochondrial genome (Brown et al. 1979) stimulated further interest in the evolution of organelle DNA.

The mammalian mitochondrial genome consists of a circular, double-stranded DNA about 15,000–17,000 bp long, approximately 1/10,000 of the smallest animal nuclear genome. It contains mostly unique (i.e., nonrepetitive) sequences: 13 protein-coding genes, two rRNA genes, 22 tRNA genes, and a control region that contains sites for replication and transcription initiation. The genome is structurally very stable, as is evident from the small variation in genome size and gene arrangement among mammalian species.

In sharp contrast, plant mitochondrial genomes exhibit a high structural variability. They undergo frequent rearrangements, duplications, and deletions (Palmer 1985b, 1991). For this reason, the genome size varies from 40,000 to 2,500,000 bp. The mitochondrial genome of the liverwort (*Marchantia polymorpha*) is 186,608 bp in length. It contains 3 rRNA genes, 29 genes for 27 species of tRNA, and 30 open reading frames (ORFs) for functionally known proteins (16 ribosomal proteins, 3 subunits of H^+-ATPase, 3 subunits of cytochrome *c* oxidase, apocytochrome *b* protein, and 7 subunits of NADH ubiquinone oxidoreductase). In addition, 32 ORFs of unknown function were predicted. Therefore, this genome contains 94 possible genes.

The chloroplast DNA (cpDNA) or genome in vascular plants is circular and varies in size from about 120,000 to 220,000 bp, with an average size of 150,000 bp (see Palmer 1991). Much of the variation in size is the result of variation in the length of an inverted repeat, and the genome is known to be structurally stable. The completely sequenced cpDNA from tobacco (a dicot) is 155,844 bp long, including an inverted repeat of 25,339 bp. The genome contains about 113 genes in total (Sugiura 1992): 21 ribosomal proteins, 4 rRNAs, 30 tRNAs, 4 RNA polymerase subunits, 1 translation initiation factor, 1 putative intron maturase, 29 genes for photosynthesis functions, 11 genes for chlororespiration (NADH dehydrogenase subunits), 1 gene for a subunit of C1p protease, 1 gene for a subunit of acetyl-CoA carboxylase, and 10 conserved ORFs of unknown function. The gene contents in the cpDNAs of the rice and the liverwort are similar to that of the tobacco cpDNA.

The synonymous rate of substitution in mammalian mitochondrial genes has been estimated to be 5.7×10^{-8} substitutions per synonymous site per year (Brown et al. 1982). This is about 10 times the value for synonymous substitutions in nuclear protein-coding genes. The rate of nonsynonymous substitution varies greatly among the 13 protein-coding genes, but it is always much higher than the average nonsynonymous rate for nuclear genes. The reason for these high rates of substitution in mammalian mitochondria seems to be a high rate of mutation relative to the nuclear rate. The high mutation rate might be due to (a) a high turnover rate of mt genome and (b) a high concentration of mutagens resulting from the metabolic functions performed by the mitochondria. On the other hand, the stringency of purifying selection against nonsynonymous mutations seems to be of the same order of magnitude as that operating on nuclear genes.

Early studies based on a few gene sequences or on restriction-enzyme mapping suggested that chloroplast genes have lower rates of nucleotide substitution than mammalian nuclear genes (Curtis and Clegg 1984; Palmer 1985b) and that plant mitochondrial DNA evolves slowly in terms of nucleotide substitution, though it undergoes frequent sequence rearrangements (Palmer and Hebron 1987). These results have been confirmed by more extensive analyses of DNA sequences (Wolfe et al. 1987, 1989a).

Table 7.7 shows a comparison of the substitution rates in the three genomes of higher plants. The average numbers of substitutions per nonsynonymous site (K_A) in the chloroplast and mitochondrial genomes are similar, but the average number of substitutions per synonymous site (K_S) in the chloroplast genome is almost three times that in the mitochondrial genome for the comparison between monocot and dicot species, and it is two to three times higher for the comparisons between maize and rice. The average synonymous substitution rate in plant nuclear genes is four to five times that in chloroplast genes. Thus, the synonymous substitution rates in plant mitochondrial, chloroplast, and nuclear genes are in the approximate ratio of 1:3:12.

If we take the divergence time between maize and rice to be 50–70 million years (Stebbins 1981; Chao et al. 1984), the nuclear data in Table 7.7 indicate an average synonymous rate of $4.1–5.7 \times 10^{-9}$ substitutions per site per year. This is similar to the rates of synonymous substitution seen in mammalian nuclear genes (Table 7.1).

Interestingly, the rate of nucleotide substitution does not correlate with the rate of structural changes in the genome of organelles. In mammals, the mitochondrial DNA evolves very rapidly in terms of nucleotide substitutions, but the spatial arrangement of genes and the size of the genome are fairly constant among species. In contrast, the mitochondrial genome of plants undergoes frequent structural changes, but the rate of nucleotide substitution is extremely low. In chloroplast DNA, both the rates of nucleotide substitution and structural evolution are very low. The lack of correlation between the rates of nucleotide substitution and the rates of structural evolution suggests that the two processes occur independently.

TABLE 7.7 **Comparison of the rates of nucleotide substitution in plant chloroplast, mitochondrial, and nuclear genes[a]**

Genomes	K_S	L_S	K_A	L_A
A. Comparison between monocot and dicot species				
Chloroplast genes	0.58 ± 0.02	4,177	0.05 ± 0.00	14,421
Mitochondrial genes	0.21 ± 0.01	1,219	0.04 ± 0.00	4,380
B. Comparison between maize and rice				
Nuclear genes	0.57 ± 0.01	8,898	0.07 ± 0.00	30,702
Chloroplast genes	0.12 ± 0.00	7,872	0.02 ± 0.00	28,518
Mitochondrial genes	0.05 ± 0.01	1,845	0.02 ± 0.00	6,357

From Wolfe et al. (1987, 1989a) and K. H. Wolfe (unpublished).

[a]K_S, number of substitutions per synonymous site. K_A, number of substitutions per nonsynonymous site between genes. L_S, number of synonymous sites. L_A, number of nonsynonymous sites.

VIRAL GENES

Many viral genomes evolve rapidly because of a high mutation rate (Holland et al. 1982). The rate of mutation in a virus depends on the polymerase or polymerases it uses to synthesize its genome. Polymerases are classified into four classes, depending on whether they use DNA or RNA templates and on whether they produce DNA or RNA products (see Levy et al. 1994). RNA-dependent RNA polymerases use RNA as a template and catalyze the synthesis of RNA polymers. RNA-dependent DNA polymerases are known as reverse transcriptases because they use RNA templates but produce, in a reverse sense, DNA polymers. The two other classes, DNA-dependent DNA and RNA polymerases, use DNA templates and produce, respectively, DNA and RNA polymers. RNA-dependent polymerases are more error prone than DNA-dependent polymerases because they lack the proofreading ($3' \rightarrow 5'$ exonuclease) activity. For this reason, the mutation rates are in general much higher in RNA viruses than in DNA viruses; note, however, that some DNA viruses, e.g., the hepatitis B virus, go through an RNA stage and reverse transcription, so that their mutation rates are also high. Table 7.8 shows a compilation of estimates of mutation rates for (1) RNA viruses, which are here defined as viruses that exclusively replicate by RNA and do not go through a DNA stage in their life cycle, (2) retroviruses, which are RNA viruses but go through a DNA stage by reverse transcription, and (3) DNA viruses.

TABLE 7.8   Rates of spontaneous mutation per replication

Organism	Mutation rate
A. RNA viruses	
Phage *QB*	1.5×10^{-3}
Poliovirus[a]	0.1–1.6×10^{-4}
Vesicular stomatis virus	3.2×10^{-4}
Influenza virus A	$>7.3 \times 10^{-5}$
B. Retroviruses	
Spleen necrosis virus	0.5–3.6×10^{-5}
Moloney murine leukemia virus	0.4–$>6.6 \times 10^{-6}$
Rous sacoma virus	4.6×10^{-5}
C. DNA-based microbes	
Bacteriophage M13	7.2×10^{-7}
Bacteriophage λ	7.7×10^{-8}
Bacteriophage T2	2.7×10^{-8}
Bacteriophage T4	2.0×10^{-8}
Escherichia coli[b]	5.4×10^{-10}
Saccharomyces cerevisiae	2.3×10^{-10}
Neurospora crassa	2.0×10^{-11}

From Drake (1991, 1993).
[a]Excluding the case of 2.3×10^{-3}.
[b]Average excluding the case of 7.9×10^{-9}.

Although the estimates are rough, it is clear that the mutation rates in RNA viruses are much higher than those in DNA viruses (e.g., bacteriophages). For example, the mutation rate for phage $Q\beta$, which is an RNA virus, is about 10,000 times higher than that in bacteriophage M13. Although both viruses have the same host bacterium (i.e., *Escherichia coli*), phage M13 benefits from proofreading and mismatch repair, two antimutagenic mechanisms that are apparently absent for phage $Q\beta$. The mutation rates for retroviruses are approximately one order of magnitude lower than those for RNA viruses. The mutation rates for *E. coli* and fungi are 10^6 to 10^7 times lower than those for RNA viruses.

Table 7.9 presents a list of the rates of nucleotide substitution per year in viral genes. Both influenza A and hepatitis C viruses are RNA viruses and their rates of synonymous substitution are very high. The retroviruses HIV-1 (the human immunodeficiency virus type 1), the Moloney murine sarcoma virus and the Moloney murine leukemia virus also have high rates of synonymous substitution. In particular, the rate in HIV-1 is comparable to that in influenza A virus. Since the mutation rates in retroviruses are in general one order of magnitude lower than those in RNA viruses (Table 7.8), the high substitution rate in HIV-1 might be due to a very short "generation time" (replication cycle of the viral genome); note that the rate of substitution depends not only on the rate of mutation per replication but also on the generation time of the virus. Although hepatitis B virus (HBV) is a DNA virus, it goes through a stage of reverse transcription. However, its rate of synonymous substitution is 100 times lower than those of

TABLE 7.9   Rates of synonymous and nonsynonymous substitutions of viral genes and nuclear genes

Organism	Gene	Substitutions per site per year		Ref[b]
		Synonymous	Nonsynonymous	
A. RNA viral genes				
Influenza A virus	Hemagglutinin	13.10×10^{-3}	3.59×10^{-3}	1
Hepatitis C virus	E	6.29×10^{-3}	0.32×10^{-3}	2
HIV-1	*gag*	9.70×10^{-3}	1.70×10^{-3}	3
Moloney murine sarcoma virus	v-*mos*	2.75×10^{-3}	0.82×10^{-3}	1
Moloney murine leukemia virus	*gag*	1.16×10^{-3}	0.54×10^{-3}	4
B. DNA viral genes				
Hepatitis B virus	P	4.57×10^{-5}	1.45×10^{-5}	5
Herpes simplex virus type 1	Genome[a]	3.50×10^{-8}		6
C. Nuclear genes				
Mammals	c-*mos*	5.23×10^{-9}	0.93×10^{-9}	1
Mammals	α globin	3.94×10^{-9}	0.56×10^{-9}	7

[a]Computed from restriction enzyme analysis of the genome.

[b]1, Gojobori et al. (1990); 2, Ina et al. (1994); 3, Li et al. (1988a); 4, Gojobori and Yokoyama (1985); 5, Orito et al. (1989); 6, Sakaoka et al. (1994); 7, Li et al. (1985b).

retroviruses. It is possible that the replication of the HBV genome does not always depend on reverse transcription or that the replication frequency of the HBV genome is not as high as that of retroviruses (Orito et al. 1989; Gojobori et al. 1990). Herpes simplex virus is a DNA virus. The synonymous rate in this virus was estimated from restriction analysis of the entire genome and so might be 3 to 4 times that shown in Table 7.9. However, it is clear that its rate of substitution is about 100,000 times lower than those in RNA viruses. As is well known, the rate in mammalian nuclear genes is at least one million times lower than those in RNA viruses (Table 7.9).

NONRANDOM USAGE OF SYNONYMOUS CODONS

Because of the degeneracy of the genetic code, most of the 20 amino acids are encoded by more than one codon (Chapter 1). Since synonymous mutations do not cause any change in amino acid sequences and since natural selection was thought to operate predominantly at the protein level, synonymous mutations were proposed as candidates for selectively neutral mutations (Kimura 1968b; King and Jukes 1969). However, if all synonymous mutations are indeed selectively neutral, the synonymous codons for an amino acid should be used with more or less equal frequencies, unless the base composition is very skewed. As DNA sequence data accumulated, however, it became evident that the usage of synonymous codons is distinctly nonrandom in both prokaryotic and eukaryotic genes (Grantham et al. 1980). In fact, in many yeast and *E. coli* genes the bias in usage is highly conspicuous. For example, 21 of the 23 leucine residues in the *E. coli* outer membrane protein II (*omp*A) are encoded by the codon CUG, though there exist five other codons for leucine. Such a large bias cannot be explained by nonrandom mutation. How to explain the widespread phenomenon of nonrandom codon usage became a controversial issue (see Li et al. 1985a), to which, fortunately, some clear answers seem to be emerging.

An observation that has been helpful for understanding the phenomenon of nonrandom usage is that genes in an organism or in related species generally show the same pattern of choices among synonymous codons (Grantham et al. 1980). Thus, mammalian, *E. coli*, and yeast genes fall into distinct classes of codon usage. Grantham et al. (1980) therefore proposed the **genome hypothesis**, postulating that the genes in any given genome use the same coding strategy with respect to choices among synonymous codons, i.e., the bias in codon usage is species-specific. The genome hypothesis turns out to be true in general, though there is considerable heterogeneity in codon usage among genes in a genome (see below).

A simple measure of nonrandom usage of synonymous codons in a gene is the Relative Synonymous Codon Usage (RSCU; Sharp et al. 1986). The RSCU value for a codon for a particular amino acid is the observed number of occurrences of that codon in the gene divided by the number of occurrences expected under the assumption of equal usage of all codons for that amino acid. For a given amino acid, let n be the number (from 1 to 6) of synonymous codons for that amino acid and X_i be the number of occurrences of the ith codon for that amino acid. Then, under the assumption of equal usage, the expected occurrences of a codon for that amino acid is $(1/n) \sum X_i$ and so the RSCU$_i$ for the ith codon is given by

$$\mathrm{RSCU}_i = \frac{X_i}{\dfrac{1}{n}\displaystyle\sum_{i=1}^{n} X_i} \tag{7.3}$$

If the synonymous codons of an amino acid are equally used in a gene, their RSCU values are equal to 1. Deviations from equal usage can arise if mutation among the four nucleotides is not random or if natural selection occurs among synonymous codons. For other measures of codon usage bias, see Sharp and Li (1986), Shields et al. (1988), Wright (1990), and Long and Gillespie (1991).

Codon Usage in Unicellular Organisms

Studies of codon usage in *E. coli* and yeast have greatly increased our understanding of the factors that affect the choice of synonymous codons. Post et al. (1979) found that *E. coli* ribosomal-protein genes preferentially use synonymous codons that are recognized by the most abundant tRNA species. They suggested that the preference resulted from natural selection because using a codon that is translated by an abundant tRNA species will increase translational efficiency and accuracy. Their finding prompted Ikemura (1981, 1982) to gather data on the relative abundances of tRNA species in *E. coli* and the yeast *Saccharomyces cerevisiae*. He showed that, in both species, a positive correlation exists between the relative frequencies of the synonymous codons in a gene and the relative abundances of their cognate tRNA species. The correlation is very strong for highly expressed genes. For instance, in *E. coli* the most abundant of the four leucine tRNAs is tRNA$_1^{Leu}$, which recognizes the CUG codon, and the *omp*A gene, which is very highly expressed, uses predominately this codon for leucine (see above).

Figure 7.6 shows schematically the correspondence between the frequencies of the six leucine codons and the relative abundances of their cognate tRNAs. In *E. coli*, tRNA$_1^{Leu}$ is the most abundant leucine tRNA species, and in highly expressed genes, CUG (the codon recognized by this tRNA) is indeed much more frequently used than the other five codons. On the other hand, in yeast, the most abundant leucine tRNA species is tRNA$_3^{Leu}$, and the codon recognized by this tRNA (UUG) is the predominant codon. In contrast, in genes with low levels of expression, the correspondence between tRNA abundance and the use of the respective codon is much weaker in both species (Figure 7.6).

The importance of translational efficiency in determining the codon-usage pattern in highly expressed genes is further supported by the following observation (Ikemura 1981). It is known that codon-anticodon pairing involves **wobbling** at the third position. For example, U in the first position of anticodons can pair with both A and G. Similarly, G can pair with both C and U. On the other hand, C in the first anticodon position can only pair with G at the third position of codons, and A can only pair with U. Wobbling is also made possible by the fact that some tRNAs contain modified bases at the first anticodon position, and these can recognize more than one codon. For example, inosine (a modified adenine) can pair with any of the three bases U, C, and A. Interestingly, most tRNAs that can recognize more than one codon exhibit differential preferences for them. For example, 4-thiouridine (S^{4}U) in the wobble position of an anticodon can recognize both A and G in the wobble position; however, it has a marked preference for

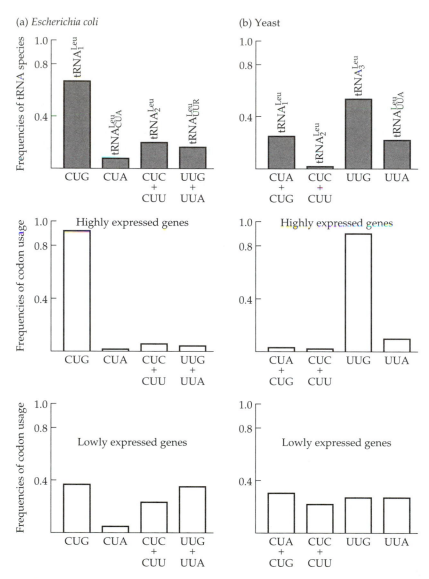

Figure 7.6 Diagram illustrating the relationship between the relative frequency of codon usage for leucine (open bars) and the relative abundance of the corresponding cognate tRNA species (shaded bars) in (a) *Escherichia coli* and (b) *Saccharomyces cerevisiae*. The plus signs (e.g., between codons CUC and CUU for *E. coli*) indicate that each of these pairs of codons is recognized by a single tRNA species (e.g., tRNA$_2^{Leu}$ for CUC and CUU in *E. coli*). From Li and Graur (1991).

A-ending codons over G-ending ones. Such a preference should be reflected in highly expressed genes. The two codons (AAA and AAG) for lysine in *E. coli* are recognized by a tRNA molecule that has S^{4}U in the wobble position of the anticodon, and, indeed, in the *E. coli ompA* gene, 15 of the 19 lysine codons are AAA, and only four are AAG.

Table 7.10 shows part of an extensive compilation of codon usage by Sharp et al. (1988). For each group of synonymous codons, if the usage is equal, the RSCU

TABLE 7.10 Codon usage in four species[a]

Amino acid	Codon	Escherichia coli		Saccharomyces cerevisiae		Drosophila		Human	
		High	Low	High	Low	High	Low	G + C	A + T
Leu	UUA	0.06	1.24	0.49	1.49	0.03	0.62	0.05	0.99
	UUG	0.07	0.87	5.34	1.48	0.69	1.05	0.31	1.01
	CUU	0.13	0.72	0.02	0.73	0.25	0.80	0.20	1.26
	CUC	0.17	0.65	0.00	0.51	0.72	0.90	1.42	0.80
	CUA	0.04	0.31	0.15	0.95	0.06	0.60	0.15	0.57
	CUG	5.54	2.20	0.02	0.84	4.25	2.04	3.88	1.38
Val	GUU	2.41	1.09	2.07	1.13	0.56	0.74	0.09	1.32
	GUC	0.08	0.99	1.91	0.76	1.59	0.93	1.03	0.69
	GUA	1.12	0.63	0.00	1.18	0.06	0.53	0.11	0.80
	GUG	0.40	1.29	0.02	0.93	1.79	1.80	2.78	1.19
Ile	AUU	0.48	1.38	1.26	1.29	0.74	1.27	0.45	1.60
	AUC	2.51	1.12	1.74	0.66	2.26	0.95	2.43	0.76
	AUA	0.01	0.50	0.00	1.05	0.00	0.78	0.12	0.64
Phe	UUU	0.34	1.33	0.19	1.38	0.12	0.86	0.27	1.20
	UUC	1.66	0.67	1.81	0.62	1.88	1.14	1.73	0.80
Met	AUG	1.00	1.00	1.00	1.00	1.00	1.00	1.00	1.00

From Sharp et al. (1988).

[a]For each group of synonymous codons, the sum of the RSCU (relative synonymous codon usage) values equals the number of codons in the group. For example, there are six codons for leucine, and so the sum of the RSCU values for these six codons should be 6. Under equal usage, the RSCU for each codon in a group should be 1, and so the degree of deviation from 1 indicates the degree of bias in usage. "High" and "low" denote genes with high and low levels of expression. For humans, "G + C" means high-GC regions, and "A + T" means high-AT regions.

value of each codon should be 1 (see above). This is clearly not so in the majority of cases. Moreover, in both *E. coli* and yeast, the codon usage bias is much stronger in highly expressed genes than in lowly expressed ones. A simple explanation (Sharp and Li 1986; Bulmer 1988) for this difference is that, in highly expressed genes, selection for translational efficiency and accuracy is sufficiently strong, so that codon usage bias is pronounced. In lowly expressed genes, on the other hand, selection is weak, so that the usage pattern is affected by mutation pressure and random genetic drift and, therefore, is less skewed. This explanation is known as the **selection-mutation-drift** hypothesis (see Bulmer 1991), which postulates that codon usage patterns are determined by the balance in a finite population between selection for the optimal codon for each amino acid and mutation together with random drift allowing the persistence of nonoptimal codons. In this theory, nonoptimal codons are present in lowly expressed genes because selection against them is weak. This view is different from the **expression-regulation** hypothesis, which postulates that nonoptimal codons are used in lowly expressed genes as a mechanism for keeping their expression low (e.g., Grosjean and Fiers 1982; Konigsberg and Godson 1983). An argument against the latter hypothesis is that it would be more efficient to modulate the level of gene expression by changing the strength of the promoter or the ribomsome binding site. See Bulmer (1991) for further discussion of the issue.

A theoretical study shows that strong usage bias occurs in a gene if the selective disadvantage (s) against nonoptimal codons is larger than $2/N$, where N is the effective size of a haploid population (Li 1987). Since the effective size for most unicellular organisms is likely to be larger than 10^6, s may need to be only of the order of 10^{-6} or even smaller for strongly biased codon usage to occur; indeed, Hartl et al. (1994) obtained an estimate of $s = 7 \times 10^{-9}$ for the *gnd* gene in *E. coli*. Thus, the strong codon usage bias observed in highly expressed genes in unicellular organisms may actually require only very small selective advantages for optimal codons. On the other hand, the fact that codon usage is not strongly biased in lowly expressed genes suggests that the selective differences between optimal and nonoptimal codons in these genes are extremely small.

In conclusion, in *E. coli* and yeast, the choice of synonymous codons is constrained by tRNA availability and other factors related to translational efficiency. These constraints will result in purifying selection, thus slowing down the rate of synonymous substitution (Ikemura 1981; Kimura 1983). In fact, it has been shown that the rate of synonymous substitution in enterobacterial genes is negatively correlated with the degree of codon usage bias (Sharp and Li 1986). Therefore, the phenomenon of nonrandom usage of synonymous codons may not be taken as evidence against the neutral theory of molecular evolution, since it can be explained in terms of the principle that stronger selective constraints result in lower rates of evolution (Chapter 2; Kimura 1983).

Codon Usage in Multicellular Organisms

Codon usage bias also varies widely among *Drosophila* genes (e.g., Shields et al. 1988; Moriyama and Hartl 1993; Sharp and Lloyd 1993). The bias is strongly associated with the GC content at third codon positions, which, in turn, has been found to be positively correlated with the GC content in their adjacent introns (Kliman and Hey 1994), though the latter correlation was not detected in earlier studies (Moriyama and Hartl 1993; Carulli et al. 1993). Since most parts of introns are apparently subject to only weak or no selective contraints, their GC content should be largely determined by the mutation pressure. Thus, a simple explanation for the positive correlation between the GC contents at third codon positions and in their adjacent introns is that the GC content at third codon positions (and therefore also codon usage) are affected by the mutation pressure. The effect is, however, weak, especially in genes with high codon usage bias (see also Sharp and Lloyd 1993). For fourfold degenerate sites, Spearman's rank-order correlation is 0.139 for the high-bias class of genes and 0.396 ($P < 0.001$) for the low-bias class. Thus, variation in mutation pattern cannot explain the GC variation at fourfold degenerate sites among high-bias genes and can account for only 16% of this variation among low-bias genes (Kliman and Hey 1994). For twofold degenerate sites the correlations are only slightly higher: $r = 0.272$ ($P < 0.02$) for high-bias genes and $r = 0.399$ ($P < 0.001$) for low-bias genes. In conclusion, the large variation in the GC content at degenerate sites appears to be mainly due to variation in the strength of natural selection, though the contribution of variation in mutation pattern is not negligible. Indeed, in high-bias genes there appears to be a strong preference for codons ending with C (Shields et al. 1988; Moriyama and Gojobori 1992; Moriyama and Hartl 1993). Available data indicate that those genes with high GC contents at degenerate sites are more highly expressed than those with low GC contents (Table 7.10; Shields et al. 1988). Other interesting observations are that the GC content in pseudogenes is reduced compared with their functional counterparts (Shields et al. 1988; Moriyama and Gojobori 1992) and that codon usage bias is less strong in *Drosophila melanogaster* than in *D. simulans*, as a result of reduction in population size in the former species (Akashi 1995). These observations suggest that, as in the case of unicellular organisms, selection for translational efficiency plays an important role in determining the choice of synonymous codons in *Drosophila*. In fact, Akashi (1994) has provided evidence for selection for translational accuracy. Apparently, the effective population size in *Drosophila* is large enough for selection to be effective in producing strong codon usage bias in highly expressed genes. Moreover, as in the case of bacteria, the rate of synonymous substitution has been shown to be negatively correlated with codon usage bias (Sharp and Li 1989; Shields et al. 1988; Moriyama and Gojobori 1992; Moriyama and Hartl 1993). Since the high GC content at degenerate sites is largely due to preference for C-ending codons at the expense of G-ending codons, the rate of synonymous substitution is negatively correlated with the C content but positively correlated with the G (A) content at the third codon positions (Figure 7.7).

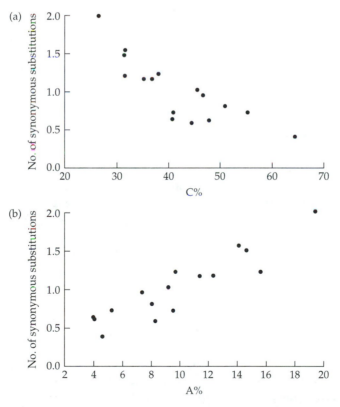

Figure 7.7   Relationships between the base content at the third codon position and the number of synonymous substitutions per site for the comparisons between *Drosophila melanogaster* and *D. pseudoobscura* genes. From Moriyama and Gojobori (1992).

In many human genes, codons tend to end in either G or C (i.e., to have a high GC content at the third position), whereas other genes have a low third-position GC content (Table 7.10). However, there are several reasons why this bias may not be strongly related to the level of gene expression. First, the α- and β-globin genes have different GC contents at the third position of codons (high and low, respectively), although they are both expressed in the same tissues (erythrocytes) in approximately equal quantities and, therefore, should have the same level of expression. Second, in chicken, the frequency with which codons are used does not correlate well with tRNA availability (Ouenzar et al. 1988). Finally, the GC content at the third codon position is strongly correlated with the GC level in introns and in both flanking regions (Chapter 13; Bernardi and Bernardi 1985; Aota and Ikemura 1986). For example, the α-globin gene is high in GC and resides in a high-GC region, whereas the β-globin gene is low in GC and resides in a low-GC region (Chapter 13; Bernardi et al. 1985). Thus, it appears that the codon-usage bias in a human gene is to a large extent determined by the GC content in the region that contains the gene. As will be discussed in Chapter 13, whether the GC content in a region is determined by natural selection or mutational bias is a controversial issue. However, since the GC content at the third position of codons in a gene tends to be higher than in introns and in surrounding regions (Chapter 13; Aota and Ikemura 1986), it is possible that the codon usage pattern in human genes is affected to some extent by natural selection. Furthermore, as mentioned earlier, a simple explanation for the positive correlation between nonsynonymous and synonymous rates in mammalian genes is that the degree of codon usage constraints varies among genes, i.e., genes subjected to strong codon usage constraints tend to have both low synonymous and nonsynonymous rates.

ADAPTIVE EVOLUTION OF LYSOZYME IN COW, LANGUR, AND HOATZIN

It is usually very difficult to provide experimental evidence that a certain molecular change is advantageous or has an adaptive value. Instead, evolutionists often rely on sequence comparison to infer whether convergent or parallel changes have occurred under similar conditions. Lysozyme has been a favorite protein for this purpose.

Lysozyme is a ubiquitous bacteriolytic enzyme found in virtually all animals. Its catalytic function is to cleave the β(1-4) glycosidic bonds between N-acetyl glucosamine and N-acetyl muramic acid in the cell walls of eubacteria. By virtue of its catalytic function and its expression in body fluids, such as saliva, serum, and tears, lysozyme usually serves as a first-line defense against bacterial invasion. In foregut fermenters, i.e., in animals in which the anterior part of the stomach functions as a chamber for bacterial fermentation of ingested plant matter, lysozyme is also secreted in the posterior parts of the digestive system and is used to free nutrients from within the bacterial cell.

Foregut fermentation has independently arisen twice in the evolution of placental mammals, once in the ruminants (e.g., cows) and once in leaf-eating colobine monkeys (e.g., langurs). In both cases, lysozyme, which in other mammals is not normally secreted in the stomach, has been recruited to degrade the

TABLE 7.11 **Pairwise comparison of lysozyme sequences among different species[a]**

Species	Species				
	Langur	Baboon	Human	Rat	Cow	Horse
Langur		14	18	38	32	65
Baboon	0		14	33	39	65
Human	0	1		37	41	64
Rat	0	1	0		55	64
Cow	5	0	0	0		71
Horse	0	0	0	0	1	

From Stewart and Wilson (1987).
[a]The numbers above the diagonal are the numbers of amino acid differences between species and those below the diagonal are the numbers of uniquely shared residues between species.

cell walls of bacteria, which carry on the fermentation in the foregut. Stewart and Wilson (1987) compared the amino acid sequences of lysozyme (~130 residues) from cows, langurs, baboons, humans, rats, and horses. They noted that there are five uniquely shared amino acids between cows and langurs (Table 7.11; Figure 7.8), compared to only one amino acid uniquely shared by cows and horses. Since cows and langurs diverged much earlier than the separation of the langur lineage from the human and baboon lineages, the uniquely shared amino acids in these two species were suggested to be the result of a series of adaptive parallel substitutions that occurred independently in both lineages.

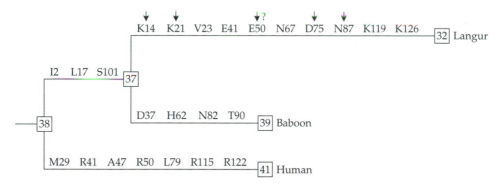

Figure 7.8 Parallel or convergent amino acid replacements in cow and langur lysozymes. The lengths of the lineages are proportional to the numbers of amino acid replacements along them. Each replacement is denoted by a one-letter abbreviation of the resultant amino acid (Table 1.1). Arrows point to five replacements in the lineage leading to langurs that occurred in parallel or convergence to those in the cow lineage. The question mark at E50 indicates that the sequence data from hoatzin lysozyme suggests that this replacement may not be adaptive (see text). The numbers of amino acid differences from cow stomach lysozyme (not shown) are in boxes. After Stewart and Wilson (1987).

Recently, Kornegay et al. (1994) sequenced the cDNA of the hoatzin lysozyme to see whether parallel substitutions similar to those in the cow and langur lysozymes have occurred in this lysozyme, because hoatzin is the only known avian foregut fermenter and has separated from the mammals about 300 million years ago. Since the hoatzin lysozyme has a deletion of a proline at position 50 (depending on the alignment), the glutamic acid (E) residue at this position uniquely shared by the cow and langur lysozymes (Figure 7.8) may not represent an adaptive substitution. What was unexpected is that the residues at positions 75 and 87 in the hoatzin lysozyme are the same as those [aspartic acid (D) and asparagine (N)] in the cow and langur lysozymes but different from those in the pigeon lysozyme. In addition, the hoatzin lysozyme has E instead of lysine (K) at positions 14 and 21. Since these two positions are at the surface of the enzyme, Kornegay et al. (1994) argued that having E in these positions may be more suitable for the less basic nature and lower pH profile of the stomach environment. In fact, position 14 is occupied by E in most ruminant stomach lysozymes other than those of the cow (Irwin and Wilson 1990; Jollés et al. 1990). Kornegay et al. made a similar argument for the E residue instead of K at position 126 (Figure 7.8; Table 7.12).

The adaptive structural changes proposed by Stewart and Wilson (1987) and Kornegay et al. (1994) are summarized in Table 7.12. They proposed that the residues at positions 14, 21, 75, 87, and 126 in the cow, langur, and hoatzin lysozymes represent adaptive substitutions. They provided evidence that some of these substitutions contribute to a better performance of lysozyme at low pH values. Furthermore, they suggested that the low arginine content confers increased

TABLE 7.12   Structural adaptations evident in hoatzin and mammalian stomach lysozymes[a]

Characteristic	Lysozyme type			
	Hoatzin stomach	Mammalian stomach	Chicken egg-white	Pigeon egg-white
Low pH optimum	+	+	–	–
Isoelectric point	~6	6.2–7.7	11.2	~10.6
Total arginines	5	3–6	11	10
Adaptive residues				
E/K14[b]	+	+	–	–
E/K21	+	+	–	–
D75	+	+	–	–
N87	+	+	–	–
E/K126	+	+	–	–

From Kornegay et al. (1994).

[a]Mammalian stomach lysozymes used were from three true ruminants (cow, goat, and axis deer) and the langur monkey. Adaptive replacements are summarized in the single-letter amino acid code and numbered according to Figure 7.8. Properties of chicken and pigeon lysozymes are used to represent nonstomach lysozymes.

[b]E/K means either E or K.

resistance to inactivation by fermentation products such as diacetyl and to proteolytic digestion by trypsin.

ADAPTATION OF HEMOGLOBIN TO HIGH ALTITUDES

Hemoglobin has played a prominent role in the study of molecular evolution for
several reasons. First, hemoglobin and its cousin myoglobin were the first proteins whose three-dimensional structures were worked out by X-ray crystallog-
raphy. This knowledge has greatly facilitated the study of the relationship
between structural constraint and rate of amino acid substitution (e.g., Table 7.5;
Dickerson and Geis 1969, 1983). Second, its abundance in the blood of animals
makes it easy to purify sufficient quantities for protein sequencing work, so it
was one of the first proteins that had been sequenced in many species. The
abundance of sequence data makes detailed comparative analyses feasible (e.g.,
Zuckerkandl and Pauling 1965; Goodman et al. 1975). Third, it is a protein that
everyone is familiar with. For these reasons, hemoglobin has always been a
favorite example for illustrating the principles of molecular evolution. In Table
7.5 it was used to illustrate the inverse relationship between stringency of structural constraints and rate of amino acid substitution in different regions of a protein. In this section it will be used as an example of adaptive changes to particular environments. In Chapter 10 hemoglobin and myoglobin will be used as an
example for the evolution of duplicate genes. For these purposes, it is helpful to
have a summary of the structure and function of hemoglobin; for more detailed
descriptions, see Fermi and Perutz (1981), Dickerson and Geis (1983), Perutz
(1983), and Bunn and Forget (1986).

Structure and Function of Hemoglobin

Hemoglobin consists of two α and two β chains. In humans and almost all other
mammals the α and β chains are 141 and 146 amino acids long, respectively. The α
chain contains seven and the β chain eight helical segments, designated A to H;
they are interrupted by nonhelical segments designated AB, BC, and so on (Figure
7.9). A short nonhelical amino-terminal segment, called NA, precedes the first
helix, and another (carboxy-terminal), called HC, follows the last helix.

The four subunits form a tetrahedron with a twofold axis of symmetry, so that
if any part of the molecule is rotated by 180 degrees, it will superimpose on a segment of identical structure (Figure 7.10). The molecule can be viewed as formed
by two fairly rigid $\alpha\beta$ dimers. If the four subunits are designated by α_1, α_2, β_1, and
β_2 and the two dimers by $\alpha_1\beta_1$ and $\alpha_2\beta_2$, then the two interfaces between unlike
subunits can be defined as $\alpha_1\beta_1$ and $\alpha_1\beta_2$. Because of symmetry, the $\alpha_1\beta_1$ interface
is identical to $\alpha_2\beta_2$ and the $\alpha_1\beta_2$ interface is identical to $\alpha_2\beta_1$ (Figure 7.10).

Uptake and release of oxygen are accompanied by small changes in tertiary
structure of the segments surrounding the hemes and by a large change in quaternary structure involving a rotation of one $\alpha\beta$-dimer relative to the other by 15
degrees (Figure 7.10). The rotation that occurs on dissociation of oxygen widens
the cavity between the two β chains so that the cationic groups that form its lin-
ing can bind organic phosphates. The two quaternary structures are referred to as
the oxy or relaxed (R) and the deoxy or tense (T) structures.

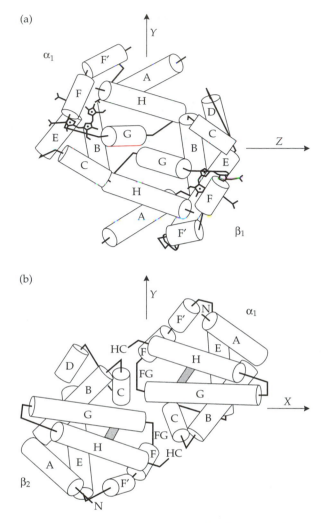

Figure 7.9   Three-dimensional orientation of the subunits in charboxyhemoglobin. The letter on each cylinder designates specific regions of α helix. The top panel (a) shows the $\alpha_1\beta_1$ dimer and the heme groups of the α and β chains. The bottom panel (b) shows the $\alpha_1\beta_2$ interface. From Baldwin (1980).

As the oxygen carrier in human blood, hemoglobin must be able to load and unload oxygen at appropriate partial oxygen pressures (PO_2) (see Bunn and Forget 1986). Hemoglobin suits this physiologic role because it has the sigmoid oxygen-dissociation (saturation) curve (Figure 7.11). This sigmoid curve reflects the fact that hemoglobin binds oxygen in a cooperative fashion, that is, the binding of the first oxygen molecule facilitates the binding of subsequent molecules. In the air sacs of the lung, PO_2 is about 95 mm Hg and the hemoglobin in red cells becomes 97% saturated with oxygen. Following circulation through capillary beds, the mixed venous PO_2 is about 40 mm Hg and hemoglobin becomes only about 75% saturated with oxygen, so that 97% − 75% = 22% of the oxygen would

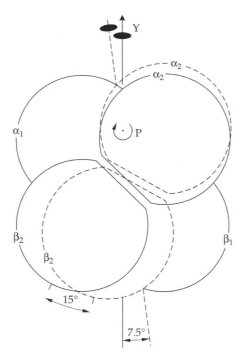

Figure 7.10 Change in the position of one dimer ($\alpha_2\beta_2$) relevant to its partner dimer ($\alpha_1\beta_1$) as a result of ligation of the heme groups. Deoxyhemoglobin is shown by a solid line and the liganded hemoglobin by an dashed line. With reference to $\alpha_1\beta_1$, which is fixed, there is considerable rotation (15 degrees) of the $\alpha_2\beta_2$ dimer upon heme ligation. In addition, there is a 7.5-degree rotation of the twofold (dyad) axis. From Fermi and Perutz (1981).

have been unloaded. Should each of the four heme groups of hemoglobin bind oxygen independently of the others, the oxygen dissociation curve would have a hyperbolic shape, like that of myglobin (Figure 7.11). Such a curve would be most unsuitable for oxygen transport, for in going from an arterial PO_2 of 95 mm Hg to a mixed venous PO_2 of 40 mm Hg, the hemoglobin is still 93% saturated (Figure 7.11), so that only 97% – 93% = 4% of the oxygen would be unloaded.

The oxygen affinity of hemoglobin decreases with decreasing pH, so that the oxygen-dissociation curve for hemoglobin is shifted to the right with increasing acidity (Figure 7.12). This is known as the (alkaline) Bohr effect, which facilitates the release of oxygen in tissue where acidity indicates the need for oxygen. Oxygen binding is influenced by chemical factors, known as heterotropic ligands, such as H^+, CO_2, Cl^-, and D-2,3-biphosphoglycerate (2,3-DPG). All these ligands reduce the oxygen affinity of hemoglobin by combining preferentially to the T (deoxy) structure, which has a high affinity for ligands but a low affinity for oxygen. The oxygen affinity of hemoglobin also decreases with rising temperature.

The oxygen affinity of hemoglobin can be conveniently expressed by the term P_{50}, the oxygen tension at which hemoglobin is half saturated (see Bunn and Forget 1986). The higher the affinity of hemoglobin for oxygen, the lower the P_{50}

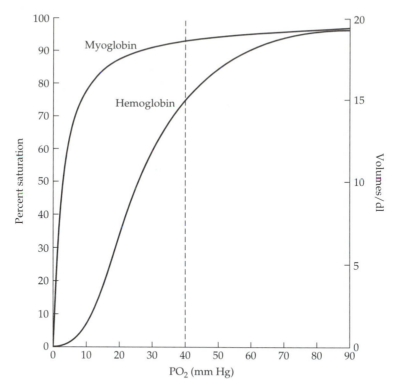

Figure 7.11 Oxygen-binding curves of hemoglobin (whole blood) and myoglobin at 37°C and pH 7.4. The ordinate on the right applies to whole blood (15 g Hb/dl). From Bunn and Forget (1986).

value and vice versa. Thus, P_{50} is inversely related to oxygen affinity. The P_{50} value for human blood at physiologic pH (7.4) and temperature (37°C) is normally 26 ± 1 mm Hg.

Llama Hemoglobin

Some animals living at high altitudes have adapted to life under hypoxic conditions (i.e., a low oxygen pressure) owing to changes in their hemoglobin sequences. The adaptation of llama (*Lama glama*), alpaca (*L. pacos*), guanaco (*L. guanacoë*), and vicuna (*L. vicugna*) to life in the Andes, South America, is such an example (Kleinschmidt et el. 1986; Piccinini et al. 1990). The blood of these mammals has a high affinity for oxygen compared to their lowland relatives of the genus *Camelus*. This oxygen affinity is achieved by a low affinity for 2,3-DPG. In the deoxy form of most mammalian hemoglobins, 2,3-DPG is bound to the central cavity between the two β chains. The amino acids involved in these contacts are: β1Val, β2His, β82Lys, and β143His in human hemoglobin, where β refers to the β chain and the number to the residue position in the chain. A suitable substitution in one of these positions can cause a reduced binding of 2,3-DPG and thus increase the oxygen affinity by destabilizing the deoxy (T) hemoglobin. In llama, alpaca, guanaco, and vicuna, β2His is replaced by Asn (Table 7.13). The

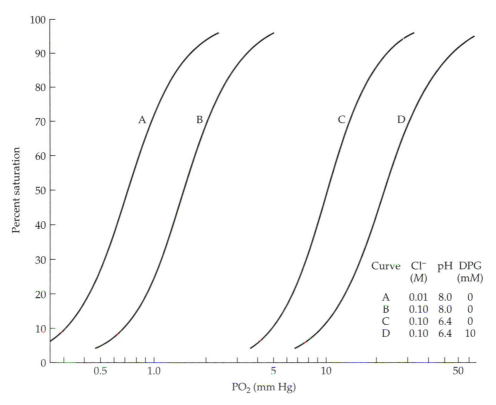

Figure 7.12 Effect of ionic strength, pH, and 2,3-DPG on oxygen affinity of human hemoglobin A at 20°C. From Bunn and Forget (1986).

side chain of Asn is shorter than that of His, and because of this shortening Asn cannot form a contact with the phosphates of 2,3-DPG (see Perutz 1983). The resulting higher oxygen affinity enables these animals to live in regions up to 4000 m. Interestingly, β2Asn is also found in the Indian and African elephant (Braunitzer et al. 1984), allowing them to climb up to 4500 m. The vicuna, the largest wild animal permanently residing in altitudes between 4000 and 5000 m, has an additional substitution (α130Ala $\rightarrow$ Thr) affecting chloride binding, which in combination with β2Asn causes the highest blood oxygen affinity of all mammals investigated so far.

A few other differences are also observed among the *Camelus* and *Lama* sequences (Table 7.13). Kleinschmidt et al. (1986) suggested that the internal non-polar isoleucine at α10 in llama, alpaca, and guanaco does not affect the hemoglobin structure. They also suggested that although position α122 is involved in the $\alpha_1\beta_1$ contact, the substitution His $\rightarrow$ Asp in llama does not seem to influence the $\alpha_1\beta_1$ contact, for the carboxyl group of aspartic acid might form a salt bridge with the amino group of β30Arg. The substitutions α8Thr $\rightarrow$ Ala, α23Glu $\rightarrow$ Asp, α120Ser $\rightarrow$ Ala, and β76Asn $\rightarrow$ Ser between the genera *Camelus* and *Lama* concern external residues that normally do not affect the structure and the oxygen binding properties of the hemoglobin molecule.

TABLE 7.13   Amino acid substitutions in Camelidae hemoglobins

Species	α chain						β chain		
	8	10	23	120	122	130	2	76	135
Lama glama	Ala	Ile	Asp	Ala	Asp	Ala	Asn	Ser	Ala
L. pacos	Ala	Ile	Asp	Ala	His	Ala	Asn	Ser	Ala
L. guanacoë	Ala	Ile	Asp	Ala	His	Ala	Asn	Ser	Ala[a]
L. vicugna	Ala	Val	Asp	Ala	His	Thr	Asn	Ser	Ala
Camelus ferus	Thr	Val	Glu	Ser	His	Ala	His	Asn	Ala
C. dromedarius	Thr	Val	Glu	Ser	His	Ala	His	Asn	Ala

From Kleinschmidt et al. (1986) and Piccinini (1990).
[a]In one animal only Ala was found, while in another animal Ala and Ser were found in the ratio of 1:1.

Bird Hemoglobins

The extensive protein sequencing work of Braunitzer and colleagues has provided many excellent examples of adaptation of bird hemoglobins to high altitudes (see Hiebl et al. 1989 and references therein). Three examples are discussed below.

The first example is the barheaded goose (*Anser indicus*) hemoglobin (Oberthür et al. 1980). These geese live and hatch their young at the Tibetan lakes (altitudes 4000–6000 m) but winter in the plains of northwest India. They migrate semiannually across the Himalayan Mountains without obvious stops en route for acclimatization. During migration flocks have been observed crossing directly over Mt. Everest at an altitude of 9200 m. The hemoglobin sequence of the barheaded goose was compared with those of the greylag goose, the Canadian goose, and the mute swan, which are its closest lowland relatives. The barheaded goose hemoglobin differs from the greylag goose hemoglobin by only four amino acids, the smallest number of replacements found among all the pairwise comparisons:

	$\alpha 18$	$\alpha 63$	$\alpha 119$	$\beta 125$
Greylag goose	Gly	Ala	Pro	Glu
Barheaded goose	Ser	Val	Ala	Asp

The substitution at $\alpha 18$ affects an external residue, the site $\alpha 63$ lies in a surface crevice, and $\beta 125$Glu is placed so that the hydrophilic side chain would protrude into the surrounding water (see Fermi and Perutz 1981). None of these replacements is likely to affect the function of hemoglobin. In contrast, $\alpha 119$Pro touches leucine $\beta 55$ at the $\alpha_1\beta_1$ contact and its replacement by alanine leaves a two-carbon gap so that the interaction between $\alpha 119$ and $\beta 55$ can no longer occur. This loss of interatomic stabilization of T structure favors the existence of the high-affinity R structure. For this reason, Perutz (1983) suggested that the substitution $\alpha 119$Pro → Ala alone is responsible for the high affinity of the barheaded goose hemoglobin.

The second example is the hemoglobin of the Andean goose (*Chloephaga melanoptera*) (Hiebl et al. 1987), which lives in the Andes at altitudes of 5000–6000 m. This hemoglobin differs from that of the greylag goose by nine replacements in the α chain and seven in the β chain. Nevertheless, Hiebl et al. (1987) postulated that its high oxygen affinity is caused by the single replacement $\beta55$Leu $\rightarrow$ Ser, which leaves a two-carbon gap at the contact with $\alpha119$Pro at the $\alpha_1\beta_1$ interface, the same proline that is replaced by alanine in the barheaded goose hemoglobin.

To test the above hypothesis that in both cases a single mutation was responsible for the high-oxygen affinity of the mutant hemoglobin (Hb), Jessen et al. (1991) used site-directed mutagenesis to produce two human Hb mutants: $\alpha119$Pro $\rightarrow$ Ala and $\beta55$Met $\rightarrow$ Ser. They compared the oxygen affinities of these two mutant Hbs with those of native human Hb and goose Hbs (Table 7.14). The increase in oxygen affinity ($\Delta\log P_{50}$) produced by either substitution was about 0.22, which was actually greater than the difference in oxygen affinity (0.15) between the Hbs of the barheaded and the greylag geese. They therefore concluded that the gap left by either substitution at the $\alpha_1\beta_1$ contact was a sufficient cause for the increase in oxygen affinity observed in nature. They also crystallized the mutant with $\beta55$Ser and determined its three-dimensional structure by X-ray analysis. They found that the major disturbances were restricted to the immediate neighborhood of the amino acid substitution. This observation explained why the mutant Hb was fully functional and strengthened the view that the increase in oxygen affinity was caused exclusively by the gap introduced at the $\alpha_1\beta_1$ contact.

The barheaded and Andean goose hemoglobins represent two types of adaptation: transitory and permanent hypoxic stress. The blood of these geese contains two hemoglobin components, one major, Hb A, and one minor, Hb D. Both hemoglobins have the same β chain but differ in the α chains, denoted by α^A and α^B, which are encoded by two duplicated α loci. In the Andean goose the substitution $\beta55$Met $\rightarrow$ Ser is in the β chain and so affects both Hbs. Thus, this single substitution results in an effective increase of oxygen affinity of whole blood, which seems to suit well the Andean goose, for this species lives permanently in the High Andes. On the other hand, in the barheaded goose hemoglobin the substitution $\alpha119$Pro $\rightarrow$ Ala, being in the α^A chain, affects only Hb A, so that Hb A has a high oxygen affinity compared to Hb D. Since barheaded geese have to cover altitutdes from near sea-level to elevations as high as 9200 m on their migratory flight over Mt. Everest, they need hemoglobins to adjust relatively quickly to different environmental conditions. Blood with hemoglobins of graded affinities can meet these requirements more effectively due to the larger variability in the

TABLE 7.14 Oxygen affinities of human and goose hemoglobins

Hemoglobin	$P_{50}(mmHg)$	$\log P_{50}$	$\Delta \log P_{50}$
Human HbA	5.8	0.76	
Human HbPα119A	3.3	0.53	0.23
Human HbMβ55S	3.4	0.54	0.22
Greylag goose	2.8	0.45	
Bar-headed goose	2.0	0.30	0.15

From Jessen et al. (1991).
Experimental conditions were as follows: pH 7.2, 0.1 M Cl$^-$, 25°C.

fine tuning of oxygen uptake and delivery resulting from the expression of two different α globin chains.

The third example, the hemoglobins of Rüppell's griffon (*Gyps rueppellii*), is even more remarkable (Hiebl et al. 1988, 1989). The fact that a Rüppell's griffon was hit by an aircraft at an altitude of 11,300 m over Abijan, Ivory Coast in west Africa shows that these vultures can ascend to extremely high altitudes. The blood of Rüppell's griffon is extraordinary in that it contains four hemoglobins instead of the usual two components in other birds. The four hemoglobins (Hb A, Hb A', Hb D, and Hb D') have the same β chain but differ in the α chain, implying the existence of four α loci, α^A, $\alpha^{A'}$, α^D, and $\alpha^{D'}$. Since the β chain of the four hemoglobins is identical to those of the European black vulture and the white-headed vulture, which ascend to altitudes of only about 4500 m, the α chains must account for the differences in the tolerance of hypoxic stress among these birds.

The α chain sequences of these birds show differences in the *respiratory box*, $\alpha34$ to $\alpha38$, which is of great structural and functional importance because it contains three $\alpha_1\beta_1$ and two $\alpha_1\beta_2$ contacts. At $\alpha34$ ($\alpha_1\beta_1$ contact) the substitution Ile $\rightarrow$ Thr has occurred in the Hb A of Rüppell's griffon (Table 7.15). With $\alpha34$Thr, a hydrogen bond can be formed to $\beta125$Asp so that the low-affinity T structure is stabilized. For this reason, the oxygen affinity of Hb A is lower than those of Hb A', Hb D, and Hb D'. A similar explanation holds at $\alpha38$ ($\alpha_1\beta_2$ contact): Rüppell's griffon Hb D and D' have the substitution Pro $\rightarrow$ Gln, which increases the oxygen affinity because the equilibrium between the T and R structures is shifted towards the R structure. A few differences are observed outside the respiratory box (Table 7.15). A change at position 12 is unlikely to exert much influence on function because this position is on the surface and is highly variable among species (Hiebl et al. 1988). Whether or not the change Cys $\rightarrow$ Asn at position 104 affects the oxygen affinity of $\alpha^{D'}$ is not clear because while $\alpha104$ is not involved in any contact, Cys at $\alpha104$ has been well conserved and Asn has been found only in the α chain of the opossum. Position 123 is at the $\alpha_1\beta_1$ contact and the substitution

TABLE 7.15   Amino acids from the α globins of Rüppell's griffon (RG), the European black vulture (BV) and the white-headed vulture (WV) at positions that show variation among sequences

α chain	Respiratory box			Other positions		
	34	38	12	104	123	137
RG α^A	Thr (–)[a]	Pro (–)	Asn	Cys	Ala	Thr
RG $\alpha^{A'}$	Ile (+)	Pro (–)	Asn	Cys	Ala	Thr
BV α^A	Ile (+)	Pro (–)	Thr	Cys	Ala	Thr
WV α^A	Ile (+)	Pro (–)	Thr	Cys	Ala	Ser
RG α^D	Ile (+)	Gln (+)	Thr	Cys	Ser	Ala
RG $\alpha^{D'}$	Ile (+)	Gln (+)	Thr	Asn	Ser	Ala
BV and WV α^D	Ile (+)	Pro (–)	Ala	Cys	Ala	Ala

Data from Hiebl et al. (1988) and references therein.

[a] – denotes a decrease and + denotes an increase in the oxygen affinity of the hemoglobin expressing the indicated amino acid.

Ala → Ser in α^D and $\alpha^{D'}$ weakens the hydrophobic interaction between Ala and β34Val by the hydrophilic Ser. However, this substitution was suggested to be of minor importance (Hiebl et al. 1988). The changes at position 137 are not yet understood; however, since all αA chains of birds, except for the white headed vulture, have Thr at this position, whereas all αD chains have Ala, the exchange Thr → Ala is unlikely to have special influence (T. Kleinschmidt, personal communication). In summary, the major changes occur only in the respiratory box, and for Rüppell's griffon three levels of affinity are predicted (Table 7.15): Hb A < Hb A' < Hb D/D'. The advantages of a low-affinity hemoglobin (greater oxygenation of the tissues) and a high-affinity component (improved arterial oxygenation) are combined. At extreme high altitudes the high-affinity Hb D/D' serve to take up oxygen while at low altitudes the intermediate and low affinity Hb A'and Hb A ensure sufficient oxygenation. Hiebl et al. (1988) suggested that this three-stage cascade system constitutes the molecular basis for flight at altitutdes of 11,300 m, at which the oxygen tension is only about 25% of that at sea level.

It is clear from the above discussion that a good knowledge of the structure and function of hemoglobin greatly facilitates the identification of changes that are important for the adaptation of life to high altitudes. In all the above examples it seems that the adaptation can be attributed to one or at most a few changes. The authors have therefore suggested that most of the other changes are functionally neutral. The same suggestion has been made by Perutz (1983) in an extensive review of adaptation of hemoglobin in different species. Of course, one should be wary that small functional differences are experimentally difficult to detect but may be of significance in the long history of evolution.

CHAPTER 8

Molecular Clocks

T HE MOLECULAR CLOCK HYPOTHESIS POSTULATES that for any given macromolecule (a protein or DNA sequence) the rate of evolution is approximately constant over time in all evolutionary lineages (Zuckerkandl and Pauling 1965). This hypothesis has stimulated much interest in the use of macromolecules in evolutionary studies, for two reasons (see Introduction). First, if macromolecules evolve at constant rates, they can be used to date species-divergence times and other types of evolutionary events, similar to the dating of geological times using radioactive elements. Moreover, phylogenetic reconstruction is much simpler under constant rates than under nonconstant rates (see Chapter 5). Second, the degree of rate variation among lineages may provide much insight into the mechanisms of molecular evolution (e.g., see Kimura 1983; Gillespie 1991). As noted in Chapter 2, under neutral mutation the rate of evolution is equal to the rate of mutation. Therefore, if the neutral mutation hypothesis is true and if the rate of neutral or nearly neutral mutation in a protein has not changed with time, the rate of evolution in that protein should be nearly constant. Thus, a large change in substitution rate may indicate a large change in evolutionary factors. For example, a large increase in the rate of evolution in a protein in a particular lineage may indicate adaptive evolution, relaxation of functional constraints (or loss of function), or a large reduction in effective population size in that lineage. This issue will be discussed in more detail later.

Although the concept of a molecular clock has had a strong impact on the study of evolution, it has always been controversial (see the Introduction). Indeed, there has been a wide range of views on this issue. In one extreme, Ochman and Wilson (1987) suggested the existence of a universal clock of synonymous substitution that is applicable to all organisms. In the other extreme, Goodman (1976, 1981) and his associates (Czelusniak et al. 1982) denied even the existence of approximate constancy. Also, there has been a strong controversy on whether generation time can have a significant effect on the rate of molecular evolution. Classical genetic studies indicated that mutation rates are more comparable among organisms when measured in terms of generation than in terms of absolute time. For this reason, the molecular clock should run faster in organisms with a short generation time, for they will go through more generations per unit time than do organisms with a long generation. This has been known as the generation-time effect hypothesis. However, while DNA–DNA hybridization data supported this hypothesis (Laird et al. 1969; Kohne 1970; Kohne et al. 1972), protein sequence data seemed to indicate a nearly constant rate in terms of absolute time (Sarich and Wilson 1973; Wilson et al. 1977). To date the two opposing views still persist.

The rapid accumulation of DNA sequence data in recent years presents an unprecedented opportunity for studying various issues related to the molecular clock hypothesis. DNA sequence data allow a closer examination of the hypothesis than do protein sequence data (e.g., the existence of both coding and noncoding regions in DNA sequences) and can be interpreted more directly than DNA hybridization and immunological distance data. For this reason, many authors (e.g., Miyata et al. 1982; Li et al. 1985a, 1987a; Wu and Li 1985; Britten 1986; Koop et al. 1986) have used DNA sequence data to examine the molecular clock hypothesis. The discussion below will be based on DNA sequence data. For DNA hybridization studies, readers may refer to Catzeflis et al. (1987), Sibley and Ahlquist (1987), Caccone and Powell (1989), and Sibley et al. (1990).

THE RELATIVE-RATE TEST

The controversy over the molecular clock hypothesis often involves disagreements on dates of species divergence. To avoid this problem, Sarich and Wilson (1973) proposed a test that does not require knowledge of divergence times. This test, called the **relative-rate test**, is illustrated in Figure 8.1a. Suppose that we want to compare the rates in lineages A and B. Then, we use a third species, C, as a reference. The reference species should have branched off earlier than the divergence between species A and B. For example, to compare the rates in the human and orangutan lineages we can use a monkey species as a reference.

From Figure 8.1a, it is easy to see that the number of substitutions between species A and C, K_{AC}, is equal to the sum of substitutions that have occurred from point O to point A (K_{OA}) and from point O to point C (K_{OC}). That is,

$$K_{AC} = K_{OA} + K_{OC} \tag{8.1a}$$

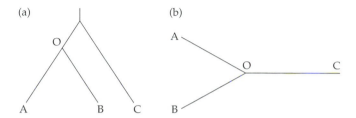

Figure 8.1   (a) The rooted tree for species A, B, and C, assuming that C is the known outgroup. O denotes the common ancestor of species A and B. (b) An unrooted tree. The root of the tree is unknown and O denotes the common node connecting the three lineages.

Similarly,

$$K_{BC} = K_{OB} + K_{OC} \tag{8.1b}$$

$$K_{AB} = K_{OA} + K_{OB} \tag{8.1c}$$

Since K_{AC}, K_{BC}, and K_{AB} can be directly estimated from the nucleotide sequences (Chapter 4), we can easily solve the three equations to find the values of K_{OA}, K_{OB}, and K_{OC}:

$$K_{OA} = (K_{AC} + K_{AB} - K_{BC})/2 \tag{8.2a}$$

$$K_{OB} = (K_{AB} + K_{BC} - K_{AC})/2 \tag{8.2b}$$

$$K_{OC} = (K_{AC} + K_{BC} - K_{AB})/2 \tag{8.2c}$$

We can now decide whether the rates of substitution are equal in lineages A and B by comparing the value of K_{OA} with that of K_{OB}. The time that has passed since species A and B last shared a common ancestor is, by definition, equal for both lineages. Thus, according to the molecular clock hypothesis, K_{OA} and K_{OB} should be equal, that is, $d = K_{OA} - K_{OB}$ should not be statistically different from 0. From Equations 8.2a and 8.2b, we obtain $K_{OA} - K_{OB} = K_{AC} - K_{BC}$. Therefore,

$$d = K_{AC} - K_{BC} \tag{8.3}$$

A simple way to test whether an observed d value is significantly different from 0 is to compare it with the standard error; for example, if the absolute value of d is larger than two times the standard error, it may be considered significant at the 5% level. Since $\text{Var}(x - y) = \text{Var}(x) + \text{Var}(y) - 2\text{Cov}(x, y)$, Equation 8.3 implies

$$V(d) = V(K_{AC}) + V(K_{BC}) - 2\text{Cov}(K_{AC}, K_{BC}) \tag{8.4}$$

Noting that $K_{AC} = K_{OA} + K_{OC}$ and $K_{BC} = K_{OB} + K_{OC}$ and that K_{OA} and K_{OB} are independent of each other because lineages A and B evolve independently, we have

$$V(d) = V(K_{AC}) + V(K_{BC}) - 2V(K_{OC}) \tag{8.5}$$

(Nei et al. 1985). Under the one-parameter model of nucleotide substitution, $V(K_{AC})$ and $V(K_{BC})$ can be obtained from formula (4.3) and $V(K_{OC})$ can be obtained as follows. From Eq. (4.2) we have

$$K_{OC} = -\frac{3}{4} \ell n \left(1 - \frac{4}{3} p_{OC} \right)$$

Since K_{OC} is given by (8.2c), we can estimate p_{OC} by

$$p_{OC} = \frac{3}{4} \left[1 - \exp \left(-\frac{4}{3} K_{OC} \right) \right] \tag{8.6}$$

Replacing p in Formula 4.3 by p_{OC}, we obtain an approximate formula for $V(K_{OC})$.

For the two-parameter model, $V(d)$ can be derived in a similar manner (Wu and Li 1985), though the derivation is more complicated. For coding regions, one can treat synonymous and nonsynonymous substitutions separately and derive the mean and variance of d for each type of substitution using the procedure above and the formulas for the mean and variance of K_{ij} given in Chapter 4 (see Wu and Li 1985).

Sometimes we are not certain about the phylogenetic relationships among the three lineages under study. In this case, we may still test the molecular clock hypothesis. We use the unrooted tree in Figure 8.1b because we do not know the root of the tree. However, Formulas 8.2a–c can still be used to calculate the branch lengths from the common node O to each of the present-day sequences. To be favorable to the molecular clock hypothesis, we assume that the lineage with the largest branch length is the outgroup. For example, if K_{OC} is the largest, then species C is assumed to be the outgroup and the test can be done as above. If K_{OB} is the largest, then species B is taken as the outgroup, and by exchanging the subscripts B and C in the above formulas, we can test the molecular clock hypothesis in a similar manner. Note, however, that a failure to reject the hypothesis may not be taken as evidence for the hypothesis because we assume the most favorable condition for testing the hypothesis. Obviously, a rejection of the hypothesis can be taken as evidence against the hypothesis.

The above test is simple and will be used in this book. Tajima (1993) has recently proposed a simple alternative test. Muse and Weir (1992) have developed a maximum likelihood test, which is likely to be more powerful but computationally more demanding. The test given in Equation 8.5 has been extended to the case of more than two lineages by Li and Bousquet (1992).

A LOCAL CLOCK IN MICE, RATS, AND HAMSTERS

Since the molecular clock hypothesis is controversial, the first question to ask is, "Does there exist a molecular clock in any group of organisms?" (Such a clock is known as a local clock.) The best organisms to look for the existence of a local clock are a group of organisms with similar physiology and life histories such as generation time. The muroid rodents (e.g., mice and rats) and their relatives would be such a group for which there is abundance of DNA sequence data. Indeed, there have been two lines of evidence suggesting that the clock-behavior is approximately maintained in this group of rodents. First, the DNA–DNA hybridization studies of Brownell (1983) and Catzeflis et al. (1987) revealed a constant rate among lineages in the murine (mouse and rat) family and the microtine (hamster) family. Second, in an analysis of nucleotide sequences using the relative-rate test with human sequences as references, Li et al. (1987a) found nearly equal rates in the mouse and rat lineages. We review below the study of O'hUigin and Li (1992), who used extensive sequence data from mice, rats, and hamsters.

First, let us compare the substitution rates in the mouse and rat lineages, using the hamster lineage as a reference. The number of substitutions per synonymous site (K_S) is 30.3% between mouse and hamster and 31.1% between rat and hamster (Table 8.1). The difference (0.8%) is not statistically significant because it is smaller than the standard error of K_S (1.0%). So, the synonymous rates in the mouse and rat lineages are nearly equal. The difference in the nonsynonymous rate (K_A), i.e., $d = 2.9\% - 2.7\% = 0.2\%$, is equal to two times the approximate standard error of K_A and may be considered statistically significant. Thus, the nonsynonymous rate seems to be slightly faster in the mouse lineage than in the rat lineage.

TABLE 8.1   **Numbers of nucleotide substitutions per 100 sites between species**[a]

Species pair	Synonymous sites		Nonsynonymous sites	
	K_S	L^b	K_A	L^b
Mouse–rat	18.0 ± 0.7	4,229	1.8 ± 0.1	15,217
Mouse–hamster	30.3 ± 1.0	4,229	2.9 ± 0.1	15,217
Rat–hamster	31.3 ± 1.0	4,229	2.7 ± 0.1	15,217
Mouse–human	53.4 ± 1.5	4,229	5.2 ± 0.2	15,217
Rat–human	51.6 ± 1.5	4,229	5.0 ± 0.2	15,217
Hamster–human	52.3 ± 1.5	4,229	5.1 ± 0.1	15,217

From O'hUigin and Li (1992).
[a]Computed by Li et al.'s (1985b) method.
[b]Number of sites compared.

mouse   rat   hamster

Second, we compare the substitution rates in the mouse (or rat) and hamster lineages, using the human lineage as a reference. The K_S value is 53.4% between mouse and human and 52.3% between hamster and human, so the difference 1.1% is smaller than the approximate standard error 1.5% and is not statistically significant. Similarly, the difference between the two lineages in the rate of nonsynonymous substitution is also not significant. The same conclusion can be drawn when the mouse lineage is replaced by the rat lineage (Table 8.1). Thus, the mouse, rat and hamster lineages have evolved at nearly equal rates in terms of nucleotide substitution.

In conclusion, there appears to be an approximate molecular clock in these rodents, at least in terms of synonymous substitution. This clock may be used to date divergence times among these rodents. For example, since the K_S value is 18.0% between mouse and rat and is about 31.0% between mouse–rat and hamster, the hamster lineage is estimated to have branched off 0.31/0.18 = 1.7 times earlier than the mouse–rat divergence.

LOWER RATES IN HUMANS THAN IN MONKEYS

There has been a long-standing controversy over the hominoid rate-slowdown hypothesis, which postulates that the rate of molecular evolution has become slower in hominoids (humans and apes) after their separation from the Old World (OW) monkeys. This hypothesis, proposed by Goodman (1961) and Goodman et al. (1971), was based on rates estimated from immunological distance and protein sequence data. Sarich and Wilson (1967) and Wilson et al. (1977) contended that the slowdown was an artifact, owing to the use of an erroneous paleontological estimate of the ape–human divergence time. They conducted relative-rate tests using both immunological distance data and protein sequence data and concluded that there was no evidence for a hominoid slowdown. A similar conclusion was drawn from DNA hybridization studies (Kohne et al. 1972; Sibley and Ahlquist 1984); however, more recent DNA hybridization studies have produced conflicting conclusions (Sibley and Ahlquist 1987; Caccone and Powell 1989). On the other hand, comparative analyses of DNA sequence data by Koop et al.

(1986), Li and Tanimura (1987), and Li et al. (1987) provided strong support for the hominoid slowdown hypothesis and the hypothesis was accepted by many molecular evolutionists. However, Easteal (1991) has argued that the slowdown occurred only in the η-globin pseudogene, because when this pseudogene was removed from comparison, the rate of nucleotide substitution in the OW monkey lineage was no longer significantly higher than that in the human lineage.

To resolve this controversy, Seino et al. (1992) and Ellsworth et al.(1993; unpublished) obtained more sequence data. In Table 8.2, K_{13} and K_{23} are the distance between an OW monkey and a New World (NW) monkey and between the human and a NW monkey, respectively. For the introns compared $K_{13} - K_{23}$ is positive, except that $K_{13} - K_{23}$ is 0 for the ε–globin and interferon-α receptor introns and is slightly negative for the lipoprotein lipase intron. When all introns are considered together $K_{13} - K_{23}$ is significantly greater than 0, implying that the rate in the OW monkey lineage is signicantly faster than that in the human lineage. The same conclusion is obtained from the flanking sequence data (Table 8.2). Thus, there is indeed evidence for the hominoid-slowdown hypothesis.

The intron sequence data suggest that the OW monkey lineage evolves 1.3 times faster than the human lineage, which is similar to that estimated (1.4) from the η-globin data. The flanking sequence data suggest that the rate ratio is larger than 2. However, since the latter data set is small, the ratio estimated from this set may not be reliable. Further data are needed to see whether the ratio varies among different DNA regions.

HIGHER RATES IN RODENTS THAN IN PRIMATES

From DNA hybridization data, Laird et al. (1969) and Kohne (1970) estimated the substitution rates between mouse and rat and between human and chimpanzee and concluded that the former rate is much higher than the latter. They attributed the higher rate in rodents to a shorter generation time, i.e., the generation-time effect. Sarich and Wilson (1973) argued that this difference in rate was based on questionable assumptions about the divergence times between species. In order to avoid controversies over the assumptions of divergence times, Wu and Li (1985) used artiodactyls and carnivores as outgroups to compare the rates of substitution in the rodent and primate lineages, but the validity of the outgroups was questioned (Easteal 1985).

To avoid the above criticism, Gu and Li (1992) used the chicken as an outgroup. They used amino acid sequences instead of DNA sequences because the chicken and mammalian lineages diverged about 300 million years ago, so that it is difficult to obtain reliable estimates of the divergences at synonymous sites. Let N_R be the number of residues at which the rodent and human sequences are different but the human and chicken sequences are identical and let N_H be the number of residues at which the human and rodent sequences are different but the rodent and chicken sequences are identical. For example, if the residues at a particular site are alanine in human and chicken but valine in rodent, then $N_R = 1$ and $N_H = 0$ for that site. However, if the residues at a site are different for the three sequences (e.g., alanine in human, valine in mouse, and leucine in chicken), then that site is noninformative and is not included in the comparison. For each protein N_R (N_H) is the sum of the N_R (N_H) values over the informative sites of the

TABLE 8.2 Differences in the number of nucleotide substitutions per 100 sites and the relative rates of substitution between the Old World monkey (species 1) and human (species 2) lineages, with the New World monkey (species 3) as a reference

Sequence	Nucleotides compared	K_{12}[a]	K_{13}[a]	K_{23}[a]	$K_{13} - K_{23}$	Rate ratio[b]
η-Globin pseudogene[c]	8,781	6.7	11.8	10.7	1.1 ± 0.3**	1.4
Introns						
IGF2	1,589	6.4	15.8	14.2	1.6 ± 0.8*	1.7
ε-globin	928	4.9	11.5	11.5	0.0 ± 0.8	1.0
Insulin	862	9.7	17.0	15.9	1.1 ± 1.3	1.3
Mast-cell carboxypeptide	1,275	5.5	13.3	12.5	0.8 ± 0.8	1.3
Carbonic anhydrase 7	501	7.2	11.1	9.7	1.5 ± 1.4	1.5
Interferon α receptor	885	7.6	14.0	14.0	0.0 ± 1.1	1.0
Apolipoprotein C3	1,270	8.7	18.5	16.9	1.6 ± 1.0	1.5
Lipoprotein lipase	1,168	7.9	13.6	13.8	−0.3 ± 1.0	1.0
Total	8,478	7.1	14.7	13.9	0.8 ± 0.3**	1.3
Flanking and untranslated regions						
ε-globin	388	5.3	13.5	10.6	2.9 ± 1.4*	3.4
Insulin	548	9.8	15.8	12.6	3.2 ± 1.5*	2.0
Total	936	7.9	14.9	11.7	3.1 ± 1.1**	2.3

Data from Bailey et al. (1991), Porter et al. (1995), and Ellsworth et al. (1993 and unpublished).

[a]K_{ij} = number of substitutions per 100 sites between species i and j.

[b]The ratio of the rate in the Old World monkey lineage to the rate in the human lineage.

[c]Excluding *Alu* sequences.

*Significant at the 5% level.

**Significant at the 1% level.

sequences. Under the assumption of equal substitution rates in the human and rodent lineages, N_R should be statistically equal to N_H for each protein. Among the 54 proteins compared, 35 proteins show a faster rate in the rodent lineage (i.e., $N_R > N_H$), 12 show a faster rate in the human lineage ($N_H > N_R$), and the rest (7 sequences) show an equal rate (Table 8.3). A sign test indicates a significantly faster rate in rodents than in humans ($P < 0.005$). When all the proteins are pooled together, $N_R = 600$ is clearly significantly larger than $N_H = 416$ ($P < 0.001$). Therefore, there is strong evidence for an overall faster substitution rate, from the common primate-rodent ancestor, in the lineage to mouse or rat (the rodent lineage) than in the lineage to humans (the human lineage).

Gu and Li also compared the frequencies of insertions and deletions (gaps) in the human and rodent sequences using the chicken sequences as references. They used two measures. One measure is simply the number of gaps in each sequence. This measure ignores the gap length. The other measure is the total number of amino acids involved in the gaps (i.e., the sum of gap lengths) in the sequence; the number in the rodent sequence is denoted by G_R and that in the human sequence by G_H. This measure is more reasonable because it gives higher weights to larger gaps. A binomial test was conducted for the gap data. Because the number of gap events in each sequence is small, no individual sequence shows a significant difference between the two lineages. When the data are pooled, in terms of the sum of gap lengths, $G_R = 385$ is significantly larger than $G_H = 108$ ($P < 0.001$), though in terms of the number of gaps, the difference (44 vs. 31) is insignificant ($0.05 < P < 0.1$; Table 8.3). As noted above, the sum of gap lengths is a more reasonable measure than the number of gaps. We may therefore conclude that the rate of insertion/deletion is higher in the rodent lineage than in the human lineage.

Comparing the rates of nucleotide substitution in the human and rodent lineages using marsupial genes as references, Easteal and Collet (1994) concluded that the rate of silent substitution did not differ between the rodent and human lineages, though the rate of nonsynonymous substitution was significantly higher in the rodent lineage than in the human lineage. They took the near equality of estimated silent rates in rodents and humans as evidence that the mutation rate is the same in rodents and primates, i.e., a molecular clock exists at silent

---

TABLE 8.3   Number of amino acid substitutions attributable to the rodent sequence (N_R) and to the human sequence (N_H), and the sum of lengths of gaps attributable to the rodent sequence (G_R) and to the human sequence (G_H)

No. of proteins[a]			Substitutions		Gaps	
$N_R > N_H$	$N_R < N_H$	$N_R = N_H$	N_R	N_H	G_R	G_H
35	12	7	600	416	385(44)[b]	108(31)
$P<0.005$			$P<0.001$		$P<0.001$	$(0.05<P<0.1)$

From Gu and Li (1992).
[a]Total number of proteins compared = 54; total number of residues compared = 16,365.
[b]The number in parentheses refers to the number of gaps.

TABLE 8.4 **Numbers of transition substitutions (A_i) and transversion substitutions (B_i) per 100 nondegenerate, twofold degenerate, and fourfold degenerate sites in comparisons between human, rodent, and marsupial genes**

Sites	Taxa compared		
	Human-rodent	**Human-marsupial**	**Rodent-marsupial**
Nondegenerate			
A_0	4.4 ± 0.3	7.3 ± 0.4	7.6 ± 0.4
B_0	4.3 ± 0.3	8.7 ± 0.4	9.7 ± 0.4
Twofold			
A_2	27.4 ± 1.6	48.9 ± 2.8	54.4 ± 3.2
B_2	5.2 ± 0.5	8.5 ± 0.7	9.3 ± 0.7
Fourfold			
A_4	25.8 ± 1.8	44.2 ± 3.6	40.6 ± 3.6
B_4	20.6 ± 1.4	45.7 ± 2.8	52.4 ± 3.2

Data from Easteal and Collet (1994).

sites. However, as silent distances between marsupial and rodent (or human) genes are large, estimates of these distances should be taken with caution. Table 8.4 shows a summary of their data for the coding regions. Easteal and Collet noted that the number of transition substitutions per 100 sites at twofold degenerate site (A_2) is similar to that at fourfold degenerate sites (A_4) for the human–rodent and human–marsupial comparisons (27.4 vs. 25.8 and 48.9 vs. 44.2) but is much higher than the latter in the rodent–marsupial comparison (54.4 vs. 40.6). For this reason, they omitted the data for A_2, arguing that these data were biased and not as suitable as those for A_4. However, note that for fourfold degenerate sites the number of transitions per site (A_4) is about the same as the number of transversions per site (B_4) for both the human–rodent comparison and the human–marsupial comparison but is much lower than the latter in the rodent–marsupial comparison. As seen in Chapter 7, A_4 tends to be higher than B_4 in mammalian genes. Thus, in the rodent–marsupial comparison, what is peculiar is A_4 rather than A_2, because A_2 is similar to B_4 whereas A_4 is much lower than B_4. Indeed, with the exception of A_4, all A_i and B_i for the rodent–marsupial comparison are larger than the corresponding values for the human–marsupial comparison, suggesting a higher rate in the rodent lineage than in the human lineage. Note that the number of transition substitutions per site (A) is estimated by the formula

$$A = -(1/2)\ln(1 - 2P - Q) + (1/4)\ln(1 - 2Q)$$

where P and Q are the proportions of transitional and transversional differences between the two sequences (Chapter 4; Kimura 1980; Wu and Li 1985). Therefore, it is easier to obtain a reliable estimate of A when Q is small, which is the case for twofold degenerate sites, than when Q is large, which is the case for fourfold degenerate sites. For this reason, A_2 is preferable over A_4 and, as noted above, the

A_2 values point to a higher synonymous rate in the rodent lineage than in the human lineage. Moreover, since the number of transversional substitutions (B) is given by the simple formula

$$B = -(1/2)\ell n \, (1 - 2Q)$$

(Kimura 1980; Wu and Li 1985), it is easier to estimate B than A. Let us therefore consider B_4. For B_4, the difference between the rodent–marsupial and human–marsupial comparisons is 6.7%, which is significant because the standard error is 2.7%. The above results are consistent with the view of a higher rate in the rodent lineage than in the human lineage for synonymous substitutions as well as for nonsynonymous substitutions.

Since at the early stage of the primate–rodent divergence the rates in the two lineages should have been very similar, the rate differences between the two lineages should have occurred mainly in more recent times. This appears to be the case. In order to estimate the substitution rate between two species we must know their divergence time. However, divergence times are usually not well established, so we consider a range of estimates and an intermediate date for every pair of species compared (Table 8.5). It is commonly thought (Gingerich 1984; Pilbeam 1984) that the divergence between the human and the Old World (OW) monkey lineages came after middle Oligocene times (some 30 Myr ago) and before early Miocene times (some 20 Myr ago), so we use a range of 20 to 30 Myr with an intermediate date of 25 Myr. The divergence time between the New World (NW) monkeys and the OW monkeys has been generally thought to be between 35 and 45 Myr ago (Gingerich 1984; Pilbeam 1984; Fleagle et al. 1986) and we use a range of 30 to 45 Myr with an intermediate of 35 Myr ago. The divergence time between mouse and rat has been estimated to be between 8 and 14 Myr ago (Jacobs and Pilbeam 1980), but Wilson et al. (1977) argued that it can be anywhere between 5 and 35 Myr ago. We use a range of 10 to 30 Myr with an intermediate date of 15 Myr. Table 8.5 shows the estimates of the substitution rate in introns. The intermediate estimates are 1.4×10^{-9}, 2.1×10^{-9} and 4.8×10^{-9} for the human–OW monkey, human–NW monkey, and mouse–rat comparisons, so the rate may be 2 to 4 times higher in rodents than in higher primates. However, these comparisons should be taken with caution because they involve assumptions of divergence times and are based on limited sequence data.

TABLE 8.5   Rates of substitution per site per year in introns

Species pair	L (bp)	Percent divergence	Time (10⁶ yrs)	Rate (10⁻⁹)
Human vs. OW monkeys	8,478	7.1	25 (20–30)	1.5 (1.2–1.8)
Human vs. NW monkeys	8,478	14.7	35 (30–45)	2.1 (1.6–2.5)
Mouse vs. rat	4,038	14.4	20 (10–30)	4.8 (2.4–7.2)

From Li, Ellsworth, and Krushkal (unpublished).

Notation: OW = Old World; NW = New World.

MALE-DRIVEN SEQUENCE EVOLUTION

Since Haldane (1935, 1947), it has been commonly believed that the mutation rate is much higher in the human male germ line than in the female germ line because the number of germ-cell divisions per generation is much larger in males than in females. However, direct estimation of mutation rates was difficult, and data supporting this view were limited, deriving mainly from sex-linked genetic diseases (see Vogel and Motulsky 1986). Thus, it was unclear how high the ratio (α) of male to female mutation rates is. Another problem with this type of data is that no distinction between indels (insertions and deletions) and substitution mutations was made because the molecular changes causing the disease were not examined. Here we are more interested in substitution mutations.

An alternative approach proposed by Miyata et al. (1987) provides a simple means for estimating α. We consider a slightly different formulation. Let u_m and u_f be the mutation rates per nucleotide site per generation in the male and female germ lines, respectively, and $\alpha = u_m/u_f$. Note that an autosomal sequence is derived from the father and the mother with equal probability. So, the mutation rate per generation for an autosomal sequence is $A = (u_m + u_f)/2$. For a Y-linked sequence the mutation rate is $Y = u_m$ because it must come from the father. For an X-linked sequence the mutation rate is $X = (1/3)u_m + (2/3)u_f$ because it is, on average, carried two-thirds of the time by the mother and only one-third of the time by the father. Therefore, the expected mutation ratio of the Y chromosome to autosomes is

$$Y/A = 2\alpha/(1+\alpha)$$

the mutation ratio of the X chromosome to autosomes is

$$X/A = (2/3)(2+\alpha)/(1+\alpha)$$

and the mutation ratio of the Y chromosome to the X chromosome is

$$Y/X = 3\alpha/(2+\alpha)$$

Miyata et al. (1987) noted that as α increases from 1 to infinity, Y/A increases from 1 to 2 while X/A decreases from 1 to $2/3 \approx 0.67$. From the sequences of a Y-linked and an autosomal pseudogene in humans they estimated $Y/A = 2.2$, and from the synonymous rates in X-linked and autosomal genes from humans and rodents they estimated $X/A = 0.60$. Since the two estimates exceed the upper and lower limits for the ratios, respectively, α is infinitely large. As this implies that the contribution of females to mutation is negligible, Miyata et al. concluded that the evolution of DNA sequences in mammals is male-driven.

This approach is ingenious, but the assumption that synonymous changes are selectively neutral may not hold. If synonymous changes are in fact subject to some, even very weak, selective constraints, then selection is more effective in reducing the rate in X-linked genes than in autosomal genes because in males an X-linked gene is present in a single copy and in females one of the two copies may be inactivated. A second problem with the data of Miyata et al. is that the genes they used were not homologous, but the mutation rates for nonhomologous sequences may be different, because the mutation rate depends on both the nucleotide composition (e.g., G and C are on average more mutable than A and T; see Chapter 1) and neighboring nucleotides (Bulmer 1986). The autosomal and Y-linked pseudogenes they used were indeed homologous and probably not subject to selection, but the data was limited. So, the reliability of their estimate was not certain. In fact, limited data (Lanfear and Holland 1991; Hayashida et al. 1992; Pamilo and Bianchi 1993) from the coding regions of mammalian X-linked and Y-linked zinc finger protein (*Zfx* and *Zfy*) genes suggested that the rate at synonymous sites is only two times higher in *Zfy* than in *Zfx*, instead of the predicted ratio $Y/X = 3$ for large α.

To obtain a better estimate Shimmin et al. (1993) sequenced the last intron (~1 kb) of the human, orangutan, baboon and squirrel monkey *Zfx* and *Zfy* genes, which are highly homologous. Table 8.6 shows the corrected percent divergence between sequences. For each species pair, the Y sequences are more divergent than the X sequences. For example, for the human-orangutan pair the divergence is only 2.1% between the two X sequences but 4.0% between the two Y sequences. As a more detailed analysis, they computed the length of each branch of the four-species tree (Figure 8.2). Clearly, in each branch the Y sequence has evolved faster than the X sequence, though the ratio Y/X varies considerably among branches, probably largely because the number of substitutions in each branch is relatively small. When all branches are considered together, they obtained $Y/X = (14.95 + 2.95 + 1.15 + 1.75 + 2.25)/(5.95 + 1.75 + 0.45 + 1.10 + 1.00) = 2.25$.

The above Y/X ratio leads to an estimate of $\alpha \approx 6$ from the relation $Y/X = 3\alpha/(\alpha + 2)$. This indicates that in higher primates the mutation rate is considerably higher in the male germ line than in the female germ line. However, the α value is far lower than suggested by Miyata et al. (1987). The study by Ketterling et al. (1993) on the germ line origins of mutation in families with hemophilia B (which is caused by defects in the gene for clotting factor IX) gave an estimate of $\alpha \approx 3.5$. Although these two estimates have large standard errors, they strongly suggest that α is, at most, of the order of 10.

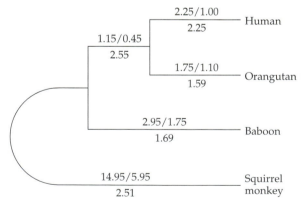

Figure 8.2. The numbers above each branch are branch lengths estimated from the Y-linked and the X-linked intron sequences (Table 8.6), respectively, and the number below each branch is the ratio of the preceding two numbers. The least-squares method was used for the estimation. From Shimmin et al. (1993).

TABLE 8.6 Number of nucleotide substitutions per 100 sites between intron sequences

Sequences	X intron				Y intron		
	Human	*Orangutan*	*Baboon*	*Squirrel monkey*	*Human*	*Orangutan*	*Baboon*
Orangutan X	2.1 ± 0.6						
Baboon X	3.0 ± 0.7	3.5 ± 0.8					
Squirrel monkey X	7.6 ± 1.2	7.3 ± 1.1	7.7 ± 1.2				
Human Y	30.6 ± 2.8	31.1 ± 2.8	31.1 ± 2.8	32.2 ± 2.9			
Orangutan Y	31.6 ± 2.9	32.6 ± 2.9	32.1 ± 2.9	33.3 ± 3.0	4.0 ± 0.8		
Baboon Y	30.2 ± 2.8	31.2 ± 2.9	30.7 ± 2.8	31.6 ± 2.9	6.3 ± 1.1	5.9 ± 1.0	
Squirrel monkey Y	32.4 ± 2.9	34.0 ± 3.0	33.2 ± 2.9	32.8 ± 2.9	18.4 ± 2.0	17.8 ± 1.9	17.9 ± 2.0

From Shimmin et al. (1993).

The mean and standard error are estimated by Tajima and Nei's (1984) method, which takes into account unequal frequencies of the four nucleotides. Gaps are not included in the comparison.

TABLE 8.7   Number of nucleotide substitutions per site between intron sequences

| Sequence | X intron | Y intron | |
	Mouse Zfx	Mouse Zfy1	Mouse Zfy2
Rat *Zfx*	0.128 ± 0.015		
Mouse *Zfy2*		0.071 ± 0.010	
Rat *Zfy*		0.172 ± 0.016	0.192 ± 0.017

From Chang et al. (1994).

Gaps are not included in the comparison.

As in the case of synonymous changes, introns may be subject to selective constraints and this can distort the rate ratio (see Charlesworth et al. 1987, 1994 for a theoretical study of the ratio of the substitution rates in Y- and X-linked genes under selection). In the present case, however, selection is unlikely to be more effective against the Y intron than against the X intron, because *Zfy* and *Zfx* are expressed in males whereas only *Zfx* is expressed in females, so that *Zfy* may be dispensable whereas *Zfx* may not.

Since the above α value was estimated from primate sequences, it may not be applicable to other mammals. In particular, the generation times in rodents are much shorter than those in higher primates, so that the number of germ-cell divisions may not be much higher in the male germ line than in the female germ line. To see the effect of generation time on α, Chang et al. (1994) sequenced the last introns of *Zfy* and *Zfx* genes in mouse and rat. There are two *Zfy* genes in the mouse, though there is only one *Zfy* gene in the rat. Since the distance between the two mouse *Zfy* introns is significantly shorter than that between either of them is to the rat *Zfy* intron (Table 8.7), the two mouse genes are probably derived from a duplication after the mouse–rat split. The average number of substitutions between the rat intron and the two mouse introns is $(0.172 + 0.192)/2 = 0.182$ and so $Y/X = 0.182/0.128 = 1.42$. From the equation $Y/X = 3\alpha/(\alpha + 2)$, we obtain $\alpha = 1.8$. In addition, Chang and Li (1995) sequenced the Y- and X-linked *Ube*1 genes and pseudogenes and estimated $\alpha = 2.0$. These estimates suggest that the α value (~2) in mice and rats is considerably lower than that in higher primates. So, there appear to be generation-time effects on α.

CAUSES OF RATE VARIATION AMONG LINEAGES

Three factors have been proposed as the major causes of rate variation among evolutionary lineages. First, the efficiency of the DNA repair system may differ among lineages. This hypothesis was proposed by Britten (1986) to explain the difference in substitution rate between the primate and rodent lineages. There is evidence from cultured cells for supporting this hypothesis. Unfortunately, no *in vivo* data seem to exist and it is difficult to evaluate the importance of this factor.

Second, the generation-time effect hypothesis postulates a higher rate of evolution in organisms with a short generation time than in organisms with a long generation time, because in any arbitrary unit of time short-generation organisms will go through more generations and therefore more rounds of germ-cell divisions (DNA replications) (Laird et al. 1969; Kohne 1970). As noted earlier, the rate of substitution is higher in monkeys than in humans and is even higher in rodents. These observations are consistent with the generation-time effect hypothesis because among these three groups of organisms, the generation time is longest in humans, intermediate in monkeys, and shortest in rodents. The hypothesis is further supported by the observation that the rate of substitution in the chloroplast genome is more than five times higher in grasses than in palms (Gaut et al. 1992).

The generation-time effect hypothesis implicitly assumes that errors during DNA replication are the major source of mutation. This assumption can be tested by comparing the sex-ratio of mutation rate (α) and the sex-ratio of the number of germ-cell divisions (c). Using published data on gametogenesis, Li (unpublished) has estimated the numbers of germ-cell divisions per generation in females (n_f) and males (n_m) and the ratio $c = n_m/n_f$. In the development of a female mouse, the number of DNA replications in the germ line (i.e., from zygote to mature egg) is $n_f \approx 27$. In the development of male mouse, the number of germ-cell divisions from zygote up to the formation of stem spermatogonia is ≈ 29. Spermatogenesis requires 10 further DNA replications to the production of spermatids, initiating on day 6 after birth and occurring on average every 8.6 days in adult male mice. The span of high reproductivity in laboratory male mice is approximately from 2 to 8 months. If we assume that the average reproductive age of male mice in the wild is 5 months, then a stem spermatognia would have gone through $(5 \times 30)/8.6 \approx 17$ divisions and the total number of DNA replications from zygote to age 5 months is $n_m \approx 29 + 10 + 17 = 56$. Therefore, $c \approx 56/27 = 2.1$. If the average reproductive age is 2 or 8 months, c becomes 1.6 and 2.4, respectively.

For the rat, $n_f \approx 28$. The number of cell divisions in the male germ line to form stem spermatogonia is ≈ 31, the spermatogenesis cycle occurs every ~12.9 days, and the period of maximal fertility of laboratory rats occurs between 100 and 300 days after birth. If we assume that the average reproductive age of males in the wild is 7 months, a stem spermatogonium would have gone through $(7 \times 30)/12.9 = 16$ divisions, and the total number of DNA replications in the male germ line is $n_m \approx 31 + 10 + 16 = 57$ and $c \approx 57/28 = 2.0$.

In humans, $n_f \approx 33$. The data for n_m are more scanty than those for rodents and so the estimate of n_m is less reliable. The number of cell divisions from zygote to stem spermatogonia at puberty was estimated to be ~40. Spermatogenesis requires 5 further DNA replications, and the spermatogenesis cycle occurs every ~16 days or 23 cycles per year. If the average reproductive age of males is 20 years, then the number of DNA replications for stem spermatogonia from puberty (age 13) to 20 is $(20 - 13) \times 23 \approx 160$, $n_m \approx 40 + 5 + 160 = 205$, and $c \approx 6.2$. If the average reproductive age of males is 15, then $c \approx 2.8$.

Although these estimates are rough, they are similar to the estimates of $\alpha = 2$ in mice and rats and $\alpha = 3$ to 6 in higher primates. This rough agreement suggests that errors in DNA replication during germ-cell divisions are indeed the primary source of mutation and that the contribution of replication-independent factors such as oxygen radicals to mutation is considerably less important.

Third, the metabolic-rate hypothesis postulates a higher rate of evolution for organisms with a high metabolic rate than for organisms with a low metabolic rate (Martin and Palumbi 1993). Martin and Palumbi (1993) found a negative correlation between adult body size and substitution rate in the η-globin pseudogene in higher primates and in the cytochrome b gene in mammals. They also reviewed published genetic distances obtained from restriction analysis of mtDNA and fossil dates for various vertebrates and found that the same general trend holds (Figure 8.3). As body size is unlikely to control the rate of DNA evolution, its relationship with substitution rate probably comes from its correlation with generation time and metabolic rate. Although generation time and metabolic rate can be used to explain some patterns of rate heterogeneity equally well, Martin and Palumbi (1993) noted that differences in metabolic rate could also explain some important exceptions to the generation-time model. For example, whales have a slow substitution rate relative to primates, despite their shorter generation times (Schlotterer et al. 1991; Baker et al. 1993). They noted also that the separation of mtDNA rates into two broad groups comprising the homeotherms and the poikilotherms (Figure 8.3) is incompatible with a simple generation time hypothesis. There are, however, exceptions to the metabolic rate hypothesis. For example, birds appear to have slower rates than do mammals (Kessler and Avise 1985; Shields and Wilson 1987), though birds have higher metabolic rates than do mammals, and the substitution rates are about the same in rats and mice, despite the fact that the body size of rats is more than twice that of mice.

The metabolic rate hypothesis is based on the argument that the rate of DNA damage is proportional to the metabolic rate because metabolism produces oxygen radicals, which are highly reactive molecules with free electrons that can damage DNA directly by attacking the sugar-phosphate backbone or the base, or indirectly via lipid peroxidates. This hypothesis may hold for mtDNA because

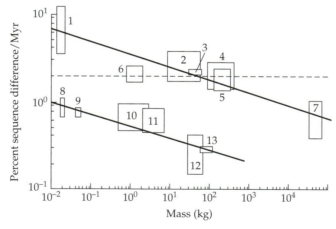

Figure 8.3 Relationship between rate of mtDNA sequence divergence (% change per million years) and body size (in kg) for various vertebrates. Data points: 1, mice; 2, dogs; 3, human-chimpanzee; 4, horses; 5, bears; 6, geese; 7, whales; 8, newts; 9, frogs; 10, tortoise; 11, salmon; 12, sea turtles; 13, sharks. Boxes represent the range of rates and body sizes for a given taxon. Solid lines are drawn to pass through the boxes. Dashed line represents the hypothesis of rate constancy. From Martin and Palumbi (1993).

about 95% of oxygen in cells is consumed by mitochondria, so that oxidative damage is greatest to mtDNA (Richter et al. 1988); this may partially explain why rates of nucleotide substitution are greater in mtDNA than in nuclear DNA (Martin and Palumbi 1993). However, whether it holds for nuclear DNA remains to be seen because of the low consumption of oxygen in nucleus. The nuclear DNA sequence data used to support the metabolic rate hypothesis were limited to the η–globin pseudogene, and this analysis involved assumptions about generation times and body sizes. Relative-rate tests on DNA–DNA hybridization data from birds (Mooers and Harvey 1994) did not support the metabolic rate hypothesis but showed generation-time effects.

It should be emphasized that the above three alternative hypotheses are not mutually exclusive and that each single hypothesis may not explain all observations. Rather, the three factors may all contribute to rate variation among lineages. It is their relative importance that is uncertain and requires more research.

OVERDISPERSED CLOCKS

Zuckerkandl and Pauling (1965) used the Poisson distribution to study the number of amino acid substitutions in a sequence. If the substitution process follows the Poisson process with a constant rate of λ substitutions per unit time, then both the expected number of substitutions in a time interval t and the variance are λt. Thus, under this simple model the variance/mean ratio, R, is 1. If the substitution process deviates from the Poisson model, R is expected to be larger than 1. Using hemoglobin and cytochrome c sequences from mammals, Ohta and Kimura (1971) found that R was approximately 2. Thus, the substitution process did not seem to follow the Poisson process with a constant rate. As R is often called the dispersion index, the clock is said to be overdispersed if R is larger than 1. Because under strictly neutral mutations, the substitution process follows the Poisson process with the substitution rate λ equal to the mutation rate u, a significant deviation of R from 1 has usually been taken as evidence against the rate-constancy assumption and the neutral mutation hypothesis. However, as will be discussed below, there are factors that can inflate the R value.

Kimura (1983) noted that a star phylogeny in which all lineages diverged at the same time is an ideal situation for studying the variability of the molecular clock, i.e., for testing $R = 1$, because under the assumption of rate constancy the number of amino acid (or nucleotide) substitutions is expected to be the same in all lineages and the variance of the number of substitutions among lineages can be easily computed. Kimura (1983) analyzed the α-globin sequences from human, mouse, rabbit, dog, horse, and bovine, which were assumed to conform to a star phylogeny, and concluded that the molecular clock hypothesis could not be rejected from these data because $R = 1.26$ was not significantly larger than 1.

In a series of papers Gillespie (1986a,b; 1987) developed a more sophisticated theory for studying R and applied it to analyze protein and DNA sequence data. He found that R is roughly in the range from 2.5 to 3.5, so that the clock appears to be overdispersed. However, the results may not be taken as evidence against the neutral mutation hypothesis for two reasons. First, the assumption of a star phylogeny may not hold, i.e., some lineages may have in fact evolved for a longer time than others. Second, the mutation rate or the generation time may not have remained the same in all lineages, i.e., there may be generation-time effects. Gillespie (1989) called these two effects collectively the lineage effect and proposed to take this effect into account by using different weights for different lineages. In the case of three lineages, the situation is quite simple. Let N_i be the number of substitutions in the ith lineage. The weight w_i for the ith lineage is chosen so that

$$E(N_1/w_1) = E(N_2/w_2) = E(N_3/w_3)$$

with the following constraint

$$w_1 + w_2 + w_3 = 3$$

In this model one can write

$$E(N_i) = w_i \mu; V(N_i) = w_i \sigma^2$$

and show that

$$R = V(N_i)/E(N_i) = \sigma^2/\mu$$

which is equivalent to the ideal case of a star phylogeny and is expected to be 1 if the substitution process in each lineage is a Poisson process with a constant rate.

Gillespie (1989) analyzed the coding sequences of 20 genes from human, mouse (or rat), and bovine. He treated synonymous and nonsynonymous substitutions separately and considered three different weights: (1) equal weights, (2) the weights obtained from nonsynonymous substitutions, and (3) the weights obtained from synonymous substitutions. In Table 8.8, the R values corresponding to the three weights are denoted by R_e, R_n, and R_s, respectively. Let us first consider nonsynonymous substitutions. For the equal-weight case, $w_p = w_r = w_a = 1$, where the subscripts p, r, and a refer to the primate, rodent, and artiodactyl lineages, respectively, and the R value varies from 0.01 to 32.22, with an average of 8.26. For the nonsynonymous weights, $w_p = 0.84$, $w_r = 1.28$, and $w_a = 0.89$ and the average R value is indeed reduced, but only from 8.26 to 6.95. For synonymous substitutions, R varies from 1.00 to 57.45 with an average of 14.41 for the equal-weight case but the average reduces to 8.67 for the synonymous-weight case ($w_p = 0.63$, $w_r = 1.61$, and $w_a = 0.76$).

All the above analyses neglected two effects: (1) the effect of corrections for multiple substitutions at the same site and (2) the correlation between comparisons, e.g., in the comparison of the two pairs (A, B) and (A, C), lineage A is used twice. To take the two effects into consideration, Bulmer (1989) derived a correction factor for the Jukes-Cantor model and found that it becomes serious when the degree of sequence divergence is 25% or higher. Gillespie (1989) conducted a simulation study and found that in the case of nonsynonymous substitution the correction factor was indeed not important and, as the above ratio of 6.95 remained about the same, he took it as evidence against the neutral mutation hypothesis. On the other hand, in the case of synonymous substitution, the correction factor was quite large, so that the above ratio of 8.67 could not be taken as evidence against the neutral mutation hypothesis.

TABLE 8.8 Estimates of *R* for nonsynonymous and synonymous substitutions at 20 loci

Locus	Nonsynonymous[a]		Synonymous[a]	
	R_e	R_n	R_e	R_s
Prolactin	26.94	12.21**	10.33	1.02
Parathyroid	8.03	3.47*	16.96	4.58
Proenkephalin	8.57	3.73*	30.34	9.05*
Proglucagon	1.58	2.13	16.99	9.39*
α-globin	0.23	0.32	18.21	4.68
β-globin	7.30	3.41*	4.72	0.73
Thyrotropin B	2.80	2.26	6.77	4.89
POMC	15.10	6.44**	9.84	1.46
Growth hormone	32.22	43.82**	3.45	17.10**
GPHA	22.43	27.74**	1.00	2.52
Luteinizing B	3.21	6.55**	11.74	2.16
Relaxin	3.38	0.13	13.08	0.28
Interleukin-2	12.80	0.85**	57.45	17.19**
Signal peptides	9.98	3.25*	18.11	7.52*
CCK	2.36	0.31	5.00	1.20
ACHRG	1.52	0.61	13.40	3.71
UPA	0.01	5.11**	21.38	0.25
ANF	1.94	1.64	10.05	2.17
Beta crystallin	1.38	2.50	5.40	0.36
Na,K-ATPase	3.49	4.46*	13.95	2.48
Average	8.26	6.95	14.41	4.64

From Gillespie (1989).

[a]Subscripts of *R*: e, n, and s refer to the use of equal weights, nonsynonymous weights, and synonymous weights, respectively.

*Significant at the 5% level.

**Significant at the 1% level.

Goldman (1994) used computer simulation to study Kimura's (1983) and Bulmer's (1989) estimators and concluded that *R* may be a poor statistic for analyzing the accuracy of Poisson models. In particular, he argued that even for the cases where *R* was found by Gillespie (1989) to deviate far beyond 1, the *R* value became nonsignificant if the assumption of a star phylogeny was removed but a new phylogeny was inferred by the maximum likelihood method with a constant rate. However, assuming a constant rate of evolution is equivalent to assuming that the root of the tree is on the longest branch. In other words, it presumes that the longest branch is not due to a faster rate of evolution but due to an earlier divergence and a longer time of evolution. Under this assumption, one may not be able to reject the molecular hypothesis, even if it actually does not hold.

In Table 8.8, in the case of nonsynonymous substitution, the largest deviations from 1 occurred for prolactin, growth hormone, and GPHA. When these three genes are removed from comparison, the average *R* value reduces to 2.77 and the deviation of *R* from 1 becomes considerably smaller. Ohta (1995) extended the analysis to include 49 genes, including prolactin, growth hormone and GPHA, and found an average R value of 5.60, which is somewhat smaller than that (6.95) for the 20 genes in Table 8.8.

The above results suggest that while the variation in the synonymous clock among lineages is mainly due to lineage effects, the variation in the nonsynonymous clock is subject to both lineage effects and episodic effects, the former being systematic whereas the latter are erratic. It has been controversial as to how the overdispersed clock for nonsynonymous substitution arose (see Gillespie 1988; Takahata 1988, 1991). Ohta (1995) thinks that the overdispersion is caused by the interplay between fluctuation of population size and weak selection. She argues that the population size of a species fluctuates in time and the effectiveness of natural selection varies accordingly. Once in a while, such as at the time of speciation, the population size becomes small and the effect of random drift predominates, so that slightly deleterious mutations may become fixed in the population. Later, mutations that compensate the fixed, slightly deleterious ones may occur and become fixed in the population. Thus, fluctuations in population size and the prevalence of slightly deleterious mutation cause irregularities in the clock. On the other hand, Gillespie (1989) thinks that the population size of a species rarely becomes small enough for the random fixation of slightly deleterious mutations to occur and that the episodic effects are mainly caused by environmental changes, so that different mutations are favored at different times—since environmental changes are episodic, so are events of nucleotide substitution. As both views involve many assumptions, the controversy is not resolved.

RATIOS OF SYNONYMOUS AND NONSYNONYMOUS RATES IN DIFFERENT LINEAGES

A quantity that is considerably easier to compute than the dispersion index is the ratio of the synonymous rate to the nonsynonymous rate. Ohta (1993, 1995) proposed to use this ratio to test her nearly neutral mutation hypothesis. She considered this ratio in the primate, artiodactyl, and rodent lineages. As noted above, rodents have a shorter generation time and are expected to have a faster molecular clock than do the other two lineages. However, since selection against slightly deleterious mutations should be more effective in rodents than in primates and artiodactyls because rodents have a larger population size and since nonsynonymous changes are more likely to be exposed to selection than synonymous changes, the generation-time effect is expected to be stronger for the synonymous rate than for the nonsynonymous rate. In other words, under the nearly neutral mutation hypothesis the ratio (r) of the synonymous rate to the nonsynonymous rate is expected to be larger in the rodent lineage than in the other two lineages.

To test the above prediction Ohta (1993, 1995) analyzed 49 genes. As seen from Figure 8.4, the ratio r is $0.137/0.037 = 3.7$ in the primate lineage, $0.184/0.047 = 3.9$ in the artiodactyl lineage, and $0.355/0.62 = 5.7$ in the rodent lineage. Clearly, the ratio is highest in the rodent lineage and Ohta's prediction appears to hold.

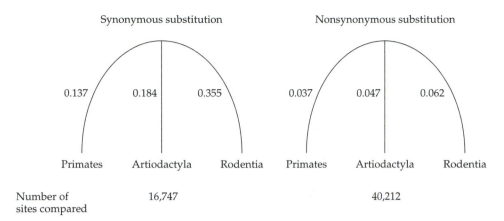

Figure 8.4 Numbers of nucleotide substitutions per site on the primate, artio-dactyl, and rodent lineages. From Ohta (1995).

CHAPTER 9

DNA Polymorphism in Populations

T HE EXTENT AND THE MECHANISM OF MAINTENANCE of genetic variability in natural populations have always been a subject of great interest to population geneticists and evolutionists. As the debate over the evolutionary forces governing protein polymorphism was inconclusive (see Introduction and Chapter 2), it has moved to the DNA level. DNA sequence data represent the highest level of genetic resolution and allow the development of more powerful statistical approaches than do protein polymorphism data. For example, the theory of gene genealogy and coalescence has been developed mainly for analyzing DNA polymorphism data. Although DNA technologies have not yet been extensively used in population genetics studies, available data have already provided considerable insight into the maintenance of DNA variation. In this chapter we review both experimental data and theoretical developments. For a more extensive treatment on this topic, readers may consult Avise (1994) and Golding (1994).

MEASURES OF DNA POLYMORPHISM

In Chapter 2 we considered gene diversity as a measure of genetic variability in a population. The gene diversity at a locus (h) is the probability that two sequences randomly chosen from the population are different; for a randomly mating population, h is also the heterozygosity. Although this measure has been commonly used for studying protein polymorphism, particularly for electrophoretic data, it may not be suitable for DNA sequence data, because the extent of genetic variation at the DNA level in nature is quite extensive. Indeed, when long sequences are considered, each sequence in the sample may be different from the other sequences, so that h is close to 1. When this is the case, h cannot discriminate among different loci or populations and is therefore no longer an informative measure of polymorphism.

For a pair of DNA sequences, it is more informative to consider the number of nucleotide differences between the two sequences rather than just whether they are different. Thus, for DNA sequence data, a more appropriate measure of polymorphism in a population is Π, the average number of nucleotide differences between two sequences randomly chosen from the population. When n sequences are taken randomly from the population, Π is estimated by

$$\Pi = \frac{1}{[n(n-1)/2]} \sum_{i<j} \Pi_{ij} \tag{9.1}$$

where Π_{ij} is the number of nucleotide differences between the ith and jth sequences and $n(n-1)/2$ is the number of possible pairs. Alternatively, we may first compute the frequencies of different alleles and estimate Π by

$$\Pi = \frac{n}{n-1} \sum_{i,j} x_i x_j \Pi_{ij} \tag{9.2}$$

where x_i and x_j are the frequencies of the ith and jth alleles in the sample, respectively, and $n/(n-1)$ is the correction factor for sampling bias. It can be shown that Equations 9.1 and 9.2 give the same result.

A mutation model that is commonly used to study Π is the **infinite-site** model, which assumes that the number of nucleotide sites on the sequence is so large that each new mutation occurs at a site that has not been mutated before (Kimura 1969). Under this model and under the assumption of random mating, Watterson (1975) showed that the mean of Π is given by

$$E(\Pi) = \theta \tag{9.3}$$

and Tajima (1983) showed that the variance of Π is given by

$$V(\Pi) = \frac{n+1}{3(n-1)}\theta + \frac{2(n^2+n+3)}{9n(n-1)}\theta^2 \tag{9.4}$$

where $\theta = 4N_e u$, N_e is the effective size of a diploid population, and u is the mutation rate per sequence per generation; Equation 9.4 assumes no recombination between sequences. Note that the mean is independent of the sample size n, whereas as n increases to infinity, the variance decreases to $\theta/3 + 2\theta^2/9$, which is the component of the variance that is due to random drift.

The Π value is likely to be larger for a long sequence than for a short one. In fact, Π increases with L because $u = Lv$ (and hence $\theta = 4N_e u$) increases with L, where v is the rate of mutation per site per generation and L is the length of the sequence (i.e., the number of nucleotide sites on the sequence). For this reason, Π is not convenient for making comparisons of DNA sequence variation among loci, because L is likely to vary among loci. This drawback, however, can be easily removed by standardizing Π by L. That is, we consider

$$\pi = \Pi/L \tag{9.5}$$

Nei and Li (1979) called π the **nucleotide diversity**, which is the number of nucleotide differences per site between two randomly chosen sequences. From Equation 9.5, one can easily show that $V(\pi) = V(\Pi)/L^2$.

Another commonly used measure of DNA polymorphism is the number of segregating nucleotide sites (K) in a sample of sequences; a **segregating site** is a site that shows variation among the sequences in the sample. Under the infinite-site model and under random mating, Watterson (1975) showed that the mean and variance of K are given by

$$E(K) = a\theta \tag{9.6}$$

$$V(K) = a\theta + b\theta^2 \tag{9.7}$$

where

$$a = 1 + \frac{1}{2} + \cdots + \frac{1}{n-1} \tag{9.8}$$

$$b = 1 + \frac{1}{2^2} + \cdots + \frac{1}{(n-1)^2} \tag{9.9}$$

Obviously, K depends on the sequence length L but this dependence can be removed by considering $k = K/L$. The advantage of k over π is that it has a smaller stochastic variance, but a drawback is that k depends on the sample size n.

A third measure of DNA polymorphism is the number of alleles (i.e., different sequences) in the sample (n_a). The mutation model commonly used to study this quantity is the **infinite-allele** model, which assumes that each mutation creates a new allele not currently present in the population. Under this mutation model and under random mating, Ewens (1972) showed that the mean of n_a is given by

$$E(n_a) = \theta \left(1 + \frac{1}{\theta+1} + \frac{1}{\theta+2} + \cdots + \frac{1}{\theta+n+1} \right) \tag{9.10}$$

Since n_a is dependent on both the sample size and the sequence length, it is less convenient than π for making comparisons among loci or populations.

Note that all the above formulas for the means and variances are derived under the assumptions that mutations are neutral and that the population is at equilibrium.

Let us now consider how to compute π, K, and n_a from sequence data by using the 11 sequences of the alcohol dehydrogenase (*Adh*) gene in *Drosophila melanogaster* obtained by Kreitman (1983). The aligned sequences were 2379 nucleotides long, i.e., $L = 2379$. Figure 9.1 shows the sites where nucleotide differences were observed among the 11 sequences. Since there were 43 such sites (Figure 9.1), $K = 43$; deletions and insertions were disregarded. It is easy to see that there are nine different alleles, i.e., $n_a = 9$. One of the alleles is represented by three sequences, while the rest are each represented by one sequence. Thus, the frequencies $x_1, \ldots, x_8$ are each 1/11 and the frequency x_9 is 3/11. We now calculate the proportion of different nucleotides for each pair of alleles. For example, alleles *1-S* and *2-S* differ from each other by 3 nucleotides out of 2379 or $\pi_{12} = 3/2379 = 0.13\%$. The π_{ij} values for all the pairs in the sample are listed in Table 9.1. By using Equation 9.2 with Π_{ij} replaced by π_{ij}, the nucleotide diversity is estimated to be $\pi = 0.0065$, so that we expect two randomly chosen sequences of 1000 base pairs to differ, on average, at 6.5 sites. Six of the alleles studied were slow-migrating electrophoretic variants (*S*), and five were the fast (*F*) type; *S* and *F* differ by only one amino acid. When the nucleotide diversity for each of the two electrophoretic classes is calculated separately, we obtain $\pi = 0.0056$ for the *S* sequences and $\pi = 0.0029$ for the *F* sequences. The twofold difference in π suggests that the *S*

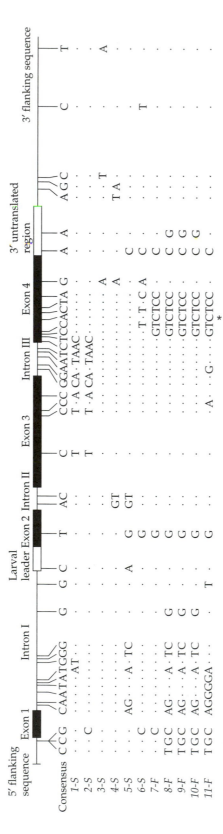

Figure 9.1   Polymorphic nucleotide sites among 11 sequences of the alcohol dehydrogenase gene in *Drosophila melanogaster*. Only differences from the consensus sequence are shown. Dots indicate identity with the consensus sequence. The asterisk in exon 4 indicates the site of the lysine for threonine substitution that is responsible for the fast/slow mobility differences between the two electrophoretic alleles. From Li and Graur (1991), which was modified from Hartl and Clark (1989).

TABLE 9.1 Pairwise percent nucleotide differences (per 100 sites) among 11 sequences of the alcohol dehydrogenase locus in *Drosophila melanogaster*[a]

Sequence	1-S	2-S	3-S	4-S	5-S	6-S	7-F	8-F	9-F	10-F
1-S										
2-S	0.13									
3-S	0.59	0.55								
4-S	0.67	0.63	0.25							
5-S	0.80	0.84	0.55	0.46						
6-S	0.80	0.67	0.38	0.46	0.59					
7-F	0.84	0.71	0.50	0.59	0.63	0.21				
8-F	1.13	1.10	0.88	0.97	0.59	0.59	0.38			
9-F	1.13	1.10	0.88	0.97	0.59	0.59	0.38	0.00		
10-F	1.13	1.10	0.88	0.97	0.59	0.59	0.38	0.00	0.00	
11-F	1.22	1.18	0.97	1.05	0.84	0.67	0.46	0.42	0.42	0.42

From Nei (1987). Data from Kreitman (1983).

[a]Total number of compared sites is 2379. *S* and *F* denote the slow and fast migrating electrophoretic alleles, respectively.

electrophoretic variant is considerably older than the *F* variant, so that the *S*-type sequences have on average diverged more at the DNA level.

At the present time, DNA polymorphism data remain limited, even for humans and *Drosophila*; *Drosophila* data will be discussed later in this chapter. Table 9.2 shows the data from humans analyzed by Li and Sadler (1991), who used published (genomic or cDNA) sequence pairs (for 49 genes) that have been verified for sequencing accuracy. The data were mainly from white Americans and so the nucleotide diversity shown in Table 9.2 may be considered as the average value for the white American population. The average level of nucleotide diversity (π) is low, on average lower than 1 nucleotide difference per 1000 sites between two random sequences. The π value for Y-linked sequences might be even lower, because a world-wide sample of 38 sequences from the last intron (~739 bp) of the *ZFY* gene revealed no variation (Dorit et al. 1995).

For the regions studied in Table 9.2, the highest one is observed at the fourfold degenerate sites and is only 0.11%. The second highest level is observed at the twofold degenerate sites; the synonymous and nonsynonymous components are 0.05% and 0.01%, respectively, and the total diversity is 0.06%. The third highest level, $\pi = 0.04\%$, is in the 3' untranslated (UT) region, though it is not significantly higher than the values in the 5' UT regions and at the nondegenerate sites, both being 0.03%. These values are well correlated with the selective constraints in different regions of the gene; e.g., the nondegenerate sites and the fourfold degenerate sites are, respectively, subject to the most and least stringent constraints and have, respectively, the lowest and the highest level of nucleotide diversity.

TABLE 9.2 Nucleotide diversity (π) in humans

Noncoding regions		Coding regions			
			Twofold degenerate		
5' UT[a]	3' UT[a]	Nondegenerate	Nonsynonymous	Synonymous	Fourfold degenerate
3,624[b]	19,769[b]	34,869[b]	10,787[b]	10,787[b]	8,537[b]
0.0003 ± 0.0003	0.0004 ± 0.0001	0.0003 ± 0.0001	0.0001 ± 0.0001	0.0005 ± 0.0002	0.0011 ± 0.0004

From Li and Sadler (1991).
[a]UT, untranslated.
[b]Total number of nucleotides compared.

GENE GENEALOGY AND THE COALESCENT THEORY

This section reviews basic concepts of gene genealogy and the coalescent theory, which has proved to be a powerful approach to the statistical analysis of DNA polymorphism data.

Consider a sample of n sequences of a DNA region from a population. If the DNA region is completely linked so that no recombination occurs between sequences, then the n sequences are connected by a single phylogenetic tree, i.e., a genealogy (Figure 9.2). In the genealogy, the n sequences can be traced back first to $n - 1$ ancestral sequences, next to $n - 2$ ancestral sequences, and so on until finally to a single common ancestral sequence. Using a different terminology, we can say that the n sequences coalesce first to $n - 1$ ancestral sequences, then to $n - 2$ ancestral sequences, and so on until finally to a single common ancestral sequence. The mathematical theory that treats the genealogical branching process backward in time as a **coalescence process** is called the **coalescent theory**. This powerful theory for studying various quantities pertaining to gene genealogies was initiated by Kingman (1982), Hudson (1982), and Tajima (1983). In the following we shall consider a diploid random mating population in the absence of natural selection and migration and for simplicity, we assume that the effective size N_e is equal to the actual population size N. For more complex situations involving selection, recombination or migration, readers may refer to Kaplan et al. (1988, 1991), Takahata and Nei (1990), and Hudson (1993).

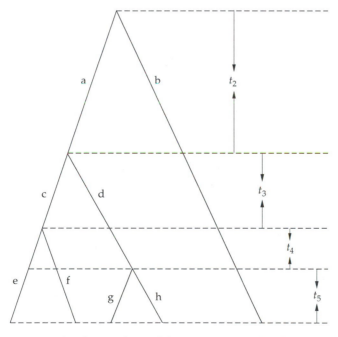

Figure 9.2 An example of genealogy of five sequences with the root of the genealogy at the top. t_m ($m = 2, \ldots, 5$) is the time (number of generations) required for the coalescence from m to $m - 1$ sequences. The dashed lines signify the partition of the genealogy into $m - 1$ parts by the m branching nodes. From Fu and Li (1993a).

An essential quantity in the coalecent theory is the time duration (t_m) required for the coalescence from m sequences to $m - 1$ sequences. Let us start with the simplest case: two sequences randomly drawn from the population, i.e., $m = 2$. The probability that the two sequences are derived from a common ancestral sequence in the preceding generation is $1/(2N)$ because there are $2N$ sequences in the population and because sampling is with replacement, i.e., the same gene can be repeatedly sampled. Obviously, the probability that they are derived from two different sequences in the preceding generation is $1 - 1/(2N)$. Therefore, the probability that the two sequences are derived from a common ancestral sequence 2 generations ago is $[1 - 1/(2N)][1/(2N)]$ and the probability that the two sequences are derived from a common ancestral sequence $t + 1$ generations ago is

$$[1 - 1/(2N)]^t[1/(2N)] \approx [1/(2N)]\exp[-t/(2N)] \tag{9.11}$$

From this distribution one can show that the mean and variance of t_2 are

$$E(t_2) = 2N \tag{9.12}$$

$$V(t_2) = 4N^2 \tag{9.13}$$

In general, the time t_m for the coalescence from m to $m - 1$ sequences follows the exponential distribution with parameter $m(m - 1)/M$:

$$g(t_m) = \frac{m(m-1)}{M}\exp\left[\frac{-m(m-1)}{M}t_m\right] \tag{9.14}$$

where $M = 4N_e$. The mean and variance of t_m are

$$E(t_m) = \frac{M}{m(m-1)} \tag{9.15}$$

$$V(t_m) = \left[\frac{M}{m(m-1)}\right]^2 \tag{9.16}$$

The genealogy of n sequences has $2(n - 1)$ branches. A branch is said to be **external** if it directly connects to an external node (i.e., a node at the tip of the tree), otherwise it is said to be **internal**. Since there are n external nodes (sequences), n of the $2(n - 1)$ branches are external and $n - 2$ branches are internal. In Figure 9.2, b, e, f, g, and h are external and a, c, and d are internal. Note that we consider a rooted tree, so that branch a is internal. The motivation of distinguishing between internal and external branches is that this will be useful later when we consider detection of natural selection from sequence data.

Let J, I, and L be, respectively, the total time length of all branches, the total time length of internal branches and the total time length of external branches. Fu and Li (1993a) showed that

$$E(L) = M \tag{9.17}$$

which is independent of the sample size n. Thus, regardless of the number of sequences sampled, the expected total time length of external branches is always $M = 4N_e$ generations. The expectations of I and J are

$$E(I) = (a - 1)M \tag{9.18}$$

$$E(J) = E(I) + E(L) = aM \tag{9.19}$$

where a is given by Equation 9.8.

Let us now consider mutations. We assume that the rate of mutation per sequence per generation is u and that the number of mutations in a sequence in a time period t follows the Poisson distribution with parameter ut. Let η_e and η_i be the total numbers of mutations in the external and internal branches, respectively, and let $\eta = \eta_i + \eta_e$ be the total number of mutations that occurred in the entire genealogy of n sequences. Fu and Li (1993a) showed that

$$E(\eta_e) = \theta \tag{9.20}$$

$$V(\eta_e) = \theta + c\theta^2 \tag{9.21}$$

$$E(\eta_i) = (a - 1)\theta \tag{9.22}$$

$$V(\eta_i) = (a - 1)\theta + \left[b - \frac{2a}{n-1} + c\right]\theta^2 \tag{9.23}$$

$$E(\eta) = a\theta \tag{9.24}$$

$$V(\eta) = a\theta + b\theta^2 \tag{9.25}$$

where b is given by Equation 9.9 and

$$
\begin{aligned}
c &= 1 && \text{if } n = 2, \text{ but} \\
&= \frac{2[na - 2(n-1)]}{(n-1)(n-2)}, && \text{if } n > 2
\end{aligned}
$$

Under the infinite-site model, which assumes that every mutation occurs at a new site of the sequences, the number of segregating (polymorphic) sites (K) among the sequences in the sample is equal to the total number of mutations η. Therefore, the mean and variances of η should be the same as those of K and indeed, Equations 9.24 and 9.25 are identical to Equations 9.6 and 9.7.

It is interesting to know the relationships among these quantities and the quantity Π defined by Equation 9.1. For example, Figure 9.3 shows that η_e and η_i become almost independent of each other (i.e., have a low correlation coefficient) when the sample size n becomes larger than 10, whereas the correlation between η and Π is strong even when the sample size is quite large.

ESTIMATION OF POPULATION PARAMETERS

A great difficulty in population genetics is our inability to measure any of the basic parameters that are involved in the theory. For example, we do not know the long-term effective population size of any species or the mutation rate u for any gene or sequence. Fortunately, the situation is improving as a result of the introduction of various molecular techniques into population genetics. In particular, the feasibility of obtaining a fairly large number of DNA sequences from a population has enabled us to obtain better estimates of certain parameters such

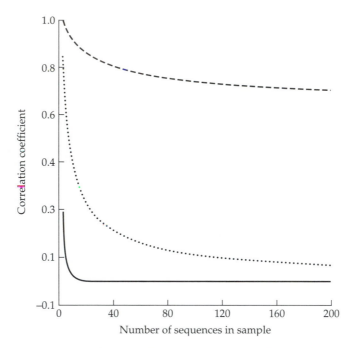

Figure 9.3 Correlation coefficients. The solid, dotted, and dashed curves show the correlation coefficients between η_e and η_i, between η_e and Π, and between η and Π, respectively. From Fu and Li (1993a).

as θ. As can be seen from preceding sections, θ is a parameter that appears in many formulas. Indeed, θ plays a prominent role in the stochastic theory of population genetics and there has been much interest in the technical aspects of estimating θ. Because the two parameters N_e and u usually do not appear separately in a formula, we are not able to estimate them separately but can estimate only their product through $\theta = 4N_e u$. Of course, if we know N_e or u, we can compute the other from θ. We discuss below methods for estimating θ. It should be noted that in estimating θ one should use sequences that are subject to no selective constraints, so that the sequence variation is not affected by natural selection.

In the past the two commonly used estimators of θ are due to Watterson (1975) and Tajima (1983). As can be seen from Equation 9.6,

$$\theta_W = K/a \tag{9.26}$$

should be an unbiased estimator of θ and this has been known as Watterson's estimator. As noted earlier, Equation 9.6 was derived under the infinite-site model. Equation 9.24 indicates that in the general case K should be replaced by η, the total number of mutations in the genealogy of the sequences in the sample. Tajima's estimator is based on Equation 9.3 and is given by

$$\theta_T = \Pi \tag{9.27}$$

It is known that Watterson's estimator is more efficient than Tajima's estimator because it has a smaller variance; compare Equations 9.4 and 9.7.

The above two estimators make little use of the genealogical relationships among the sequences in the sample. Strobeck (1983) developed a theory that takes into account the genealogical relationships between the alleles in a sample, however, he considered only the cases of two and three alleles. Felsenstein (1992a) considered any number of sequences and suggested that much improvement in efficiency over Watterson's estimator can be achieved by incorporating the genealogical relationships among sequences. Fu and Li (1993b) investigated the maximum amount of improvement that might be achieved in practice.

Let us consider the hypothetical genealogy in Figure 9.2. The tree is divided into segments by the horizontal lines through each of the branching nodes (or tips). Let η_j be the total number of mutations that occurred between the $(j-1)$-th node and the jth node, $j = 2, 3, \ldots, n$. Suppose that in an ideal situation, we can observe or infer without error the η_j values. Then, using the coalescent theory and the maximum likelihood method, Fu and Li (1993b) showed that the optimal estimator θ_m is the solution of the following equation for θ

$$\sum_{m=2}^{n} \frac{\eta_m + 1}{\theta + m - 1} = \frac{\eta}{\theta} \tag{9.28}$$

and that the variance of this estimator is

$$V_{\min} = \theta \left[\sum_{m=2}^{n} \frac{1}{\theta + m - 1} \right]^{-1} \tag{9.29}$$

Since θ_m is based on $\eta_2, \ldots, \eta_n$ whereas θ_W is based only on η (i.e., the sum of $\eta_2, \ldots, \eta_n$), θ_m uses more information and should be more efficient than θ_W. Indeed, Fu and Li (1993b) showed that

$$V(\theta_m) \leq V(\theta_W)$$

The ratio $V(\theta_m)/V(\theta_W)$ indicates the relative efficiency of θ_W to θ_m. Figure 9.4 shows that this ratio is much smaller than 1, if θ is large. Thus, θ_m can be much more efficient than θ_W and there is considerable room for improvement. For this reason, there is currently much effort to improve the estimation of θ. Felsenstein (1992b), Griffiths and Tavaré (1994), and Kuhner et al. (1995) use the maximum likelihood approach, while Fu (1994a) proposes an approximate approach that is computationally less demanding than the likelihood approach.

Fu's estimator is based on the following observation. Since in practice the η_j's cannot be estimated without error or, in other words, we have less information than knowing precisely $\eta_2, \ldots, \eta_n$, we cannot have a better estimator than θ_m, i.e., a variance smaller than $V_{\min}$. Thus, if one can obtain an estimator that has a variance very close to $V_{\min}$, then that estimator should be close to the best estimator. Fu (1994a) showed that by using the simple UPGMA method to reconstruct a genealogy and to infer the number of mutations on each branch of the genealogy from the sequence data one can in fact obtain an estimate of θ with a variance very close to $V_{\min}$. A computer program is available from Dr. Y.-X. Fu.

Note that we have so far assumed no recombination. If the rate of recombination per sequence per generation (r) is low, so that $4N_e r$ is of the order of 1, the effect of recombination is not serious and the above approach by Fu is still

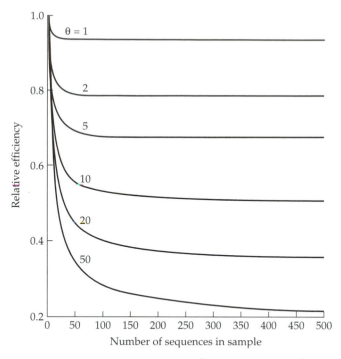

Figure 9.4   Efficiency of Watterson's estimate $\hat{\theta}_W$ of θ relative to $\hat{\theta}_m$. The six curves from top to bottom correspond respectively to $\theta = 1, 2, 5, 10, 20,$ and 50. The efficiency is calculated by $\text{Var}(\hat{\theta}_m)/\text{Var}(\hat{\theta}_W)$ where $\text{Var}(\hat{\theta}_W)$ and $(\hat{\theta}_m)$ are given by Equations 9.7 and 9.29, respectively. From Fu and Li (1993b).

applicable. However, if the recombination rate is higher, the effect needs to be taken into account; Fu (1994b) has developed a new estimator for this purpose.

DETECTION OF NATURAL SELECTION

In molecular population genetics or evolutionary studies, one often assumes that a DNA region is subject to no natural selection, so that all mutations in the region are selectively neutral. For example, noncoding regions such as intergenic regions or introns are usually assumed to be subject to no selective constraints, and in coding regions synonymous mutations are often assumed to be neutral. The issue here is how to test the neutrality assumption by sampling a number of sequences from the population. Note that selection here means any type of selection, including negative selection.

Tajima (1989) proposed to use the two different estimates of θ, $\theta_T = \Pi$ and $\theta_W = K/a$, to detect selection; for the two estimates, see Equations 9.26 and 9.27. His test statistic is

$$D\frac{\Pi - K/a}{\sqrt{V(\Pi - K/a)}} \tag{9.30}$$

where V stands for variance. The rationale for this test is as follows. Since K ignores the frequencies of variants, it is strongly affected by the existence of deleterious alleles, which are usually kept at low frequencies. In contrast, Π is not much affected by the existence of deleterious alleles because it is dependent on the frequencies of variants with a small contribution from low-frequency alleles. Thus, if the sample includes some (usually slightly) deleterious alleles, the estimate of θ based on K is likely to be larger than that based on Π, i.e., D in Equation 9.30 should have a negative sign. The presence of overdominant selection tends to have the opposite effect, because alleles with intermediate frequencies increase Π considerably but have little effect on K, i.e., D should have a positive sign. Therefore, the difference $\Pi - K/a$ can be used to detect the presence of selection. The denominator is intended to normalize the statistic. Tajima (1989) suggested using a beta distribution as an approximation of the distribution of D so that critical values of the test can be obtained, but a more accurate approach is to use Monte Carlo simulation to obtain the critical values of the statistic (Fu and Li 1993a).

One can construct a test similar to the above one by using any pair of estimates of θ as long as the variance of their difference can be calculated. However, in order for such a test to be useful, the two estimates must be sufficiently different when selection is present. Because the expectations of η_e and $\eta_i/(a-1)$ are both equal to θ under neutral mutation and because they are likely to be different when selection is present, Fu and Li (1993a) proposed the following test:

$$G = \frac{\eta_e - \eta_i/(a-1)}{\sqrt{V[\eta_e - \eta_i/(a-1)]}} \tag{9.31}$$

As is well known, deleterious mutations are usually quickly eliminated from the population, and those that are present in the sample are most likely to have arisen recently. Recent mutations are close to the tips (external nodes) in the genealogy (see Golding et al. 1986) and therefore are mostly included in the value of η_e. In contrast, mutations in the internal branches are most likely to be neutral and $\eta_i/(a-1)$ is not strongly affected by the presence of selection. Therefore, these two estimates of θ should be useful for testing for the presence of selection. An advantage of G over D is that η_e and η_i are weakly correlated, whereas Π and $\theta_W = K/a$ are strongly correlated (Figure 9.3).

Since Π is less affected than η_e by the presence of selection, an obvious variant of test G is

$$F = \frac{\Pi - \eta_e}{\sqrt{V(\Pi - \eta_e)}} \tag{9.32}$$

The critical values of the G and F statistics have been provided by Fu and Li (1993a), who also gave an example of the application of these tests to DNA sequence data.

Simonsen et al.'s (1995) simulation study of the relative performances of different tests under a selective sweep suggested that Tajima's test given by Equation 9.30 tends to be more powerful than the others. However, Charlesworth et al.'s (1995) result from limited simulations under background selection suggested the superiority of Fu and Li's (1993a) tests over Tajima's test. (See the section on the Joint Effects of Linkage and Selection for the definitions of selective sweep and background selection.) Readers may also refer to the maximum likelihood approach by Golding and Felsenstein (1990), which is a between-species method but can also be applied to within-species polymorphism data.

TESTING THE NEUTRAL MUTATION HYPOTHESIS

The tests presented in the preceding section are for testing the hypothesis that all mutations in a DNA region are selectively neutral, but not for testing the neutral mutation hypothesis. The latter hypothesis postulates that the majority of mutations that have contributed significantly to the genetic variation in natural populations are neutral or nearly neutral (Chapter 2; Kimura 1968a, 1983). This assumption is much weaker than the assumption that all mutations are neutral. Indeed, the neutral mutation hypothesis assumes that the most prevalent type of selection is purifying (negative) selection (Kimura 1983). Thus, for example, even if all the rare variants at a locus are deleterious, the neutral mutation hypothesis still holds as long as the majority of the more common variants are selectively neutral or nearly so; note that in this case the hypothesis that all mutations are neutral does not hold. Clearly, we need different tests for testing the neutral mutation hypothesis.

According to the neutral mutation hypothesis, both the variation within populations and the differences between populations are mainly due to neutral or nearly neutral mutations. In other words, polymorphism is a transient phase of molecular evolution, and the rate of molecular evolution is positively correlated with the level of within-population variation (Kimura and Ohta 1971a). This prediction had previously been used to examine protein polymorphism and evolution (Chakraborty et al. 1978; Skibinski and Ward 1982). Recently, tests based on using within- and between-population DNA sequence comparisons have been developed (Kreitman and Aguadé 1986; Hudson et al. 1987; McDonald and Kreitman 1991; Sawyer and Hartl 1992). We discuss two such tests below.

The Hudson-Kreitman-Aguadé Test

This test by Hudson, Kreitman, and Aguadé (1987) is known as the **HKA test**. It is to test whether the levels of within- and between-population DNA variation are positively correlated, as predicted by the neutral mutation hypothesis.

Consider m regions (loci) in species 1 and 2. Suppose that at each of the m regions n_1 and n_2 random sequences have been obtained from species 1 and 2, respectively. Let K_{1i} be the number of segregating sites at locus i in the sample of n_1 sequences from species 1; K_{2i} is similarly defined for species 2. Let D_i be the number of nucleotide differences at locus i between a random sequence from the sample from species 1, and a random sequence from the sample from species 2. The observations K_{1i} and K_{2i} ($i = 1, \ldots, m$) are measures of within-species variation and the D_i are measures of between-species divergence. The HKA test is to test whether the latter quantities are statistically consistent with the former.

The HKA model makes the same assumptions (e.g., neutrality, equilibrium, and no recombination) as those in the derivation of Equations 9.6 and 9.7. In addition, it assumes that all loci are unlinked (or evolve independently), that species 1 and 2 are at equilibrium at the time of sampling with effective population sizes N and Nf, respectively, and that the two species were derived t generations ago from a single ancestral population that was at equilibrium with an effective size of $N(1 + f)/2$, i.e., the average of the population sizes of species 1 and 2. The test statistic is

$$X^2 = \sum_{i=1}^{m} \left[K_{1i} - \hat{E}(K_{1i}) \right]^2 / \hat{V}(K_{1i}) + \sum_{i=1}^{m} \left[K_{2i} - \hat{E}(K_{2i}) \right]^2 / \hat{V}(K_{2i})$$

$$+ \sum_{i=1}^{m} \left[D_i - \hat{E}(D_i) \right]^2 / \hat{V}(D_i) \tag{9.33}$$

where $\hat{E}$ and $\hat{V}$ denote estimates of expectation and variance, respectively. The estimated expectation and variance of K_{1i} can be obtained by Equations 9.6 and 9.7, i.e.,

$$E(K_{1i}) = a_1 \theta_i \tag{9.34}$$

$$V(K_{1i}) = a_1 \theta_i + b_1 \theta_i^2 \tag{9.35}$$

where a_1 and b_1 are given by Equations 9.8 and 9.9 with n replaced by n_1. The equations for the expectation and variance of K_{2i} are the same as above except that θ_i and n_1 are replaced by $f\theta_i$ and n_2, respectively. The estimated expectation and variance of D_i are given by

$$E(D_i) = \left[t + (1+f)/2 \right] \theta_i \tag{9.36}$$

$$V(D_i) = E(D_i) + \left[\theta_i (1+f)/2 \right]^2 \tag{9.37}$$

which were obtained by Li (1977) and Gillespie and Langley (1979). The above equations involve θ_i, f, and t, which can be estimated from the following system of $m + 2$ equations:

$$\sum_{i=1}^{m} K_{1i} = a_1 \sum_{i=1}^{m} \hat{\theta}_i \tag{9.38}$$

$$\sum_{i=1}^{m} K_{2i} = \hat{f} a_2 \sum_{i=1}^{m} \hat{\theta}_i \tag{9.39}$$

$$\sum_{i=1}^{m} D_i = \left[\hat{t} + (1+\hat{f})/2 \right] \sum_{i=1}^{m} \hat{\theta}_i \tag{9.40}$$

$$K_{1i} + K_{2i} + D_i = \hat{\theta} \left[\hat{t} + (1+\hat{f})/2 + a_1 + \hat{f} a_2 \right], \qquad i = 1, \ldots, m-1 \tag{9.41}$$

This system of equations is obtained from Equations 9.34–37 and the analogous equations for $E(K_{2i})$, by summing different combinations of the equations and replacing the expectations of the random variables with the observed values. From computer simulations, Hudson et al. (1987) found that if n_1, n_2, and t are sufficiently large, X^2 is approximately χ^2 distributed with $2m - 2$ degrees of freedom; the degree of freedom is $2m - 2$ because $m + 2$ parameters are estimated from the above system of equations and should be deducted from the $3m$ degrees of freedom in Equation 9.33. In essence, the HKA test is a goodness-of-fit test, i.e., one first uses Equations 9.38–41 and the observations K_{1i}, K_{2i}, and D_i to estimate the parameters θ_i's, f, and t under the assumption of neutrality, and then puts the estimated parameters into Equation 9.33 to see if there is any significant deviation from the prediction of the neutral mutation hypothesis. The

HKA test is conservative for rejecting selective neutrality because of the assumptions of complete linkage within loci and no linkage between loci.

The HKA test has been applied to sequence data from three contiguous regions: the 5′ flanking region of *Adh*, the *Adh* coding region and the *Adh–dup* coding region in *Drosophila melanogaster* and *D. simulans* (Kreitman and Hudson 1991). The *Adh–dup* gene is 3′ to the *Adh* gene and is related to *Adh* by an ancient tandem duplication (Schaeffer and Aquadro 1987). The data consist of 11 sequences across the above three regions from *D. melanogaster* and 1 sequence from *D. simulans*. Since no within-species variation can be computed from the single sequence available from *D. simulans*, the HKA test is modified to

$$X^2 = \sum_{i=1}^{m} \left[K_i - \hat{E}(K_i) \right]^2 / \hat{V}(K_i) + \sum_{i=1}^{m} \left[D_i - \hat{E}(D_i) \right]^2 / \hat{V}(D_i)$$

where K_i refers to the number of segregating sites in region i in the sample from *D. melanogaster*. For this reason, the degree of freedom becomes $m - 1$. The above test is based on the number of segregating sites and is denoted as "Seg" in Table 9.3. An alternative test is to substitute the average pairwise difference (Pwd) for the number of segregating sites; Pwd is Π as defined in Equation 9.1. This test may be more sensitive to among-region differences when the observed polymorphism frequency distributions differ across regions. However, since the expected variance of Pwd exceeds the expected variance of the number of segregating sites (K), Pwd has less power to reject the null hypothesis when the frequency distributions are the same among regions.

TABLE 9.3   HKA tests for silent-site differences in three regions

Region	Test[a]	Within species Obs.	Within species Exp.	Between species Obs.	Between species Exp.	$\hat{\theta}$[b]	t (time)	X^2	P
5′ Flanking	Seg	30	37.5	78	57.7	12.8			
vs. *Adh*		20	12.5	16	19.2	4.3	4.5	3.7	0.054
5′ Flanking	Pwd	11.4	15.1	78	59.3	15.0			
vs. *Adh*		7.6	4.0	16	15.7	4.0	3.9	3.8	0.05
5′ Flanking	Seg	30	27.2	78	71.2	9.3			
vs. *Adh-dup*		13	15.8	50	41.7	5.4	7.7	1.4	0.25
5′ Flanking	Pwd	11.4	9.2	78	70.1	9.2			
vs. *Adh-dup*		3.4	5.5	50	42.3	5.5	7.7	1.8	0.18
Adh	Seg	20	12.0	16	19.9	4.1			
vs. *Adh-dup*		13	21.0	50	34.8	7.2	4.9	5.4	0.02
Adh	Pwd	7.6	3.4	16	16.9	3.4			
vs. *Adh-dup*		3.4	7.6	50	38.1	7.6	5.0	6.7	0.01

From Kreitman and Hudson (1991).
[a]Seg, number of segregating sites; Pwd, average pairwise number of differences.
[b]$\theta = 4Nv$, where v is the rate of mutation *per site* per generation, not the rate of mutation *per sequence* per generation. The ^ indicates estimated θ.

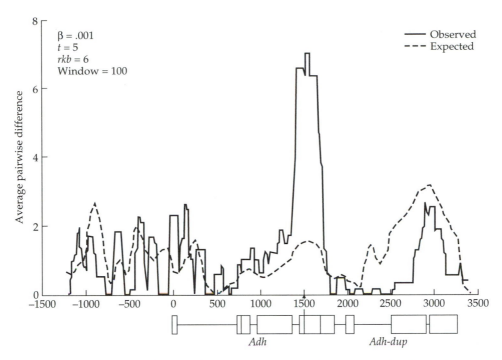

Figure 9.5   Sliding window of observed and expected polymorphism levels in a sample of 11 *Drosophila melanogaster* alleles. The average pairwise number of nucleotide differences in a window of 100 silent sites is plotted at each nucleotide position. The arrow at position 1490 marks the site of the *Adh^f/Adh^s* amino acid replacement polymorphism. See Kreitman and Hudson (1991) for more detailed information about the sliding-window method. From Kreitman and Hudson (1991).

Kreitman and Hudson (1991) considered silent sites, which are defined as positions at which mutations do not change the polypeptide sequence encoded by the gene. In the 5′ flanking region (1243 bp), every site is silent. In a coding region a site can be only partially silent (synonymous), and the number of silent sites is equal to the number of possible silent nucleotide changes in the region divided by 3; the number of silent sites is 319.2 for *Adh* (903 bp) and 599.8 for *Adh–dup* (1293 bp). Table 9.3 shows the pairwise tests of the three regions: 5′ flanking, *Adh*, and *Adh–dup*. The 5′ flanking region vs. *Adh* test and the *Adh* vs. *Adh–dup* test are both significant at the 5% level, whereas the 5′ flanking region vs. *Adh–dup* test is not. These results implicate the *Adh* coding region as the cause of departure from neutrality. The departure occurred because the within-species polymorphism in the *Adh* coding region was higher than expected (Table 9.3). This pattern fits the suggestion that the *Adh* fast (*Adh^f*) and slow (*Adh^s*) electrophoretic alleles have been maintained by balancing selection, because levels of neutral polymorphism are expected to be elevated at sites sufficiently tightly linked to a balanced polymorphism. Indeed, Figure 9.5 shows that the level of polymorphism is highest at position 1490, the site where the threonine–lysine difference distinguishing the *Adh^f* and *Adh^s* alleles, and that the polymorphism level is also very high at sites closely linked to the balanced polymorphism. Thus, the high polymorphism at silent sites in the *Adh* coding region is apparently not due to selection at these sites *per se* but is maintained because of their linkage to a balanced polymorphism.

The McDonald-Kreitman Test

McDonald and Kreitman (1991) proposed a simple method to contrast the patterns of within-species polymorphism and between-species divergence at synonymous (silent) and nonsynonymous (replacement) sites in the coding region of a gene. Obviously, in a coding region the synonymous sites and the replacement sites should have the same evolutionary history because they are tightly linked. Therefore, if polymorphism and divergence at both types of site are due to neutral mutations, the ratio of replacement to synonymous differences between species should be the same as the ratio of replacement to synonymous polymorphisms within species. A significant difference between the two ratios can therefore be used to reject the neutral mutation hypothesis.

Consider two samples of sequences from the coding region of a gene from species 1 and 2. A nucleotide site in the sequences is said to be polymorphic if it shows any variation in one or both samples but is said to have a fixed difference between the two samples if it shows no variation in each sample but is fixed at different nucleotides in the two samples. For example, suppose that for 12 and 10 sequences taken from species 1 and 2, the nucleotides at position X are T,T,T,T, T,T,T,T,T,T,T,T in the 12 sequences from species 1 and G,G,G,G,G,G,G,G,G,G,G in the 10 sequences from species 2, and that the nucleotides at position Y are G,G,G,G,G,G,G,G,G,G,G,G,G in species 1 and G,A,G,G,G,G,T,G,G,G in species 2. Then position X has a fixed difference, whereas position Y has 2 polymorphic sites. For two samples of sequences, let F_R and F_S be, respectively, the numbers of fixed replacement and synonymous differences and P_R and P_S be, respectively, the numbers of replacement and synonymous polymorphic sites. Then the McDonald-Kreitman test uses a 2×2 contingency table with F_R and F_S in one row and P_R and P_S in the other row (see Table 9.4). If there is no selective difference between synonymous and replacement sites, then the test should be nonsignificant. See Sawyer and Hartl (1992) for a theoretical treatment.

McDonald and Kreitman (1991) applied the above test to DNA sequences of the coding region of the alcohol dehydrogenase (*Adh*) gene from *Drosophila melanogaster*, *D. simulans*, and *D. yakuba* (Table 9.4). The ratio of replacement changes to synonymous changes that are fixed between species is significantly greater than the ratio of replacement to synonymous polymorphisms. Indeed, the former is $7/17 = 0.41$, whereas the latter is only $2/42 = 0.05$; Fisher's test and the G-test (see Sokal and Rohlf 1981) both give a probability of $P < 0.006$. A minor problem in this study is the rule that the between-species variation at a site is neglected if that site is polymorphic in any of the three species. This rule tends to underestimate between-species variation (Graur and Li 1991). For example, at nucleotide position 1590, all the alleles in *D. melanogaster* have T, half the alleles in *D. simulans* have T, and the other half have C, and all the alleles in *D. yakuba* have C. At this position, McDonald and Kreitman assigned a single synonymous polymorphism within *D. simulans* but no difference between species. However, under their assumption of monomorphism in the common ancestral species, a substitu-

TABLE 9.4 **Numbers of replacement and synonymous fixed differences between species and of polymorphisms within species**

	Differences	
Type of change	*Fixed*	*Polymorphic*
Adh **gene**[a]		
Replacement	7	2
Synonymous	17	42
G6pd **gene**[b]		
Replacement	21	2
Synonymous	26	36

[a]McDonald and Kreitman (1991): 12, 6, and 24 sequences from *Drosophila melanogaster*, *D. simulans*, and *D. yakuba*, respectively.
[b]Eanes et al. (1993): 32 and 12 sequences from *D. melanogaster* and *D. simulans*, respectively.

tion must have occurred between *D. melanogaster* and *D. yakuba*. Of course, recounting a few such cases does not change the conclusion because the *P* value is rather small. Another caveat is that only 6 sequences were sampled from *D. simulans* and so a fixed difference may become a polymorphic difference if the sample size is increased (Graur and Li 1991). Nevertheless, the high ratio of fixed replacement to fixed synonymous changes is difficult to explain under the neutrality assumption. Rather, it would be easier to explain the observation under the assumption that most of the observed fixed replacement differences were due to adaptive fixation of selectively advantageous mutations. A mutation that became fixed by selection would have been a polymorphism for a shorter time period than would a mutation that became fixed by random drift. For this reason, an adaptive substitution is less likely to appear as a polymorphism than is a neutral substitution (see Chapter 2). Interestingly, Sawyer and Hartl (1992) estimated that a small selective advantage ($\sim 1.5 \times 10^{-6}$) would be sufficient to explain the observed data.

An even more extreme case was found by Eanes et al. (1993) for the glucose-6-phosphate dehydrogenase (*G6PD*) gene in *D. melanogaster* and *D. simulans* (Table 9.4). The ratio of fixed replacement differences to fixed synonymous differences is $21/26 = 0.81$, whereas the ratio of replacement polymorphisms to synonymous polymorpisms is only $2/36 = 0.06$; the G-test gives $P \ll 0.001$. This implies a tenfold excess of replacement changes over that expected if the *G6PD* gene were evolving in a strictly neutral fashion. The function of *G6PD* in *D. melanogaster* is well established as the initial enzymatic step in the pentose shunt pathway, and the rapid amino acid change in *G6PD* has been suggested to be due to selective response to varying pentose shunt function in changing evironments (Eanes et al. 1993; Eanes 1994).

Another case of deviation from the neutrality expectation is the *jingwei* gene in *D. melanogaster* and *D. simulans* (Long and Langley 1993). In addition, at the *zeste* locus, there is a higher proportion of replacement differences among fixed sites than among polymorphic sites (Hey and Kliman 1993), though a statistical test would lack power because no polymorphic replacement difference was found. On the other hand, the patterns at *PGI*, *6PGD*, *GAPDH*, and *GPDH* (see the review by Eanes 1994) and at *yp2* and *per* (Hey and Kliman 1993) reflect molecules that are either very constrained or appear to be changing in a neutral fashion when the McDonald-Kreitman test is applied. As the number of loci studied is still limited, it is too early to draw a general conclusion on the proportion of genes whose polymorphism-divergence pattern cannot be explained by the neutralist view (see the review by Brookfield and Sharp 1994).

Note that all the above tests assume that synonymous mutations are neutral, but this is apparently not the case for *Drosophila* genes (see Chapter 7). However, Charlesworth (1994) suggested that the McDonald-Kreitman test is conservative, because nonsynonymous mutations are subject to stronger negative selection than are synonymous mutations. Akashi (1995) discussed how codon usage bias may be taken into account when conducting the McDonald-Kreitman test.

JOINT EFFECTS OF LINKAGE AND SELECTION

The neutralist prediction of a positive correlation between intraspecific variation and interspecific divergence can be violated by the presence of positive Darwinian (directional) selection. When directional selection drives an advantageous mutation through a population to fixation, much of the neutral variation at linked sites is eliminated during the process. This phenomenon, known as the "hitchhiking effect" or "selective sweep," was first analyzed by Maynard Smith and Haigh (1974) and later by Kaplan et al. (1989), Stephan et al. (1992), and others. This effect is expected to increase with the strength of linkage. Thus, if advantageous substitutions occur at an appreciable rate, the level of standing variation should be low in regions with a severely restricted rate of recombination. In general, a positive correlation should exist between the level of polymorphism and the rate of recombination in a region. The hitchhiking process, however, is not expected to affect the rate of divergence between species at linked neutral sites (Birky and Walsh 1988; Hudson 1990), though it affects the rate at linked sites under selection. Thus, in contrast to the neutralist prediction, positive selection can lead to uncoupling of levels of polymorphism and divergence. The uncoupling can also occur by the presence of balancing selection, which can increase the level of polymorphism at linked loci.

Aguadé et al. (1989) seem to have been the first to use the hitchhiking effect as an indirect means to detect the action of directional selection at the DNA level. The *yellow* gene and the *achaete-scute* (*ASC*) region are located near the tip of the X chromosome in *D. melanogaster*, which has a very low rate of recombination. A survey of restriction-map variation at the *yellow-ASC* (*y-ASC*) region in 64 lines of *D. melanogaster* from nature showed reduced nucleotide variation, although no apparent reduction was observed in other populations (Beech and Leigh-Brown 1989; Eanes et al. 1989; Macpherson et al. 1990). In fact, in that survey only 9 polymorphisms were detected out of 176 restriction sites (approximately 2112 nucleotide-site equivalents) and 6 out of the 9 polymorphisms were present only once in the sample; given that the average π for the genome is 0.05, the probability of observing such a sample is only 2×10^{-6} (Hudson 1990). A population-bottleneck effect was considered unlikely because the same sample was analyzed for other regions of the X chromosome with no evidence of reduction in levels of variation. An even greater reduction in polymorphism in the *achaete* region was found in the sibling species *D. simulans*: no nucleotide variation was detected in a sample of 103 lines when 390 site equivalents were studied (Martín-Campos et al. 1992). The reduction in variation in both species is obviously not due to purifying selection (selective constraints), because the divergence in this region between the two species is comparable to those in other regions: 6% compared to 7% and 6% silent divergences at the 5'*Adh* and *hsp82* regions, respectively. From these lines of evidence, it is reasonable to conclude that the reduction in nucleotide variation in the *y–ASC* region is due to a hitchhiking effect; see further data in Aguadé and Langley (1994). The same conclusion was drawn for the *cubitus interruptus Dominant* locus located on the nonrecombining fourth chromosome: sequencing of a 1.1-kb region of the gene in 10 natural lines of *D. melanogaster* and 9 of *D. simulans* had uncovered only one polymorphism (Berry et al. 1991). In addition, no variation was observed at the *su-f* region in *D. melanogaster* and *D. simulans* (Langley et al. 1993) and in the *fw* locus of *D. ananassae* (Stephan and Mitchell 1992).

Much interest in the issue of polymorphism vs. linkage was generated by Begun and Aquadro (1992), who showed that for 20 gene regions from across the genome of *D. melanogaster* the level of nucleotide diversity is positively correlated with the regional rate of recombination. This could not be explained by variation in mutation rates or in functional constraints because there was no correlation between recombination rate and DNA sequence divergence between *D. melanogaster* and *D. simulans*. They concluded that the correlation resulted from hitchhiking (selective sweeps) associated with the fixation of advantageous mutants. The positive correlation is supported by further data from the X and third chromosomes (Aguadé and Langley 1994; Aquadro et al. 1994). Figure 9.6 shows that the amount of DNA variation is positively correlated with the rate of recombination in each of these two chromosomes.

Charlesworth et al. (1993) pointed out that, in addition to the hitchhiking effect, selection against deleterious mutations can also reduce nucleotide variation at linked sites. This type of selection, called **background selection**, and hitchhiking essentially involve the same process, namely, the elimination of neutral or nearly neutral variants as a result of selection at linked sites. Background selection, like positive selection, reduces the effective population size at nearby sites. For an autosomal region with no recombination and with a total mutation rate of U per generation and selection s against mutant homozygotes, and hs against heterozygotes (h denotes degree of dominance), the effective size in a random mating population is reduced from N_e to $f_0 N_e$, where

$$f_0 \approx \exp\left[-U / (2hs)\right] \tag{9.42}$$

For an X-linked region with the same parameters and with no recombination,

$$f_0 \approx \exp\left[-1.5U / (2hs + s)\right] \tag{9.43}$$

Therefore, the expected nucleotide diversity (π) at a neutral site is reduced from $4N_e v$ to $4f_0 N_e v$, where v is the mutation rate per site. The background effect decreases rapidly with increasing recombination frequency. Based on their simulation study and data on U, s, hs, and recombination rates in *D. melanogaster*, Charlesworth et al. (1993) estimated that the reduction in π in this species can be

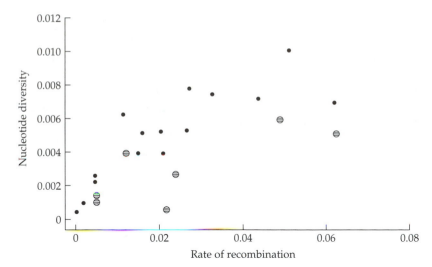

Figure 9.6. Scatterplot of nucleotide diversity versus rate of recombination (adjusted coefficient of exchange) for 15 third-chromosome gene regions and seven X-linked gene regions, all from populations from the United States. Third-chromosome and X-linked genes are represented by solid and hatched circles, respectively. Levels of variation for X-linked genes have been adjusted to make them comparable to autosomal genes by multiplying the former by 4/3. Rates of recombination have been adjusted by 1/2 for autosomes, and by 2/3 for X chromosomes, to compensate for the lack of recombination in males and assuming a 1:1 sex ratio. From Aquadro et al. (1994).

very high. Background selection may be sufficient for explaining relatively mild reductions in π in certain regions, though it may not account for drastic reductions such as that observed in the *su-f* region (Langley et al. 1993).

It is difficult to distinguish between hitchhiking and background selection effects because they lead to many similar observable consequences for the pattern of neutral variability (Charlesworth et al. 1993). Aquadro et al. (1994) suggest that one way to evaluate the relative contributions of hitchhiking and background selection is to compare the levels of DNA variation in regions of equivalent recombination on the X chromosome and autosomes. Because X-linked genes are hemizygous in males, recessive advantageous mutations are fixed more rapidly on the X chromosome than on an autosome (Charlesworth et al. 1987), the hitchhiking effect on nucleotide variation is likely to be stronger on X-linked genes than on autosomal genes. So, X-linked loci should tend to have lower levels of variation than autosomal loci. In contrast, background selection is a mutation-selection equilibrium model in which recessive deleterious mutations are eliminated more efficiently from the X chromosome than the autosome, so that the X chromosome has a larger proportion of mutation-free gametes compared to the autosomes. [This prediction follows from Equations 9.42 and 9.43 because $1/(2hs) > 1.5/(2hs + s)$ as long as $h < 1$.] For this reason, X-linked genes are expected to have higher levels of variation (all else being equal), after correcting for different effective population size for X-linked and autosomal genes. Limited data (Figure 9.6; Aquadro et al. 1994) suggest that, for genes experiencing similar recombination rates, sequence variation is generally lower for X-linked genes than for autosomal genes, supporting the notion that the hitchhiking effect is an important cause of the correlation between recombination and variation. However, the data is too limited to draw a firm conclusion.

The issue, therefore, remains unsettled. Gillespie (1994) argued that directional selection is usually too weak for the hitchhiking effect to be effective. Aguadé and Langley (1994) and Braverman et al. (1995) noted that, under the selective-sweep hypothesis, tightly linked loci should tend to have low frequency alleles so that Tajima's D (Formula 9.30) should tend to have a negative value, but this does not seem generally to be the case; see also Charlesworth et al. (1995). Further research is needed to evaluate the relative importances of selective sweeping and background selection.

How much selection is required to explain the observed relationship between polymorphism and recombination? This is a difficult question and estimates have so far been obtained only under the assumption that the effect of background selection is negligible. Based on the slope of variation vs. recombination frequency in *D. melanogaster*, Wiehe and Stephan (1993) developed a theory for estimating the intensity of selective sweeps, which is defined as $2N_e s v_a$, where s is the selection coefficient for heterozygotes and v_a is the number of selected substitutions per nucleotide site. Using the tightly linked regions in the data of Begun and Aquadro (1992), Wiehe and Stephan (1993) estimated $2N_e s v_a$ to be $> 1.3 \times 10^{-8}$. Using available data from the third chromosome of *D. melanogaster*, Aquadro et al. (1994) obtained an estimate of $2N_e s v_a = 5.4 \times 10^{-8}$. If $N_e = 2 \times 10^6$ and $s = 0.01$, then $v_a = 1.4 \times 10^{-12}$. If the mutation rate per site is $v = 10^{-9}$, then about 0.14% of the mutations are selectively advantageous. This number is very small but becomes larger if s is smaller, which is likely to be the case.

MHC POLYMORPHISM

In this and the next section we shall consider some examples where the presence of natural selection was detected by methods not described in the preceding sections.

The **major histocompatibility complex** (**MHC**) is a large multigene family whose products control the immune system's capacity to recognize foreign proteins (see Klein 1986). MHC molecules bind foreign antigens (peptides) and display them on the cell surface. Such an assembly can be recognized by T cell receptors—the recognition activates the T lymphocytes, and the activated lymphocytes initiate appropriate immune responses to destroy the invaded cell or the invader itself. In the case of grafted tissue, the T cells recognize as foreign the MHC molecules of the donor, if they are different from those of the host, and the immune attack is then directed against the transplant. However, if the MHC molecules are recognized as "self," no immune response is initiated. This is the origin of the term "histocompatibility."

The human MHC, usually referred to as the **human leukocyte antigen** (**HLA**) system, is located on chromosome 6 in a region that is more than 4 million bp in length and contains more than 100 genes (see Klein et al. 1993). Only some of these genes encode molecules involved in the presentation of peptides to T lymphocytes. Others include genes that control the degradation of proteins into peptides and transport of the peptides across membranes; genes that specify other components of the immune system; genes that subserve functions that are either unknown or unrelated to the immune response; and nonfunctional genes. The MHC genes proper are classified into two classes, I and II; the two classes are evolutionarily related, though distantly.

It has long been realized that some MHC loci are highly polymorphic because tissue grafts are invariably rejected by the host whenever they are contributed by an unrelated donor. This has indeed been borne out by biochemical and molecular data. For example, heterozygosity at the human *HLA-A* and *-B* loci is about 90% in many populations, and the number of alleles detected in a sample of 200 individuals ranges approximately from 15 to 30 (see Klein 1986; Roychoudhury and Nei 1988). Similarly, the number of major alleles at the mouse *H2-K* and *-D* (or *H-2K* and *-2D*) loci are 54 and 55, respectively (Klein 1986). How such high polymorphism is maintained has been a subject of debate for decades. Recently there has been significant progress towards the resolution of this controversy.

One simple analysis that can provide insight into the mechanism of maintenance of polymorphism is Watterson's (1978) homozygosity test, which compares the observed homozygosity in a sample of size n containing k alleles to the expected homozygosity conditional on n and k under neutrality. If purifying selection is operating, the observed value is likely to be higher than the expected value (owing to the presence of low frequency alleles), whereas if balancing selection (e.g., overdominant selection, frequency-dependent selection) is operating, the reverse is expected to be true. Hedrick and Thomson (1983) compared the observed homozygosity in 22 samples at both the *HLA-A* and *-B* loci with the neutral expectation. In all cases the observed homozygosity is lower than the expected value, and they concluded that some form of balancing selection was responsible for the high polymorphism.

Hughes and Nei (1988) reasoned that if there is any positive Darwinian selection at MHC loci, it must occur at the **antigen recognition site (ARS)**, also known as the peptide binding region. A class I MHC molecule has three extracellular domains (α_1, α_2, and α_3), a transmembrane portion, and a cytoplasmic tail (Figure 9.7). The α_3 domain associates noncovalently with β_2-microglobulin, a polypeptide encoded by a gene that lies outside the MHC but is distantly related to the MHC genes. The ARS of the class I MHC molecule is located in the α_1 and α_2 domains and consists of 57 amino acid residues.

To see if natural selection occurs at the ARS, Hughes and Nei (1988) compared the pattern of nucleotide substitution at the ARS with the pattern in regions outside the ARS. In each region, they computed the number of substitutions per synonymous site (K_S) and the number of substitutions per nonsynonymous site (K_A) between alleles. In the absence of selection on a DNA region, K_S and K_A should be more or less the same. If there is purifying selection, K_S is expected to be higher than K_A, because selection would occur at the amino acid level. By contrast, if there is positive Darwinian selection, K_A is expected to be higher than K_S. Table 9.5 shows the mean K_S and K_A values for all sequence comparisons at several class I loci. In the case of the human *HLA-A* locus, K_A is much higher than K_S in the ARS, the difference being significant at the 0.1% level when the standard errors computed by Nei and Jin's (1989) method are used. In the other regions, K_S is higher than K_A. The same pattern of nucleotide substitution is also observed for the *HLA-B* and *HLA-C* loci. It is interesting to see that K_A in the ARS is signifi-

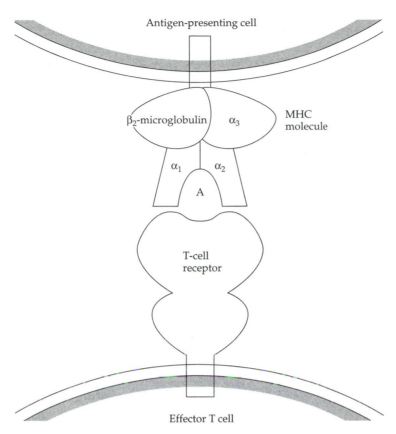

Figure 9.7 A class I MHC molecule presenting an antigen to a T cell. The antigen A is bound in the antigen recognition site of the MHC molecule. The antigen recognition site is composed of amino acids from subunits α_1 and α_2; subunits α_3 and β_2-microglobulin form most of the remainder of the molecule. The T-cell receptor, the variability of which is generated somatically to produce unique clones, makes specific contact with both the MHC molecule and antigen. From Hedrick et al. (1991).

cantly higher than K_S in any region. This clearly indicates that amino acid substitutions in the ARS are elevated by positive Darwinian selection. By contrast, K_A is much smaller than K_S in the non-ARS regions, suggesting that amino acid substitutions are generally subject to purifying selection in these regions.

Table 9.5 also includes the results of allelic comparisons of the mouse $H2$-K and -L loci. The K_A in the ARS is again substantially higher than the K_S in any region. (The unusually low value of K_S for α_3 may have been caused by interlocus genetic exchange; Hughes and Nei 1988.) Therefore, the pattern of nucleotide substitution for the mouse MHC loci is essentially the same as that for the human MHC loci. The same conclusion holds for class II loci (Hughes and Nei 1989).

Another extraordinary feature of the MHC polymorphism is the high nucleotide diversity among MHC alleles (see the review by Klein and Figueroa 1986). In other genetic systems, alleles at a given locus usually differ by a few nucleotide substitutions at most (see Table 9.2). In the MHC, some alleles differ by 100 or more substitutions. This observation means that these alleles have diverged from each other a long time ago. In fact, nucleotide sequencing work has revealed

TABLE 9.5 Mean numbers of nucleotide substitutions per synonymous site (K_S) and per nonsynonymous site (K_A) expressed as percentages between alleles from the same class-I MHC polymorphic loci in humans (HLA-A, -B, and -C) and mice (H2-K and -L)[a]

Locus (number of sequences)	Antigen recognition site (ARS) (N = 57)		Remaining codons in α_1 and α_2 (N = 124, 125)[b]		Domain α_3 (N = 92)	
	K_S	K_A	K_S	K_A	K_S	K_A
Human						
HLA-A (5)	3.5	13.3***	2.5	1.6.	9.5	1.6**
HLA-B (4)	7.1	18.1**	6.9	2.4*	1.5	0.5
HLA-C (3)	3.8	8.8	10.4	4.8	2.1	1.0
Overall means	4.7	14.1***	5.1	2.4	5.8	1.1**
Mouse						
H2-K (4)	15.0	22.9	8.7	5.8	2.3	4.0
H2-L (4)	11.4	19.5	8.8	6.8	0.0	2.5**
Overall means	13.2	21.2*	8.8	6.3	1.2	3.6**

From Hughes and Nei (1988).
[a]The K_S and K_A values were computed by Nei and Gojobori's (1986) method I. The difference between K_S and K_A is significant at the 5% level (*), the 1% level (**), or the 0.1% level (***).
[b]N, Number of codons compared. N is 124 for the mouse and 125 for the human.

alleles that are shared by species that diverged millions of years ago (see the review by Klein et al. 1993). For example, Figueroa et al. (1988) found that two allelic lineages marked with deletions/insertions at the Aβ1 locus are shared by mice and rats, which diverged at least 10 million years ago. Such polymorphisms are obviously more ancient than the species themselves and are called **transspecies polymorphisms** (Klein 1986). In such a case, the divergence of the allelic lineages predates the divergence of the species. Figure 9.8 shows such an example from the human HLA-A and -B loci and the chimpanzee C-A and -B loci. The human allele H-A11 is closer to the chimpanzee allele C-A108 than to any other human alleles at the A locus, whereas the human allele H-A24 is closer to the chimpanzee allele C-A126 than to any other human alleles. This comparison implies that the two human alleles H-A11 and H-A24 diverged before the human-chimpanzee split and so did their counterparts in the chimpanzee, i.e., alleles C-A108 and C-A126. Thus, a pair of polymorphic allelic lineages is shared by humans and chimpanzees.

Many other examples of transspecies polymorphism at MHC loci have been found in primates and rodents (e.g., McConnell et al. 1988; Lawlor et al. 1988; Mayer et al. 1988; Fan et al. 1989). A most surprising case is that certain human MHC alleles appear to have diverged before the separation of prosimian and anthropoid primates more than 65 million years ago (Klein 1986). The antiquity and the large number of MHC alleles in humans strongly suggest that the lineage leading to modern humans has never gone through any severe population bottleneck (see Klein et al. 1993).

In summary, there are four extraordinary features of MHC polymorphism (Nei and Hughes 1991): (1) the extent of polymorphism is extremely high, (2) the number of nucleotide differences between alleles is unusually large, (3) in the ARS the rate of nonsynonymous substitution is higher than the rate of synonymous substitution, and (4) allelic lineages may persist in a population for tens of millions of years. Any hypothesis for the origin and maintenance of MHC polymorphism must be able to explain all these features.

The antiquity of MHC alleles provides strong evidence against selective neutrality. A pair of neutral alleles at a locus may persist in a population for $2N_e$ generations but not much more (Takahata and Nei 1990), where N_e is the effective population size. If we assume that $N_e = 10^5$ for rodent species as suggested by the extent of protein polymorphism (Nei and Graur 1984) and that there are two generations in a year in nature, then $2N_e$ generations correspond to 10^5 years, which is too short compared to the observed antiquity of rodent MHC alleles. Moreover, the neutral mutation hypothesis cannot explain the third feature, i.e., $K_A > K_S$ in the ARS.

The overdominant selection hypothesis can adequately explain all four features (Nei and Hughes 1991). As a form of balancing selection, overdominant selection can maintain a large extent of polymorphism and, as noted above, it can explain the obervation of $K_A > K_S$ in the ARS. Takahata and Nei (1990) and Takahata (1990) showed that with overdominant selection the coalescence time for alleles in a population can be tens of millions of years under reasonable selection, population, and mutation parameters. As a consequence, large divergences between alleles can occur. The overdominance hypothesis was originally proposed by Doherty and Zinkernagel (1975), who showed that different MHC molecules bind different foreign antigens and argued that a heterozygote having two different MHC molecules would have a better chance to bind foreign antigens and thus would be more resistant to infectious diseases than a homozygote with

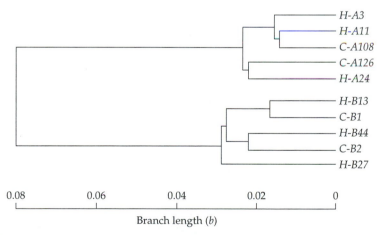

Figure 9.8 Phylogenetic tree for alleles from the class I MHC loci *A* and *B* in the human (*HLA-A* and *-B*) and the chimpanzee (*CHA-A* and *-B*). Human alleles are denoted by *H-* and chimpanzee alleles by *C-*. This tree was obtained by UPGMA, and *b* represents the branch length measured in the number of nucleotide substitutions per site. From Nei and Hughes (1991).

only one type of MHC molecule. For further biological support for this hypothesis, see Nei and Hughes's review (1991).

Frequency-dependent selection was one of the most popular early hypotheses for the maintenance of MHC polymorphism. The most popular model of frequency-dependent selection assumes that host individuals carrying a recently arisen mutant allele have a selective advantage because pathogens will not have had time to evolve the ability to infect host cells carrying a new mutant antigen (Snell 1968; Bodmer 1972). Hughes and Nei (1988) noted that this model will result in a constant turnover of alleles in the population because old alleles lose resistance to pathogens and will become lost by random drift. While this model is expected to generate a higher value of K_A than K_S, it can explain neither the high degree of polymorphism nor the long persistence of MHC alleles; this indeed has been shown theoretically (Takahata and Nei 1990). Another form of frequency-dependent selection assumes minority advantage. Since this form of selection can enhance the retention of alleles as does overdominant selection, it can also explain the above four features of MHC polymorphism. Takahata and Nei (1990), however, argued that while it is possible that a new mutant allele has a selective advantage over old ones because there are no pathogens adapted to it, it is difficult to imagine why an allele should regain a selective advantage when it becomes old and again rare in the population.

Hill et al. (1991) have provided evidence that in West Africa specific class I and class II alleles are associated with resistance to *Plasmodium falciparum* malaria and suggested that their data favor the hypothesis of frequency-dependent selection rather than overdominant selection. Hughes and Nei (1992) argued that the study of Hill et al. was not a test of the overdominance hypothesis because this hypothesis depends on the exposure of the population to two or more pathogens, but Hill et al. did not consider two or more strains of *P. falciparum* and did not study other pathogens. Moreover, among individuals having the protective DRw13.02 haplotype, there is a higher proportion of heterozygotes among those with severe malarial anaemia than with mild controls. This observation suggests that the fitness of DRw13.02 homozygotes is higher than that of DRw13.02 heterozygotes, so that the selection is directional (i.e., not frequency dependent) and thus cannot maintain polymorphism. In response, Hill et al. (1992) argued that their study might be a test of the overdominance hypothesis because *P. falciparum* shows striking polymorphism in many of its immunodominant antigens and because most Gambian children with malaria harbor more than one antigenically distinct parasite strain. They argued further that the finding that alleles with frequencies of 0.13–0.15 are protective is not incompatible with minority-advantage models, for these alleles might have been even more protective when they were less common. As the issue remains controversial, further studies are needed to distinguish between the two types of natural selection for maintaining MHC polymorphism.

Other hypotheses such as a high mutation rate in the MHC complex, gene conversion, mating preference, and maternal-fetal incompatibility either lack solid evidence or may not explain the above four observations adequately (Nei and Hughes 1991). For example, neither a high mutation rate nor gene conversion can explain why the rate of nonsynonynous substitution is higher than that of synonymous substitution in ARS (but not in other regions).

The genetic systems governing self-incompatibility in plants exhibit many features similar to those of the MHC system. In many plant species, self-incompatibility is controlled by a single locus, the *S* locus (De Nettancourt 1977). In gametophytic self-incompatibility, pollen deposited on a stigma fails to germinate and grow into the style of the pistil if the *S* allele expressed by the pollen matches one of the two alleles in the pistil. The *S* system shares three striking features with the MHC system (Ioerger et al. 1990; Clark 1993). First, it carries a very large number of alleles; e.g., a rare *Oenothera* species with a population size of only ~500 individuals contained 45 distinct *S* alleles (Emerson 1939; Lewis 1949). As Wright (1939) showed theoretically, the self-incompatibility system favors the introduction of new alleles because the probability of avoiding a pistil bearing the same allele as the pollen is greatest for rare alleles. This reduces the rate of loss of new alleles from a population, as compared to the case of neutral alleles, and for this reason the *S* locus can maintain many alleles. Second, it exhibits transspecies polymorphisms. Self-incompatibility is a type of balancing selection and so an allele can persist for a very long time. According to Ioerger et al.'s (1990) estimates from nucleotide sequence data, some allelic lineages predated the divergence of three species in Solanacea (an ornamental tobacco plant, a wild petunia, and a wild potato) some 27 to 36 million years ago. Third, it shows extremely high within-species allelic diversity. Indeed, the amino acid sequence similarity between two alleles in a species can be lower than 40%; the high allelic divergence follows from the antiquity of alleles. The sporophytic self-incompatibility system such as that in *Brassica* also shows similar features as the gametophytic system (Nasrallah and Nasrallah 1989; Boyes et al. 1991), though it appears to be controlled by multiple loci.

PROTEIN AND DNA POLYMORPHISMS IN AMERICAN OYSTERS: CONTRASTING PICTURES OF GEOGRAPHIC DIFFERENTIATION

Population geneticists have always been interested in geographically structured populations, because the pattern of genetic differentiation among different parts of such a population may provide insight into the mechanism of maintenance of genetic polymorphism and insight for understanding how genetic differentiation may occur because of environmental variation or isolation by distance (Wright 1943). In the 1970s and early 1980s, starch gel electrophoresis of soluble proteins was a popular tool for conducting such studies. One example was the study of electrophoretic variation in proteins encoded by nuclear genes in American oysters (*Crassostrea virginica*) sampled from Massachusetts to Texas (Buroker 1983). Surprisingly, the study revealed a near uniformity of allele frequencies throughout this range (Figure 9.9), despite the fact that it stretches over 4000 miles. The result was attributed to high gene flow due to "the rather long planktonic stage of larval development, since this species has the ability to disperse zygotes over great distances when facilitated by tidal cycles and oceanic currents" (Buroker 1983).

A very different pattern, however, was observed in a later study using restriction analysis of mtDNA (Reeb and Avise 1990), which showed a major genetic "break" distinguishing all sampled populations north versus south of the mid-Atlantic coast of Florida (Figure 9.9). The latter finding was in striking concordance with a previous restriction analysis of mtDNA in the horseshoe crab (Saunders et al. 1986). In both mtDNA studies, numerous restriction site changes were involved, and the estimated mean nucleotide sequence divergence between the north and south regions was high, about 2.6% and 2.0%, respectively.

There are several possible explanations for the apparent inconsistency between nuclear and cytoplasmic structures in oysters (Reeb and Avise 1990; Karl and Avise 1992). The discrepancy could be due to biological or demographic factors, such as a higher rate of interpopulation gene flow mediated by sperm rather than by eggs, or directional selection favoring different mtDNA haplotypes in the two regions. Alternatively, it could be due to a slower rate of evolutionary change in allozyme frequencies, or balancing selection at multiple allozyme loci.

To distinguish between the above two classes of competing hypotheses, Karl and Avise (1992) conducted an analysis of restriction fragment length polymorphism (RFLP) in single-copy nuclear (scn) DNA. They designed PCR primers suitable for amplifying each of four anonymous nuclear loci (a locus is anonymous if its function and genomic location are unknown). Nuclear DNA was isolated from each of 277 oysters collected at 9 locations between Massachusetts and Louisiana. Restriction site polymorphisms from all four loci revealed a pronounced genetic break along the eastern coast of Florida, as was true for the mtDNA study (Figure 9.9).

The similar pattern of geographic differentiation registered in mtDNA and scnDNA indicates that biological factors differentially affecting mitochondrial and nuclear genomes cannot account for the original discrepancy between the mtDNA and allozyme data sets. Thus, the geographic uniformity in allozymes is probably not due to greater dispersal of male gametes or to the slow rate of evolutionary sorting of nuclear lineages from an ancestral gene pool, because effects of such processes should also be reflected in the distributions of scnDNA alleles. Neither is the population genetic break in mtDNA attributable to directional selection on mtDNA, unless the kinds of selection favoring different mtDNA haplotypes in the Atlantic versus the Gulf extend to scnDNA alleles as well. Indeed, it is difficult to imagine a selective force that could have molded in such a consistent fashion the mtDNA break across species as different as the oyster, horseshoe crab, fishes, reptile, and bird that have been assayed (see the review by Avise 1992). One remaining possibility is that several of the allozyme loci have been under uniform balancing selection, and thus, do not record the population subdivision suggested by the mtDNA and scnDNA data (Karl and Avise 1992). This explanation, of course, requires that the same type of balancing selection operates in both the Atlantic and the Gulf populations. At any rate, either balancing selection must have acted on some allozyme loci, or diversifying selection must have acted on mtDNA and some scnDNA loci. This deduction underlines the need for caution in inferring population structure and gene flow from any single gene or perhaps even class of marker loci (Avise 1994).

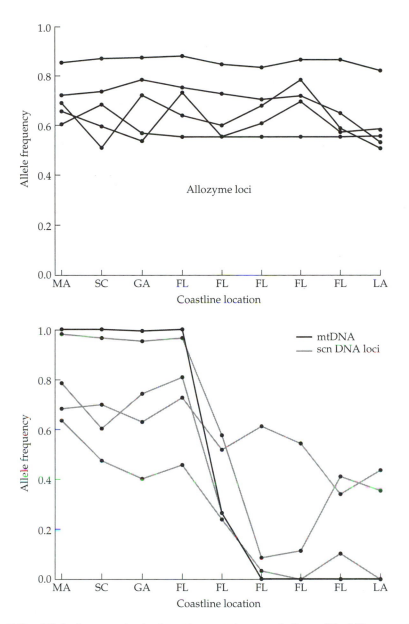

Figure 9.9 Allele frequencies in American oyster populations. (Top) Frequencies of the most common alleles at five polymorphic allozyme loci; (Bottom) five DNA loci along a coastline transect running from Massachusetts (MA) through South Carolina (SC), Georgia (GA), Florida (FL), and Louisiana (LA). The allozyme loci are *Est*1, *Lap*1, *6Pgd*, *Pgi*, and *Pgm*; loci assayed at the DNA level are mtDNA (the solid line) and each of four anonymous single-copy nuclear genes (shaded lines). From Avise (1994).

CHAPTER 10

Evolution by Gene Duplication and Domain Shuffling

T HE EVOLUTIONARY SIGNIFICANCE OF GENE DUPLICATION was first recognized by Haldane (1932) and Muller (1935), who suggested that a redundant duplicate of a gene may acquire divergent mutations and eventually emerge as a new gene. However, few examples of duplicate genes were discovered before the advent of biochemical and molecular biology techniques. The development of protein sequencing methods in the 1950s provided a tool for the study of long-term evolution, and by the late 1950s the α and β chains of hemoglobin were recognized to have been derived from gene duplication (Rhinesmith et al. 1958; Braunitzer et al. 1961; Konigsberg 1961). Later, isozyme, DNA–RNA hybridization, and cytogenetic studies added much further evidence for the frequent occurrence of gene duplication in evolution (e.g., Markert 1964; Ritossa and Spiegelman 1965; Becak et al. 1966; Ritossa et al. 1966; Harris et al. 1967; Wolf et al. 1969). Using data from these various types of studies, Ohno (1970) put forward the view that gene duplication is the only means by which a new gene can arise. Although other means of creating new genes or new functions are now known and are discussed later in this chapter, Ohno's view remains largely valid.

There is now ample evidence that gene duplication is the most important mechanism for generating new genes and new biochemical processes that have facilitated the evolution of complex organisms from primitive ones. For example, the genome of vertebrates contains many gene families that are not found in invertebrates (see Doolittle 1985, 1995), and many gene duplications apparently occurred in the early evolution of chordates (see Ohno 1970; Miyata et al. 1994). Gene duplication is also important for generating many copies of genes of the same function, thereby enabling the production of a large quantity of RNAs or proteins. Furthermore, internal (partial) gene duplication plays a major role in increasing the functional complexity of genes in evolution. Many proteins of present-day organisms show internal repeats of amino acid sequences, and the repeats often correspond to the functional or structural domains of the proteins (see Barker et al. 1978). This observation suggests that the genes coding for these proteins were formed by internal gene duplication. This type of duplication allows the improvement of the function of a protein by increasing the number of active sites or the aquisition of an

269

additional function by modifying a redundant segment. Many complex genes in present-day organisms might have evolved from small, simple primordial genes via internal duplication and subsequent modification.

The discovery of split genes prompted Gilbert (1978) to suggest that recombination in introns provides a mechanism for the exchange of exon sequences between genes, a phenomenon that has been commonly known as **exon shuffling.** Many examples of exon shuffling have been found (see, e.g., Doolittle 1985; Patthy 1985, 1994), indicating that this mechanism has played a significant role in the evolution of eukaryotic genes with new functions. However, an exon-shuffling event is likely to be biologically significant only if it includes a structural or functional domain. Moreover, a domain-shuffling event may occur without the involvement of any exon (see Doolittle 1995). Thus, it is perhaps more meaningful to talk about domain shuffling rather than exon shuffling.

TYPES OF DNA DUPLICATION

A duplication may involve (1) part of a gene, (2) a single gene, (3) part of a chromosome, (4) an entire chromosome, or (5) the whole genome. These types of duplication are called, respectively, (1) partial or internal gene duplication, (2) complete gene duplication, (3) partial chromosomal duplication or partial polysomy, (4) chromosome duplication, aneuploidy, or polysomy, and (5) genome duplication or polyploidy. The first four categories are also known as regional duplications, because they do not affect the entire haploid set of chromosomes.

Ohno (1970) argued that genome duplication has generally been more important in evolution than regional duplication, because in the latter case only parts of the regulatory system of structural genes may be duplicated—such an imbalance may disrupt the normal function of the duplicated genes. However, as will be discussed later, partial and complete gene duplications apparently also have played a very important role in evolution; for example, this importance is suggested by the fact that genes controlling the same function are often closely linked. Partial polysomy, if it involves many genes, and polysomy are likely to cause a severe imbalance in gene product, and their chance of being incorporated into the population is small. Indeed, experiments in *Drosophila* have shown that the viability of terminal partial polysomy declines with the length of the triplicated region, becoming generally lethal when more than one-half of an autosomal arm is present in three doses (Lindsley et al. 1972). Also, it is well known that in man trisomies of larger chromosomes are lethal conditions, and even those of the smaller ones result in sterility; e.g., trisomy for chromosome 21 causes Down's syndrome. Thus, polysomy and partial polysomy are probably not important in evolution. Genome duplication was probably common in animal evolution before sex chromosomes became differentiated. In organisms with well-differentiated sex chromosomes such as mammals, birds, and reptiles, genome duplication disrupts the mechanism of sex determination and is likely to be quickly eliminated from the population; thus, it probably has contributed little in the evolution of bisexual organisms. In contrast, in plants genome duplication is a widespread phenomenon (see, e.g., Lewis 1980; Stebbins 1980). Chromosome duplication and genome duplication as mechanisms of increasing the size of a genome will be discussed in Chapter 13.

DOMAIN DUPLICATION AND GENE ELONGATION

A protein **domain** is a well-defined region within a protein that either performs a specific function, such as substrate binding, or constitutes a stable, compact structural unit within the protein that can be distinguished from all the other parts. The former is referred to as a **functional domain**, and the latter, a **structural domain** or **module** (Gō and Nosaka 1987). Defining the boundaries of a functional domain is often difficult because functionality is in many cases conferred by amino acid residues that are located in different regions of the polypeptide. A structural module, on the other hand, consists of a continuous stretch of amino acids.

The above distinction is important when considering possible evolutionary mechanisms by which multidomain proteins have come into existence. If a functional domain coincides with a module, its duplication will increase the number of functional segments. In contrast, if functionality is conferred by amino acid residues scattered among different modules, the effects of a duplication of a single module may not be functionally desirable. The internal repeats found in many proteins often correspond to either structural modules or functional domains residing within a single module (Barker et al. 1978).

Several possible relationships may exist between the structural domains and the arrangement of exons in the gene (Figure 10.1). In some genes, a more or less exact correspondence exists between the exons of the gene and the structural domains (Figure 10.1a, b); this relationship exists for many globular proteins (Gō 1981). When such a relation exists, duplication of an exon will result in duplication of the corresponding structural domain. In the majority of cases, however, a more complex relationship exists between protein domains and exons, such as shown in the situations in Figure 10.1c–e. In the case of Figure 10.1c, duplication of an exon results in duplication of two domains, whereas in the case of Figure 10.1e, duplication of an exon only duplicates part of a domain, thus being more likely to disrupt the protein function and less likely to survive.

A survey of genes in eukaryotes shows that internal duplications have occurred frequently in evolution. This increase in gene size, or **gene elongation**, is one of the most important steps in the evolution of complex genes from simple ones. Theoretically, elongation of genes can also occur by other means. For example, a mutational change converting a stop codon into a sense codon can also elongate the gene (Chapter 1). Similarly, insertion of a DNA segment into one of the exons or the occurrence of a mutation obliterating a splicing site will achieve a similar result. These types of molecular changes, however, would most probably disrupt the function of the elongated gene, because the added regions would consist of an almost random array of amino acids. Indeed, in the vast majority of cases, such molecular changes have been found to be associated with pathological manifestations. For instance, the hemoglobin abnormalities Constant Spring and Icaria resulted from mutations turning the stop codon into codons for glutamine and lysine, respectively, thus adding 30 additional residues to the α chains of these variants (Weatherall and Clegg 1979). By contrast, duplication of a structural domain is less likely to be problematic. Indeed, such a duplication can sometimes even enhance the function of the protein produced, for example, by increasing the number of active sites. A well-known example is the haptoglobin $\alpha 2$ allele in humans, which was formed by a nonhomologous crossing-over

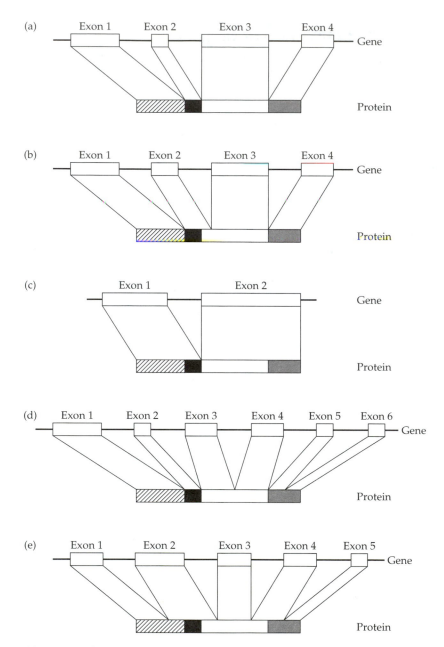

Figure 10.1   Possible relationships between the arrangement of exons in a gene and the structural domains of the protein it encodes: (a) each exon corresponds exactly to a structural domain; (b) the correspondence is only approximate; (c) an exon encodes two or more domains; (d) a single structural domain is encoded by two or more exons; and (e) lack of correspondence between exons and domains. The four structural domains of the protein are designated by different boxes (hatched, solid black, white, and shaded). From Li and Graur (1991).

within different introns of two $\alpha1$ genes, probably a slow and a fast electrophoretic variant ($\alpha1S$ and $\alpha1F$). The internal duplication nearly doubled the length of the polypeptide (changing from 84 to 143 amino acids) and it seems to increase the stability of the haptoglobin-hemoglobin complex and the efficiency in rendering the heme group of hemoglobin susceptable to degradation (Black and Dixon 1968). The $\alpha2$ gene is probably of recent origin, at least more recent than the human-chimpanzee split, but it has a fairly high frequency (30 – 70%) in Europe and in parts of Asia (Mourant et al. 1976). It is likely that the $\alpha2$ allele will replace the $\alpha1$ alleles.

The examples below are to illustrate the point that gene elongation during evolution has depended largely on the duplication of domains.

The $\alpha2$ Type I Collagen Gene

A remarkable example of gene elongation by duplication is given by the $\alpha2$ type I [$\alpha2(I)$] collagen gene. Collagen is the main supportive protein of skin, bone, cartilage, and connective tissue in vertebrates. The major part of the mature collagen consists of three polypepetide chains, called α chains, arranged in a long rodlike helical configuration. This region contains about 338 repeats of the amino acid triplet Gly-X-Y, where X and Y are often proline and hydroxyproline. The chicken $\alpha2(I)$ collagen gene has been sequenced (Yamada et al. 1980; Boedtker et al. 1985); it has a length of approximately 38 kilobases and contains 52 exons, among the largest numbers of exons ever observed in a gene. The triple-helical coding region of the pro-$\alpha2(I)$ collagen gene is composed of 42 triple-helical exons and 2 "junction" exons. Of the 42 triple-helical exons, 5 contain 45 bp; 23, 54 bp; 5, 99 bp; 8, 108 bp; and 1, 162 bp—remarkably, all in multiples of the 9 bp that code for the triplet Gly-X-Y! For example, exon 8 is 54 bp long and encodes 18 repeats of Gly-X-Y as shown below:

GGT CCT CCT GGG TTT CAA GGT GTT CCT GGT GAG CCT GGT GAA CCT
Gly Pro Pro Gly Phe Gln Gly Val Pro Gly Glu Pro Gly Glu Pro

GGT CAA ACA
Gly Gln Thr

It is quite possible that all of these exons were derived from an ancestral exon of 54 bp by multiple duplications and recombinations (Figure 10.2). Many of the 54-bp exons are contiguous and could be the result of a series of exon duplications. On the other hand, a 99-bp exon could have arisen from an unequal crossing-over between two exons of 54 bp (Figure 10.2). One might argue that exons of 108 bp arose by precise deletion of the intervening sequence and fusion of two 54-bp exons. However, the precise deletion of an intron appears to have occurred rarely in evolution, though a number of such cases have been found, e.g., in a rat insulin gene (Lomedico et al. 1979). It seems more likely that an unequal crossing-over between a 99-bp exon and a 54-bp exon gave rise to a 108-bp exon and a 45-bp exon (Figure 10.2). At any rate, because of its regular repetitive structure, partial gene duplication can occur easily in this gene. Such duplication may not disrupt the gene function if the duplicated part is a multiple of the 9-bp sequence coding for Gly-X-Y, which seems to be necessary for forming a helical structure.

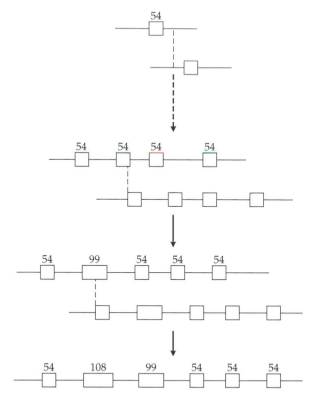

Figure 10.2 A hypothetical scheme for the evolution of the exons in the DNA region coding for the helical region in the α2(I) collagen polypeptide. The number on the top of a block denotes the number of base pairs (bp) in the exon. The dotted line denotes the place of unequal crossing-over. An unequal crossing-over between two exons of 54 bp can give rise to an exon of 99 bp and an exon of 9 bp, and an unequal crossing-over between an exon of 99 bp and an exon of 54 bp can give rise to an exon of 108 bp and an exon of 45 bp. From Li (1983).

Apolipoprotein Genes

The apolipoproteins are the protein components of plasma lipoproteins, which are the major carriers of cholesterol, phospholipids, and triglycerides in vertebrate blood. In mammals, nine different types of apolipoproteins (A-I, A-II, A-IV, B, C-I, C-II,C-III, D, and E) have been identified and characterized (see Li et al. 1988b for a review). The major function of the apolipoproteins is lipid binding so as to make lipoproteins water soluble. Most apolipoproteins have, in addition, acquired specialized functions in lipid metabolism. For example, apoC-II is the physiological activator of lipoprotein lipase, an enzyme that hydrolyzes triglycerides to release fatty acids, which are either used for energy or are re-esterified and stored in adipocytes. The *apoA-I*, *apoA-II*, *apoC-I*, *apoC-II*, *apoC-III*, and *apoE* genes have the same genomic structure, each containing a total of three introns at the same locations (Figure 10.3). The *apoA-IV* gene has lost the first intron, but has retained the other two. Thus, these seven genes clearly belong to one multigene family. Note, however, that the *apoB* and *apoD* genes have very different genomic structures and probably do not belong to the same family (see Li et al. 1988b).

From analysis of amino acid sequence data, Barker and Dayhoff (1977), Fitch (1977a), and McLachlan (1977) found that *apoA-I* contains multiple repeats of a 22 amino acid unit (22-mer) each of which is a tandem array of two 11-mers. The 22-mer repeat unit has been suggested to be a structural element that builds an amphipathic α helix with a proline at the amino end (Segrest et al. 1974; McLachlan 1977). An important feature of the amphipathic helix model (Figure 10.4) is the presence of two clearly defined faces, one hydrophobic, that is inserted between the fatty-acyl chains of the phospholipid molecules, and the other hydrophilic, that interacts with the phospholipid head groups and the aqueous phase. Another feature of the model is that the negatively charged residues, Glu and Asp, tend to occupy positions along the center of the polar face, while the positively charged residues, Lys and Arg, are located on the lateral edges of the polar face.

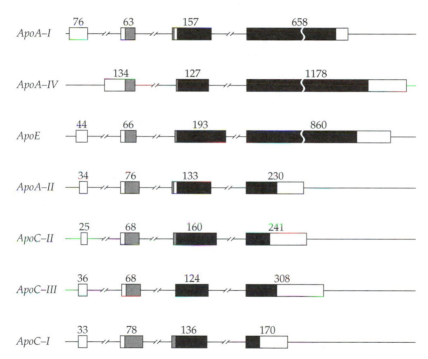

Figure 10.3 Structural organization of the human apolipoprotein A-I, A-II, A-IV, C-I, C-II, C-III, and E genes. Transcription is from left to right. The wide bars represent the exons, and the thin line represents the 5′ flanking region, introns, and 3′ flanking region of the respective genes. The wide bars are divided into several regions: the open bars at the two ends represent the 5′ and 3′ untranslated regions; the shaded bars, the signal peptide regions; and the solid bars, the mature-peptide regions of the respective genes. In *apoA-I* and *apoA-II*, the prosegment is represented by a narrow open bar between the signal peptide and mature peptide region. The numbers above the exons indicate the length (number of nucleotides) of the exons. The lengths of the exons are drawn to scale, except for the last exons in *apoA-I, A-IV,* and *E*. From Chan et al. (1990).

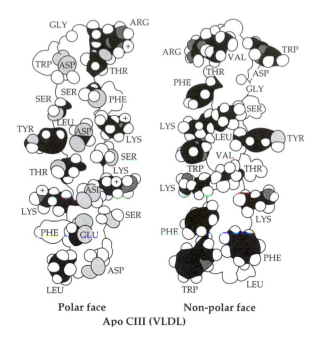

GLY   ARG
TRP ASP   THR
SER   SER   PHE
TYR   LEU ASP   LYS
THR   SER   LYS
LYS   ASP   SER
PHE   GLU
LEU   ASP

ARG   VAL   TRP
THR   ASP
PHE   GLY
SER
LYS   LEU   TYR
VAL
TRP   THR
LYS
LYS
PHE   PHE
TRP   LEU

Polar face   **Non-polar face**
Apo CIII (VLDL)

Figure 10.4   Space-filling models of the amphipathic helical region of apoC-III. The amphipathic sequence, residues 40–67, has been built as an α helix and the polar and nonpolar faces are shown. From Segrest et al. (1974).

DNA-sequencing work has provided better data for defining the repeats in *apoA-I* (Karanthanasis et al. 1983; Luo et al. 1986) and has further revealed similar repeat patterns in *apoA-II, apoA-IV, apoC-I, apoC-II, apoC-III,* and *apoE* (Boguski et al. 1984; Luo et al. 1986). The results are summarized in Figure 10.5. In each gene the repeats start from the first codon in exon 4 and end close to the end of this exon. Most of the repeats are full 22-mers, each of which is composed of two half-repeats (11-mers) and is divided into groups A and B. The main reason for considering the 22-mer as the fundamental unit of repeat is that most 11-mers are more similar to the 11-mer one unit removed than to the adjacent 11-mer (Figure 10.5; Fitch 1977a).

Some remarkable features emerge from the alignment shown in Figure 10.5. Most prolines (P) appear in the first column of group A. Columns 6 and 10 of both groups are mostly occupied by hydrophobic amino acids (M, methionine; V, valine; L, leucine; I, isoleucine; F, phenylalanine; Y, tyrosine; W, tryptophan), the rest being indifferent amino acids (G, glycine; A, alanine; S, serine; T, threonine; N, asparagine; Q, glutamine; H, histidine; C, cysteine). In both groups, column 3 is predominantly occupied by hydrophobic amino acids, columns 4 and 5 by acidic amino acids (D, aspartic acid; E, glutamic acid), and column 9 by basic amino acids (R, arginine; K, lysine); in column 11 there is no clear pattern. In group A, column 2 is predominantly occupied by hydrophobic amino acids, but in group B, this column has no predominant pattern. Thus, there are major similarities and only minor differences between groups A and B.

The large variation in size among the soluble apolipoproteins (see Figure 10.3) used to be a puzzle because they were apparently derived from a common ancestral gene. It is now clear that the size variation is largely due to differences in the number of repeats in the last exon, i.e., exon 4.

In view of the fact that most of these apolipoproteins have diverged before the bird–mammal split, i.e., more than 270 million years ago, it is remarkable that

Figure 10.5   Internal repeats in exons 4 of human apolipoproteins A-I, A-IV, E, A-II, C-III, C-II, and C-I. Most of the repeats are 22-mers (22 residues long), each of which is made up of two 11-mers, and the other repeats are 11-mers. The amino acids are divided into five groups: (1) the proline (P) group; (2) the acid group: aspartic acid and glutamic acid (D and E); (3) the basic group: arginine and lysine (R and K); (4) the hydrophobic group: methionine, valine, leucine, isoleucine, phenylalanine, tyrosine, and tryptophan (M, V, L, I, F, Y, and W); and (5) the indifferent group: glycine, alanine, serine, threonine, asparagine, glutamine, histidine, and cysteine (G, A, S, T, N, Q, H, and C). A column containing 16 or more amino acids of the same group (except for the indifferent group) is said to possess that character, and the amino acids of that group are boxed. Numbers along the left and right margins are residue numbers in the mature peptide. From Li et al. (1988b).

they still share so much similarity in repeat pattern. Thus, the 22-mer repeat unit appears to be a conserved structural and functional unit. However, the existence of many copies of this repeat in a protein has allowed the acquisition of new functions. For example, apoE can bind to both the apoB\apoE low-density lipoprotein receptor and a specific apoE receptor and, in addition to its function in lipoprotein metabolism, it may be involved in neural regeneration, immunoregulation, and modulation of growth and differentiation (see the review by Mahley 1988).

Prevalence of Domain Duplication

Table 10.1 shows further examples of gene elongation by duplication. All involve one or more domain duplications, and some of the sequences (e.g., ferredoxin, serum albumin, and tropomyosin α chain) were evidently derived from doubling or multiplication of a primordial sequence.

In each of the above examples, the duplication event can easily be inferred from sequence homology. There may be many other complex genes that have also evolved by internal gene duplication, but the duplicated regions have become so diverged that sequence homology between them is no longer discernable. For example, the variable and constant region domains of immunoglobulin genes show no significant sequence homology, but their tertiary structures suggest that the two kinds of domains were derived from a common ancestral domain (see Hood et al. 1975). Thus, internal duplications in protein evolution are probably much more ubiquitous than the empirical data have indicated.

Occasionally, duplication of a domain may enable a gene to perform a new function. For example, it seems that the coenzyme binding domain of several dehydrogenases, which binds nicotinamide adenine dinucleotide (NAD), was derived from duplication of a primordial mono-nucleotide binding domain (Rossman et al. 1975). Moreover, two duplicated domains may diverge in function and enable the gene to perform a new or more complex function. For instance, the variable and constant region domains of immunoglobulin genes were probably derived from a common ancestral domain, but they now have distinct functions—antigen recognition by the variable regions and effector functions by the constant regions (Leder 1982). Many complex genes might have arisen in this manner.

FORMATION OF GENE FAMILIES AND ACQUISITION OF NEW FUNCTIONS

A complete gene duplication produces two identical copies. How they will evolve varies from case to case. The copies may, for instance, retain their original function, enabling the organism to produce a larger quantity of certain RNA species or proteins. Or, one copy may be incapacitated by the occurrence of deleterious mutations and become a functionless pseudogene (see the next section). More importantly, however, gene duplication may result in the emergence of genetic novelties or new genes. This will happen if one of the duplicates retains its original function, while the other accumulates molecular changes such that, in time, it can perform a rather different task.

Repeated genes are divided into two types: variant and invariant repeats. Invariant repeats are identical or nearly identical in sequence to one another. In

TABLE 10.1 Examples of proteins with internal domain duplications

Sequence (organism)	Length of protein[a]	Length of repeat	Number of repeats	Percent repetition
$\alpha_1\beta$-glycoprotein (human)	474	91	5	96
Angiotensin I-converting enzyme (human)	1306	357	2	55
Calbindin (human, bovine)	260	43	6	99
Calcium-dependent regulator protein (human)	148	74	2	100
Epidermal-growth-factor precursor (human)	1217	40	3	10
Ferredoxin (*Clostridium pasteurianum*)	55	28	2	100
Fibronectin (human)	2324	40	12	21
Guanosine cyclic 3',5'-phosphate-dependent protein kinase (human)	670	120	2	36
Hemopexin (human)	439	207	2	94
Hexokinase (human)	917	447	2	97
Immunoglobulin γ chain C region (human)	329	108	3	98
Immunoglobulin ε chain C region (human)	423	108	4	100
Interleukin-2 receptor (human)	251	68	2	54
Interstitial retinol-binding protein (human)	1243	303	4	98
Lactase-phlorizin hydrolase (human)	1927	480	3[b]	79
Lymphocyte-activation gene 3 (LAG-3) protein (human)	470	138	2	59
Ovoinhibitor (chicken)	472	64	7	95
P-glycoprotein (human)	1280	609	2	95
Parvalbumin (human)	108	39	2	72
Plasminogen (human)	790	79	5	50
Proglucagon (rat)	161	36	3	67
Pro-von Willebrand factor (human)	2791	586	3	
		30	2	
		117	2	
		354	4[b]	86
Protease inhibitor, Bowman-Birk type (soybean)	71	28	2	79
Protease inhibitor, submandibular-gland type (rodent)	115	54	2	94
Ribonuclease/angiogenin inhibitor (human)	461	57	8	99
Ribosomal protein L35 (rat)	122	11	2	18
Serum albumin (human)	584	195	3	100
Tropomyosin α chain (human)	284	42	7	100
Twitchin (*Caenorhabditis elegans*)	6049	100	31	
		93	26	91
Villin (human)	826	360	2	87
Vitamin D-dependent calcium-binding protein (bovine)	260	54	3	63

Data from Barker et al. (1978) and Graur and Li (unpublished).
[a]Number of amino acid residues.
[b]Plus one truncated repeat.

several cases, the repetition of identical sequences can be shown to be correlated with the synthesis of increased quantities of a gene product that is required for the normal function of the organism. Such repetitions are referred to as **dose repetitions**. Dose repetitions are quite common whenever a metabolic need for producing large quantities of specific RNAs or proteins arises (Ohno 1970). Representative examples include the genes for rRNAs and tRNAs (see below), which are required for translation, and the genes for histones, which constitute the chief protein component of chromosomes and are synthesized in large quantities during cell division (Elgin and Weintraub 1975). Another good example of dosage repetition is the amplification of an esterase gene in a California *Culex* mosquito, leading to insecticide resistance (Mouchäs et al. 1986).

Variant repeats consist of copies of a gene that, although similar to each other, differ in their sequence to a lesser or greater extent. Representatives of variant repeats are various types of isozymes (see below). Variant repeats may differ to some extent in function or regulation, so that they contribute to the fine-tuning of physiological functions of an organism.

All the genes that belong to a certain group of repeated sequences in a genome are referred to as a **gene family** or **multigene family**. Members of a gene family usually reside in close proximity to each other on the same chromosome. In some cases, some functional or nonfunctional family members may be located on other chromosomes.

When duplicate genes become very different from each other in either function or sequence, it may no longer be convenient to assign them to the same gene family. The term **superfamily** was coined by Dayhoff (1978) in order to delineate between closely related and distantly related proteins. Accordingly, proteins that exhibit at least 50% similarity to each other at the amino acid level are considered to be members of a family, while homologous proteins exhibiting less than 50% similarity are considered to be members of a superfamily. For example, the α globins and the β globins are classified into two separate families, and together with myolgobin they form the globin superfamily (see later). However, the two terms cannot always be used strictly according to Dayhoff's criteria. For example, human and carp α-globin chains exhibit only a 46% sequence similarity, which is below the limit for assigment to the same gene family. For this reason, the classification of proteins into families and superfamilies is determined not only according to sequence resemblance, but also by considering auxiliary evidence pertaining to functional similarity or tissue specificity. The genomes of higher eukaryotes contain numerous superfamilies of genes, e.g., the collagens, the actins, the immunoglobulins, and the serine proteases. In contrast to the case of a gene family, the members in a superfamily usually have diverged a long time ago and may have quite distinct functions. A well-known example is that of trypsin and chymotrypsin, which diverged about 1500 Myr ago (Barker and Dayhoff 1980). These two digestive enzymes have acquired distinct functions: trypsin cleaves polypeptide chains at arginine and lysine residues, whereas chymotrypsin cleaves polypeptide chains at phenylalanine, tryptophan, and tyrosine residues. Examples of gene superfamilies will be presented in later sections.

The number of genes in a gene family varies widely. Some genes are repeated within the genome a few times, and these are referred to as "lowly repetitive." Others may be repeated 50 times or more within the genome and are called "highly repetitive." Between these two extremes are moderately repetitive genes. In the following section, rRNA and tRNA genes will be used to illustrate highly repetitive invariant genes. Lowly repetitive genes will be represented by isozymes and the color-sensitive pigment proteins.

RNA-Specifying Genes

Table 10.2 shows the numbers of rRNA and tRNA genes for a variety of organisms. The mitochondrial genome of mammals contains only one 12S and one 16S rRNA gene. This is apparently sufficient for the mitochondrial translation system, because the genome has only 13 protein-coding genes (Anderson et al. 1981). The mycoplasmas, which are the smallest self-replicating prokaryotes, contain two sets of rRNA genes. The genome of *Escherichia coli* is 4–5 times larger and it contains 7 sets of rRNA genes. The number of rRNA genes in yeast is approximately 140, and the numbers in fruit flies and humans are even larger. *Xenopus laevis* has a larger genome size and a larger number of rRNA genes than humans. Thus, there exists a good correlation between the number of rRNA genes and genome size. This is also true for the number of tRNA genes (Table 10.2).

A conspicuous exception to the above rule is that *Tetrahymena* has a genome size considerably larger than that of *Saccharomyces cerevisiae* but has only one set of rRNA genes. This is, however, the copy number in the germinal nucleus, i.e., the micronucleus. In the derivation of macronuclei from the micronucleus the copy number is amplified to about 200 (Yao et al. 1974). Thus, a large number of rRNAs can be produced in a cell, because macronuclei serve as somatic nuclei during vegetative growth and are responsible for virtually all of the transcriptional activity during growth and division.

Highly repetitive genes such as the rRNA genes are generally very similar to each other in sequence. One factor responsible for the homogeneity may be purifying selection, because such genes may have very specific functional or structural requirements. However, homogeneity often extends to regions having no known function (e.g., untranscribed regions), and so the maintenance of

TABLE 10.2   *Numbers of rRNA and tRNA genes per haploid genome in various organisms*

Genome source	Number of rRNA genes[a]	Number of tRNA genes	Approximate genome size (bp)
Human mitochondrion	1	22	1.7×10^4
Mycoplasma genitalium	2	33	5.8×10^5
Escherichia coli	7	~100	4×10^6
Saccharomyces cerevisiae	~140	320–400	1.3×10^7
Tetrahymena thermophila	1	ND[b]	2×10^8
Drosophila melanogaster	130–250	~750	2×10^8
Human	~300	~1,300	3×10^9
Xenopus laevis	400–600	~7,800	8×10^9

Updated from Li (1983).

[a]For rRNA genes, the values refer to the number of complete sets of rRNA genes.

[b]ND = not determined.

homogeneity necessitates invoking other mechanisms such as unequal crossing-over and gene conversion (see Chapter 11).

Isozymes

In addition to invariant repeats, the genomes of higher organisms contain numerous multigene families whose members have diverged to various extents in regulation and function. Good examples are families of genes coding for isozymes such as those for lactate dehydrogenase, aldolase, creatine kinase, pyruvate kinase, and carbonic anhydrase (see the Isozymes series edited by M. C. Rattazzi et al. 1982). **Isozymes** are enzymes that catalyze the same biochemical reaction but may differ from each other in tissue specificity, developmental time, electrophoretic mobility, or biochemical properties. They are encoded by different loci, usually duplicated genes, as opposed to **allozymes**, which are distinct forms of an enzyme encoded by different alleles at a single locus.

The study of multilocus isozyme systems has greatly enhanced our understanding of how cells with identical genetic materials can differentiate into hundreds of different types that constitute the complex body organization of vertebrates. Although isozymes still serve the same function, different members of the family may become better adapted for utilization of substrates in different tissues and this increases the fine tuning of cell physiology.

A well-known example of isozymes is the two genes encoding the A and B subunits of lactate dehydrogenase (LDH) in vertebrates (Markert 1964). These two subunits form five tetrameric isozymes, A_4, A_3B, A_2B_2, AB_3, and B_4, all of which catalyze either the conversion of lactate into pyruvate in the presence of the coenzyme nicotinamide adenine dinucleotide (NAD^+), or the reverse reaction in the presence of the reduced coenzyme (NADH). It has been suggested that LDH-B_4 and the other isozymes rich in B subunits, which have a high affinity to NAD^+, function as a true lactate dehydrogenase in aerobically metabolizing tissues such as heart, whereas LDH-A_4 and the isozymes rich in A subunits, which have a high affinity for NADH, are especially geared to serve as a pyruvate reductase in anaerobically metabolizing tissues such as skeletal muscle (Everse and Kaplan 1975; Nadal-Ginard and Markert 1975). Figure 10.6 shows the developmental sequence of LDH production in the heart. The more anaerobic the heart is, specifically in the early stages of gestation, the higher the proportion of LDH isozymes rich in A subunits will be. Thus, the two genes have become specialized to different tissues and to different developmental stages. As the two subunits are present in almost all vertebrates studied to date, the duplication that produced the two genes probably occurred before or during the early stage of vertebrate evolution.

An interesting feature of the above example is that the two subunits can form heteromultimers, which increase the versatility of the enzyme. Many other examples of multimeric enzymes that are composed of polypeptides encoded by duplicated genes can be found among multilocus isozymes (see Edwards and Hopkinson 1977).

Color-Sensitive Pigment Proteins

Humans, apes, and Old World monkeys possess three color-sensitive pigment proteins. The blue pigment is encoded by an autosomal gene, while the red and green pigments are each encoded by an X-linked gene (Nathans et al. 1986). The

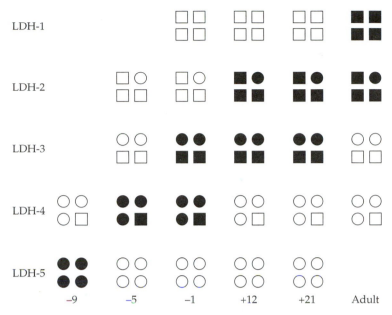

LDH-1

LDH-2

LDH-3

LDH-4

LDH-5

−9   −5   −1   +12   +21   Adult

Figure 10.6 Developmental sequences of five lactate dehydrogenase (LDH) isozymes in the rat heart. Negative and positive numbers at the bottom denote days before or after birth, respectively. Squares indicate B subunits and circles indicate A subunits. Solid symbols indicate quantitatively predominant forms. Notice the shift from A to B subunits during ontogenesis. From Li and Graur (1991).

amino acid sequences of the red and green pigments are 96% identical, but their similarity with the blue pigment is only 43%. The blue pigment gene and the ancestor of the green and red pigment genes diverged about 600 million years ago (Yokoyama and Yokoyama 1989). In contrast, the close linkage and high homology between the red and green pigments point to a very recent gene duplication. Because New World monkeys have only one X-linked pigment gene, whereas Old World monkeys and humans have two or more, the duplication probably occurred in the ancestor of apes and Old World monkeys after their divergence from the New World monkeys about 35–40 million years ago. As a consequence of this duplication, humans, apes, and Old World monkeys can distinguish three colors (i.e., they are **trichromatic**).

Interestingly, in some New World monkey species, e.g., squirrel monkeys, there are three highly polymorphic alleles at the X-linked locus (Jacobs and Neitz 1986). Two of these alleles have spectral sensitivity peaks similar to those of human red and green color pigments, respectively, while the third one has an intermediate spectral peak. For this reason, a female monkey heterozygous for two of the three alleles is trichromatic, like a normal human, though male monkeys, who carry only one X chromosome, are **dichromatic**, like a color-blind human, i.e., they can only distinguish between blue and green or blue and red, but not between green and red. Thus, in the case of humans and Old World monkeys, trichromatic vision is achieved by a mechanism akin to isozymes (i.e., two distinct proteins encoded by different loci), while in heterozygous female squirrel monkeys, the same end is achieved through the use of two "allozymes" (i.e.,

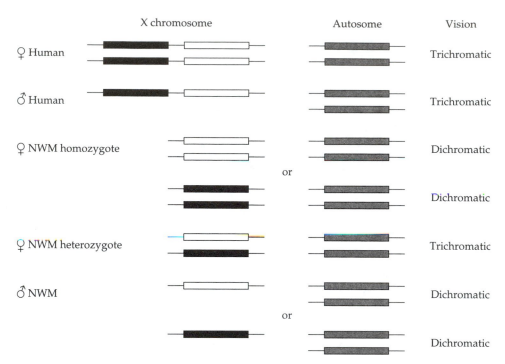

Figure 10.7 Molecular basis of trichromatic vision in humans and New World monkeys (NWM). Although there is only one X-linked color photopigment gene in NWM, a female heterozygous for two alleles is trichromatic, like a normal human. Note that male New World monkeys cannot achieve trichromatic vision. The solid, empty, and shaded boxes denote the red-, green-, and blue-photopigment genes, respectively. From Li and Graur (1991).

two distinct allelic forms at a single locus) (Figure 10.7). If trichromatic vision confers a selective advantage on its carriers, as commonly thought, then the long-term maintenance of the three color-sensitive alleles at a locus in New World monkeys is achieved through overdominant selection (Chapter 2). For more details about the evolution of the triallelic system in New World monkeys, see Shyue et al. (1995).

NONFUNCTIONALIZATION OF REDUNDANT DUPLICATES

A duplicate gene may accumulate deleterious mutations and become nonfunctional, as long as the other duplicate gene is functioning normally (Haldane 1933). Indeed, this is a much more likely fate for a redundant duplicate gene than evolving into a new gene, because deleterious mutations occur far more often than advantageous ones (Ohno 1972). The nonfunctionalization of a duplicate gene produces a pseudogene. Pseudogenes thus produced are called **unprocessed** pseudogenes, as opposed to processed pseudogenes, which will be discussed in Chapter 12.

Table 10.3 lists the structural defects found in several globin pseudogenes. Most of these pseudogenes contain multiple defects such as frameshifts, premature stop codons, and obliteration of splicing sites or regulatory elements, so that it is difficult to identify the mutation that was the direct cause of gene silencing. In a few cases, identification of the "culprit" is possible. For example, human globin pseudogene $\psi\zeta1$ contains only a single major defect, a nonsense mutation, which is probably the direct cause of nonfunctionalization. (The notation ψ is used to distinguish a pseudogene from its functional counterpart.) Some pseudogenes, such as the $\psi\beta^x$ and $\psi\beta^z$ in the goat β-globin multigene family, have been derived from the duplication of an existing pseudogene (Cleary et al. 1981).

Considering the pattern of increase in genome size from bacteria to mammals, Nei (1969) predicted that the genomes of higher organisms would contain a large number of nonfunctional genes. There is now indeed ample evidence that loss of duplicate genes has occurred repeatedly in the long history of the evolution of life. For example, electrophoretic studies suggest that salmonid and catostomid fishes, which are of tetraploid origin, have lost duplicate gene expression in about 50% of the cases (e.g., Allendorf et al. 1975; Ferris and Whitt 1977; Ferris et al. 1979; Buth 1979). Furthermore, pseudogenes have been found in almost every gene family that has been examined in detail, e.g., the 5S rRNA genes (Jacq et al. 1977), the α- and β-globin gene clusters (Nishioka et al. 1980; Proudfoot 1980; Little 1982), and the immunoglobulin genes (Huang et al. 1981).

The rate of loss of a duplicate gene from a population or the rate of fixation of a pseudogene in a population depends on population size as well as on the rate of null mutation. This is because the process occurs mainly by random genetic drift. In an extremely large population, the effect of genetic drift is negligible, so that defective duplicate genes may never be fixed in the population (Fisher 1935; Nei and Roychoudhury 1973). In finite populations, fixation of pseudogenes can occur and many authors have studied the rate of fixation, assuming that there are two duplicate loci and that mutation occurs irreversibly from the normal to the nonfunctional or null state (Bailey et al. 1978; Kimura and King 1979; Li 1980; Watterson 1983; and references therein). In the case of no linkage, Watterson (1983) showed that the mean time until fixation of the null allele at one of the two loci is roughly given by

$$T = N \log N - N\psi(2Nv) + 2.53N \tag{10.1}$$

where $\psi(.)$ denotes the digamma function, v is the rate of mutation from the normal to the null state per gene per generation, and N is the effective population size. It is assumed that the population is initially free of null mutants and that double-null homozygotes are lethal but all other genotypes are normal.

Table 10.4 shows some simulation results with the same assumptions as above. There are several interesting properties. First, tight linkage ($r = 0.001$) has only a minor effect on T, as long as the population size is not extremely large. Second, T is small in a small population but increases with population size. This is because as population size increases the effect of random drift becomes weaker and selection becomes effective. When N is 10^6, it takes approximately 10 million generations for a pseudogene to be fixed in the population. Third, in a small population T is largely determined by v, the null mutation rate. Indeed, when N is of

TABLE 10.3   Defects (+) in globin pseudogenes[a]

Pseudogene	TATA box	Initiation codon	Frame shift	Premature stop
Human $\psi\alpha1$		+	+	+
Human $\psi\zeta1$				+
Mouse $\psi\alpha3$	+		+	+
Mouse $\psi\alpha4$			+	
Mouse $\beta h3$	?	+	+	+
Goat $\psi\beta^X$	+		+	+
Goat $\psi\beta^Z$	+		+	+
Rabbit $\psi\beta2$			+	+

From Li (1983).

[a]A plus indicates the existence of a particular type of defect; a question mark indicates the possibility of the defect.

the order of 100, T is roughly given by $1/(2v)$. For example, when $v = 10^{-5}$, $N = 100$, and $r = 0.5$, T is approximately 50,900 generations. In a large population, however, an increase in v has only a minor effect on T. For instance, if $N = 10^5$ and $r = 0.5$, T decreases from 1.16×10^6 to 0.97×10^6 when v increases from 10^{-5} to 10^{-4}. Fourth, the standard deviation is as large as the mean. Thus, the fixation time is subject to large random errors.

TABLE 10.4   Mean time (T) until fixation of a null allele at one of two duplicate loci

v[a]	N[a]	r[a]	T (in generations)
10^{-5}	10^2	0.5	50,900 (50,900)[b]
		0.001	50,700 (51,000)
	10^3	0.5	62,400 (57,300)
		0.001	57,000 (51,200)
	10^4	0.5	163,000 (133,000)
		0.001	134,000 (104,000)
	10^5	0.5	1,160,000 (934,000)
		0.001	936,000 (685,000)
	10^6	0.5	10.8×10^6 (8.7×10^6)
		0.001	9.3×10^6 (7.0×10^6)
10^{-4}	10^2	0.5	5,870 (5,500)
	10^3	0.5	14,000 (11,600)
	10^4	0.5	127,000 (82,900)
	10^5	0.5	970,000 (756,000)

From Li (1980).

[a]v, null mutation rate; N, effective population size; r, recombination value.
[b]The values in parentheses are standard deviations.

TABLE 10.3 *(continued)*

Pseudogene	Essential amino acid	Splice GT/AG	Stop codon	AATAAA
Human $\psi\alpha1$	+	+	+	+
Human $\psi\zeta1$				+
Mouse $\psi\alpha3$		+		
Mouse $\psi\alpha4$	+			
Mouse $\beta h3$	+	+	?	?
Goat $\psi\beta^X$	+	+	+	+
Goat $\psi\beta^Z$	+	+	+	+
Rabbit $\psi\beta2$	+	+		

The preceding computation shows that the rate of fixation of pseudogenes is highly dependent on population size and mutation rate. At the present time, however, there is still no data to test the details of these relationships. In practice, the rate of fixation would depend on several other factors such as the effect of null alleles on the fitness of heterozygotes (Takahata and Maruyama 1979; Li 1980), attainment of disomic segregation at meiosis in the case of tetraploidization (Li 1980), and differentiation of the regulatory systems of the duplicate genes (Ferris and Whitt 1979). All these factors would complicate the relationships among T, N, and v in Table 10.4. In general, however, these factors have an effect of prolonging the fixation time (Li 1982).

In a multigene family, the gene number may fluctuate from time to time as a result of unequal crossing-over. If the number becomes larger than the optimal number, some of the genes will be free to become pseudogenes. In this case, the mean time for the first pseudogene to become fixed in the population is expected to be considerably shorter than that for the case of two duplicate genes, because any of the multiple genes can become nonfunctional. This seems to be in agreement with actual data, because most of the multigene families include some pseudogenes at present.

DATING GENE DUPLICATIONS

Two genes are said to be **paralogous** if they are derived from a duplication event, but **orthologous** if they are derived from a speciation event. For example, in Figure 10.8, genes α and β were derived from duplication of an ancestral gene and are therefore paralogous, while gene α from species 1 and gene α from species 2 are orthologous, and so are gene β from species 1 and gene β from species 2.

We can estimate the date of duplication, T_D, from sequence data, if the rate of substitution in genes α and β is known. The rate of substitution can be estimated from the number of substitutions between the orthologous genes in conjunction

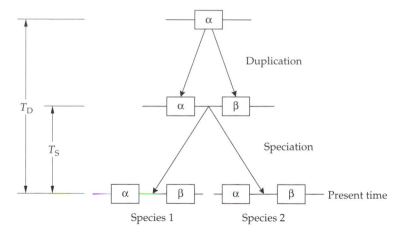

Figure 10.8 Model for estimating the time of a gene duplication event (T_D). Two genes α and β were derived from a duplication event T_D time units ago in an ancestral species. The species then split into two species, 1 and 2, T_S time units ago. The α genes in species 1 and 2 are orthologous, and so are the β genes, but the α genes are paralogous to the β genes. From Li and Graur (1991).

with knowledge of the time of divergence, T_S, between species 1 and 2 (Figure 10.8). An estimate of T_D can be obtained as follows.

For gene α, let K_α be the number of substitutions per site between the two species. Then, the rate of substitution in gene α is estimated by

$$r_\alpha = K_\alpha / (2T_S) \tag{10.2}$$

The rate of substitution in gene β, r_β, can be obtained in a similar manner. The average substitution rate for the two genes is given by

$$r = \left(r_\alpha + r_\beta\right)/2 \tag{10.3}$$

To estimate T_D, we need to know the number of substitutions per site between genes α and β ($K_{\alpha\beta}$). This number can be estimated from four pairwise comparisons: (1) gene α from species 1 and gene β from species 2, (2) gene α from species 2 and gene β from species 1, (3) both genes from species 1, and (4) both genes from species 2. From these four estimates we can compute the average value ($\overline{K}_{\alpha\beta}$) for $K_{\alpha\beta}$, from which we can estimate T_D as

$$T_D = \overline{K}_{\alpha\beta} / (2r) \tag{10.4}$$

Note that in the case of protein-coding genes, by using the numbers of synonymous and nonsynonymous substitutions separately, we can obtain two independent estimates of T_D. The average of these two estimates may be used as the final estimate of T_D. However, if the number of substitutions per synonymous site between genes α and β is large, say larger than 1, then the number of synonymous substitutions cannot be estimated accurately, and so synonymous substitutions may not provide a reliable estimate of T_D. In such cases, only the number of nonsynonymous substitutions should be used.

In the above, we have assumed rate constancy. This assumption can be tested by the four pairwise comparisons mentioned above. The assumption fails if an approximate equality does not hold among the four comparisons. As will be discussed in the next chapter, problems due to concerted evolutionary events may also arise and complicate the estimation of T_D.

Another method for dating gene-duplication events is to consider the phylogenetic distribution of genes in conjunction with paleontological data pertinent to the divergence date of the species in question. For example, all vertebrates with the exception of jawless fish (Agnatha) encode α- and β-globin chains. There are two possible explanations for this observation. One is that the duplication event producing the α and β globins occurred in the common ancestor of the Agnatha and the other vertebrates, but all the Agnatha species have lost one of the two duplicates. This is possible but not likely, because such a scenario would require that the losses occurred independently in many evolutionary lineages. The other explanation is that the duplication event occurred after the divergence of jawless fish from the ancestor of all other vertebrates but before the radiation of the other vertebrates from each other (450–500 million years ago). This latter explanation is thought to be more plausible, and the duplication date is commonly taken to be 450–500 million years ago (Dayhoff 1972; Dickerson and Geis 1983).

Obviously, the above methods can only provide us with rough estimates of duplication dates, and all estimates should be taken with caution. Note that in comparison with estimation of species divergence dates, estimation of duplication dates is more difficult because in the former case many genes can be used whereas in the latter only the sequence data of the duplicate genes can be used, so that the estimate is subject to larger standard errors.

THE GLOBIN SUPERFAMILY OF GENES

In this section the globin superfamily and in the next section the homeobox genes will be used as two examples to illustrate how divergent duplicate genes may contribute to the physiological or structural complexity of higher organisms.

The globin superfamily has experienced all the possible evolutionary pathways that can occur in families of duplicated genes: (1) retention of original function, (2) acquisition of new function, and (3) loss of function in some duplicates. In humans, the globin superfamily consists of three families: the myoglobin family, whose single member is located on chromosome 22, the α-globin family on chromosome 16, and the β-globin family on chromosome 11 (Figure 10.9).

The three families of globin genes produce two types of functional proteins: myoglobin and hemoglobin. The two proteins diverged about 600–800 million years ago (Figure 10.10; Dayhoff 1972; Doolittle 1987) and have become specialized in several respects. In terms of tissue specificity, myoglobin became the oxygen-storage protein in muscles, whereas hemoglobin became the oxygen carrier in blood. In terms of quarternary structure, myoglobin retains a monomeric structure, while hemoglobin has become a tetramer composed of two α and two β chains. In terms of function, myoglobin evolved a higher affinity for oxygen than did hemoglobin, while the function of hemoglobin has become more refined and diversified (see Bunn and Forget 1986; Stryer 1988). Mammalian hemoglobin,

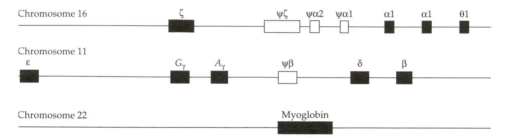

Figure 10.9   The chromosomal arrangement of the three gene families belonging to the globin superfamily of genes in humans: the α-globin family on chromosome 16; the β-globin family on chromosome 11; and myoglobin on chromosome 22. Solid black boxes denote functional genes, while empty boxes denote pseudogenes. From Li and Graur (1991).

for instance, has the capabilities of (1) transporting H^+ and CO_2 in addition to O_2, (2) binding four oxygen molecules cooperatively, (3) responding to tissue acidity (the Bohr effect), and (4) regulating its oxygen affinity through the level of organic phosphate in blood (see Chapter 7). Apparently, the heteromeric structure has facilitated the refinement of the function of hemoglobin. In fact, isolated α chains are monomers and do not have properties (2)–(4) and although β chains can form tetramers, they do not show cooperativity and thus, like the α chain and myoglobin, would not be a suitable oxygen carrier in blood.

The hemoglobin in humans and the vast majority of vertebrates is made up of two types of chains, one encoded by an α family member, the other by a member of the β family. As discussed above, the α and β families diverged about 450-500 million years ago (Figure 10.10). Since jawless fish contain only one type of monomeric hemoglobin, polymerization of hemoglobin in vertebrates probably occurred close to the time of the α–β divergence.

In humans, the α family consists of four functional genes: ζ, $\alpha1$, $\alpha2$, and $\theta1$ (Figure 10.9). It also contains three pseudogenes: $\psi\zeta$, $\psi\alpha1$, and $\psi\alpha2$. The β family consists of five functional genes: ϵ, $^G\gamma$, $^A\gamma$, β, and δ, and one pseudogene, $\psi\beta$. The two families have diverged in both physiological properties and ontological regulation. In fact, distinct globins appear at different developmental stages: $\zeta_2\epsilon_2$ and $\alpha_2\epsilon_2$ in the embryo, $\alpha_2\gamma_2$ in the fetus, and $\alpha_2\beta_2$ and $\alpha_2\delta_2$ in adults; the time at which $\theta1$ is expressed is not known. Furthermore, differences in oxygen-binding affinity have evolved among these globins. For example, the fetal hemoglobin $\alpha_2\gamma_2$ has a higher oxygen affinity than either adult hemoglobin, enabling the fetus to receive enough oxygen from the maternal blood (the high affinity occurs because the γ chain has serine at position 143 instead of histine as in the β chain) (see Wood et al. 1977; Bunn and Forget 1986). This phenomenon exemplifies again the fact that gene duplication can result in refinements of a physiological system.

Among the α family members the embryonic type ζ is the most divergent, having branched off more than 300 million years ago (Figure 10.10). The $\theta1$ globin branched off about 260 million years ago. Because the divergence time between the two α genes is uncertain, only the α_1 gene is shown in the figure. The α_1 and α_2 genes have almost identical DNA sequences and produce an identical polypeptide. This similarity would seem to indicate a very recent divergence time. However, the similarity could also be the result of concerted evolution (Zimmer et al. 1980), a phenomenon that will be discussed in Chapter 11. The two genes are present in humans and all the apes and so could have diverged from each other more than 20 million years ago.

Among the β family members, the adult types (β and δ) and the nonadult types (γ and ϵ) diverged about 155–200 million years ago (Efstratiadis et al. 1980). The ancestor of the two γ genes diverged from the ϵ gene about 100–140 million years ago. The duplication that created $^G\gamma$ and $^A\gamma$ occurred after the separation of the human and Old World monkey lineage from the New World monkey lineage, about 35 million years ago (Shen et al. 1981). The divergence between the δ and β genes was previously estimated to be 40 million years ago (Dayhoff 1972; Efstratiadis et al. 1980), but more recent DNA sequence data suggest that it may have occurred even before the eutherian radiation, about 80

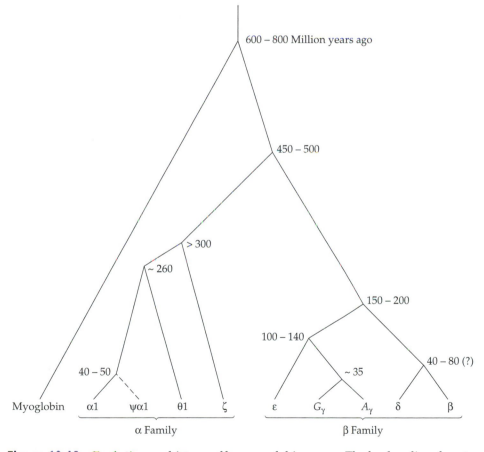

Figure 10.10 Evolutionary history of human globin genes. The broken line denotes a pseudogene lineage. Only one of the two α-globin genes is shown in the figure, because the date of their divergence from each other is uncertain. From Li and Graur (1991).

million years ago (Goodman et al. 1984; Hardison and Margot 1984). We note from the above discussion that, in both families, there is a good correlation between the time of divergence and the degree of functional or regulatory divergence between genes.

HOMEOBOX GENES

Homeotic (*Hox*) genes are genes in which mutations can transform part of a body segment, or an entire body segment, into the corresponding part of another segment. For example, in the **Antennapedia** mutant of *Drosophila* the antennae on the head of the fly are transformed into an additional pair of the second legs. The first homeotic mutant, bithorax, was discovered by Bridges and Morgan (1923) and later shown by E. B. Lewis to be part of a gene cluster, the bithorax complex (see Lewis 1978). The first molecular cloning of bithorax genes was done by Bender et al. (1983). Later, McGinnis et al. (1984) and Scott and Weiner (1984) simultaneously discovered that homeotic genes share a common sequence element of 180 bp. This element has been known as the **homeobox**, and all genes containing this box, including nonhomeotic genes, are called homeobox genes. The discovery of the homeobox has generated much excitement, for it not only facilitates the study of the function of homeotic genes, but also serves as a probe for fishing out and identifying other homeobox genes (see Gehring 1994). As homeotic genes are master control genes that specify the body plan and regulate development of higher organisms, understanding their regulation, structure, and function is key to the study of genetic control of development in higher organisms (see Gehring et al. 1994). These genes are also essential for understanding the evolution of body plans, which has been a major issue in evolution.

Today, more than 350 homeobox genes and their relatives have been identified and sequenced in fungi, plants, and animals, and they can be grouped into at least 30 different classes (see Kappen et al. 1993; Bürglin 1994); in the past many gene names were very confusing but a new nomenclature has been adopted (Scott 1992). The evolution of homeobox genes has been studied by many authors (e.g., Kappen et al. 1993; Bürglin 1994; Garcia-Fernàndez and Holland 1994; and references in Bürglin 1994 and Zhang and Nei 1996). Here we shall focus on the Antennapedia (*Antp*)-class homeobox genes, which are the best studied and are of special importance, because they specify the body segments along the anterior-posterior axis of the animal embryo and their developmental patterning.

In vertebrates (with the possible exceptions of primitive vertebrates; see Carroll 1995), there are four clusters of *Antp*-class genes located on four different chromosomes (Figure 10.11). The genes within and between the clusters are related and can be classified into 13 cognate gene groups (Kappen and Ruddle 1993); however, none of the clusters has all 13 cognate genes. The amphioxus (*Branchiostoma floridae*), which is thought to be a sister group of vertebrates, has one cluster of at least 10 cognate genes. There are eight *Antp*-class homeobox genes in *Drosophila melanogaster*, which are located on the same chromosome but in two separate clusters, the Antennapedia complex and the bithorax complex. The nematode (*Caenorhabditis elegans*) has one cluster of four *Antp*-class homeobox genes. For more details about these gene clusters, see references in Carroll (1995).

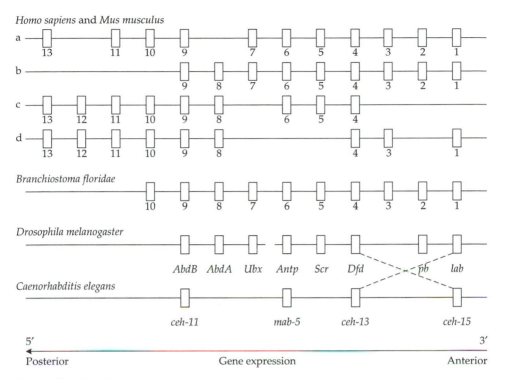

Figure 10.11 Genomic organizations of Antennapedia-class homeobox genes in the human (*Homo sapiens*), mouse (*Mus musculus*), amphioxus (*Branchiostoma floridae*), *Drosophila* (*D. melanogaster*), and nematode (*Caenorhabditis elegans*). There is an inversion of genes *ceh-13* and *ceh-15* in the nematode. The arrow indicates the order of gene expression. Moving along the clusters in the 3'-to-5' direction, each successive gene expresses later in the developmental process and more posterior along the anterior-posterior axis of the animal embryo. From Zhang and Nei (1996).

The orthologous and paralogous relationships of these genes are believed to be as given in Figure 10.11. Interestingly, the temporal order of expression of the genes in each cluster is from 3' to 5', i.e., along the 3'-to-5' direction, each successive gene expresses later in the developmental process and more posterior along the anterior-posterior axis of the animal embryo.

There are studies of *Antp*-class homeobox genes in many other animals such as hydras (Murtha et al. 1991; Naito et al. 1993), flatworms (Bartels et al. 1993), annelids (Dick and Buss 1994; Snow and Buss 1994), crustaceans (Cartwright et al. 1993), acorn worms (Pendleton et al. 1993), lampreys (Pendleton et al. 1993), and others. For simplicity, these genes are not included in the following discussion.

The following discussion is based mainly on Zhang and Nei (1996); for other studies, see Kappen et al. (1989), Schughart et al. (1989), and Schubert et al. (1993). It should be noted that the relationships inferred in any of these studies are not very certain because the homeobox sequence, which has been the basis of the inference, is only 180 bp or 60 codons long. Therefore, the inferences discussed below should be taken with caution, especially for those groupings with a low bootstrap value.

Figure 10.12 shows an inferred phylogenetic tree of 98 *Antp*-class homeo-domains from the five species in Figure 10.11. The genes in the tree are clustered into two major groups, I and II. Group I consists of subgroups A, B, and C; this subgrouping is tentative because the bootstrap proportions (BP), except that for subgroup B, are not high. Subgroup A includes cognate genes 1 and 2 of the mouse, human, and amphioxus; *pb* (proboscipedia) and *lab* (labial) of *Drosophila*; and *ceh-13* of the nematode. Inclusion of *ceh-13* in this subgroup is reasonable, because *ceh-13* and *ceh-15* of the nematode have apparently been inverted and *ceh-13* seems to be orthologous to cognate gene 1 (Bürglin and Ruvkin 1993). Sub-group B includes cognate gene 3 of the mouse, human, and amphioxus. Sub-group C consists of cognate genes 4–8 of the mouse, human, and amphioxus; *Dfd* (Deformed), *Scr* (Sex combs reduced), *Antp*, *AbdA* (Abdominal-A), and *Ubx* (Ultrabithorax) of *Drosophila*; and *mab-5* and *ceh-15* of the nematode. *ceh-15* and *mab-5* C are supposed to be orthologous with cognate gene groups 4 and 6, respectively (Kappen and Ruddle 1993), but they form a cluster separate from other genes in subgroup C. This cluster should be taken with caution because it has a low BP value. In the figure, cognate groups 5–8 do not form monophyletic groups, but this result could be due to sampling errors.

Group-II genes are divided into subgroups D and E, though each grouping is supported by only a low BP value. Subgroup D includes cognate genes 9 and 10 of the mouse, human, and amphioxus. The *Hox-9* gene of the amphioxus clusters with cognate group 10 genes, but this may be due to sampling errors, especially in view of the low BP value for this clustering. Also, whether cognate group 10 should be in subgroup D or E is uncertain, because when only human and mouse genes were used in the analysis, this group was included in subgroup E, not in subgroup D as shown in Figure 10.12 (Zhang and Nei 1996). With this caveat in mind, Figure 10.12 suggests that subgroup E consists of cognate genes 11–13 of the mouse and human, and *AbdB* (Abdominal-B) of *Drosophila*. The *Drosophila AbdB* gene is supposed to belong to cognate gene group 9, and in view of the low BP value for the E subgroup, its inclusion in subgroup E seems to be due to sampling errors. The nematode gene *ceh-11*, which was previously aligned with cognate group 9 genes (Bürglin et al. 1991; Wang et al. 1993), seems to be an outgroup to all other *Antp*-class genes; however, this is uncertain because it is supported by a low BP value.

Figure 10.12 suggests that the divergence of nematode *ceh-13*, *Drosophila-lab*, and chordates cognate group 1 genes postdated the divergence of cognate gene groups 1 and 2 and that the last common ancestor of the nematode, *Drosophila*, and chordates already had cognate gene groups 1 and 2. Similarly, Figure 10.12

Figure 10.12 Phylogenetic tree of 98 *Antp*-class homeobox genes of the human, ▶ mouse, amphioxus, *Drosophila*, nematode and two outgroup genes (nematode *ceh-5* and *ceh-19*). The tree is constructed by the neighbor-joining method. The numbers for interior branches are bootstrap percentages. Percentages less than 50 are not shown. The branches are measured in terms of the proportional difference of two amino acid sequences with the scales given below the tree. Each pair of the human and mouse homologous genes are merged into a single lineage if the cluster is supported by a bootstrap value higher than 85%. After Zhang and Nei (1996).

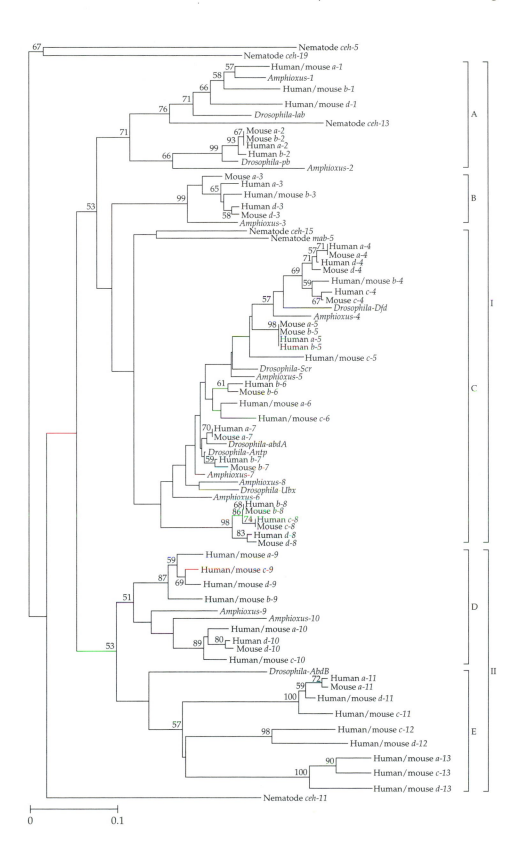

suggests that the last common ancestor of the nematode, *Drosophila*, and chordates also had cognate gene 3, the ancestral gene of subgroup C, and the ancestral gene of group II. In other words, there were at least five cognate genes in the last common ancestor of Pseudocoelomates (nematode in Figure 10.12) and Coelomates (insects and chordates). In the nematode, however, cognate genes 2 and 3 seem to have been lost later, and the ancestral genes of cognate genes 4–8 apparently gave rise to *mab-5* and *ceh-15*. The positions of *ceh-15* and *ceh-13* are inverted according to the sequence comparison and functional analysis of Bürglin and Ruvkun (1993), probably due to a chromosome inversion in the nematode.

Figure 10.12 further suggests that the divergence of subgroups D and E occurred earlier than that of the amphioxus and vertebrate cognate group 10 genes. Therefore, if the tree is correct on this point, the amphioxus must have had cognate group 11–13 genes previously. These genes have important functions in the limb formation in the mouse. It is possible that they have been lost, consistent with the fact that the fins of the amphioxus, which may be homologous to the limbs of tetrapods, are poorly developed. Or it is possible that these genes have not yet been found in the amphioxus.

Figure 10.13 shows a scenario proposed by Zhang and Nei (1996) for the evolution of the *Antp*-class genes through gene duplications. In the beginning, duplication of a single ancestral gene produced the ancestral genes of groups I and II, and the mutational change of these genes initiated the step for the anterior and posterior parts of the animal embryo to be controlled by separate genes. Then, a series of duplications of the ancestral group I gene generated subgroups A, B, and C. Subgroups A and B are involved in the segmentation of the anterior part, whereas subgroup C genes control the intermediate part along the anterior-posterior axis of the animal embryo. In the case of group II genes, the first duplication seems to have produced the ancestral genes of cognate groups 9–10 and the others, and the subsequent duplications generated the

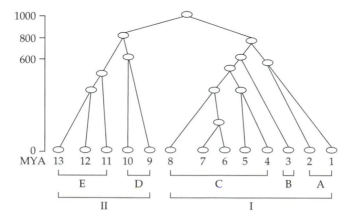

Figure 10.13 Evolutionary scenario of the 13 cognate groups of the *Antp*-class homeobox gene family. Branch lengths are not proportional to the evolutionary time. I and II are two major groups and A, B, C, D, and E are five subgroups. MYA, million years ago. From Zhang and Nei (1996).

remaining cognate genes. An ancestral cluster of at least 10 cognate genes was probably formed before the divergence between the amphioxus and vertebrate lineages. Gene clusters a, b, c, and d in vertebrates were probably generated by two events of genome duplication in the early stage of vertebrate evolution (Kappen et al. 1993; Kappen and Ruddle 1993), and later a few genes in each cluster were lost.

On the basis of amino acid divergence and under the assumption of a molecular clock, the time of divergence between groups I and II has been estimated to be about 1000 million years ago (Figure 10.13), suggesting that the *Antp*-class homeobox gene family evolved in the very early stage of metazoan evolution. Similarly, subgroups D and E and subgroups B and C were estimated to have diverged about 800 and 600 million years ago, respectively. These estimates suggest that the five subgroups were formed by 600 million years ago, predating the divergence of Pseudocoelomates and Coelomates.

Of course, many of the above inferences are uncertain and much further research is needed to clarify the evolutionary history of these genes. Also, the above discussion is concerned mainly with the evolutionary history of the *Antp*-class homeobox genes. For an excellent review of the relationships between the evolution of homeotic genes and the evolution of body plans and body parts, see Carroll (1995), who suggested that the generation of arthropod and tetrapod diversity has largely involved regulatory changes in the expression of homeotic genes and the evolution of interactions between homeotic genes and the genes they regulate, though changes in the number and function of homeotic genes have also been important. Presently, the *Antp*-class homeobox genes are the best studied, yet our understanding of their evolution is limited. In view of the existence of a large number of homeobox genes, their evolution is a subject that will require many years of effort by both molecular biologists and evolutionists.

DOMAIN AND EXON SHUFFLING

There are two types of **domain shuffling**: **domain duplication** and **domain insertion**. Domain duplication refers to duplication of one or more domains in a gene and so is a type of internal duplication, which was discussed earlier in the context of gene elongation. Domain insertion is the process by which structural or functional domains are exchanged between proteins or inserted into a protein. Both types of shuffling have been important in the evolution of genes, especially for the emergence of new genes. Here, we discuss the insertion of a domain from one gene into another, with the consequent production of mosaic or chimeric proteins (Doolittle 1985; Patthy 1985, 1991, 1994).

In the past "exon shuffling" instead of "domain shuffling" has been commonly used but the latter seems to be a better or, at least, broader term (see Doolittle 1995). Shuffling of a domain is more likely to succeed than shuffling of an exon, because a domain is a functional or structural unit whereas an exon is often not. Of course, an exon may be successfully shuffled even if it does not encode a domain, but such a shuffling may have little functional or evolutionary significance. Note further that in the case where a multiexonic domain is shuffled, domain rather than exon is the unit of shuffling. Moreover, domain shuffling can occur without the involvement of any exon. In fact, no intron has yet been found in prokaryotic protein coding genes, but many cases of domain shuffling have already been discovered in bacteria (see the review by Doolittle 1995). Nevertheless, the existence of introns has greatly promoted domain shuffling, especially in vertebrate genes. For this and for the historic reasons, we shall also discuss exon shuffling.

There has been a controversy regarding the following questions: How many kinds of domain exist and what was their origin? One view is that there were only a few small polypeptides in early stages of life and that most contemporary proteins are descended from them via duplication and modification (Doolittle 1981, 1995). The opposing view is that a large number of polypeptides existed early in the history of living systems and that all proteins, past and present, are the result of shuffling these primordial structural units (Dorit et al. 1990, Dorit and Gilbert 1991). This view is based on the "intron early" notion (Gilbert 1978), which assumes that primitive proteins were encoded by "minigenes" that were spliced together, with coding regions (exons) being separated by introns. The intron early hypothesis is a subject of intense debate and will be discussed in Chapter 13.

Mosaic Proteins

One of the first mosaic proteins discovered was tissue plasminogen activator (TPA) (Figure 10.14). Plasminogen is converted by TPA into its active form, plasmin, which dissolves fibrin, a soluble fibrous protein in blood clots. The conversion is greatly accelerated by the presence of fibrin itself, the substrate of plasmin. Fibrin binds both plasminogen and TPA, thus aligning them for catalysis. This mode of molecular alignment allows plasmin production only in the proximity of fibrin, thus conferring fibrin specificity to plasmin. By contrast, urokinase (UK), a urinary plasminogen activator, lacks fibrin specificity. A comparison of the amino acid sequences of TPA and UK showed that TPA contains a 43-residue sequence at its amino-terminal end that has no counterpart in UK (Banyai et al. 1983). This segment can form a finger-like structure (Figure 10.14a) and is homologous to the finger domains responsible for the fibrin affinity of another protein, fibronectin (FN) (Figure 10.14b), which is a large glycoprotein present in plasma and cell surfaces that promotes cellular adhesion. Deletion of this segment leads to a loss of the fibrin affinity of TPA. The homology of TPA with FN is restricted to this finger domain. Thus, domain (exon) shuffling was probably responsible for the acquisition of this domain by TPA from either FN or a similar protein.

TPA also contains a segment homologous to epidermal growth factor (EGF) and the growth-factor-like regions of other proteins, such as Factor IX and Factor X, which are blood-clotting enzymes in the blood-coagulation pathway. In addition, the carboxy-terminal regions of TPA are homologous to the proteinase parts of trypsin and other trypsin-like serine proteinases, which are enzymes that hydrolyze proteins into peptide fragments. Finally, the nonproteinase parts of TPA contain two structures similar to the kringles of plasminogen. (A kringle is a cysteine-rich sequence that contains three internal disulfide bridges and forms a pretzel-like structure resembling the Danish cake bearing this name.) Thus, during its evolution, TPA captured at least four DNA segments from at least three

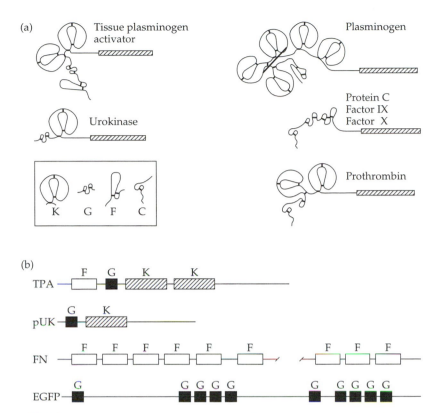

Figure 10.14   (a) Various structural modules observed in tissue plasminogen activator (TPA) and some other proteins. K = kringle module, G = growth-factor module, F = finger module, and C = vitamin K-dependent calcium-binding module. The crosshatched bars represent the protease regions homologous to trypsin. (b) The origin of the modules acquired through exon insertion in the tissue plasminogen activator (TPA) protein. UK, urokinase; EGFP, epidermal growth factor precursor; FN, fibronectin. (a) After Patthy (1985); (b) from Li and Graur (1991).

other genes: plasminogen, epidermal growth factor, and fibronectin (Figure 10.14b). Moreover, the junctions of these acquired units coincide precisely with the borders between exons and introns, thus, lending further credibility to the idea that exons have indeed been transferred from one gene to another. For more examples of domain shuffling, see Table 10.5, Doolittle (1985, 1995), and Patthy (1985, 1987, 1991, 1994).

Phase Limitations on Exon Shuffling

For an exon to be inserted into an intron of a gene without causing a frameshift in the reading frame, the phase limitations of the receiving gene must be respected (Patthy 1987, 1994). In order to understand this constraint, let us consider the different types of introns in terms of their possible positions relative to the coding regions. Introns residing between coding regions are classified into three types according to the way in which the coding region is interrupted. An intron is of **phase 0** if it lies between two codons, of **phase 1** if it lies between the first and sec-

---

TABLE 10.5   Some animal proteins that have sequence segments in common

---

A. EGF type
  Epidermal-growth-factor precursor
  Tumor growth factors
  Low-density lipoprotein receptor
  Factor IX
  Factor X
  Protein C
  Tissue plasminogen activator
  Urokinase
  Complement C9
  Notch protein (*Drosophila*)
  Lin-12 (Nematode)
  Thrombospondin

B. Fibronectin "finger"
  Fibronectin
  Tissue plasminogen activator

C. C9 type
  Complement C9
  Low-density lipoprotein receptor
  Notch (*Drosophila*)
  Lin-12 (Nematode)

D. Proprotease "kringle"
  Plasminogen
  Tissue plasminogen activator
  Urokinase
  Prothrombin

E. β-2 type
  β_2-glycoprotein
  Complement factor B
  Complement factor H
  Factor XIII b-chain

F. Lectin globular domain
  Pulmonary-surfactant protein
  Hepatic lectin
  Asialoglycoprotein receptor
  Cartilage proteoglycan
  Humural lectin (Sarcophaga)

From Doolittle (1985), with further examples from Doolittle (1987) and Patthy (1987).

---

ond nucleotides of a codon, and of **phase 2** if it lies between the second and third nucleotides of a codon (Figure 10.15). Exons are grouped into classes according to the phases of their flanking introns. For example, the middle exon in Figure 10.15b is flanked by a phase-0 intron at its 5′ end and by a phase 1 intron at its 3′ end and is said to be of **class 0-1**. An exon that is flanked by introns of the same phase at both ends is called a **symmetrical exon**, otherwise it is **asymmetrical**. For example, the middle exon in Figure 10.15a is symmetrical. Of the nine possible classes of exons, three are symmetrical (0-0, 1-1, and 2-2), and six are asymmetrical.

Only symmetrical exons can be inserted into introns. For example, in Figure 10.16a the insertion of a 0-1 exon into a phase-0 intron causes a frameshift in all the subsequent exons. Moreover, insertion of symmetrical exons is also restricted. A 0-0 exon can be inserted only into introns of phase 0, an 1-1 exon can be inserted only into introns of phase 1, and a 2-2 exon can be inserted only into introns of phase 2. For example, Figure 10.16b shows that the insertion of a 0-0 exon into an intron of phase 1 causes a frameshift in the inserted exon and in all exons on the 3′ side, while Figure 10.16c shows that the insertion of a 0-0 exon into an intron of phase 0 causes no frameshift. Interestingly, the vast majority of known movable exons are of the 1-1 class (Patthy 1991).

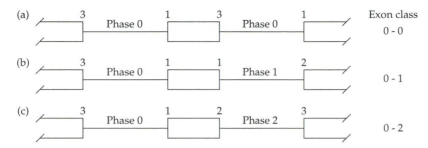

Figure 10.15   Phases of introns and classes of exons. Exons are represented by boxes. The number at the exon/intron junction indicates the codon position of the last nucleotide of the exon, while the number at the intron/exon junction indicates the codon position of the first nucleotide of the exon. Only three of the nine possible exon classes are shown. From Li and Graur (1991).

ALTERNATIVE PATHWAYS FOR PRODUCING NEW FUNCTIONS

In addition to gene duplication and domain shuffling, there are other mechanisms for producing new genes or polypeptides. Four such mechanisms are discussed in the next four sections.

Alternative splicing

Alternative splicing of the primary RNA transcript can result in the production of different polypeptides from the same DNA segment. This phenomenon has been found in numerous eukaryotic genes and in a number of eukaryotic transposons and animal viruses (see Smith et al. 1989).

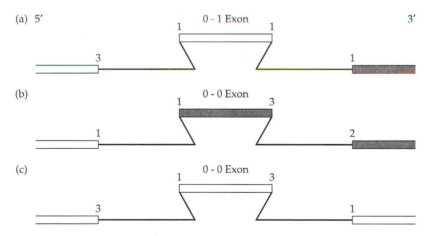

Figure 10.16   Consequences of exon insertion into introns. Shaded boxes indicate frameshifts. (a) Insertion of a 0-1 asymmetrical exon into a phase 0 intron; (b) insertion of a 0-0 symmetrical exon into a phase-1 intron; (c) insertion of a 0-0 symmetrical exon into a phase-0 intron. The insertions in (a) and (b) are abortive insertions. From Li and Graur (1991).

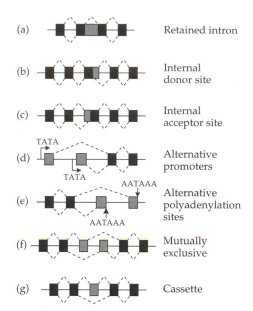

(a) Retained intron

(b) Internal donor site

(c) Internal acceptor site

(d) TATA / TATA — Alternative promoters

(e) AATAAA / AATAAA — Alternative polyadenylation sites

(f) Mutually exclusive

(g) Cassette

Figure 10.17 Modes of alternative splicing. Constitutively spliced exons are shown as black boxes and alternatively spliced exons as shaded boxes. Splicing pathways are shown by the diagonal lines. Promoters and poly(A) sites are denoted TATA and AATAAA, respectively. From Smith et al. (1989).

Figure 10.17 shows the classification of the modes of alternative splicing summarized by Smith et al. (1989):

a. **Retained introns**: Perhaps the simplest form involves introns that can either be spliced or retained in the mature RNA. Failure to splice the intron can result in the insertion of a peptide segment if the open reading frame is maintained. Alternatively, premature termination of the reading frame or frameshift may result in functionally different products or may effectively shut off gene expression if the unspliced RNA produces no functional product. Examples of such retained introns are found in the genes for fibronectin and *Drosophila* P element transposase (see references in Smith et al. 1989).

b. **Alternative 5′ donor sites and 3′ acceptor sites**: The use of alternative donor or acceptor sites for a given exon results in the excision of introns of different lengths with complementary variations in exon size. Again, this can give rise to insertion of small alternative peptide segments, frameshifting, or premature termination of reading frames. Such use of competing splice sites is common in viral transcription units such as the adenovirus Ela and SV40 T/t units, and is also used in eukaryotic genes such as the *Drosophila* ultrabithorax and transformer genes.

c. **Alternative promoters and cleavage/polyadenylation sites**: In some instances, alternative 5′ and 3′ end sequences arise through the differential utilization of promoters or cleavage/poly(A) sites, each associated with their own exons. This can lead to variability in the amino acid sequence of the N or C terminal of protein, respectively, if the variable segments extend into the protein-coding region. The variations in 5′ and 3′ untranslated sequences may lead to differential regulation of gene expression via effects on RNA stability, transport, or translational efficiency. Alternative promoter selection dictates the splicing pattern of the myosin light chain 1/3 (MLC1/3) gene, while examples of alternative 3' end exons with associated poly(A) addition sites are found in the α and β tropomyosins, the immunoglobulin μ heavy chain, and the calcitonin/CGRP genes.

d. **Internal mutually exclusive exons**: Internal exons are sometimes used mutually exclusively. One member of the pair is always spliced into the mRNA, but never both. The two exons are thus never spliced together, nor are both skipped. These exons code for alternative peptide segments within an otherwise constant flanking amino acid sequence. Such exons are found in the genes for the contractile proteins α and β tropomyosin, troponin-T, and myosin light chain 1/3, and in the pyruvate kinase gene.

e. **Cassette exons**: Some cassette exons can be included or excluded independently of other exons, and usually the same reading frame is maintained whether the exon is spliced in or out, though frameshift or premature termination can result from inclusion. When several cassette exons are present in a gene a very high degree of diversity can be generated, as in the NCAM and troponin-T genes. The observed hypervariability within the N-terminal region of troponin-T region is based on the presence of five cassette exons at the 5' end of the gene, these in conjuction with a pair of mutually exclusive exons in theory allow the generation of up to 64 different isoforms.

Alternative splicing has often been used as a means of developmental regulation. A very intriguing situation is seen in several genes involved in the process of sex determination in *Drosophila melanogaster*. At least three genes, *Sexlethal* (*Sxl*), *transformer* (*tra*), and *doublesex* (*dsx*), are spliced differently in males and females (Figure 10.18). In the case of *dsx*, the gene has six exons, and exons 1, 2, 3, and 4 are used in the female, whereas exons 1, 2, 3, 5, and 6 are used in the male. In the cases of *Sxl* and *tra*, the products of the alternative splicing in males contain premature termination codons and are therefore nonfunctional. For example, exon 3 in *Sxl* contains an in-frame stop codon, but the mRNA in females does not contain this exon.

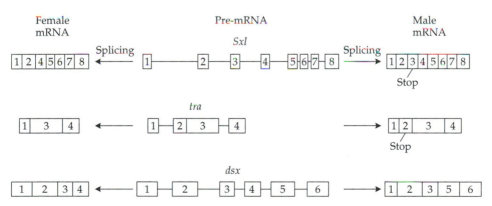

Figure 10.18 The patterns of splicing of the *sexlethal* (*Sxl*), *transformer* (*tra*), and *doublesex* (*dsx*) genes in *Drosophila melanogaster* females (left) and males (right). "Stop" indicates a termination codon that renders the mature mRNA nonfunctional. From Baker (1989).

A special instance of alternative splicing is illustrated by the case of intron-encoded proteins (Perlman and Butow 1989). In such cases, the intron contains an open reading frame that encodes a protein or part of a protein that is completely different in function from the protein encoded by the flanking exons. In some cases, the open reading frame is an extension of the upstream exon, e.g., intron a14α in the yeast mitochondrial gene *cox I* (Figure 10.19a). In other cases the intron includes not only a free-standing protein-coding gene, but also the necessary signals for transcription initiation and termination (Figure 10.19b). Intron a14α is intriguing, since it encodes a maturase required for its proper self-splicing from the pre-mRNA. This maturase also functions as an endonuclease in DNA recombination.

The evolution of alternative splicing requires that an alternative splice junction site be created *de novo*. Since splicing signals are usually 5–10 nucleotides long, it is possible that such sites are created with an appreciable frequency by mutation. Indeed, many such examples are known in the literature. For example, Figure 10.20 illustrates a case in which a synonymous substitution in a glycine codon turned a coding region into a splice junction. In cases of pathological manifestations such as the β[+]-thalassemia in Figure 10.20, the new splice site is used more frequently than the old splice site (i.e., most of the mRNA synthesized after such a mutation occurs is of the altered type). Such a mutation will obviously have deleterious effects and is not expected ever to become fixed in the population. However, if the newly created splice site is much weaker, then most mRNA will be of the original type and only small quantities of the new mRNA will be made. Such a change will not obliterate the old function, and yet will create an opportunity to produce a new protein.

Overlapping Genes

It has been found that a DNA segment can code for more than one gene by using different reading frames. This phenomenon is widespread in viruses, organelles, and bacteria. Figure 10.21a shows the genetic map of φX174, which is a single-stranded DNA bacteriophage. Several overlapping genes are observed. For example, gene *B* is completely contained inside gene *A*, while gene *K* overlaps gene *A* on the 5′ end and overlaps gene *C* on the 3′ end. A more detailed illustration of the latter case is given in Figure 10.21b.

Overlapping genes can also arise by the use of the complementary strands of a DNA sequence. For example, the genes specifying tRNA[Ile] and tRNA[Gln] in the human mitochondrial genome are located on different strands, and there is a

Figure 10.19 Examples of intron-encoded proteins: (a) an open reading frame (shaded) that is an extension of the upstream exon (empty box) (e.g., intron a14α in the yeast mitochondrial gene *cox I*); (b) a free-standing open reading frame, the transcription initiation and termination signals of which reside within the intron (e.g., the intron of *sun Y* gene in bacteriophage T4). From Perlman and Butow (1989).

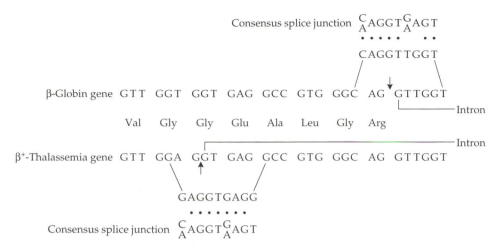

Figure 10.20   The nucleotide sequences at the border between exon 1 and intron 1 in the β-globin gene from a normal individual and a patient with β⁺-thalassemia. The mutated nucleotide is shown in boldfaced type. The arrows indicate the splicing sites. Each of the splice junctions is compared with the sequence of the consensus splice junction, and dots denote identity of nucleotides. From Li and Graur (1991).

three nucleotide overlap between them that reads 5′–CTA–3′ in the former and 5′–TAG–3′ in the latter (Anderson et al. 1981).

The question arises as to how overlapping genes may come into existence during evolution. To answer this question, we note that open reading frames abound throughout the genome. Therefore, it is possible that potential coding regions of considerable length exist in either a different reading frame of an existing gene or on the complementary strand. Because only three of 64 possible codons are termination codons, even a random DNA sequence might contain open reading frames hundreds of nucleotides long. If by chance such a reading frame contains an initiation codon and a transcription initiation site, or if such sites are created by mutation, an additional mRNA will be transcribed and subsequently translated into a new protein. Whether the new product has a beneficial function or not is another matter, but if it does, the trait may become fixed in the population.

The rate of evolution is expected to be slower in stretches of DNA encoding overlapping genes than in similar DNA sequences using only one reading frame. The reason is that the proportion of nondegenerate sites is higher in overlapping genes than in nonoverlapping genes, thus reducing the proportion of synonymous mutations out of the total number of mutations (Miyata and Yasunaga 1978).

Gene Sharing

An extremely intriguing situation of creating a new function is the recruitment of a gene product to serve an additional function without any changes in its amino acid sequence. This phenomenon has been termed **gene sharing** (Piatigorsky et al. 1988). Gene sharing means that a gene acquires and maintains a second function without duplication and without loss of the primary function. Gene sharing

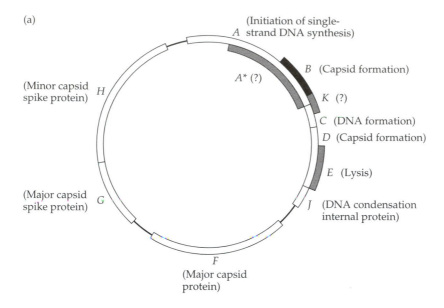

(b)

K protein: M   S   R   K   I   I   L   I   K   Q   E   L   L   L   L   V   Y   E   L
A protein: ...S   D   E   S   K   N   Y   L   D   K   A   G   I   T   T   A   C   L   R   I
C protein:
DNA sequence: TCTGATGAGTCGAAAAATTATCTTGATAAAGCAGGAATTACTACTGCTTGTTTACGAATT

K protein: N   R   S   G   L   L   A   E   N   E   K   I   R   P   I   L   A   Q   L   E
A protein: K   S   K   W   T   A   G   G   K   *
C protein: M   R   K   F   D   L   S   L   R   S   S   R
DNA sequence: AAATCGAAGTGGACTGCTGGCGGAAAATGAGAAAATTCGACCTATCCTTGCGCAGCTCGAG

K protein: L   L   L   C   D   L   S   P   S   T   N   D   S   V   K   N   *
A protein:
C protein: S   S   Y   F   A   T   F   R   H   Q   L   T   I   L   S   K   T   D   A   L...
DNA sequence: AGCTCTTACTTTGCGACCTTTCGCCATCAACTAACGATTCTGTCAAAAACTGACGCGTTG

Figure 10.21   (a) Genetic map of the circular genome of the single-stranded DNA bacteriophage φX174. Note that the *B* protein-coding gene is completely contained within the *A* protein-coding gene and that gene *K* overlaps two genes, *A* and *C*. (b) Sequence of the *K* gene showing overlap with the 5′ part of the *A* gene and the 3′ part of the *C* gene. Asterisks indicate stop codons. (See Table 1.1 for the one-letter abbreviations for amino acids.) (a) Modified from Kornberg (1982) by Li and Graur (1991); (b) from Li and Graur (1991).

may, however, require a change in the regulation system of tissue specificity or developmental timing.

Gene sharing was first discovered in crystallins (Table 10.6), which are used in the eye lens to maintain transparency and proper light diffraction. The first finding was that the ε crystallin from birds and crocodiles is identical in its amino acid sequence with lactate dehydrogenase B and possesses identical LDH activity (Wistow et al. 1987). Subsequent work showed that the two "proteins" are encoded by the same gene (Hendriks et al. 1988). A second crystallin, δ, which

TABLE 10.6   **Eye-lens crystallins and their relationships with enzymes and stress proteins**

Crystallin	Distribution	Identical or [Related]
α	All vertebrates	**Small heat-shock proteins (αB)**
		[*Schistosoma mansoni* antigen]
β	All vertebrates (embryonic γ	[*Myxococcus xanthus* protein S]
γ	not in birds)	[*Physarum polycephalum* spherulin 3a]
δ	Most birds, reptiles	**Argininosuccinate lyase (δ2)**
ξ	Crocodiles, some birds	**Lactate dehydrogenase B**
ζ	Guinea pig, degu rock cavy, camel, llama	**NADPH: quinone oxidoreductase**
η	Elephant shrews	**Aldehyde dehydrogenase I**
λ	Rabbits, hares	[Hydroxyacyl CoA dehydrogenase]
μ	Kangaroos, quoll	[Ornithine cyclodeaminase]
ρ	Frogs (*Rana*)	[NAPDH-dependent reductases]
τ	Lamprey, turtle; moderately abundant in most vertebrates	α-Enolase
S	Cephalopods	[Glutathione S-transferases]
Ω	Octopus	[ALDH]
j	Cubomedusan jellyfish	?

From Wistow (1993).

exists in all birds and reptiles, has also been shown to be identical in sequence with another enzyme, argininosuccinate lyase, which catalyzes the conversion of argininosuccinate into the amino acid arginine, and these proteins are likewise encoded by the same gene (Piatigorsky et al. 1988). Similarly, τ crystallin in lampreys, bony fishes, reptiles, and birds has been shown to be identical to and encoded by the same gene as α-enolase, a glycolytic enzyme converting 2-phosphoglycerate into phosphoenolpyruvate (Piatigorsky and Wistow 1989). Thus, δ, ϵ, and τ crystallins illustrate instances of gene sharing, whereby a gene acquires additional roles without being duplicated. The α, β, and γ crystallins, on the other hand, are classical examples of proteins that evolved by means of gene duplication and subsequent sequence divergence from ancestral genes specifying different proteins (e.g., heat-shock genes, which encode proteins expressed following exposure to excessive heat).

The finding that the same polypeptide can serve both as an enzyme and as a structural protein blurs the traditional distinction between enzymes and nonenzymatic proteins. Note also that once an enzyme has been recruited to serve as a structural protein in lens, it is subject to two sets of evolutionary pressure (Piatigorsky and Wistow 1991). This constraint may not allow sequence changes that enhance its function as a crystallin but that are not of benefit to the enzymatic role. A solution to this adaptive conflict may result if a gene duplication occurs (Piatigorsky and Wistow 1991). Such an example is provided by the argininosuccinate lyase (ASL)/δ crystallin family. Of the two tandemly arranged chicken ASL/δ crystallin genes, $\delta1$ is specialized for lens expression and produces >95% of the lens δ crystallin mRNA; by contrast, the $\delta2$ gene, which appears to encode the enzymatically active ASL, produces most of the ASL/δ crystallin mRNA in nonlens tissues, although it is still much more abundant in the lens than in other tissues (Thomas et al. 1990).

RNA Editing

RNA editing is defined as the transcriptional or posttranscriptional modification of an RNA molecule that changes its coding specificity by a process other than splicing. The first example of RNA editing is the insertion and deletion of uridine (U) residues within the coding regions of many mitochondrial transcripts of kinetoplastid trypanosomes (Benne et al. 1986; Simpson 1990). Many types of RNA editing are now known (see the review by Chan 1993).

One of the common types of RNA editing is C-to-U conversion in the RNA transcript. This conversion may occur partially or completely in some tissues but not in the others, leading to different variants of a protein in different tissues or multiple variants in the same tissue. Occasionally, it can produce a new protein with a quite different function, such as in the case of the synthesis of apolipoprotein B-48 mRNA (Chen et al. 1987; Powell et al. 1987). Apolipoprotein (apo) B is the largest protein species of apolipoproteins, which are the major carriers of various lipids in human blood. ApoB is heterogeneous but exists primarily in two forms: apoB-100 and apoB-48. In humans, apoB-100 is synthesized primarily by the liver and is the major protein constituent of very low density lipoproteins (VLDL) and low-density lipoproteins (LDL), whereas apoB-48 is synthesized by the intestine and is found in chylomicrons and chylomicron remnants. ApoB-48, unlike apoB-100, does not bind to the LDL receptor and has a molecular weight about 48% of that of apoB-100, which is a huge monomer protein, having 4536 amino acid residues. Traditionally, one would have guessed that apoB-48 is either produced by posttranslational cleavage of apoB-100 or encoded by a different gene. Unexpectedly, it was found that apoB-48 is translated from an intestinal mRNA with an in-frame UAA stop codon resulting from a C-to-U change in the codon CAA encoding Gln^{2153} in apoB-100 mRNA (Chen et al. 1987; Powell et al. 1987). Thus, although there is only one apoB gene in humans and rodents, two quite different proteins can be produced from the same gene.

CHAPTER 11

Concerted Evolution of Multigene Families

THE DEVELOPMENT OF DNA REANNEALING AND HYBRIDIZATION techniques in the mid-1960s provided a convenient, though crude, means for measuring the degree of divergence between related sequences and for exploring the structure and organization of eukaryotic genomes. Application of these techniques led to two intriguing observations. First, the genome of higher organisms is composed of highly and moderately repeated sequences as well as single-copy sequences (Chapter 13; Britten and Kohne 1968). Second, denatured repetitious DNA strands from one species form duplexes more rapidly with one another than they do with homologous strands from a different species. In other words, within one species the members of a repeated sequence family may be highly similar, whereas members from closely related species may differ greatly. Edelman and Gally (1970) found this observation perplexing and suggested a new concept, namely coevolution of DNA sequences, that is, the members of a sequence family in a species evolve together, so that they maintain high similarities among themselves, as they diverge from members in other species.

An excellent example of this phenomenon was provided in a study of the ribosomal DNAs from *Xenopus,* the African toads (Brown et al. 1972). In *Xenopus* and most other vertebrates, the genes specifying the 18S and 28S ribosomal RNAs are present in hundreds of copies and are arranged in one or a few tandem arrays. Each repeat unit consists of a transcribed and a nontranscribed spacer (NTS) region, which is now commonly called the intergenic spacer (IGS) (Figure 11.1). The IGS regions within the ribosomal DNA of *Xenopus laevis* are highly homogeneous but are very different from those in *X. borealis.* The large interspecific difference indicates that the IGS regions accumulate mutations rapidly in evolutionary time. Therefore, the high intraspecific homogeneity is maintained not by selective constraints (purifying selection) but by certain mechanisms by which mutations in a repeat can spread "horizontally" to all members in the same family. For this reason, Brown et al. (1972) called this phenomenon **horizontal evolution,** in contrast to vertical evolution, which refers to the spread of a mutation to individuals in a breeding population through generations. Later the terms **coincidental evolution** (Hood et al. 1975) and **concerted evolution** (Zimmer et al. 1980) were suggested. The latter is now commonly used in the literature.

Figure 11.1 Diagramatic representation of a typical repeated unit of rRNA genes in vertebrates. The thick bar designates the repeat unit, and the arrow indicates the transcribed unit. ETS, external transcribed spacer; ITS, internal transcribed spacer; IGS, intergenic spacer. After Arnheim (1983).

This new phenomenon was so puzzling that many mechanisms have been proposed for its occurrence. These include the saltatory replication hypothesis (Britten and Kohne 1968; Buongiorno-Nardelli et al. 1972), unequal crossing-over (Edelman and Gally 1970; Smith 1974, 1976), replication slippage (Dover 1986), gene conversion (Edelman and Gally 1970; Birky and Skavaril 1976), and duplicative transposition (Dover 1982). The phenomenon has been a subject of intense research in the last two decades. A large body of data obtained by restriction enzyme analysis and DNA sequencing techniques has attested to the generality of concerted evolution in multigene families (see reviews by Ohta 1980; Dover 1982; Arnheim 1983). Moreover, theoretical studies have greatly sharpened our understanding of the population dynamics of concerted evolution (e.g., Ohta 1980; Dover 1982; Nagylaki 1988). However, our knowledge is still limited in many aspects, such as in the rate of unequal crossing-over, the rate of gene conversion in higher organisms, and the rate of concerted evolution in a family.

GENE CONVERSION AND UNEQUAL CROSSING-OVER

Gene conversion and unequal crossing-over are now thought to be the two most important mechanisms for the occurrence of concerted evolution. They are also the two mechanisms that have received extensive quantitative treatments. To better understand their roles in the evolution of multigene families, we discuss some characteristics of gene conversion and unequal crossing-over, especially the rates of occurrence, as they determine the intensity of concerted evolution.

Gene Conversion

Gene conversion is a recombination process in which two sequences interact in such a way that one is converted by the other (see Chapter 1). It is a nonreciprocal process, because one sequence is changed whereas the other is not. For the mechanism of gene conversion, readers are referred to Holliday (1964), Meselson and Radding (1975), and Szostak et al. (1983).

Figure 11.2 shows a classification of conversional interactions according to the pathways involved; the chromatids are depicted after DNA synthesis but before segregation. (Chromatids are the copies of a chromosome produced by replication.) Pathway *c* is the interaction between two alleles at the same locus and is known as classical (allelic) gene conversion because this was the first type of gene conversion studied. The other pathways are (*a*) intrachromatid, (*b*) sister-chromatid, (*d*) semiclassical (nonallelic between homologous chromosomes), and (*e*) ectopic (heterochromosomal) conversion events, which refer, respectively, to interactions between genes (*a*) on the same chromatid, (*b*) at nonhomologous loci on sister chromatids, (*d*) at nonhomologous loci on homologous chromosomes, and (*e*) on nonhomologous chromosomes (Nagylaki 1990). [Petes et al. (1991) use somewhat different terms.] Here we assume that gene conversion occurs immediately after chromosome replication, so interactions between genes at homologous loci on sister chromatids have no effect and have not been included in the above classification. Collectively, intrachromatid and sister-chromatid interactions are intrachromosomal, whereas classical, semiclassical, and ectopic interactions are interchromosomal.

Gene conversion has been demonstrated in many species of fungi, in *Drosophila,* and in *Zea mays* (see Nelson 1975; Hilliker and Chovnick 1981; Lamb and Helmi 1982; Nagylaki 1983; and references therein). However, even in the best studied organism, the yeast *Saccharomyces cerevisiae,* data on the rate and other features of gene conversion are limited. The following brief summary is taken largely from the yeast data reviewed in Petes and Hill (1988) and Petes et al. (1991). It should be cautioned that conclusions drawn from yeast data may not be applicable to other organisms; for example, compared with higher eukaryotes, yeast seems to manifest a high rate of gene conversion, though the observed high

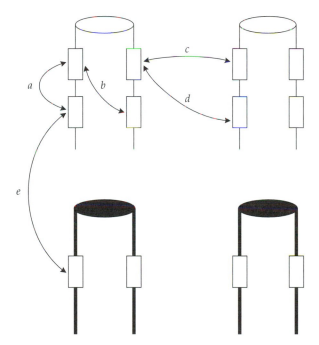

Figure 11.2 Types of gene conversion between repeated genes. Chromatids are depicted after DNA synthesis but before segregation. Repeats are indicated by rectangles and arrows indicate types of interactions. Nonhomologous chromosomes are indicated by lines of different thickness. The conversions depicted by pathways *a, b, c, d,* and *e* are termed (*a*) intrachromatid, (*b*) sister-chromatid, (*c*) classical (allelic), (*d*) semiclassical, and (*e*) ectopic (heterochromosomal). (Somewhat different terms are used by Petes and Hill 1988.) After Petes and Hill (1988) and Nagylaki (1990).

rate might be partly due to the ease of detecting gene conversion events in yeast. Four primary conclusions have been drawn from yeast studies.

First, allelic gene conversion has been observed for every locus that has been examined in detail and the frequencies vary from about 0.5% to 18% (Table 11.1; Fogel et al. 1981); an exceptionally high rate (30%) has been observed at the *his4*

TABLE 11.1   Rates of gene conversion at 30 heterozygous sites

Gene	Total	Segregations 3A:1a	Segregations 1A:3a	Rate of conversion (%)[a]	Disparity[b]
pet1	4,924	4	27	0.63	***
trp1	20,826	62	47	0.52	
mat1	23,135	93	104	0.85	
ura3	2,315	10	9	0.82	
ade6	1,589	4	9	0.82	
his5-2	2,315	14	10	1.04	
tyr1	8,391	59	44	1.23	
CUP1	18,016	123	124	1.37	
gal2	2,416	36	17	2.19	*
leu2-1	3,203	41	25	2.06	
trp5-48	2,315	23	31	2.33	
met1	1,589	18	17	2.20	
met10	892	17	16	3.70	
ura1	15,014	331	275	4.04	*
ilv3	9,487	230	239	4.94	
lys1-1	2,315	51	78	5.57	*
SUP6	892	22	33	6.17	
thr1	21,220	691	594	6.06	**
his4-4	12,533	411	303	5.70	***
ade8-10	1,118	48	30	6.98	
met13	10,454	459	474	8.92	
ade8-18	15,480	351	239	3.81	***
adc14	892	48	39	9.75	
ade7	1,028	47	27	7.20	*
his2	2,481	215	231	17.98	
arg4-4	1,188	5	13	1.52	
arg4-3	2,405	28	23	2.12	
arg4-19	5,352	87	68	2.90	
arg4-16	14,490	508	302	5.59	***
arg4-17	13,476	485	551	7.69	*
Total	221,751	4,521	3,999		***

Data from Fogel et al. (1978).

[a]Sum of 3A:1a and 1A:3a segregations divided by the total segregations.

[b]*, **, and ***: significant at the 5%, 1%, and 0.1% level, respectively.

locus (Nag et al. 1989). Conversion events were observed for every type of allele that has been examined, including point mutations, small (1–4 bp) insertions and deletions, and large insertions and deletions. In general, the frequency of gene conversion does not appear to be related specifically to the type of mutational change; however, this conclusion is based on limited data.

The rate of semiclassical gene conversion (conversion between nonhomologous loci on homologous chromosomes) has been studied by using two duplicated *his4* loci separated by a plasmid sequence (Jackson and Fink 1985). The observed frequency of ~2% (Table 11.2) is within the range of frequencies observed for allelic gene conversion (see above). A frequency of ~1% has been observed for semiclassical gene conversion between duplicated *his3* loci (Klein 1984). Intrachromosomal gene conversion has been studied also at a few loci; the rates observed are about 1 or 2% (Table 11.2; Klein and Petes 1981; Klein 1984). Limited data suggest that ectopic (heterochromosomal) gene conversion also occurs at fairly high rates (Jinks-Robertson and Petes 1985, 1986; Lichten et al. 1987). For example, in one study the conversion rate between two duplicated 1.7-kb *his3* genes (on chromosomes IX and XV) was about 0.5%, compared with an allelic rate of 1.5% (Jinks-Robertson and Petes 1985). A general conclusion is that high rates of interchromosomal and intrachromosomal gene conversion may not require very high sequence similarities.

Second, bias, or disparity, is typical. At a heterozygous locus with alleles *A* and *a*, if the two conversion events 3*A*:1*a* and 1*A*:3*a* occur with equal frequency, the conversion is said to show parity or no bias. If deviation from parity occurs, the allele whose frequency increases as a result of the deviation is said to have a **conversional advantage;** the other allele has a **conversional disadvantage.** In an

TABLE 11.2 Frequencies of different types of meiotic recombination events between two yeast duplicated *his4* separated by a plasmid sequence[a]

Event	X176 No. of tetrads (total 580)	%	X144 No. of tetrads (total 106)	%	X209 No. of tetrads (total 277)	%
Equal crossing-over	237	41	33	31		
Unequal crossing-over between homologous chromosomes	99	17	18	17		
Unequal intrachromosomal crossing-over	9	1.5	1	0.9		
Semiclassical gene conversion	14	2	3	2.8		
Unequal intrachromosomal gene conversion					3	1.1
Multiple recombination events	21	3.6	3	2.8		

From Jackson and Fink (1985).
[a]X176, X144, and X209 are different types of crosses.

examination of the data summarized in Fogel et al. (1978), Nagylaki and Petes (1982) found that gene conversion bias is a common phenomenon (see Table 11.1) and that the degree of bias appears to depend on the type of mutation (i.e., substitution mutation and deletion or insertion). In fact, approximately half of the mutant sites examined in yeast showed sufficiently large deviations from parity to be detected with a sample size ≥ 100. For small insertions (4 bp), the studies of Borts and Haber (1987) and Symington and Petes (1988) showed parity or a small bias. By contrast, the studies of Fogel et al. (1981), McKnight et al. (1981), and Pukkila et al. (1986) indicated that large (≥ 100 bp) deletions frequently show departures from parity.

Third, the amount of DNA transferred during a single gene-conversion event varies from a few base pairs to greater than 12,000 bp (see Borts and Haber 1987; Petes et al. 1991). Studying closely linked sites within the *ARG4* gene, Fogel and Mortimer (1969) concluded that the tract length of a typical conversion event is several hundred base pairs. In studies involving strains with multiple heterozygous restriction sites, the average tract lengths estimated were: 1500 bp (Borts and Haber 1987), 2300 bp (Judd and Petes 1988) and 3700 bp (Symington and Petes 1988). Occasionally, much larger conversion tracts were observed in conversion events (DiCaprio and Hastings 1976). The size of the tract depends on the strain and the position in the genome, but within a tract gene conversion is usually continuous.

Fourth, the rate of gene conversion varies from site to site and is context-dependent (Table 11.1; see Petes et al. 1991). Symington and Petes (1988) found that among 82 tetrads associated with a crossover in the 22-kb interval between *LEU2* and *CEN3*, one heterozygous marker converted in 27 tetrads whereas another marker converted only in 4 tetrads. Large (38-fold) differences in the rate of gene conversion were observed between *leu2* heteroalleles inserted at different positions in the genome (Lichten et al. 1987). In *S. pombe* a G→T mutation within the coding region of the *ade6* gene stimulates both meiotic gene conversion and crossing-over (Gutz 1971; Ponticelli et al. 1988). The effect of this mutation is context-dependent; when a 3-kb fragment containing the mutant sequence was placed in a different chromosomal location, no stimulation of meiotic recombination was observed, suggesting that some nucleotides more than 1 kb from the mutation were required for hotspot activity.

Unequal Crossing-Over

Unequal crossing-over creates a sequence duplication in one chromosome (or chromatid) and a corresponding deletion in the other; thus, unlike gene conversion, it is a reciprocal recombination process. Figure 11.3 shows an example in which an unequal crossing-over event has led to duplication of three repeats in one daughter chromosome and deletion of three repeats in the other. Obviously, unequal crossing-over can produce fluctuations in the number of repeats in a family. One good example is the rRNA genes in *Drosophila* (Ritossa and Scala 1969; Schalet 1969). A large deletion of rRNA genes can result in a "bobbed" mutant, which, in homozygous form, manifests small bristles, slow development, reduced growth and fertility, and poor viability. Conversely, a bobbed mutant can revert to normal by an increase in the number of rRNA genes.

Unequal crossing-over may occur between the two sister chromatids of a chromosome or between two homologous chromosomes at meiosis. The latter

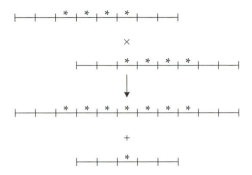

Figure 11.3 Model of unequal crossing-over. As a result of unequal crossing-over, both daughter chromosomes have an altered number of repeats and an altered frequency of the two repeat types when compared to the parental frequencies (50%). After Arnheim (1983).

type apparently can occur with high frequency. For example, a frequency of 17% was observed in a study with two duplicated *his4* loci separated by a bacterial plasmid sequence (Table 11.2); this frequency is much higher than that (2%) of semiclassical gene conversion for the same locus. The frequency (~1.5%) of unequal sister-chromatid (intrachromosomal) crossing-over is much lower than that of unequal crossing-over between homologous chromosomes but is as high as that of intrachromosomal gene conversion (Table 11.2). A similar frequency of unequal sister-chromatid crossing-over was also observed by Maloney and Fogel (1987). Unequal crossing-over may also occur between nonhomologous chromosomes; however, except for some cases such as that of rRNA genes in man and apes, this type of crossing-over is not important because it creates translocations, which generally have deleterious consequences.

EXAMPLES OF CONCERTED EVOLUTION

Numerous examples of concerted evolution of multigene families have been proposed, including the 5S DNA family in *Xenopus* (Brown and Sugimoto 1973), the γ-globin genes in primates (Jeffreys 1979; Slightom et al. 1980; Scott et al. 1984), the α-globin genes in primates (Liebhaber et al. 1981), the heatshock genes in *Drosophila* (Brown and Ish-Horowicz 1981) and the chorion gene superfamily in the silkmoth (Hibner et al. 1991). In the following we consider three examples that have been studied in detail.

X-linked Color Vision Genes in Higher Primates

As noted in the preceding chapter, Old World monkeys (OWM), apes, and humans are trichromatic because they posses three color photopigments: the short-wave (blue), middle-wave (green), and long-wave (red) pigments. The gene encoding the blue pigment is autosomal, whereas the genes for the red and green pigments are tandemly arrayed on the X chromosome (Nathans et al. 1986). The red and green pigment genes are believed to have arisen from a duplication event before the separation of the human-ape lineage from the OWM lineage, though

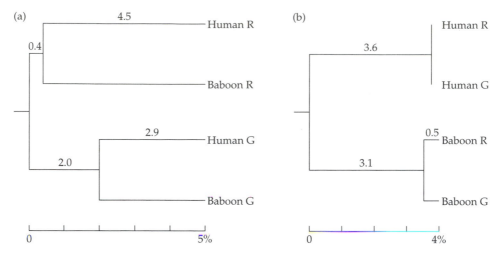

Figure 11.4   (a) Tree inferred from exons 4 and 5 of the red (R) and green (G) pigment genes of human and baboon. (b) Tree inferred from intron 4 sequences of human and baboon red and green pigment genes. In each tree, the topology was inferred by the neighbor-joining method and the root was placed in between the two clusters. The branch lengths were computed by assuming equal rates in different branches. The sequence data were from Shyue et al. (1994) and Zhou and Li (1996).

after the separation of the OWM-ape-human lineage from the New World monkeys, which posses only one X-linked and one autosomal pigment gene. Indeed, the tree (Figure 11.4a) constructed from the synonymous substitutions in exons 4 and 5 of the two genes from a human and a baboon (an OWM) is consistent with the view that the two genes diverged before the separation between the human and OWM lineages. The degrees of synonymous divergence in exons 4 and 5 are 8.1% between the two human genes and 11.5% between the two baboon genes (Zhou and Li 1996). These values are slightly higher than the average divergence between human and Old World intron sequences (7.1%, Table 8.2) and so are consistent with the above proposed duplication date.

    Given the strong evidence that the red and green pigment genes diverged before the human-OWM split, one would expect the homologous introns from the two genes to have diverged to a degree similar to the average divergence between human and OWM introns (7.1%). However, the intron 4 sequences (each ~2-kb long) from the red and green genes of a normal human were found to be identical (Shyue et al. 1994). This absence of divergence was totally unexpected, because the two genes should have diverged more than 20 million years ago, a minimum estimate for the human-OWM split. At any rate, it is significantly lower than the nonsynonymous divergence (2.0% ± 0.5%) between the two genes from the same human individual studied, contrary to the usual situation in which introns are less conserved than exons (Chapter 7). A simple explanation for this observation is that the intron 4 sequences between the two genes have been completely homogenized by gene conversion. This observation is difficult to explain by unequal crossing-over, because neither exon 4 nor exon 5 has been homogenized. In addition, the intron 2 sequences from the two genes were found to differ by only 0.3%, indicating that this intron has been almost completely homogenized by gene conversion (Shyue et al. 1994).

In the above discussion, we have simplified the presentation by neglecting the fact that a normal human X chromosome may carry more than one red or one green pigment gene (Nathans et al. 1986; Neitz and Neitz 1995). The multiple copies of the green (or red) pigment genes in humans are, however, highly similar, which can be a result of either recent origins or homogenization or both. In this case, homogenization can occur by either gene conversion or unequal crossing-over—obviously, the variation in gene copy number should have resulted from unequal crossing-over. As will be explained later, contraction and expansion in copy number in a sequence family by unequal crossinging-over can lead to homogenization.

Since the intron 4 sequences studied by Shyue et al. (1994) were from only one individual (a European), it was not known whether the homogenization occurred in that individual, or a common ancestor of Europeans, or an even earlier hominid ancestor. Zhou and Li (1996) therefore sequenced introns 4 of the red and green pigment genes from a male Asian Indian and found them to be identical too (Table 11.3). They also obtained introns 4 sequences of the two genes from a male chimpanzee and a male baboon (Table 11.3). The two chimpanzee introns differed by only 0.3% and the two baboon introns by only 0.9%. These values are again significantly lower than both the divergences in exons 4 and 5 and the intron divergences between any pair of the three species studied. Therefore, gene-conversion events have apparently also occurred in introns 4 of the two genes in

TABLE 11.3 **Mean and standard error of the number of nucleotide substitutions per 100 sites in intron 4 and in exons 4 and 5 between the red and green pigment genes**

	Intron 4	Exons 4 and 5	
Species and genes[a]	K^b	K_S[c]	K_A[c]
Human G/Human R	0.0 ± 0.0	8.1 ± 4.3	5.8 ± 2.0
Chimpanzee G/Chimpanzee R	0.3 ± 0.1	6.6 ± 3.5	5.1 ± 1.8
Baboon G/Baboon R	0.9 ± 0.2	11.5 ± 5.1	5.1 ± 1.8
Human G/Chimpanzee G	1.1 ± 0.3	4.2 ± 3.2	0.0 ± 0.0
Human R/Chimpanzee R	1.0 ± 0.3	0.0 ± 0.0	0.7 ± 0.7
Human G/Baboon G	7.3 ± 0.7	5.8 ± 3.6	1.1 ± 0.8
Human R/Baboon R	7.1 ± 0.7	9.0 ± 4.3	1.3 ± 1.0
Chimpanzee G/Baboon G	7.8 ± 0.8	6.0 ± 3.1	1.1 ± 0.8
Chimpanzee R/Baboon R	7.5 ± 0.7	9.0 ± 4.3	0.7 ± 0.7

From Zhou and Li (1996).

[a]G and R represent the green and red pigment genes, respectively.

[b]The K value is the number of substitutions per 100 sites and was computed by Kimura's 2-parameter method (Kimura 1980).

[c]The K_S and K_A values are the numbers of substitutions per 100 synonymous sites and per 100 nonsynonymous sites, respectively, and were computed by Li's methods (Li 1993).

chimpanzees and in baboons. For this reason, the tree inferred from the human and baboon intron 4 sequences groups the two human introns in one cluster and the two baboon introns in another, and is drastically different from that inferred from exons 4 and 5 (Figure 11.4).

It is possible that some of the putative gene-conversion events in intron 4 also involved either or both of exons 4 and 5, which are only about one-tenth the length of intron 4. In addition, there is evidence for gene-conversion events in exons in humans (Winderickx et al. 1993; Shyue et al. 1994) and OWMs (Ibbotson et al. 1992). Nevertheless, exons 4 and 5 remain relatively divergent between the two genes (Table 11.3). A most likely explanation of these observations is that natural selection for maintaining the distinct functions of exons 4 and 5 of the red and green pigment genes has acted against sequence homogenization in either of these exons. Homogenization of exon 4 or 5 between the red and green pigment genes would reduce the spectral sensitivity differences between the two genes, and this would be disadvatageous to its carriers, so that the mutant haplotype would eventually be eliminated from the population.

In summary, a lesson from the X-linked color pigment genes in higher primates is that concerted evolution may occur with different intensities in different regions of the genes, owing to different selective constraints in different regions.

Ribosomal RNA Genes in Humans and Apes

The rDNA multigene family in humans and apes provides a model system for studying the concerted evolution of a family of repeated genes distributed on several chromosomes (see Arnheim 1983). In humans and chimpanzees, the approximately 400 rDNA genes are distributed among nucleolus organizers located on five pairs of chromosomes (Arnheim et al. 1980). As noted in Figure 11.1, each rDNA repeat unit contains a nonconserved region, i.e., the intergenic spacer (IGS). If rDNA genes on nonhomologous chromosomes have experienced no genetic interaction, each nucleolus-organizer region would have evolved independently, and there should be high variation in the IGS region among different nucleolus organizers within individuals of a species. Thus, a high degree of homogeneity in this region among genes in different nucleolus organizers would suggest that the rDNA family has evolved in a concerted fashion.

Using restriction enzyme analysis, Arnheim et al. (1980) found that different regions of the rDNA repeat unit in humans and the apes have different characteristic evolutionary patterns. The transcribed segments have undergone few changes because the 18S and 28S RNA gene sequences have been highly conserved in evolution. In contrast, differences exist among species in the IGS region with no comparable variation existing within or among individuals of a species.

Figure 11.5 shows a summary of the data in Arnheim et al.'s (1980) analysis of human and chimpanzee rDNA genes. Each human repeat unit has a *Hpa*I site in the IGS region 3' to the 28S gene. The chimpanzee lacks this site. The fact that none of the other great apes (gorilla, orangutan, and gibbon) have a *Hpa*I site in this position suggests that the most recent common ancestor of humans and chimpanzees also lacked this site. Therefore, the *Hpa*I site most probably originated in the human lineage after the human-chimpanzee divergence and eventually became fixed in every human rDNA repeat. Some other restriction sites (e.g., *Hind*II) in the IGS region also demonstrate species-specific homogeneity

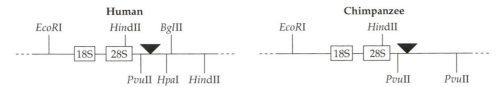

Figure 11.5 Restriction sites in human and chimpanzee ribosomal RNA genes. The restriction sites above the gene are polymorphic within the species. Those below the gene are monomorphic. The inverted triangles denote length polymorphism in the IGS. From Arnheim (1983).

(Figure 11.5). These observations indicate that in spite of the fact that the rDNA genes are distributed on five pairs of nonhomologous chromosomes, the whole family has evolved together in a concerted manner.

Krystal et al. (1981) have provided further evidence for genetic exchange between rDNA clusters on nonhomologous chromosomes. They examined the distribution of restriction sites that are polymorphic within humans (such sites are also indicated in Figure 11.5). Their results reveal a general uniform distribution of all of the polymorphisms over the chromosomes (with a few exceptions), and are consistent with a model of concerted evolution that involved exchanges between sister chromatids or homologous chromosomes and between nucleolus organizers on nonhomologous chromosomes. The exchanges could have been either gene conversions or unequal crossing-overs. The latter type of events would cause the number of rDNA genes per cluster to fluctuate randomly. In fact, estimates of the number of rDNA genes on the human nucleolus containing chromosomes by *in situ* hybridization have revealed extensive copy-number variation among the chromosomes within an individual and among the same chromosomes from different individuals (Warburton et al. 1976).

In humans (and apes) unequal crossing-over events between rDNA genes on nonhomologous chromosomes would probably cause no significant deleterious effects because the rDNA genes are located on the stalks between the short arm and satellite body of chromosomes 13, 14, 15, 21, and 22. For example, Figure 11.6 shows that an unequal crossing-over event between rDNA genes on chromosomes 14 and 21 produces a translocation involving only rDNA genes and satellite material. Of course, genetic exchange between rDNA genes on nonhomologous chromosomes can also occur via gene conversion. The relative importance of gene conversion and unequal crossing-over in the concerted evolution of rDNA genes in humans and apes cannot be evaluated from the data available at the present time.

The "360" and "500" Nongenic Repeat Families in Drosophila

The "360" and "500" repeat families are two abundant tandem families of noncoding sequences found within a group of eight sibling species of the *melanogaster* species subgroup of *Drosophila* (see Barnes et al. 1978; Strachan et al. 1982); their names signify the approximate lengths (~360 and ~500 bp) of their repeat units. Strachan et al. (1985) chose these two families to illustrate the transition stages of concerted evolution for three reasons. First, the repeats of these two

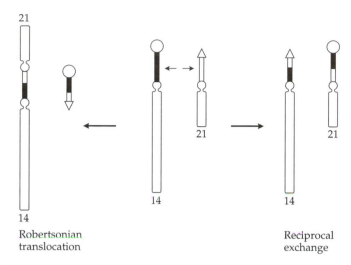

Robertsonian
translocation

Reciprocal
exchange

Figure 11.6 Cytogenetic consequences of genetic exchanges between human rDNA sequences on nonhomologous chromosomes. The satellite segments of chromosomes 14 and 21 are indicated by an open circle and an open triangle, respectively. The rDNA sequences for both chromosomes are immediately below the satellites and are represented as a filled (chromosome 14) or open (chromosome 21) rectangle. The centromeres of the chromosomes are shown as a constriction between the long and short chromosomal arms. In a reciprocal exchange, only the rDNA and satellite segments move to a nonhomologous chromosome. The breaks in the rDNA might also be resolved by the fusion of chromosomes 14 and 21, yielding a Robertsonian translocation that is dicentric and an acentric rDNA and satellite-containing fragment that would be lost. In order to simplify the figure, exchanges involving two chromosomes are shown. In fact, these interactions would be expected to take place after DNA replication. Because homologous chromosomes are also paired, each bivalent involved contains four chromatids. From Arnheim (1983).

families do not seem to be subject to any discernible natural selection, so that the dynamics of spread of mutant variants due to concerted evolution can be separated from that due to natural selection. Second, their phylogenetic distribution allows a detailed examination of the various stages of concerted evolution. Third, the two families have been evolving in parallel and so their patterns of concerted evolution can be contrasted.

Strachan et al. (1985) randomly selected clones of members of each family from each of the following species: *Drosophila melanogaster, D. mauritiana, D. simulans, D. orena, D. teissieri,* and *D. yakuba.* They sequenced many clones and computed the intraspecies and interspecies sequence divergences (proportions of nucleotide differences including deletions and insertions). As shown in Table 11.4, the interspecies divergences are approximately one order of magnitude greater than the within-species divergences, except for those divergences between *D. mauritiana* and *D. simulans* (both families) and between *D. yakuba* and *D. teissieri* (the 500 family). These results show concerted evolution in each family.

To detect the various stages of concerted evolution, Strachan et al. (1985) examined in detail the patterns of variation at each nucleotide position among

TABLE 11.4 **Intraspecific (diagonal) and interspecific sequence divergences (%)[a]**

	Mau	Sim	Ere	Ore	Tei	Yak
Mau	3.4/6.4	5.4	n.d.	31.5	33.4	34.8
Sim	8.1	4.5/7.8	n.d.	31.6	33.2	34.7
Ere	42.5	41.8	–	n.d.	n.d.	n.d.
Ore	–	–	–	4.8/3.4	28.8	35.0
Tei	34.2	34.2	38.7	–	8.5/14.8	35.3
Yak	34.1	33.8	40.5	–	12.0	1.0/3.4

From Strachan et al. (1985).
[a]Figures above the diagonal refer to the 360 family, while those below the diagonal refer to the 500 family. Figures on the diagonal are within-species variations; the figures left and right of the / refer to the 360 and 500 families, respectively. Dashes signify the inability to detect the 500 family in *Drosophila orena*. n.d. = not determined. Species abbreviations: Mau, *D. mauritiana*; Sim, *D. simulans*; Ere, *D. erecta*; Ore, *D. orena*; Tei, *D. teissieri*; and Yak, *D. yakuba*.

all clones of each family of the two pairs of species: *D. mauritiana* versus *D. simulans* and *D. yakuba* versus *D. teissieri*. They classified the patterns into six classes as shown in Figure 11.7. The class-1 pattern is characterized by complete homogeneity across all clones randomly sampled from a pair of species (e.g., positions 121 to 124 in Figure 11.7b). This class represents an absence of mutations (or their spread), in the supposed progenitor bases; this is, of course, only an approximate statement because homogeneity in a limited sample of sequences does not imply homogeneity in the population. Class 2 represents rare mutations (or low levels of subsequent spread), resulting in the appearance of a minority of clones with a new mutation in a given position in one species while the other species remains homogeneous for the progenitor base at the corresponding position (e.g., position 158 in Figure 11.7b). Class 3 covers those cases in which no decision can be made between minority and majority frequencies in that a mutation and the progenitor base are in approximately equal frequencies, while the other species is homogeneous for one of the two bases. No examples of this were found in the 360 family, although such examples were found in the 500 family. Class 4 includes those positions in which a mutation, which is apparently absent in one of the species, has replaced the progenitor base in the majority of members in the other species (e.g., position 193 in Figure 11.7b). This interpretation assumes that the base that is in the minority in one species and that is homogeneous in the other species is the progenitor base. Class 5 represents positions in which the two species are internally homogeneous for bases that are diagnostically different for each of the species (e.g., position 151 in Figure 11.7b). This class represents the classical observation of concerted evolution, i.e., the final outcome of the dual process of intrafamily homgenization and population fixation of mutations that arose independently in one or both species. All subsequent mutations beyond this point are represented by class 6; no such example is shown in Figure 11.7b.

(a)

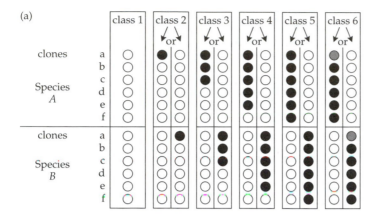

(b)

```
        130        140        150        160        170        180        190
CON  TATTGGTTGT AAGGAGTACA AAATGGTACT NCTTTTGCTC TCTGACCATT TTTAGTCAAG TTATAGCCAA AANA
mau 1 .........  .........  .........  T........  .........  .........  .........  ..C.
    2 .....T..AC .......T.  ...C...T.  T........  GG......C. A....C....  .........  ..A.
    3 .........  .........  .........  T........  .........  T........  .........  ..C.
    4 .........  .........  .........  T........  .........  .........  .........  ..C.
    5 .........  .........  .........  T........  .........  .........  .........  ..C.
    6 .........  .........  .........  T........  .........  .........  .........  ..C.
sim 1 ......C... ..T.G.....  .........  C........  .........  .........  .........  ..A.
    2 .........  .........  .......C C  C........  .........  .........  .........  ..A.
    3 .........  .........  .A.......  C......G.  .........  .........  .........  ..AC
    4 .........  .........  .........  C........  .........  .........  .........  ..A*
    5 .........  .........  .........  C........  .........  .G....... .........  ..AT
    6 .........  .......T.  .........  CA.......  .........  .........  .........  ..A.
    7 .....C.... .....A....  .........  C........  .........  .........  ...C....  ..AC
    8 .........  .........  .........  C........  .........  .........  .........  ..A.
    9 .........  .........  .........  C........  .........  .........  .........  ..A.
```

Figure 11.7 (a) Graphic representation of transition stages during the spread of new mutations. Classes 1–6 represent the patterns of distribution of mutations at individual nucleotide positions across clones a–f in two species. Open circles, solid circles, and shaded circles represent nucleotide differences that can be A, T, G, or C. In classes 2–5 only two types of bases are found at a given position across all clones of a pair of species. (b) Alignment of partial sequences of clones representing the 360 families of *Drosophila mauritiana* (mau) and *D. simulans* (sim). Numbers on the top of the sequence denote the nucleotide positions in the complete sequence. * denotes deletion, · is the same as the nucleotide in the overall consensus represented as CON, and N denotes no unambigous consensus nucleotide. From Strachan et al. (1985).

The data for the two species pairs are shown in Table 11.5. The following two conclusions can be drawn from this table:

First, for each of the two families, a greater number of class-5 and -6 nucleotide positions (representing a more extensive spreading of variant repeats) is found between *D. yakuba* and *D. teissieri* than between *D. mauritiana* and *D. simulans*. For example, for the 360 family there are 55 class-5 positions between *D. yakuba* and *D. teissieri* but only 5 class-5 positions between *D. mauritiana* and *D. simulans*. This observation is consistent with the assumption of a relatively constant rate of concerted evolution because *D. yakuba* and *D. teissieri* diverged earlier than did *D. mauritiana* and *D. simulans*.

TABLE 11.5 Numbers (proportions) of the six classes of nucleotide sites at individual positions across all clones in the 360 and 500 family sequences of two closely related pairs of *Drosophila* species[a]

	360 family		500 family	
Classes	D. mauritiana vs. D. simulans	D. yakuba vs. D. teissieri	D. mauritiana vs. D. simulans	D. yakuba vs. D. teissieri
1	278 (76.8)	113 (56.4)	348 (72.9)	424 (76.1)
2	72 (19.9)	21 (10.3	104 (21.8	74 (13.3)
3	–	2 (1.0)	–	3 (0.6)
4	6 (1.7)	4 (2.0)	2 (0.4)	10 (1.8)
5	5 (1.4)	55 (27.5)	2 (0.4)	38 (6.8)
6	1 (0.3)	9 (4.5)	1 (0.2)	9 (1.6)

From Strachan et al. (1985).

[a]For the definition of the six classes of sites, see Figure 11.7. Numbers refer to the absolute number of positions falling into each class. Numbers in parentheses represent percentages obtained by dividing the absolute numbers of positions within each class by the number of nucleotide positions available for comparison. In the case of the 360 family of *D. teissieri*, a large deletion event resulted in there being only ~200 nucleotide positions available for comparison with the 360 family of *D. yakuba*. Those mutation events that could not be assigned an unambiguous location were excluded from the analysis. For example, in the cases of runs of a particular nucleotide, deletions and insertions representing the same nucleotide are topographically ambiguous.

Second, the rate of spread of mutations is apparently faster in the 360 family than in the 500 family in all four species. For example, for the pair of *D. yakuba* and *D. teissieri*, the proportion of class-5 nucleotide sites is 27.5% for the 360 family but only 6.8% for the 500 family. This difference in rate is expected because the 360 family is confined to the X chromosome (see Brutlag 1980), whereas the 500 family members are distributed on all four pairs of chromosomes (Strachan et al. 1982).

The major mechanism for the concerted evolution in these two families has been suggested to be unequal crossing-over, because this is in keeping with the experimental proof of unequal exchanges within tandem arrays of rDNA that are embedded between some of the arrays of the 360 and 500 families on the X and Y chromosomes (see references in Strachan et al. 1985). However, there were indications that gene conversions have also contributed to the concerted evolution of these two families (Strachan et al. 1985).

CONCERTED EVOLUTION BY GENE CONVERSION

Edelman and Gally (1970) appear to have been the first to suggest that gene conversion can homogenize repeated genes in a genome. Birky and Skavaril (1976) initiated a theoretical investigation of this problem by using computer simulation. Later, Ohta (1977) used an analytic method (the diffusion theory) to study the extinction time of a variant repeat. The mathematical theory of gene conversion has since been greatly extended; e.g., Ohta (1984, 1985, 1986), Slatkin (1986), Walsh (1986, 1987, 1988), Kaplan and Hudson (1987), Basten and Weir (1990), and Nagylaki (1990). We consider only some simple situations here.

To understand how gene conversion can homogenize repeated genes, let us start with the case of tandem repeats and follow the evolution of a single chromosome lineage. If the family consists of only two repeats, then a single conversion event can result in homogeneity. In a larger family the situation is more complicated, but as illustrated in Figure 11.8, repeated cycles of gene conversion will eventually lead to homogeneity, if there is no mutation. Nagylaki and Petes (1982) developed a theory for the fixation probability and the mean time to fixation or loss of a variant repeat. They showed that a small conversional bias (advantage or disadvantage) can have a dramatic effect on the fixation probability of a new variant and that gene conversion can act sufficiently rapidly to be an important mechanism for producing and conserving sequence homogeneity.

The above treatment neglected random genetic drift and the effects of mutation and crossing-over. To incorporate these factors, we assume that the population is diploid and mates randomly, with effective size N, that every mutation cre-

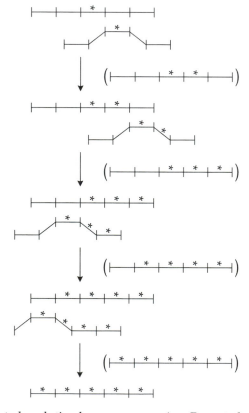

Figure 11.8   Concerted evolution by gene conversion. Repeated cycles of gene conversion events (denoted by *) cause the duplicated genes on each chromosome to become progressively more homogenized. The sequence in parentheses denotes the sequence selected for the next round of gene conversion. The process does not affect the number of repeated sequences on each chromosome. Modified from Arnheim (1983)

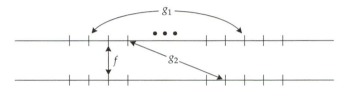

Figure 11.9 Three identity probabilities. f is the probability that two genes at the same locus on two homologous chromosomes are identical. g_1 is the probability that two distinct genes on the same randomly chosen gamete are identical. g_2 is the probability that two genes at different loci on two randomly chosen gametes are identical. From Ohta (1982).

ates a new allele (the rate of mutation is u per gene per generation), and that (equal) crossing-over is independent of gene conversion and at most one crossover occurs per generation in the entire multigene family in an individual.

Let us consider the case of interchromosomal gene conversion, which is mathematically simpler than that of intrachromosomal gene conversion. We assume unbiased gene conversion because conversion bias complicates the mathematics greatly. The family consists of n tandem repeats and all loci are equivalent and exchangeable. See Nagylaki (1984a) for a more detailed discussion of the model.

In the presence of mutation, complete homogeneity can no longer be attained, but there will be a balance between the homogenizing effect of gene conversion and the diversifying effect of mutation. Following Ohta (1982, 1983a,b), we consider three probabilities of identity (Figure 11.9): (i) f, the probability that two genes at the same locus, chosen at random from two distinct gametes, are identical; (ii) g_1, the probability that two distinct genes on the same randomly chosen gamete are identical; and (iii) g_2, the probability that two genes at different loci on different gametes, chosen at random, are identical; subscripts 1 and 2 signify whether one or two gametes (chromosomes) are involved. Thus, f represents the expected homozygosity at a locus and $h = 1 - f$, the expected heterozygosity, is a measure of intralocus genetic variability in the population; g_1 is an index of homogeneity between repeats within a chromosome; and g_2 incorporates both intralocus and interlocus variation. Nagylaki (1984a) showed that at equilibrium the above probabilities of identity are given by

$$\hat{f} = 2(nu + \lambda)/(nD) \tag{11.1a}$$

$$\hat{g}_1 = \hat{g}_2 = 2\lambda/(nD) \tag{11.1b}$$

where

$$D = 2u + 2\lambda/n + 8Nu(u + \lambda) \tag{11.1c}$$

and λ is the conversion rate per repeat per generation. Note that $\hat{g}_1$ and $\hat{g}_2$ are equal and that all three identity probabilities are independent of the crossover frequency.

Table 11.6 shows the values of $\hat{f}$ and $\hat{g}_1 = \hat{g}_2$ for various parameter values. It is assumed that each allele mutates to a new allele at the rate of 10^{-6} per generation, i.e., $u = 10^{-6}$. The case of no gene conversion (i.e., $\lambda = 0$) is given for comparison; in this case, $\hat{g}_1 = \hat{g}_2 = 0$ because mutations at different loci are assumed to be differ-

TABLE 11.6   Probabilities of identity at equilibrium

N	n	$\lambda = 0$	$\lambda = 10^{-5}$		$\lambda = 10^{-3}$	
		$\hat{f}$	$\hat{f}$	$\hat{g}_1 = \hat{g}_2$	$\hat{f}$	$\hat{g}_1 = \hat{g}_2$
10^3	10	0.996	0.978	0.489	0.962	0.952
	50	0.996	0.965	0.161	0.840	0.800
	100	0.996	0.962	0.087	0.733	0.666
	500	0.996	0.959	0.019	0.428	0.286
10^4	10	0.962	0.820	0.410	0.716	0.709
	50	0.962	0.732	0.122	0.344	0.328
	100	0.962	0.714	0.065	0.216	0.196
	500	0.962	0.699	0.014	0.070	0.046
10^5	10	0.714	0.313	0.156	0.201	0.199
	50	0.714	0.214	0.036	0.050	0.047
	100	0.714	0.200	0.018	0.027	0.024
	500	0.714	0.188	0.004	0.007	0.005

N = effective population size; n = number of repeats; λ = the conversion rate per repeat per generation; $\hat{f}$ = probability that two genes at the same locus, chosen randomly from distinct gametes, are identical; $\hat{g}_1$ = probability that two distinct genes on the same gamete, chosen at random, are identical; $\hat{g}_2$ = probability that two genes chosen randomly from different loci on different gametes are identical. In all cases, the mutation rate is 10^{-6} per repeat per generation.

ent. Gene conversion tends to homogenize repeated genes, and so the interlocus identities $\hat{g}_1$ and $\hat{g}_2$ increase with λ, the conversion rate. Indeed, if λ is much larger than nu, then $\hat{g}_1$ and $\hat{g}_2$ can be almost as large as $\hat{f}$, the intralocus identity. For example, for $n = 10$, $N = 10^3$, and $\lambda = 10^{-3}$, we find $\hat{g}_1 = \hat{g}_2 = 0.95$ and $\hat{f} = 0.96$. However, if N and n are large, $\hat{f}$, $\hat{g}_1$, and $\hat{g}_2$ can become quite small; e.g., for $N = 10^5$, $n = 50$ and $\lambda = 10^{-3}$, $\hat{g}_1 = \hat{g}_2 = 0.047$ and $\hat{f} = 0.050$. Thus, if the population size or the gene family size is large and if the conversion rate is low, it is more difficult to maintain high sequence homogeneity in a family. Note that gene conversion reduces the intralocus identity because it introduces different alleles into a locus from other loci. Thus, $\hat{f}$ decreases as λ increases.

Nagylaki (1984a) has studied also the time of convergence to equilibrium (T). He showed that T is greater than both n/λ and $4nN$. If the conversion rate λ is much less than $1/(4N)$, then T is very close to n/λ, i.e., T is proportional to the family size n and inversely proportional to the conversion rate λ. If λ is much greater than $1/(4N)$, then T is very close to $4nN$, i.e., T is proportional to both the family size and the effective population size.

For intrachromosomal gene conversion and other cases, readers may consult Ohta (1982, 1983a,b, 1984), Ohta and Dover (1983), and Nagylaki (1984b, 1988, 1990). To date most of the theoretical studies on gene conversion have assumed no conversion bias. Since conversion bias appears to occur often and can have strong effects on the above probabilities of identity, it should be taken into account in future studies. Another factor that needs much further research is natural selection, for it has been considered in only a few studies (e.g., Slatkin 1986; Walsh 1985, 1986).

CONCERTED EVOLUTION BY UNEQUAL CROSSING-OVER

In a single chromosome lineage it is not difficult to see how unequal crossing-over can homogenize repeats in a tandem array. For example, in Figure 11.3, as a result of an unequal crossover, each of the daughter chromosomes has become more homogeneous than the parental chromosomes. If this process is continued, the numbers of wild-type and mutant repeats on a chromosome will fluctuate with time, and eventually one type of repeat will become dominant in the family. This, in fact, has been verified by both computer simulation (Black and Gibson 1974; Smith 1974) and mathematical analysis (Ohta 1976; Perelson and Bell 1977). Black and Gibson (1974) used a model in which the shift of genes of unequal matching at a crossover is always one gene, and duplication and deletion occur by equal chance of $1/2$. Their simulation showed that the number of unequal crossovers required for the loss of gene lineages increases with the family size. Perelson and Bell (1977) treated this model by a birth-death process and showed that the "crossover fixation time," which is the expected number of unequal crossovers needed until a particular type of repeat becomes fixed in the family, is approximately $2n_0(n_0 - 1)$, where n_0 is the initial number of repeats in the family. This result agrees well with that of Ohta (1976), who used the diffusion theory. Smith's (1974) simulation used a more realistic model in which the shift of unequal matching at the crossover is not restricted to one repeat but to a certain range (latitude), e.g., the family size may vary randomly in the range $n_0 - n_0/10 \sim n_0 + n_0/10$. He found that the crossover fixation time depends not only on the family size but also on the allowed latitude of shift, i.e., the larger the latitude, the shorter the crossover fixation time. For example, if the average family size is 100, the crossover fixation is faster when the allowed latitude is 95 ~ 105 than when the latitude is 99 ~ 101. However, the former is not different from the case of 75 ~ 125, suggesting that beyond a certain value, increasing the allowed latitude becomes ineffective in accelerating the crossover fixation.

When the effects of random genetic drift and mutation are taken into account, the situation becomes much more complicated. Unlike gene conversion, unequal crossing-over creates shifts in the position of repeats and also fluctuations in the family size. The effect of shift of positions is difficult to treat mathematically. Moreover, f is now difficult to define as it refers to the probability of identity of two genes at the *same* locus in two different chromosomes, and the definition of g_2 is ambiguous because it partly includes f. For these reasons, unequal crossing-over is mathematically less well studied than gene conversion. Nevertheless, in a series of studies Ohta (e.g., 1978, 1980, 1982, 1983a) has obtained approximate results for g_1 and g_2.

RELATIVE ROLES OF GENE CONVERSION AND UNEQUAL CROSSING-OVER

As a mechanism of concerted evolution, gene conversion appears to have several advantages over unequal crossing-over (see Baltimore 1981; Nagylaki and Petes 1982). First, gene conversion causes no change in gene number, whereas unequal crossing-over changes the number of repeated genes in a family. Such a change may cause dosage imbalance. For example, deletion of one of the two duplicated α-globin genes in man gives rise to the α-thalassemia-2 syndrome (Kan et al.

1975). Second, gene conversion can act as a correction mechanism not only on tandem repeats but also on dispersed repeats. In contrast, unequal crossing-over has certain restrictions when repeats are dispersed. It probably can act effectively on repeats dispersed on nonhomologous chromosomes if they are located at the ends of chromosome arms as in the case of 5S RNA genes in *Xenopus* (Brown and Sugimoto 1973) and that of rRNA genes in man and apes (Arnheim 1983). But, it will be greatly restricted if the dispersed repeats are located inside chromosomes, as in the case of rRNA genes in mice (Arnheim 1983) and in *Drosophila melanogaster* (Coen and Dover 1983). If repeats are dispersed on a chromosome, unequal crossing-over can cause deletion or duplication of the genes between the repeats. Figure 11.10 shows a hypothetical case of unequal crossing-over between two repeat clusters, resulting in the deletion of a unique gene in one chromosome and a corresponding duplication in the other chromosome. Third, gene conversion can have a preferred direction, and a small disparity can have a large effect on the probability of fixation of mutant repeats (Nagylaki and Petes 1982; Walsh 1985).

In large families of tandemly repeated sequences, however, unequal crossing-over may be equally or even more important than gene conversion for concerted evolution. First, in such families, the number of repeats can apparently fluctuate greatly without causing significant adverse effects. This is suggested by the observations that the number of rRNA genes in *Drosophila* varies widely among individuals in the same species (Ritossa and Scala 1969) and that the number of 5S RNA genes is about 24,000 copies in *Xenopus laevis* but only 9000 copies or so in *X. borealis* (Brown and Sugimoto 1973). Moreover, in humans, several families of tandem repeats that exhibit extraordinary degrees of variation have been found (Nakamura et al. 1987). Second, as suggested by the data in Table 11.2, unequal crossing-over probably occurs more often than nonallelic gene conversion. Third, a gene conversion event may involve only a small region, say, from a few bases to several hundred bases. In the examples of gene conversion in the γ-globin genes proposed by Slightom et al. (1980) and Scott et al. (1984), the DNA segments involved are all relatively short. By contrast, the number of repeats exchanged in an unequal crossing-over event can be large. In Szostak and Wu's (1980) study with yeast rRNA genes, a single unequal crossing-over event affects on average seven repeats. Obviously, the larger the number of repeats exchanged in an unequal crossing-over event is, the higher the rate of concerted evolution will be (see the last section). This advantage of unequal crossing-over may be enough to offset the advantages of gene conversion. As dis-

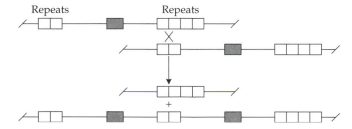

Figure 11.10   Crossing-over involving dispersed repeats (empty boxes). The shaded box denotes a unique gene. In the crossing-over event, this gene is deleted in one chromosome but duplicated in the other chromosome. From Li and Graur (1991).

cussed on earlier, unequal crossing-over seems to have played a more important role than gene conversion in the concerted evolution of the "360" and "500" repeat sequence families in *Drosophila*. These repeated sequences do not seem to have any function and so unequal crossing-over probably can occur freely within each repeat cluster. Unequal crossing-over has also been suggested to have played a much more important role than gene conversion in the concerted evolution of immunoglobulin V_H gene family in mouse (Gojobori and Nei 1984). Of course, the observed lower rate of gene conversion in this family might have been due to detection bias, for it is generally much easier to detect unequal crossing-over than gene conversion.

OTHER MECHANISMS OF CONCERTED EVOLUTION

In addition to gene conversion and unequal crossing-over, several other mechanisms have also been proposed for concerted evolution. These include the master-slave hypothesis (Callan 1967), the saltatory replication hypothesis (Britten and Kohne 1968; Buongiorno-Nardelli et al. 1972), replication-slippage or slipped-strand mispairing (Dover 1987; Walsh 1987), and duplicative transposition (Dover 1982).

Like unequal crossing-over, slipped-strand mispairing (SSM) is an expansion-contraction process that can lead to homogenization of a tandem repeat family. As noted in Chapters 1 and 13, SSM is probably more important than unequal crossing-over in the generation of tandem arrays of short repeats. Recent molecular data obtained from such techniques as Southern blot analysis and the polymerase chain reaction method have revealed the existence of numerous microsatellite DNAs in man (Chapter 13). SSM seems to have played a major role in the sequence homogenization of such loci. Of course, SSM and unequal crossing-over can operate on the same gene family. Indeed, this seems to be the case for the rRNA gene family in eukaryotes. Unlike the prokaryotic large subunit (LSU) rRNA genes, the eukaryotic LSU rRNA genes each consist of not only core regions but also expansion segments (also known as variable regions). While unequal crossing-over is probably mainly responsible for the coevolution of the whole rRNA repeats, SSM seems to be important for the coevolution of the expansion segments within a repeat (Hancock and Dover 1988).

The master-slave hypothesis postulates that in each generation new slave copies are replicated from the master sequences; thus, variant repeats are removed in each generation. This hypothesis is incompatible with the observation that neighboring repeats in the rRNA gene family in *Xenopus* can be unequal in length (Wellauer et al. 1976; see also Ohta 1980).

The saltatory-replication hypothesis proposes that a recent amplification of a sequence is the basis for intraspecific homogeneity. Under this hypothesis, the homogeneity of a family should slowly disappear, owing to accumulation of mutations, and for two species that diverged after the sequence amplification the intraspecific variation should be approximately equal to the interspecific variation (Arnheim 1983). As discussed above, this prediction is not supported by observations.

Duplicative transposition involves the duplication and movement of a sequence from one position to another (see Chapter 12). A sequence with a high propensity to duplicate with movement of one copy to another location has a good chance of spreading in a genome, and hence spreading through a sexually reproducing population (Dover 1982). As will be discussed in the next chapter, there are many such families in the genome of higher vertebrates. However, if the high homogeneity among the members of a family is due to recent expansion rather than being maintained by exchange of information among members, then the family is not really undergoing a *concerted* evolution, as there is little interaction between members.

FACTORS AFFECTING THE RATE OF CONCERTED EVOLUTION

In the above sections we discussed the effects of family size and population size on the rate of concerted evolution of a multigene family (e.g., see Table 11.6). In the following sections we shall discuss other factors.

Arrangement of Repeats

How cohesively the members of a repeated sequence family can evolve together depends on how they are arranged in the genome. There are, roughly speaking, two types of arrangement. First, the family members are highly dispersed all over the genome. For example, *Alu* repeats are interspersed with single-copy sequences throughout the genome (see Chapter 12). This type of arrangement is least favorable for concerted evolution because it greatly reduces the chance of unequal crossing-over and gene conversion and because unequal crossing-over would often lead to disastrous cytological consequences. The high similarities among *Alu* repeats is largely due to recent saltatory amplification of the family rather than concerted evolution (Chapter 12).

Second, all members of a family are clustered either in a single tandem array or in a small number of tandem arrays. As mentioned above, the rDNA family in *Xenopus* is present in a tandem array. This arrangement is most favorable for unequal crossing-over and gene conversion to operate. If the repeats are located on more than one chromosome, the rate of unequal crossing-over would be greatly reduced, unless all the clusters are at the ends of chromosome arms, as in the case of the rDNA family in humans. Moreover, the rate of gene conversion might also be reduced to some extent (see the last section of this chapter).

Functional Requirement

In terms of functional requirement, there are two extreme types of multigene families. One is that the function has an extremely stringent structural requirement, and often requires a large amount of the same gene product. The rRNA genes and the histone genes are well-known examples. The other extreme is that the function requires a large amount of diversity. The immunoglobulin genes and the major histocompatibility genes belong to this category. If all other conditions are the same for the two different types of gene families, the rate of concerted evolution is expected to be higher in the former type than in the latter type of families. Indeed, according to Gojobori and Nei's (1984) estimate, the rate is roughly 100 times higher in human rDNA family than in the mouse immunoglobulin V_H family. This large difference in rate seems to be mainly due to the difference in selection scheme between the two families (Gojobori and Nei 1984). In rRNA genes, homogeneity appears to be advantageous, and since unequal crossing-over and gene conversion both have the effect of reducing the genetic variability among multigenes, they apparently operate effectively for maintaining the homogeneity of rRNAs. In the case of V_H genes, however, an individual who has many copies of the same DNA sequence owing to unequal crossing-over or gene conversion seems to be at a disadvantage and may thus be eliminated from the population. We note further that homogeneity, on the one hand, and unequal crossing-over and gene conversion, on the other hand, would go in a cycle. Unequal crossing-over and (nonallelic) gene conversion both arise as a consequence of sequence misalignment between nonallelic repeats; there is indeed evidence from fusion genes in the human β-globin complex that the rate of unequal crossing-over increases with the length of sequence identity between duplicate genes (Metzenberg et al. 1991). Thus, the higher the homogeneity among the repeats in a family is, the higher the rates of unequal crossing-over and gene conversion will be (and vice versa). For this reason, the rates of unequal crossing-over and gene conversion in the rDNA family would be higher than those in the V_H family. This could be an important reason for the large difference in the rate of concerted evolution between the two families. It goes without saying that, in a multigene family whose stringency of functional requirements is intermediate between the above two extremes, the condition for concerted evolution would be moderately favorable.

Structure of Repeat Unit

By the structure of repeat unit we mean the numbers and sizes of noncoding regions (e.g., introns and spacer regions) and coding regions contained in a repeat unit. Since noncoding regions generally evolve rapidly, it would be difficult to maintain high homology among the repeats in a family, if each repeat contains large or many noncoding regions. As just mentioned, the rates of unequal crossing-over and gene conversion are expected to decrease with decreasing intraspecific homogeneity. Thus, the structure of repeat unit should play a role in concerted evolution.

Zimmer et al. (1980) estimated that in higher primates the rate of concerted evolution in the α-globin gene region is 50 times faster than that in the β-globin gene region. They suggested that the rate in the β-globin region has been greatly reduced because the introns and flanking sequences are highly divergent between the two β genes. It is interesting to note that the β genes have introns that are several times longer than those in the α genes and that the intergenic region between the two β genes is 2400 bases longer than that between the two α genes. Zimmer et al. (1980) suggested that the larger introns and intergenic region in β genes arose as a response to selection for reduction in the rate of unequal crossing-over, which produces a single gene state (the Lepore state) in which the major gene is under the control of the minor promoter. Because the noncoding sequences in the β region are highly divergent, they can act as barriers to unequal crossing-over. As the length of such sequences goes up, their effectiveness as barriers will rise. Note, however, that the selective advantage for a reduction in the rate of unequal crossing-over would be very small, at most as large as the amount of reduction in rate. Thus, the larger introns and intergenic region might have arisen by chance rather than by selection. It is also possible that the introns and the intergenic region were already large before the divergence of the higher primates, and this has promoted the divergence between the two β genes, including divergence in regulation. On the other hand, the introns and intergenic region in the α genes could have been small or could have been reduced since the divergence of higher primates, and as a consequence, concerted evolution has occurred at a higher rate in the α genes than in the β genes.

Natural Selection and Biased Gene Conversion

Positive natural selection can accelerate the process of concerted evolution because the rate and probability of fixation of a variant repeat that is favored by natural selection will be larger than those for a selectively neutral variant repeat. In fact, since unequal crossing-over and gene conversion can cause a large variation in the number of variant repeats per genome among individuals in the population, natural selection can become very strong among individuals in the population. By contrast, negative natural selection can effectively eliminate deleterious new variants. Therefore, positive selection can accelerate the differentiation between repeat members from different species whereas negative selection can maintain high homogeneity among repeat members within a species.

The effect of biased gene conversion on the evolution of multigene families would be similar to that of natural selection, though it may be weaker. The probability and rate of fixation of a new variant repeat is increased, if it is favored by gene conversion. On the other hand, if the predominant type of repeat is favored by gene conversion, then new variants may be eliminated quickly from the genome. Thus, like natural selection, biased gene conversion can accelerate between-species differentiation and maintain high within-species homogeneity.

Both natural selection and biased gene conversion work more effectively in large populations than in small ones because the effect of random genetic drift decreases with increasing population size.

EVOLUTIONARY IMPLICATIONS

Since concerted evolution of members of a gene family is different from the traditional notion of independent evolution of each family member, it has some profound evolutionary implications.

Spread of Advantageous Mutations

Gene conversion, unequal crossing-over, and other mechanisms of concerted evolution allow spreading of a variant repeat to all family members. This capability of horizontal spreading has profound evolutionary consequences, for an advantageous mutant repeat can replace all other repeats and becomes fixed in the family. We note that the advantage that a single variant can confer to an organism is usually very limited. The advantage would, however, be greatly amplified if the mutation spreads to many or all members in the family. Thus, through concerted evolution, a small selective advantage can become a great advantage. In this respect, concerted evolution surpasses independent evolution of individual gene family members (see Arnheim 1983; Walsh 1985).

Arnheim (1983) has compared the evolution of RNA polymerase I transcriptional control signals with that of RNA polymerase II transcriptional control signals. The former RNA polymerase transcribes only the rRNA genes whereas the latter transcribes all protein-coding genes (Chapter 1). RNA polymerase I transcription control signals appear to have evolved much faster than the signals for RNA polymerase II. For example, in cell-free transcription systems, mouse rDNA clone does not work in human cell extract, but clones of protein-coding genes from astonishingly diverse species are transcribed in heterologous systems (e.g., silk worm genes in human cell extracts and mammalian genes in yeast). Arnheim (1983) argues that in the case of RNA polymerase I transcription units, mutations that favorably affect transcription initiation could be propagated throughout the rDNA multigene family as a consequence of concerted evolution. It has also been suggested that even mutations that disfavorably affect transcription initiation to some extent can spread through the rDNA family (Arnheim 1983; Dover and Flavell 1984); since such mutations initially affect only a small number of rDNA genes and so would be subject to weak selection, they may persist long enough to allow mutations in the RNA polymerase I gene that are better able to interact with mutations in rDNA genes. Such correlated changes between the rDNA genes and the RNA polymerase I gene are a type of molecular coevolution. In contrast, in the case of RNA polymerase II transcription units, advantageous mutations affecting transcription initiation that occur in any gene would not be expected to be propagated throughout all genes, for they belong to many different families.

Retardation of Divergence of Duplicate Genes

The traditional view of the creation of a new gene has been that a gene duplication occurs and one of the two resultant genes gradually diverges and becomes a new gene (e.g., Ohno 1970). It is now clear that the process may not be so simple as previously assumed. As long as the degree of divergence between the two genes is not large, the diverged copy may be deleted by unequal crossing-over or converted to the conserved copy by gene conversion. In the former case it requires another duplication to create a redundant copy, while in the latter case divergence must start again from scratch. Thus, divergence of duplicate genes may proceed much more slowly than traditionally thought, and for this reason the chance of creation of a new gene from a redundant copy is reduced. On the other hand, gene conversion may also prevent a redundant copy from becoming nonfunctional for a long period time or, alternatively, may enable a "dead gene" (pseudogene) to be "resurrected" (see, e.g., Martin et al. 1983; Walsh 1987).

Evolutionary History of Duplicate Genes

It has been customary to assume that, following a gene duplication, the two resultant genes will diverge monotonically with time. Under this assumption it is rather simple to infer the time of the duplication event. For example, the protein sequences of human β and δ globins are more similar to each other than to rabbit $\beta 1$, or mouse β major and minor sequences (Dayhoff 1972). It has therefore been inferred that the two human genes were derived from a duplication event about 40 million years (Myr) ago (Dayhoff 1972), long after the mammalian radiation (about 80 Myr ago). The evolutionary history of duplicate genes inferred in this traditional manner may be erroneous in view of the fact that duplicate genes can correct each other. In fact, it has recently been suggested that the β and δ genes originated from a duplication event before the mammalian radiation (Hardison and Margot 1984). This suggestion is based on the observations that the large intron and the 3' untranslated region of rabbit pseudogene $\psi\beta2$ are more similar to human δ than to rabbit $\beta1$, and that the pseudogene $\psi\beta3$ in mouse is similar to δ at its 3' end (Edgell et al. 1983). This hypothesis is not supported by the observation that no obvious descendant of δ is found in the goat β-globin family (Cleary et al. 1981; Shapiro et al. 1983), but if it turns out to be true, it will provide an excellent example how gene correction events can obscure the evolutionary history of duplicate genes. In a large multigene family, gene correction events will occur more often, and it will be even more difficult to trace the evolutionary relationships of the family members.

Generation of Genic Variation

From the evolutionary point of view there is an analogy between evolution of multigene families and evolution of subdivided populations. One may regard each repeat in a multigene family as a deme in a subdivided population. Then, transfer of information between repeats would be equivalent to migration of genes or individuals between demes. It is well known that migration will reduce the genetic difference between demes but will increase the amount of genic variation, e.g., the number of alleles, in a deme. Similarly, transfer of information between repeats will reduce the genetic difference between repeats but will increase the amount of genic variation at a locus. This can be seen from Table 11.6. Note that the intralocus variation, $1 - f$, is lower in the absence of gene conversion, i.e., $\lambda = 0$, than in the presence of gene conversion. Note also that $1 - g_1$ and $1 - g_2$, the interlocus variations, are 1 in the absence of gene conversion but become lower in the presence of gene conversion.

It is well known that some major histocompatibility complex (MHC) loci are highly polymorphic in humans, mice, and certain other mammals (Klein 1986). For example, allelic gene products at a *H-2* antigen locus from different inbred mouse strains can differ extensively in amino acid sequences and hence in serological properties. More than 50 different alleles at each of the *H-2 K* and *H-2 D* loci have been observed in mice. Weiss et al. (1983) and many others have suggested that the large numbers of alleles at the *H-2* loci are mainly generated by gene conversion-like events; donor sequences could be from the *H-2* loci and from nearby loci. An alternative explanation is that these alleles have persisted in the population for a long time (Figueroa et al. 1988), probably being maintained by overdominant selection (Hughes and Nei 1988). These are two extreme views and it is possible that both gene conversion and overdominant selection have been responsible for the high genetic variation at MHC loci.

CHAPTER 12

Evolution by Transposition and Horizontal Transfer

"JUMPING GENES" (TRANSPOSABLE GENETIC ELEMENTS) were first discovered in maize by Barbara McClintock in the late 1940s. She found that genes associated with the development of color pigments in kernels could be turned on or off at abnormal times by the action of certain genetic elements, which she called "controlling elements," that apparently could move from site to site on different maize chromosomes. However, the idea of jumping genes ran contrary to the traditional view that genomes are stable, static entities in which genes can be assigned to specific loci and retain their precise chromosomal position over long periods of evolution. Since the traditional concept was not casually arrived at but came from a large body of genetic data, McClintock's observation was thought to be a rare phenomenon and not of general interest.

With the great advances in molecular biology techniques in the 1970s and 1980s, transposition of genetic elements was found to be a widespread phenomenon not only in maize but also in other organisms. DNA sequences that possess an intrinsic capability to change their genomic location are called **mobile elements** or **transposable elements** (TEs). The new data stimulated vigorous debates on the biological significance of TEs. Since TEs can "jump about" in a genome and can move genetic materials from one genomic location to another, they can produce novel gene mutations and chromosomal rearrangements that cannot be produced by other known mechanisms. For this reason, many authors (e.g., Never and Saedler 1977; Cohen and Shapiro 1980; Bingham et al. 1982; Ginzburg et al. 1984; Syvanen 1984) argued that TEs play an important role in evolution, especially in speciation events. On the other extreme, Doolittle and Sapienza (1980), Orgel and Crick (1980), Temin (1980) and others regarded TEs merely as genetic parasites. They noted that movement of TEs in a genome is more likely to produce deleterious rather than beneficial effects and argued that TEs persist in nature not because they confer advantages to their hosts but because they can multiply faster than other parts of the host genome. This great divergence in opinion notwithstanding, it is now generally recognized that the structural organization of genomes is much more fluid and prone to evolutionary changes than previously thought. Although the evolutionary implications of DNA transposition remain not fully understood, progress in both empirical and theoretical population studies of TEs has considerably increased our understanding of the spread and maintenance of TEs in populations (see Charlesworth and Langley 1991 and references therein). Moreover, molecular studies have greatly increased our understanding of the mechanism and the effects of DNA transposition (see Berg and Howe 1989; Lewin 1994).

DNA can also move between organisms. In fact, it has long been known that a bacterium can absorb DNA from its surroundings and functionally integrate the exogenous DNA into its genome, a phenomenon known as **transformation** (Griffith 1928). There are also other mechanisms by which genes can be transferred across bacterial species boundaries (see the later discussion of examples of horizontal gene transfer). Thus, cross-species (horizontal) gene transfer might have played a significant role in bacterial evolution; evidence for such a role has been increasing (see the reviews by Smith et al. 1992; Kidwell 1993; Syvanen 1994). Furthermore, recent evidence suggests that horizontal gene transfer can occur from prokaryotes to eukaryotes and vice versa. There is also strong evidence for the transfer of TEs among eukaryotes as well as among prokaryotes (see Kidwell 1993). Therefore, although the possibility of horizontal gene transfer was originally viewed by biologists with much skepticism, it may indeed be a widespread phenomenon.

TRANSPOSITION AND RETROPOSITION

Transposition is defined as the movement of genetic material from one chromosomal location to another. There are two types of transposition, distinguished by whether the TE is replicated or not. In the **nonreplicative** type of transposition, the element itself moves from one site to another (Figure 12.1a). In **replicative** (or **duplicative**) transposition, the TE is copied, and one copy remains at the original site, while the other inserts at a new site (Figure 12.1b). Thus, replicative transposition is characterized by an increase in the number of copies of the TE. Some TEs use only one type of transposition; others use both the replicative and nonreplicative pathways.

In the above types of transposition, the genetic information is carried by DNA. It is known that genetic information can also be transposed through RNA. In this mode, the DNA is transcribed into RNA, which is then reverse-transcribed into cDNA (Figure 12.1c). In order to distinguish between the two modes, the RNA-mediated mode has been termed **retroposition.** In contrast with DNA-mediated transposition, retroposition is always of the duplicative form, because it is a reverse-transcribed copy of the element, not the element itself, that is transposed. Both transposition and retroposition are found in eukaryotic and prokaryotic organisms (see Weiner et al. 1986; Temin 1989).

When a TE is inserted into a host genome, a small segment of the host DNA at the insertion site (usually 4–12 bp) is duplicated (Figure 12.1). The duplicated repeats are in the same orientation and are called **direct repeats.** This characteristic is a hallmark of transposition and retroposition.

Some TEs can transpose themselves in all cells; others are highly specific. For example, *P* elements in *Drosophila melanogaster* are usually mobile only in germ cells. The genomic locations of the recipient sites for transposition also show variation among different TEs. Some elements show an exclusive preference for a specific genomic location. For example, *IS4* incorporates itself exactly and always at the same point in the galactosidase operon of *Escherichia coli*. Others, such as bacteriophage *Mu*, can transpose themselves at random to almost any genomic

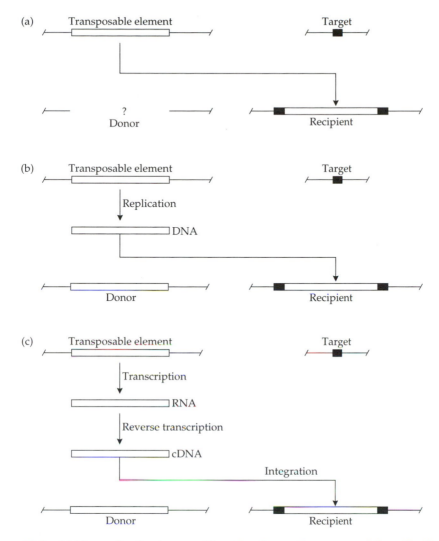

Figure 12.1 (a) Nonreplicative transposition. The element is transposed from the donor site to a target site. What will happen to the donor site is unclear. One model proposes that the ends of the donor DNA are not joined to each other and that the DNA molecule is destroyed. Another model proposes that the double-strand break is repaired by the host repair system. (b) Replicative transposition. The element is replicated and a copy is inserted at a target site, while the other copy remains at the donor site. For a more detailed explanation of nonreplicative and replicative transposition, see Lewin (1994). (c) Retroposition. The element is transcribed into RNA, which is reverse-transcribed into DNA. The DNA copy is inserted into a host genome. For an example of retroposition, see Figure 12.4. Note that both transposition and retroposition create a short repeat (black box) at each end of the newly inserted element. From Li and Graur (1991).

location. Many TEs show intermediate degrees of genomic preference. For example, 40% of all *Tn10* transposons in *E. coli* are found in the *lacZ* gene, while Drosophila P elements have an affinity for the X chromosome and also prefer to insert into sequences 5' to the coding region of genes, rather than into the coding region itself. Some TEs exhibit higher affinities for a particular type of nucleotide composition. For example, *IS1* favors AT-rich insertion sites (Devos et al. 1979). However, it should be noted that the distribution pattern of elements in a genome is affected by both natural selection and the rate of recombination (see Charlesworth and Langley 1991).

TRANSPOSABLE ELEMENTS

According to their mode of transposition and the types of genes they contain, which may not be a reflection of common ancestry, transposable elements have been classified into three types: insertion sequences, transposons, and retroelements. In addition, there are genomic sequences that have been produced by retroposition of RNAs but lack the intrinsic capability to transpose themselves. Such sequences are called retrosequences and will be discussed in later sections on retrogenes and retropseudogenes.

Insertion sequences

Bacterial insertion sequences were discovered in the late 1960s and early 1970s during investigations of expression of genes in E. coli and bacteriophage lambda (see Galas and Chandler 1989 for a review). They were originally found as unstable mutations but were subsequently shown by hybridization and heteroduplex analysis to be insertions of the same few DNA segments in different positions and orientations. The similarity of bacterial insertion sequences to the genetic elements in maize discovered by McClintock (1952) became clear when it was recognized that they are residents of the *E. coli* genome and that the observed insertion mutations were examples of their movement to new genetic locations.

Insertion sequences (IS) are the simplest transposable elements (Figure 12.2). They carry no genetic information except that which is necessary for transposition. Insertion sequences are usually 700–2500 bp in length. They have been found in bacteria, bacteriophages, and plasmids. Moreover, the maize controlling elements are also insertion sequences (IS) by this definition. Bacterial insertion sequences are denoted by the prefix IS followed by a type number. The structure of an insertion sequence, *IS1* from *E. coli* and *Shigella dysinteria*, is shown schematically in Figure 12.2a. *IS1* is approximately 770 bp in length, including two inverted nonidentical terminal repeats, 23 bp each. It contains two reading frames, InsA and InsB, which encode one or two forms of transposase, an enzyme that catalyzes the insertion of transposable elements into insertion sites. There are dozens of different types of insertion sequences in *E. coli*, and the genomes of most strains isolated from the wild contain variable numbers of each (e.g., Sawyer et al. 1987).

Transposons

In the mid 1960s it became evident that certain genes responsible for resistance to antibiotics in bacteria are capable of transferring from one DNA molecule to another. The first direct evidence that such transfer is by a process analogous to the insertion of IS elements was reported by Hedges and Jacob (1974), who postulated that the gene for ampicillin resistance is carried by a DNA element that can be transposed between molecules. They called such an element a **transposon** (Tn). Later, the term was used to denote any element that carries one or more

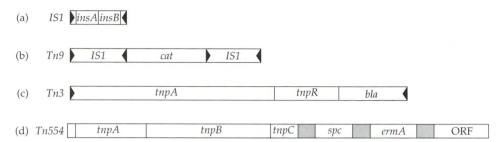

Figure 12.2   Schematic representation of four transposable elements in bacteria. Black triangles denote inverted repeats. (a) Insertion sequence *IS1* from *Escherichia coli* and *Shigella dysenteria* is flanked by imperfect inverted repeats 23 bp long. (b) Composite transposon *Tn9* from *E. coli* contains two copies of *IS1* flanking the cat gene, which encodes the protein conferring chloramphenicol resistance. (c) Transposon *Tn3* from *E. coli*, which confers streptomycin resistance, contains three genes, two of which (*tnpR* and *bla*) are transcribed on one strand, and the third (*tnpA*) on the other. *Tn3* is flanked by two perfect inverted repeats, 38 bp long. (d) Transposon *Tn554* from *Staphilococcus aureus* lacks terminal repeats and contains five genes and an open reading frame (ORF). Three of the genes encode transposases (*tnpA*, *tnpB*, and *tnpC*) and are transcribed as a unit. The *spc* and *ermA* genes confer spectinomycin and erythromycin resistance, respectively. The *spc* gene, which encodes an S-adenosylmethionine-dependent methylase, is transcribed on a different strand than the other genes. The ORF is abundantly transcribed, but it is not known whether or not it is translated. From Li and Graur (1991).

genes that encode other functions in addition to those related to transposition; such genes are known as exogeneous genes. Some authors call all transposable elements, including IS elements, transposons.

Since transposons also carry exogeneous genes they are usually larger than IS elements, about 2500–7000 bp long. In bacteria, some transposons are complex or compound transposons, so named because two complete, independently transposable insertion sequences in either orientation flank one or more exogenous functional genes (Figure 12.2b). Interestingly, in complex transposons, the entire transposon can transpose as a unit, and also one or both of the flanking insertion sequences can transpose independently.

Other bacterial transposons, as well as many eukaryotic transposons, are flanked only by short repeated sequences in various orientations and contain no insertion sequences (Figure 12.2c). Not all transposons, however, are symmetrical in structure. Some have asymmetrical ends, lacking either inverted or direct terminal repeats (Figure 12.2d). The coding regions of some transposons in animals, e.g., *P* elements in *Drosophila*, are interrupted by introns (Figure 12.3).

Figure 12.3   Schematic structure of a complete *P* element in *Drosophila melanogaster*. The element is flanked by short inverted repeats, 31 bp long, and its coding region contains four exons (white boxes) interrupted by three introns (black boxes). The element is about 2900 bp long. From Li and Graur (1991).

Transposons in bacteria often carry genes that confer antibiotic (e.g., *Tn554*), heavy-metal (e.g., *Tn21*), or heat (e.g., *Tn1681*) resistance to their carriers. Plasmids can carry such transposons from cell to cell, and as a consequence, resistance can quickly spread throughout populations of bacteria exposed to such environmental factors.

Several phages in bacteria are in fact transposons. For example, bacteriophage *Mu* is a very large transposon (~38 kb) that encodes not only the enzymes that regulate its transposition, but also a large number of structural proteins necessary to construct the packaging of its DNA.

Transposons of many types are quite widespread in the genomes of animals, plants, and fungi. *Drosophila melanogaster*, for instance, contains multiple copies of 50–100 different kinds of transposons (Rubin 1983).

Retroelements

As mentioned above, there are transposable elements that transpose by retroposition; all such elements contain a gene for the enzyme reverse transcriptase. Moreover, there are genetic entities that contain a gene for reverse transcriptase but do not have any intrinsic capability to transpose. Temin (1989) proposed to call the entire group of reverse transcriptase-bearing entities, whether transposable or not, **retroelements** (Table 12.1); the term retroids was coined earlier by Fuetterer and Hohn (1987).

Among the known retroelements, retroviruses are the most complex (see the review by Eickbush 1994). Their genome contains at least three genes: *gag, pol,* and *env* (Figure 12.4a). The *gag* gene encodes a polyprotein that is processed into several small structural proteins that are the major proteins of the nucleocapsid and matrix of the virion. The *pol* gene encodes a polyprotein that consists of a reverse transcriptase (RT); RNase H (RH), which destroys the RNA strand of the RNA; an integrase (IN), which is responsible for the integration of the double-stranded DNA intermediate made from the RNA transcript into the chromosomes of the host genome; and possibly also an aspartate proteinase (PR), which is responsible for processing the primary translation products of the *gag* and *pol* genes into the various structural and enzymatic components of the virion. The

TABLE 12.1   Classification of retroelements and retrosequences

Element	Reverse transcriptase	Transposition	Presence of LTRs[a]	Virion particles
Retron	yes	no	no	no
Retroposon	yes	yes	no	no
Retrotransposon	yes	yes	yes	no
Retrovirus	yes	yes	yes	yes
Pararetrovirus	yes	no	yes	yes
Retrosequence	no	no	no	no

From Temin (1989).
[a]LTR = Long terminal repeats.

env gene encodes the envelope protein. Many retroviruses possess additional genes; for example, the AIDS virus (HIV) possesses at least six additional genes (see Trono 1995). The coding region of a retrovirus is flanked by long terminal repeats (LTRs), which contain important control sequences, e.g., a promoter, an enhancer, and a poly(A) addition signal.

The next most complex retroelements are **retrotransposons,** which are TEs with LTRs (Table 12.1). With few exceptions, retrotransposons do not have the *env* gene for constructing virion particles and therefore cannot independently transport themselves between cells. Most retrotransposons have a gene organization similar to that of either the *gypsy* or *copia* elements. The gene content and organization of *gypsy* elements are very similar to those of retroviruses (Figure 12.4b): The three open-reading frames (ORFs) have sizes and arrangements similar to the *gag, pol,* and *env* genes of retroviruses. In fact, recent data suggest that gypsy is indeed an infectious retrovirus, and since *gypsy* is found in *Drosophila,* this finding changes the traditional view that retroviruses exist only in vertebrates (Kim et al. 1994). The other major subgroup of retrotransposons is represented by the element *copia* in *D. melanogaster.* Note that *copia* does not contain the *env* gene. Note also that the integrase domain in *copia* precedes the reverse transcriptase domain, rather than follows the RNase H domain, as in retroviruses and *gypsy* (Figure 12.4b). Retrotransposons similar to *copia* have been found in many eukaryotes including the *Ty1* elements in yeast (Ty stands for transposon yeast) and the *Ta1* of *Arabidopsis thaliana.*

Retroposons, unlike retrotransposons, do not have LTRs (Table 12.1). They are sometimes called non-LTR retrotransposons (e.g., Eickbush 1994) or nonviral retrotransposable elements (Weiner et al. 1986). Because the best known elements within this group are the long interspersed nucleotide elements (LINEs) found in mammalian genomes, they have also been called LINE-like elements. Two examples of retroposons found in *D. melanogaster,* the I factor and R2, are shown in Figure 12.4c. The retroposons are probably the most varied group of retroelements, and the I factor and R2 are not representative of any major subgroup. Most retroposons share at least a few of the features associated with the *gag* and *pol* genes of retroviruses. Most retroposons, like the I factor, contain two ORFs, while others, like R2, contain a single ORF (Figure 12.4c).

Some DNA viruses appear to use reverse transcription in replication but do not have an integrated replicative form. These viruses are called pararetroviruses (Table 12.1; Figure 12.4d) and include the hepadnaviruses (e.g., the hepatitis B virus) and cauliflower mosaic virus. Since pararetroviruses do not have the ability to insert themselves into the host genome, they actually are not qualified to be transposable elements; however, they clearly have a common evolutionary origin with the retroviruses (see later).

Retrons are the simplest retroelements. They do not have LTRs and their nucleotide sequence contains only a region coding for reverse transcriptase and one or two other regions that are transcribed. Retrons have been found in some bacterial genomes (Inouye et al. 1989; Lampson et al. 1989) as well as in the mitochondrial genome of the plant *Oenothera berteriana* (Schuster and Brennicke 1987). Retrons do not excise and therefore are integral parts of the genome. Unlike proviruses, retrons cannot construct virion particles and appear as a single copy in the genome of *myxobacteria* and *E. coli.*

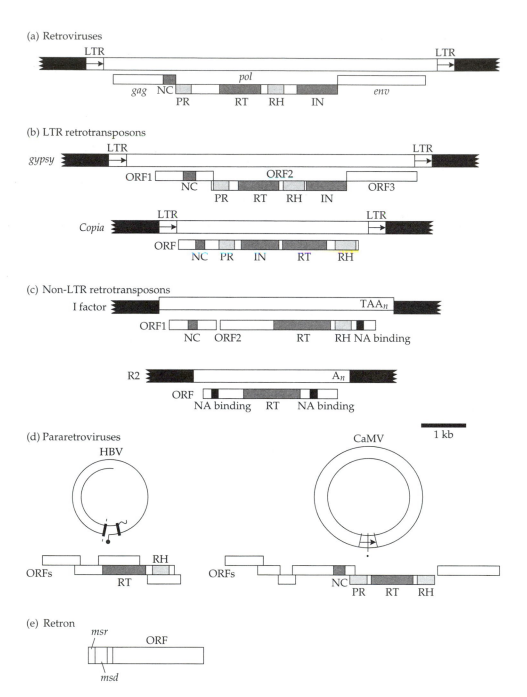

(a) Retroviruses

(b) LTR retrotransposons

(c) Non-LTR retrotransposons

(d) Pararetroviruses

(e) Retron

EVOLUTIONARY RELATIONSHIPS AMONG RETROELEMENTS

Since all retroelements are reverse transcriptase-bearing genetic elements, they are apparently related to one another. Indeed, as early as 1970, Temin postulated that retroviruses evolved from moveable genetic elements (i.e., cellular protoviruses). The elucidation of the structural features and nucleotide sequences of transposable elements (e.g., *Ty1* and *copia*) revealed clear similarity to those of

◀ **Figure 12.4**   Structural comparison of retroelements. The open-reading frames (ORFs) for each element are shown below the DNA diagram as a series of horizontal boxes. Boxes at the same level represent ORFs separated by termination codons; boxes at different levels represent ORFs in different reading frames. LTRs (long terminal repeats) are represented by arrows. (a) A consensus retrovirus is shown depicting only the *gag, pol,* and *env* genes. Additional genes are found in many retroviruses. Identified domains within the ORFs are indicated by shading. NC, nucleocapsid protein; PR, asparate protease; RT, reverse transcriptase; RH, RNase H; IN, integrase. (b) The gypsy and copia elements represent two typical retrotransposons. Gypsy has a third ORF, ORF3, which encodes the env protein. (c) Retroposons are the most varied group of retroelements and the I factor and R2 shown are not representative of any major group. NA binding denotes an identified cysteine-histidine motif with an unknown function in RNA or DNA binding. Note the absence of LTRs. (d) Pararetroviruses represented by the hepatitis B virus (HBV) and the cauliflower mosaic virus (CaMV). The black dot at the end of one strand of the HBV genome corresponds to the protein primer used to prime first-strand (cDNA) synthesis. The wavy line at the end of the other, incomplete strand of HBV corresponds to the RNA primer used to prime second-strand synthesis. Thick bars in HBV represent short direct repeats, involved in the strand-transfer reactions. (e) A retron from the myxobacterium, *Myxococcus xanthus*. The *msr* gene is transcribed into RNA. The *msd* gene is transcribed from the complementary strand and then reverse-transcribed into DNA by the reverse transcriptase encoded by the ORF of the retron. The two molecules are subsequently attached to each other via a 2′,5′-phosphodiester bond to form a branched molecule called multicopy single-stranded DNA (msDNA). After Eickbush (1994).

retroviruses (Figure 12.4) and provided a basis for the following scenario for the emergence of retroviruses (Temin 1980). Two insertion-like sequences transposed around a cellular gene for a DNA polymerase; this DNA polymerase became the ancestor of the reverse transcriptase. The two insertion sequences then formed the ancestors of the long terminal repeats (LTRs) at the end of the provirus. They also formed a transposon for the DNA polymerase gene. Two of these protoviral elements integrated around a cellular gene that was the precusor of a virion protein, for example, the core protein precursor. Deletion of an internal LTR followed by transposition of the region containing the polymerase gene and the new gene gave an element with two structral genes. Repeating this process of integration of two elements around a cellular gene that was a precusor of another virion protein (for example, the envelope protein), followed by deletion of an internal LTR, gave an element with the structure of a provirus.

In the above scheme Temin (1980) proposed that the viral reverse transcriptase evolved from DNA polymerase because at that time no cellular reverse transcriptase genes had been found in either eukaryotes or prokaryotes. Later, cellular genes for reverse transcriptase were found in myxobacteria and *E. coli* (Figure 12.4). This observation suggests that reverse transcriptase existed before retroviruses and not the reverse. Temin (1989) therefore revised his scheme as follows (Figure 12.5): The path of evolution went from retrons, which are the simplest reverse transcriptase-bearing entities, to retroposons by gaining the ability to transpose, next to retrotransposons by gaining LTRs, then to retroviruses by the acquisition of structural genes, and subsequently to pararetroviruses by loss of the ability to transpose. Of course, it is possible that some of the present-day retrotransposons have been derived from retroviruses, rather than the other way around.

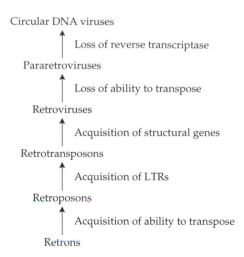

Figure 12.5 Schematic representation of possible evolutionary pathways of retroelements. From Li and Graur (1991).

Many authors (e.g., Miyata et al. 1985; Doolittle et al. 1989; Xiong and Eickbush 1988, 1990) have studied the evolutionary relationships among retroelements. Figure 12.6 shows a phylogeny of retroelements inferred from reverse transcriptase sequences (Eickbush 1994). The tree can be divided into two major branches, as indicated by the arrow on the tree. The major branch on the left half of the tree contains all the non-LTR retroelements. Within this branch, the ms-associated DNAs, the organellar elements, and the group II introns form three separate monophyletic subgroups. In addition, with the exception of the three trypanosome elements *CZAR, SLAC,* and *CRE1,* all the other retroposons form a monophyletic group, which includes the I factor, R2 (*R2Dm* and *R2Bm*) and the LINE (*L1Hs* and *L1Md*) elements. The major branch on the right half of the tree contains all the LTR retroelements. Two subgroups are worth noting. First, the copia and the *Ty1* elements form a subgroup. Since the *Ty1* and copia elements are only distantly related to other retrotransposons, it has been suggested that *Ty1* is the result of a horizontal transfer from *Drosophila* to yeast (Yuki et al. 1986). In this regard it is interesting to note that not all species of *Drosophila* carry the *copia* element, so that it is difficult to imagine that the *copia* element has been carried continuously in the genome since the divergence of fungi and animals. The close ecological dependence of *Drosophila* and yeast is in keeping with the possibility of horizontal transfer (Doolittle et al. 1989). Second, several *Drosophila* elements (gypsy, 412, etc.), yeast *Ty3,* and the caulimoviruses form a large subgroup. Since plants, fungi, and animals have long been separated from one another, this subgroup may represent at least two cases of horizontal transfer.

Since the *ms*-associated DNAs, the organellar elements, and the group II introns are the simplest retroelements and are probably of prokaryotic origin, their common ancestor may be considered the root of the tree (Eickbush 1994). Under this assumption, the branching order of the tree is consistent with Temin's (1980) view that the evolution of retroelements went from retrons to retroposons, to retrotransposons, and then to retroviruses (Figure 12.5), though this tree does not support the view that hepadnaviruses (as represented by the hepatitis B

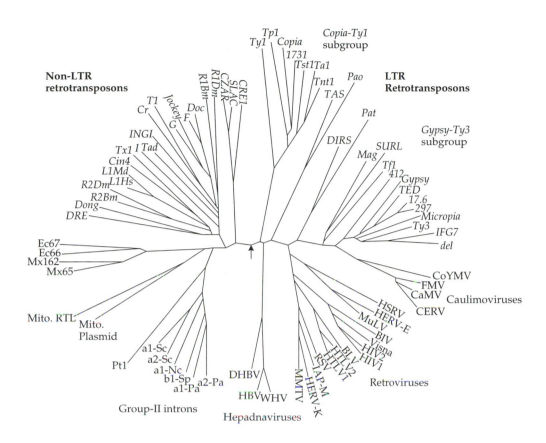

Figure 12.6 Unrooted phylogenetic tree of retroelements. The tree was derived from the amino acid sequences of the reverse transcriptase domain encoded by each retroelement. The amino acid sequence divergence of the 178 positions common to all enzymes was used to derive the tree using the neighbor-joining method. All elements that define a particular class or subclass of retroelements are indicated by italic letters. Pararetroviruses are called hepdanaviruses in the figure. Identification of the host species for each element can be found in Xiong and Eickbush (1988, 1990) and Eickbush (1994). From Eickbush (1994).

viruses) were derived from retroviruses. (As mentioned above, *gypsy* and its relatives might be retroviruses.) Note, however, that Figure 12.6 is based on very limited sequence data, so that the branching order of the subgroups and the branching order of the elements within each subgroup should be taken with caution. Further data are needed for obtaining a more reliable tree.

RETROGENES

The classification in Table 12.1 includes a class of **retrosequences,** or **retrotranscripts,** which are genomic sequences that have been derived through the reverse transcription of RNA and subsequent integration into the genome but lack the ability to produce reverse transcriptase. The template RNA is usually the transcript of a gene. Some authors call retrosequences "nonviral retroposons" (e.g., Weiner et al. 1986). A possible process of producing retrosequences is shown in Figure 12.7. If a gene is not transcribed within any germline cells, the creation of a retrosequence requires the RNA to cross cell barriers. This can happen when an RNA molecule becomes encapsulated within the virion particle of a retrovirus and is then transported to a germline cell where it is reverse-transcribed (Linial 1987). This process is referred to as **retrofection.**

Since retrosequences originated from RNA sequences, most of them bear marks of RNA processing and are, hence, also referred to as **processed retrosequences**. The diagnostic features of such sequences include: (1) lack of introns, (2) precise boundaries coinciding with the transcribed regions of genes, (3) stretches of poly-A at the 3' end (there are exceptions, as in most of snRNA pseudogenes), (4) short direct repeats at both ends, indicating that transposition may have been involved, (5) various post transcriptional modifications, such as the addition or removal of short stretches of nucleotides, and (6) chromosomal position different from the locus of the original gene from which the RNA was transcribed.

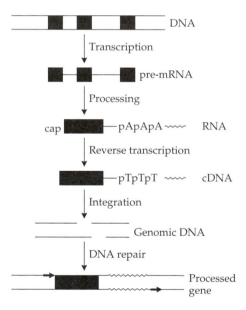

Figure 12.7 Creation of a processed retrosequence. The black boxes represent exons. The wavy lines indicate a poly-A tail in mRNA and the complementary poly-T in cDNA. The process is as follows. The DNA is transcribed into pre-mRNA, then processed to mRNA. The mRNA is reverse-transcribed into cDNA, which becomes integrated into the genomic DNA. The gaps are repaired, and so two direct short repeats flanking the inserted retrosequence are created (black horizontal arrows). If the insertion is done in a different cell from the one in which the RNA has been synthesized, the process of creating a retrosequence requires the mRNA to become incorporated into a retroviral particle and transported to the target cell. Such a process is called retrofection. From Li and Graur (1991).

A retrosequence is called a **retrogene** (a processed gene) if it is functional but a **retropseudogene** if it is nonfunctional. Examples of retrogenes are discussed below, while examples of retropseudogenes will be presented in the next section.

A retrogene is a functional retrosequence producing a protein that is identical or nearly identical to that produced by the gene from which the retrogene has been derived. There are several reasons why it is highly unlikely that a reverse-transcribed gene will retain its functionality. First, the process of reverse transcription is error prone and many differences (mutations) between the RNA template and the cDNA can occur. Second, unless a processed gene has been derived from a gene transcribed by RNA polymerase III, it usually does not contain regulatory sequences that reside in the untranscribed regions and is therefore likely to be inactive, even if its coding region is intact. The processed human metallothionein pseudogene appears to be such a case (Karin and Richards 1984). Third, a processed gene may be inserted at a genomic location that may not be adequate for its proper expression. For these reasons, a processed gene is "dead on arrival" in the vast majority of cases.

However, processed functional genes have been found. The human phosphoglycerate kinase (*PGK*) family consists of an active X-linked gene, a processed X-linked pseudogene, and an autosomal gene. The X-linked gene contains 11 exons and 10 introns. Its autosomal homologue, on the other hand, is unusual in that it has no introns and is flanked on its 3' end by remnants of a poly-A tail, strongly suggestive of a reverse transcription process. Interestingly, the autosomal *PGK* gene is expressed almost exclusively in the testes. Thus, the reverse-transcribed *PGK* gene has not only maintained an intact reading frame and the ability to transcribe and produce a functional polypeptide, but also has acquired a novel tissue-specificity (McCarrey and Thomas 1987).

The muscle-specific calmodulin gene in chicken is intronless and apparently was produced by a reverse-transcriptase-mediated event (Gruskin et al. 1987). This might also be the case for the intronless globin gene of *Chironomus* (Antoine et al. 1987). All but one of the 17–20 intronless actin genes in *Dictyostelium* are funtional (Romans and Firtel 1985), and these might also be retrogenes.

The rat and mouse preproinsulin I gene may represent an instance of a semi-processed retrogene. The gene contains a single small intron in the 5' untranslated region. In comparison, its homologue, preproinsulin II contains the same small intron, as well as an additional larger intron within the coding region for the C peptide. All preproinsulin genes from other mammals, including other rodents, contain two introns as well. Moreover, the preproinsulin I gene is flanked by short repeats, and the polyadenylation signal is followed by a short poly-A tract (Soares et al. 1985). These features suggest that preproinsulin I gene might be a semiprocessed retrogene derived from a partially processed preproinsulin II mRNA. Indeed, based on comparisons between the two preproinsulin genes, preproinsulin I appears to have been derived from an abberant mRNA transcript that was initiated at least 500 base pairs upstream of the normal cap site and from which only the first intron has been spliced out. It is likely that the aberrant transcript contained 5' regulatory sequences not normally transcribed, so that the retrogene has maintained its function following its integration into a new genomic location.

RETROPSEUDOGENES

A retropseudogene is a nonfunctional retrosequence. Most retropseudogenes are apparently derived from processed RNA transcripts (in this case they are called **processed pseudogenes**), though some seem to have been derived from unprocessed RNA transcripts (see Weiner et al. 1986). In principle, retropseudogenes can be derived from all RNA transcripts that are present in germ cells or can retroinfect germ cells. Indeed, retropseudogenes derived from RNA polymerases (POL) II and III transcripts (e.g., mRNA, snRNA, tRNA, and 7SL RNA) are known. However, so far there seems to be no example of a retropseudogene that is derived from a RNA polymerase I transcript (e.g., 18S, 28S, 5.8S, and 5S rRNAs). It is not clear why this class of RNA transcripts cannot be or are rarely retrotransposed.

Pseudogenes Derived from POL II Transcripts

Most POL II retropseudogenes are derived from mRNA sequences; the others are mainly from snRNAs (small nuclear RNAs, Chapter 1). They bear all the hallmarks of a functional retrosequence but have molecular defects that prevent them from being expressed. Many processed pseudogenes get truncated during retrofection. 5' truncation of processed mRNA is particularly common, though 3' truncation is also known. Truncation of the 5' end can occur during (1) transcription (e.g., initiation downstream of the normal initiation site), (2) RNA processing (e.g., faulty splicing), and (3) reverse transcription (e.g., failure of the enzyme to complete the reverse transcription to the 3' end of the RNA molecule, which corresponds to the 5' end of the DNA).

Pseudogenes derived from mRNAs are abundant in mammals. Table 12.2 shows a list of such pseudogenes in humans and rodents for which both the number of functional genes and the number of processed pseudogenes are known or have been estimated. Note that there is a tendency to underestimate the number of pseudogenes because old pseudogenes may have diverged in sequence from their parental gene to such an extent that they are no longer detectable by molecular probes derived from the functional homologue. Interestingly, in some cases the number of processed pseudogenes is much greater than the number of functional genes. For example, in the genome of mouse about 200 processed pseudogenes have been produced from a single gene for glyceraldehyde-3-phosphate dehydrogenase.

Although processed pseudogenes are abundant in mammals, they are relatively rare in other organisms, including chicken, amphibia, and *Drosophila*. For example, mammals have 20–30 tubulin retropseudogenes while chicken and *Drosophila* have none, though, like mammals, chicken and *Drosophila* have many α- and β-tubulin genes. It is not clear why such a conspicuous difference occurred. One hypothesis is that it is due to differences in gametogenesis between mammals and other organisms (Weiner et al. 1986). While spermatogenesis is very similar among mammals, chicken, amphibia, and *Drosophila*, mammalian oogenesis differs from that in the other organisms by prolonging the lambrush stage from birth to ovulation. This state of relatively suspended animation may last 40 years in humans, but for only a few months in amphibians, and for less than three weeks in birds, and is virtually absent in *Drosophila*. If retroposi-

TABLE 12.2 The number of retropseudogenes and the number of parental functional genes

Species	Gene	Number of genes	Number of retropseudogenes
Human	Argininosuccinate synthetase	1	14
	β-actin	1	~20
	β-tubulin	2	15–20
	Cu/Zn superoxide dismutase	1	≥4
	Cytochrome c	2	20–30
	Dihydrofolate reductase	1	~5
	Nonmuscle tropomyosin	1	≥3
	Glyceraldehyde-3-phosphate dehydrogenease	1	~25
	Phosphoglycerate kinase	2[a]	1
	Ribosomal protein L32	1	~20
	Triosephosphate isomerase	1	5–6
Mouse	α-globin	2	1
	Cytokeratin endo A	1	1
	Glyceraldehyde-3-phosphate dehydrogenease	1	~200
	Myosin light-chain	1	1
	Propiomelanocortin	1	1
	Ribosomal protein L7	1–2	≥20
	Ribosomal protein L30	2	≥15
	Ribosomal protein L32	1	16–20
	Tumor antigen p53	1	1
Rat	α-tubulin	2	10–20
	Cytochrome c	1	20–30

From Weiner et al. (1986).
[a]One of which is a retrogene.

tion in mammals in fact occurs predominantly in the female germline, retrosequences would be underrepresented on the Y chromosome.

Pseudogenes Derived from POL III Transcripts

The genomes of mammals and other higher eukaryotes contain several types of highly repetitive interspersed sequences. These sequences, originally detected as rapidly reannealing components of genomic DNA, have been divided into two major classes referred to as short and long interspersed elements (SINEs and LINEs, respectively) (Singer 1982). The SINEs typically range from 75 to 500 bp in length, while the LINEs can extend to 7000 bp. DNA-sequencing studies have indicated that the SINEs are derived from POL III transcripts while the LINEs from POL II transcripts (see Deininger 1989; Hutchison et al. 1989). LINEs are classified as retroposons because they contain a region coding for reverse transcriptase (Table 12.1 and Figure 12.4); actually, most LINEs are nonfunctional and such LINEs should be classified as retrosequences. In contrast, SINEs generally do not code for reverse transcriptase and are retrosequences, though they have often been classified as retroposons.

A typical (or generic) SINE sequence contains an internal POL III promoter, an A-rich 3′ end, and flanking direct repeats (Figure 12.8a). The promoter has a typical bipartite structure as represented by the A and B boxes. The 3′ A-rich end regions vary from less than 8 to longer than 50 bp. The flanking direct repeats are produced by duplication of target sequences at the site of integration. These repeats are generally A-rich, suggesting a preference of insertion at A-rich regions.

Many different SINE families are known (Table 12.3). The *Alu* family was the first to have been well characterized (see Schmid and Shen 1985), and so in the past other SINE families were often referred to as *Alu*-like families. The *Alu* family was originally defined as the fraction of rapidly reannealing repetitive sequences that contain a distinctive *Alu1* site (Houck et al. 1979). The human *Alu* family has all of the features of the generic SINE but is a head-to-tail dimer of two similar sequences ~130 bp long (Figure 12.8b); some monomers have also been

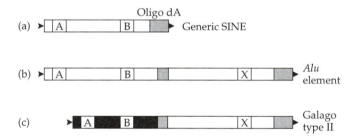

Figure 12.8 (a) Features of the generic SINE. The typical SINE sequence contains two major features: an internal RNA polymerase III promoter and an A-rich 3′ end, which is quite variable in length and in sequence among different SINE family members (shaded region). The promoter has a typical bipartite structure as represented by the A and B boxes. The element is typically flanked by short direct repeats (indicated by arrowheads) formed during its integration into the genome. (b) The *Alu* family. The human *Alu* repeat has all of the features of the generic SINE, but is a dimer. The right half of the sequence is clearly related to the left half, but has 31 extra bp (marked with an X) and does not contain an active RNA polymerase III promoter. There is a short A-rich region between the two halves of the dimer, and typically a much longer, more variable A-rich region at the 3′ end of the repeat. (c) The galago type II family. In the prosimian galago, there is a different form of composite Alu family in addition to the normal form seen in (b). The new form is a fusion of different SINE family member (a Monomer family member; blackened region), which contains a different RNA polymerase III promoter, with the right half of the Alu family member. This helps to illustrate the capacity of some SINEs (e.g., Monomer family members) to be able to fuse with and carry other sequences to new locations. From Deininger (1989).

TABLE 12.3   A collection of representative SINE families

Family	Species distribution	Copy number	% of genome
Alu	Primates	500,000	5
Alu subA	Primates	75,000	0.7
Alu types IIa & IIb	Galago	200,000	2
Monomer	Galago	200,000	0.8
B1 (rodent type II)	Rodents	80,000	0.3
B2a (rodent type IIa)	Rodents	80,000	0.7
B2b (rodent type IIb)	Mouse		
Identifier (ID)	Rat	130,000	0.5
	Mouse	12,000	0.05
	Hamster	2,500	0.01
Rabbit C	Rabbit	170,000	1.7
Artiodactyl C-BCS	Goat and cow	300,000	3
art2	Goat	100,000	1.8
Salmon SINE	Salmon	ND	?

From Deininger (1989).
ND, not determined.

found. The left half has a POL III promoter, while the right half contains a 31-bp insert inside the POL III promoter.

A related family, the type II family, is found in the genome of *Galago crassicaudatus*, a prosimian (Figure 12.8c). The right half of this family is almost identical to the right half of the *Alu* family. The left half is entirely different from the *Alu* family but also contains a POL III promoter. The type II family probably arose from a fusion of two different SINE elements.

The discovery of SINEs raised the question of their origin. The first clue in this regard was the finding that the human *Alu* family has high sequence similarity to the 7SL RNA gene (Ullu et al. 1982; Ullu and Tschudi 1984). 7SL RNA is an abundant cytoplasmic RNA that functions in protein secretion as a component of the signal recognition particle. Ullu et al. (1982) noted that approximately 100 nucleotides at the 5' end and 45 nucleotides at the 3' end of the RNA are homologous with the human *Alu* right monomer consensus sequence (Figure 12.9). The central portion of 155 nucleotides is unique to 7SL RNA and shows no similarity with the human *Alu* DNA or the rodent *Alu* equivalent, the B1 element, which is a monomer (Figure 12.9).

The next important finding concerning the origin of SINEs was the discovery that the Monomer family in the prosimian galago, which is closely related to the left-half sequence of the type II family, is related to a methionine-tRNA gene (Daniels and Deininger 1985). The two sequences exhibit nearly 70% similarity. Since the Monomer family is not present in other species, not even in other primates, it must have been derived from the tRNA gene relatively recently. There are other SINE

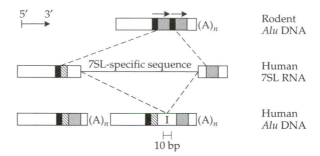

Figure 12.9 The structural relationship of human 7SL RNA to the consensus sequences of human and rodent *Alu* DNA. Homologous sequences are indicated by identical shading. Human *Alu* DNA is a head-to-tail dimer of two similar sequences ~130 bp long. The right monomer contains an insert (I) not present in the left half. The rodent *Alu*-equivalent sequence is a monomer. The mouse B1 *Alu*-equivalent consensus sequence contains an internal tandem duplication of 30 bp between positions 62 and 121. The type-I Chinese hamster *Alu*-equivalent sequence has a similar structural arrangement. Arrows above the rodent *Alu* DNA indicate the position of the 30-bp tandem duplication; $(A)_n$ denotes an A-rich sequence that follows the *Alu* sequence at 3' end. From Ullu and Tschudi (1984).

families that appear to be amplified tRNA pseudogenes (e.g., Deininger and Daniels 1986; Nagahashi et al. 1991; Okada and Ohshima 1993; Sakagami et al. 1994).

However, SINEs are obviously not directly derived from 7SL RNA or tRNA transcripts because the sequence dissimilarity between any element and the parental gene is considerable. For example, the B1 element not only has a deletion relative to the 7SL RNA gene, but also contains a small internal duplication (Figure 12.9). A possible scheme for the derivation of the *Alu* element and the B1 element is shown in Figure 12.10.

Actually, something must occur to make a sequence much more effective in multiplying itself for it to produce a SINE family (see Deininger 1989). An important requirement for the formation of a new SINE family appears to be a POL III promoter that is functional and allows strong expression in the germline, for only retroposition in the germline can result in heritable sequence amplification. An important difference between a POL II and a POL III transcript is that the latter contains an internal promoter whereas the former does not, and this could be the major reason why POL III retropseudogenes are generally much more abundant than POL II retropseudogenes (Tables 12.2 and 12.3). Another important requirement would be to have an A-rich 3' end. For a retroposition to occur, a SINE RNA molecule must be reverse-transcribed into DNA, and having an A-rich 3' end may enable the molecule to self-prime its own reverse transcription.

Evolutionary Fate of Retropseudogenes

Due to the ubiquity of reverse transcription, the genomes of mammals are literally bombarded with copies of reverse-transcribed sequences. The vast majority of these copies are nonfunctional from the moment they are integrated into the genome. Moreover, such sequences cannot be easily rescued by gene conversion, since they are mostly located at great chromosomal distances from the parental functional gene. The phenomenon of a functional locus pumping out defective copies of itself and dispersing them all over the genome has been likened to a volcano generating lava, and the process has been termed the **Vesuvian mode** of evolution (P. Leder, cited in Lewin 1981).

As soon as a retropseudogene is established as a chromosomal sequence in the genome, it is affected by two evolutionary processes (Graur et al. 1989). The first involves the rapid accumulation of point mutations. This accumulation eventually obliterates the sequence similarity between the pseudogene and its functional homologue, which evolves much more slowly. The nucleotide composition of the pseudogene will come to resemble more and more its nonfunctional surroundings, and it will eventually "blend" into its surrounding regions. This process has been called **compositional assimilation**.

The second evolutionary process is characterized by a pseudogene becoming increasingly shorter compared to the functional gene. This **length abridgment** is caused by the excess of deletions over insertions. It has been estimated that a processed pseudogene loses about half of its DNA in about 400 million years. This process is so slow that the human genome, for instance, still contains major chunks of pseudogenic DNA that were found in very distant ancestors. Obviously, these ancient pseudogenes have by now lost almost all similarity to the functional genes.

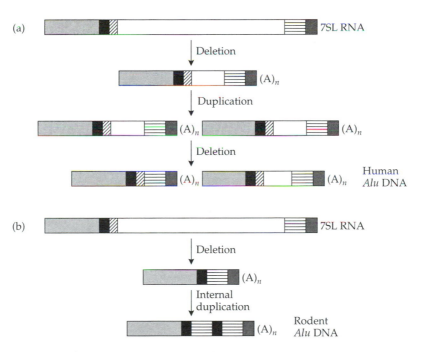

Figure 12.10 Origin of *Alu* sequences in (a) humans and other primates and (b) rodents. Different regions in the 7SL RNA genes are shaded or hatched differently to emphasize the deletions and rearrangements in the *Alu* sequences. $(A)_n$ means that A is repeated n times. Note the dimeric structure in (a) and the monomeric structure in (b). From Li and Graur (1991).

In summary, it seems that processed genes are created at a faster rate than the rate by which they are obliterated by deletions. It can therefore be concluded that abridgment is too slow a process to offset the increase in genome size following the continuous Vesuvian bombardment (Graur et al. 1989). So, the limitation on the number of pseudogenes in a genome is probably due to other factors such as natural selection.

EVOLUTION OF *Alu* SEQUENCES

The evolution of *Alu* sequences has received much attention, because the *Alu* repeats represent one of the most abundant families of retrosequences and because *Alu* sequence data are abundant (see the review by Jurka 1995). A major issue is how *Alu* repeats have become so abundant.

Noting that a typical *Alu* sequence contains an internal POL III promoter and an A-Rich 3′ end, Jagadeeswaran et al. (1981) and Van Arsdell et al. (1981) proposed a model for the transposition of *Alu* sequences (and other repetitive sequences with a similar structure). The model assumes that the complementary strand of an *Alu* sequence is transcribed into an RNA sequence, which contains the entire *Alu* sequence, including its A-rich 3′ end, and ends in a U-rich sequence characteristic of the site of termination of POL III transcription. The 3′ oligo(U) sequence of the RNA transcript pairs with the internal oligo(A) to prime cDNA synthesis, and the cDNA sequence is inserted back into the genome. In this model, retroposition is a cascade process (Deininger et al. 1992), because new sequences may become additional sources for producing *Alu* sequences. This model was supported by a phylogenetic analysis of 59 human *Alu* sequences by Bains (1986), who found that all these sequences were about equally divergent from the center of a dendrogram, i.e., there was no subgroup of *Alu* elements with significantly greater similarity between members of the subgroup than between any of these members and other *Alu* elements. Based on this observation, Bains (1986) suggested that the *Alu* family derives from a large pool of precursors (sources).

However, Slagel et al. (1987) identified a subfamily in an analysis of 22 *Alu* sequences. Four of the sequences appeared to be closely related because they not only formed one cluster in a dendrogram constructed from distances based on nucleotide mismatches but also shared a two-base deletion that was not shared by any of the other sequences in the sample. In a subsequent statistical analysis of more sequences, Willard et al. (1987) proposed a subdivision of *Alu* sequences into at least three subfamilies, each of which was founded by a distinct master sequence. This suggestion was supported by a study using the correspondence analysis (Quentin 1988), which is a powerful technique for identifying clusters.

The existence of subfamilies has also been suggested by extensive analyses of "shared differences" (Britten et al. 1988; Jurka and Smith 1988). In this approach, a consensus sequence is constructed from the sequences under study. Shared (diagnostic) differences are defined as the differences (substitutions, insertions, or deletions) shared by a subset of sequences but differing from the consensus. For example, Figure 12.11 shows the shared differences of 30 *Alu* sequences (Britten et al. 1988). The diagram focuses on a few highly shared nucleotides by rearranging the sequence first to show off the diagnostic positions and second to isolate the CpG dinucleotides that are hot spots of mutation. The consensus nucleotide is shown at the left column and the *Alu* repeats are aligned approximately in order

```
Class:                    IV          III         II

 93  T . . . . . . . . . . . . . . . . .CCCC CCCACCCCC 12  ┐
 98  C . . T . A . . . . . . .T . . . .TTTT TTTTTTTTT 15  │
218  C . . . . . . . . . - . . . - . . . GGGGG GGGGGGGGG 14  │ Diagnostic
196  G . . . . . . . . . . . - . . . CCCCC CCCCCCCCC 12  ├ for
199  G . . . . . . . . . . . - . . . TTTTT TTTTTTTTT 13  │ class III
132  A . - . . . . . . - . . . . .T . ---- ---- . ---- 15  │
152  G . . . . . . . . . . . . . . . CC.CC CCC.AA.CT  8  ┘


 64+1  - . . . . . . . . . . . . . . . . . . . CTTCTCCC.  5  ┐
 64+2  - . . . . . . . . . . . . . . . . . . . TTTT.TTTC  7  │ Diagnostic
 76  A . . . . . . . . . . . . . . . . . . . . TTTTTTTTT  9  ├ for
 86  T . . . . . . . A . . . . . . . . . . . GGGGGAGGG  8  │ class II
162  G . . . . . . . . . . . . . . . . .C . . .AAA.AAAA  7  ┘


 72  G . . . . . . . . . . . A . . . .AAAA . . . . .A . . .  6  Reversion G/A/G
153  G . . . - . . . . . . . A . . . .A.AA .AA.T.AA.  8  These 2 were CpGs
197  G . . . . . . . . . . . . . - . . . T.A.A A . . . . .AC.  4  in class II & III
------------------------------------------------------------------
  4  C . . . . . . . . .T . . .TT. .GT. . .T . . .T . . .  6
  5  G . . . . .AA . . .AAA . . . . . . .A. . . . . .A.AAA 10
 57  C . . . .TTT . . . . . .A. . T . . . . T.TTT . .T.  9  Below line is a
 58  G . . . . . . . . . .A. . .A .A.A. . . . . . .AA.A  7  sample of posi-
138  C . . . . . . . . . .T.T . . . .TTT . . . . . .TTA  7  tions with large
139  G . . . .AAA . . . . . .AA . . . . . AA.AAA . . .A 11  SD among the 50
150  C . .T.T.T. . . .T . . . .G . . . .T . . .TA. .T.  7  that are part of
151  G . . . .A . . . . . . .AAA. A. .A. AA. .CAA. . 10  CpGs.
174  C.T.T . . . . . . . . .T . . . T.T . . .T.TTT . .T 10
175  G . . . .C . . . . . . . . . . - . . . . . . T.A.-.A. .  2
206  C . . . . . . . . . . .-T. .T TT.T. .T . . .TTA.  8
207  G. .T . . .A.A.A- . . . . .AA. . . . .A. . .A  7
213  C . . . . . . . . . .T- . . .T . . . .T . . .T.TTT  7
214  G . . . . . . . .-AAA. C.A.A .AAA.A . . .  9
236  C. .T.TG. .T. . .G. .T T . . .T .T . . . . . .T  8
237  G. . .C. .A. .A. . .A.T . .AAT CA.A-AT.A  9
238  C . . . . . . .T. .T . . .- . . . .T .TA.TTGTT  8
239  G. . . .A. . .A. . .A. . . . . . . A. .A.A. . . .  6
268  C . . . . . . . . .-T.TTA. T . . . . T. .T . . . .T  7
269  G. . . .A.A.- .A. .C. .A. . . . .AA.AA. .  8
276  C . . . . . . . . . .T. .TT. . . .- . . TT. . .TTT.  8
277  G . . . . . .A.A . . . . . . AA.A. . .A.A.T.A  8
     a b c d e f g h i j k l m n o p q   r s t u v  ABCDEFGHI
```

Figure 12.11 Nucleotides in the diagnostic positions of 30 *Alu* repeats. The consensus nucleotide is shown at the left and the *Alu* repeats are aligned vertically approximately in order of divergence from the consensus. All matches to consensus are shown as periods. The sequence has been rearranged so that the diagnostic positions with the largest number of shared differences (SD) (diagnostic sites) are near the top. The numbers to the right are the number of shared differences from the consensus. Below the line are included a sample of CpG dinucleotide positions to show the lack of pattern in the sharing of mutations at these hot spots. Letters at bottom identify the individual *Alu* repeats. From Britten et al. (1988).

of divergence from the consensus. The diagram immediately identifies diagnostic positions that define three classes of *Alu* repeats, denoted as IV, III, and II. For example, at position 93 all class-IV sequences have T, while almost all class-III and –II sequences have C, and at position 76 all class-IV and –III sequences have A, while all class-II sequences have T. Class-I sequences are even more divergent than class-II sequences from the consensus sequence and are not shown in the figure.

Jurka and Milosavljevic (1991) regard each of classes I, III, and IV as one subfamily and divide class II, the major class, into three subfamilies. Recently, Britten (1994) added a fifth class to represent newly and very recently inserted elements.

Britten et al. (1988) proposed that the patterns of sharing shown in Figure 12.11 arose from a series of different source (master) genes producing each class of *Alu* repeats in turn and being sequentially replaced during evolution rather than from a set of coexistent sources producing all classes at all times. If several different sources operate concurrently, they would produce sets of *Alu* repeats of all classes, each initially identical to the respective source. Then each class would age and would include sequences with a range of divergence from each other and from their source. However, observation shows the opposite. Members of the classes that are more divergent from the consensus are more divergent from each other. For example, the average pairwise divergence between class-II members (the rightmost 9 *Alu* repeats) is 56.3 mutations (20% substitutions, insertions, and deletions, including CpG positions), whereas that between the leftmost 13 *Alu* repeats of class IV is only 31.6 mutations (11%). Apparently, the rightmost 9 *Alu* repeats were copied from earlier source sequences and inserted into the genome earlier in the past. Interestingly, the average sequence similarity with the 7SL RNA gene decreases in the order of classes I, II, III, IV, and V sequences. This observation can be explained by the scheme shown in Figure 12.12: The first source gene was the most similar to the 7SL RNA gene and subsequent source

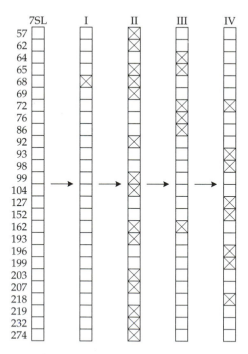

Figure 12.12   Sequence of mutations at diagnostic positions between different subfamilies of *Alu* sequences. Roman numerals indicate successive waves of subfamilies that served at various periods as the predominant source of *Alu* sequences. Substitutions distinguishing each subfamily from the preceding one are denoted by X. From Li and Graur (1991).

genes were derived from preceding ones by mutation, becoming less similar to the 7SL RNA gene. Each source gene served at one time or another as the predominant source of *Alu* sequences and was superseded by a descendant line. The successive waves of fixation did not occur in sudden bursts, but consecutive subfamilies continued to coexist within the genome for long periods of time. For example, from the degree of sequence divergence and the rate of evolution of noncoding sequences in primates, Britten (1994) estimated that most of the class-II members were produced between 50 and 30 million years ago and that most of the class-III members were produced between 46 and 26 million years ago.

The source (master) gene model is suggested to be applicable also to other SINE families and to LINE families (Deininger et al. 1992). However, whether there are only a few or many source genes active at each time remains unclear (see Matera et al. 1990; Hutchinson et al. 1993; Deininger and Batzer 1994). Since no source gene has yet been identified, the nature and function of source genes are unknown. Also, how source genes have evolved in succession is unclear.

EFFECTS OF TRANSPOSITION ON THE HOST GENOME

Transposition and retroposition can have profound effects on the size and structure of genomes. In particular, TEs have been considered the best example of "selfish DNA," which may not confer any advantage to the host but spread in the genome because they multiply faster than the host genomic sequences (Doolittle and Sapienza 1980; Orgel and Crick 1980). For this reason, transposition of TEs can significantly increase the genome size. This effect will be discussed in the next chapter. Here, we shall discuss the ways in which TEs may affect the expression and evolution of host genes and the structure and function of the host genome.

First, as mentioned in the earlier section on transposable elements, transposons in bacteria often carry genes that confer antibiotic or other resistance to their carriers and thus may enable the host species to survive in an adverse environment (see Figure 12.2).

Second, the insertion of a TE either within or adjacent to a gene may alter its temporal or spatial pattern of expression (see the reviews by Schmid and Maraia 1992; McDonald 1993, 1995; Britten 1996). This change can occur in several ways (Figure 12.13): (1) If the insertion is in the 5' flanking region, a change in expression may occur as a result of either the "readthrough" of transcripts initiated in the TE promoter or the presence of positive or negative enhancer-like sequences contained within the TE (e.g., Horowitz et al. 1984). (2) TEs inserted in some genes have been found to contain termination or polyadenylation signals that change expression patterns (Krane and Hardison 1990). (3) The insertion of a TE into the coding region of a gene can alter the pattern of gene splicing (e.g., Purugganan and Wessler 1992). Although the vast majority of such changes are likely to be deleterious, occasionally a change may be advantageous or only slightly deleterious so that it can survive. Several successful examples have been found (see the review by McDonald 1995). One such example is the mouse sex limited protein (*Slp*) gene (Stavenhagen and Robins 1988), which is a member of the histocompatibility complex and arose by tandem duplication of the fourth component of the complement *C4* gene. Although *Slp* and *C4* are highly homologous, they are markedly different in their tissue-specific expression. These regulatory differences have been shown to be attributable to an

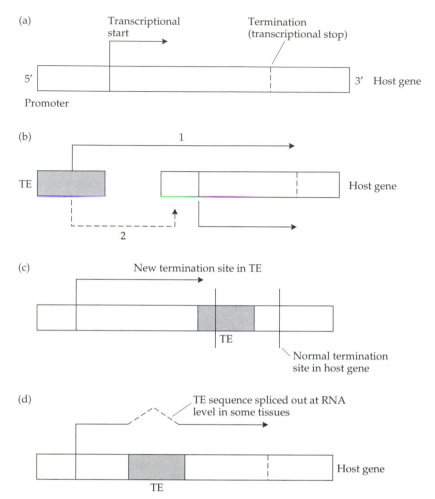

Figure 12.13 Mechanisms by which TEs may alter the regulation of gene expression. (a) The expression of wild-type genes is regulated by control sequences usually contained in a 5′ promoter region. (b) The insertion of a TE 5′ to a gene can result in an altered pattern of expression due to (1) transcriptional initiation within the inserted element and readthrough into the host gene or (2) alteration in the function of the mutant gene's own control sequences due to "enhancer" sequences contained within the TE. (c) A TE inserted into a gene may contain termination or polyadenylation signals that change patterns of gene expression. (d) A TE inserted within a gene's coding region may act like a primitive intron or may activate cryptic splice sites within the gene, resulting in altered splicing patterns in some tissues that may change temporal or spatial patterns of gene expression. From McDonald (1995).

enhancer sequence contained within a cryptic retrotransposon, which was inserted adjacent to the ancestral *Slp* gene several million years ago. As two additional examples, the human θ1 α globin has been shown to utilize a truncated *Alu* sequence as part of its promoter (Kim et al. 1989) and the *Alu* sequence in the last intron of the human CD8 gene has been found to operate as part of an enhancer (Hambor et al. 1993).

Susan R. Wessler (personal communication) has suggested that a TE-mediated change in gene expression may have a fair chance to survive in the case of a duplicate gene, because the other copy may be sufficient to maintain the normal function. As many plants are polyploids, TE transposition might have played an important role in the diversification of gene expression in plant evolution.

Third, the structure and function of the protein encoded by a gene may be altered or destroyed by the insertion of a TE into the gene. The insertion of a TE into the coding region of a gene will most probably alter the reading frame and may have drastic phenotypic effects, unless the TE can function as an intron, so that it is spliced during RNA processing (Wessler et al. 1987). Similarly, the insertion of an element into a regulatory region may incapacitate the gene. In comparison, insertion of an element into an intron or an intergenic region may have a high chance to survive. For example, there are 13 *Alu* repeats in the introns of the human thymidine kinase gene (Flemington et al. 1987), and yet there is no evidence of deleterious effects. However, occasionally such an insertion can interfere with the processing of the primary RNA transcript and affect the function of the gene. For example, the insertion of an *Alu* element into intron 5 of the *NF1* gene resulted in deletion of exon 6 during splicing of the RNA transcript and subsequently shifted the reading frame (Wallace et al. 1991). This defect led to several features of neurofibromatosis type 1, including one cutaneous neurofibroma, cervical nerve root tumours, and macrocephaly.

Fourth, TEs can promote genomic rearrangements. Inversions, translocations, duplications, and large deletions and insertions can be mediated by transposable elements. These rearrangements can take place, for example, as a direct result of transposition, i.e., by moving pieces of DNA from one genomic location to another, thus altering the milieu in which they are expressed. A more indirect effect will ensue if as a result of transposition, two sequences that had previously had little similarity with each other are now sharing a similar TE, so that crossing-over between them is possible. This type of "nonhomologous recombination" may occasionally produce a beneficial gene duplication or gene rearrangement. For example, a duplication of the entire growth hormone gene early in human evolution might have occurred via *Alu*–*Alu* recombination (Barsh et al. 1983), and the duplication giving rise to the human $^G\gamma$ and $^A\gamma$ globin genes may have resulted from recombination between short L1 elements (Maeda and Smithies 1986). In the majority of cases, however, it may result in a deleterious mutation. Figure 12.14 illustrates how an unequal crossing-over event, facilitated by the presence of multiple *Alu* sequences within the introns flanking exon 5 of the low density lipoprotein receptor gene, gave rise to a mutant gene lacking the exon (Hobbs et al. 1986). The deletion of exon 14 by the same mechanism was also observed (Lehrman et al. 1986). Patients homozygous for these deletions have a high level of cholesterol in the blood (hypercholesterolemia). Recombination between two *Alu* elements has also been shown to be responsible for the deletion of the promoter and the first exon of the adenosine deaminase gene in patients with adenosine deaminase deficiency (Markert et al. 1988). In general, genomic instability has been demonstrated for all regions containing *Alu* repeat sequences (Calabretta et al. 1982).

Fifth, insertion of TEs into members of a multigene family can reduce the rate and limit the extent of gene conversion between members and therefore promote the rate of divergence between duplicate genes (Hess et al. 1983; Schimenti

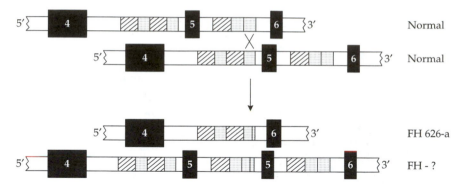

Figure 12.14   Unequal crossing-over in the low-density-lipoprotein receptor gene. Exons are indicated by solid bars and are numbered. *Alu* sequences in the introns are indicated by hatching (right arms) and by stippling (left arms); an *Alu* sequence may consist of two arms (dimeric) or one arm (monomeric). The position of the postulated crossing-over is indicated by the X. The deleted (observed) and inserted (inferred) products of the recombination are depicted as FH 626-a and FH-? below the arrow. From Hobbs et al. (1986).

and Duncan 1984). The reason is that the process of gene conversion involves pairing of homologous DNA, and nonhomologous regions created by insertion of TEs might reduce the chance of such pairing and thus the chance of gene conversion between duplicate genes.

Sixth, there is evidence that some TEs may cause an increase in the rate of mutation. For example, strains of *E. coli* that contain the transposable element *Tn10* were found to have elevated rates of mutation (Chao et al. 1983). Under most conditions this trait will be deleterious to the carrier. However, under severe environmental stress it is possible that an elevated rate of mutation might be advantageous in creating mutations, some of which may be better adapted to the new circumstances.

Since insertion of an element at a site is a rather rare event and thus can be used as a genetic marker, SINE and LINE elements are useful for evolutionary study and DNA fingerprinting (see Deininger and Batzer 1994). For example, the rice pSINE1 insertions have been found to be useful markers in the evolution and identification of different strains of rice (Mochizuki et al. 1992). In addition, SINE elements have been used to study the phylogenetic relationships among Pacific salmonids (Murata et al. 1993) and human evolution (Hammer and Horai 1995).

HYBRID DYSGENESIS

In the 1970s several groups of geneticists were surprised to observe a variety of unusual genetic phenomena when males from natural populations of *Drosophila melanogaster* were crossed with females from long-established laboratory strains (see Sved 1976; Kidwell and Kidwell 1976; Bregliano and Kidwell 1983). These phenomena include high frequencies of partial or complete sterility, male recombination (in *Drosophila* recombination is usually restricted to females), visible and lethal mutation, reversion of mutation, chromosomal rearrangement, transmission ratio distortion (including sex ratio distortion), and chromosomal nondisjunction. These correlated abnormal genetic traits have been collectively called **hybrid dysgenesis** (Sved 1976; Kidwell and Kidwell 1976). The dysgenic effects are usually asymmetrical, that is, they occur in one type of hybrid but usually not in the reciprocal hybrid. The syndrome has later been found to be caused by transposable elements.

There are two well-documented dysgenic systems in *Drosophila*, the *P-M* and *I-R* systems (see Bregliano and Kidwell 1983). We shall only discuss the *P-M* system; *P* and *M* stand for paternal and maternal, respectively. The asymmetry of hybrid dysgenesis in this system is shown in Figure 12.15. When a male from a *P* strain mates with a female from an *M* strain, the offspring are dysgenic, whereas in the reciprocal mating, the offspring are normal.

The cause of *P-M* dysgenesis is a family of transposable elements called *P* elements. In *P* strains, there are 30–50 *P* elements in the genome; however, many of them may be defective because of deletions. They are distributed throughout all chromosomes, although in some strains transposition shows a preference for the X chromosome. *M* strains do not carry *P* elements. The asymmetry of the hybrid dysgenesis system is thought to result from maternal inheritance of a *P* element-encoded repressor in F_1 progeny of *P* female × *M* male matings and the absence

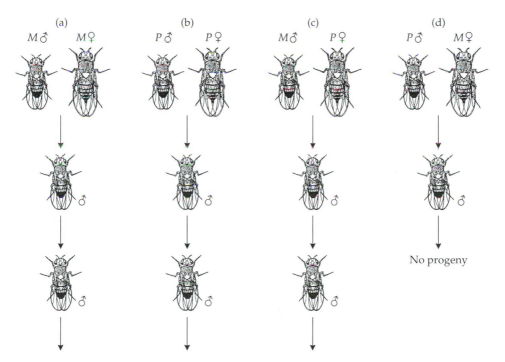

Figure 12.15 Hybrid dysgenesis in *Drosophila*. (a) Matings within *M* strains and (b) matings within *P* strains produce normal progeny. (c) Normal progeny are also produced in crosses between *M* males and *P* females. (d) In the crosses between *P* males and *M* females, dysgenic offspring are produced, many of which are sterile. From Li and Graur (1991).

of such a repressor in F_1 progeny of the reciprocal M female $\times$ P male matings. Dysgenic traits develop because of the transposition of P elements in germline cells that contain no repressors in cytoplasm, which is maternally derived. In the presence of repressors, P-element transposition may be wholly or partially inhibited. Transposition of P elements normally does not take place in somatic cells, because the third intron is not excised as it is in germline cells.

An interesting observation was made by Kidwell (1983) concerning the distribution of P-carrying strains. P characteristics were not found in any $D.$ *melanogaster* strains collected before 1950, and collections made subsequently showed increasing frequencies of P with decreasing age (Figure 12.16). A similar observation was made with the I-R dysgenic system. Two hypotheses were suggested to explain the distribution of P elements. Engels (1981) proposed that most strains in nature are of the P type but that they tend to lose the transposon in laboratory populations (the stochastic loss hypothesis). The second hypothesis postulates a recent introduction of P transposons into $D.$ *melanogaster* populations followed by a rapid spread of P elements in formerly M populations (Kidwell 1979). There are several reasons why the second hypothesis, the recent invasion hypothesis, is more plausible. First, laboratory strains lacking active P elements may be only 30 years old or younger, whereas theoretical studies (Charlesworth 1985; Kaplan et al. 1985) showed that the mean time for the stochastic loss of ele-

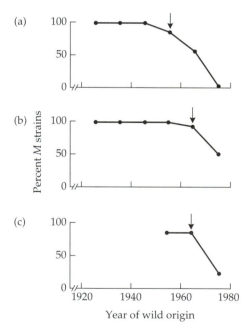

Figure 12.16 Changes in the frequencies of M strains (i.e., strains devoid of P elements) in natural populations in (a) North and South America, (b) Europe, Africa, and the Middle East, (c) Australia and the Far East. Arrows denote the first appearance of P strains. Note that the proportion of M populations decreased first in the American continents and only later decreased in the other continents. From Li and Graur (1991) after Kidwell (1983).

ments is at least the order of $1/v$, where v is the excision rate and is of the order of 10^{-4} per generation, or at most 10^{-3} per year. The stochastic-loss hypothesis is also not supported by the observation that P-carrying laboratory strains that have been monitored for almost 20 years did not lose their P characteristics. Second, there is a temporal-geographical cline in the distribution of P strains, suggesting that the invasion of the Americas occurred approximately 10 years earlier than that in Europe, Africa, and the rest of the world (Kidwell 1983; Anxolabéhère et al. 1988). Third, active P elements are highly invasive when they are introduced into susceptible populations either by crossing or by experimental germline transformation (Kiyasu and Kidwell 1984; Daniels et al. 1987; Good et al. 1989). Lastly, there is evidence that P elements have been recently acquired by *D. melanogaster* from *D. willistoni*, a distantly related species (Daniels et al. 1990). This acquisition appears to have occurred by horizontal transfer (see the last section of this chapter).

Hybrid dysgenesis has been thought for a while to represent an early stage in the process of speciation, by acting as a postmating reproductive isolation mechanism between different populations belonging to the same species (Bingham et al. 1982; Ginzburg et al. 1984). Indeed, the sterility of hybrids produced by crosses between the sibling species *D. melanogaster* and *D. simulans* is very similar to dysgenesis (e.g., rudimentary gonads, segregation distortion). There are, however, problems with this view. First, although hybrids exhibit reduced fitness and are, therefore, partially isolated in terms of reproduction, the transposition of P elements in the germline virtually ensures that most of the chromosomes transmitted to the hybrids will bear P elements, and the cytotype will eventually change to the P type too. Thus, provided the reduction in fitness of the hybrids is not too great, the P element will spread through the entire population. Indeed, the hybrid must be almost completely sterile for an effective reproductive isolation to last. Second, P elements can be transposed horizontally as an infectious agent from individual to individual (see above). Thus, an entire population may rapidly be taken over by P, so that hybrid dysgenesis is likely to last for very short periods of time in nature. Indeed, many species of *Drosophila* are known in which all individuals and all populations carry P elements or P-like elements, and yet hybrid dysgenesis does not occur in any of these species. Finally, hybrid dysgenesis seems to be restricted to *Drosophila* and may not represent a universal phenomenon in nature. Even in *Drosophila* no evidence has been found for the involvement of mobile elements as barriers to gene flow between sibling species.

MAINTENANCE OF TRANSPOSABLE ELEMENTS IN POPULATIONS

There are two opposing views regarding the widespread existence of TEs in natural populations. One view is that TEs become widespread because they confer selective advantages to their hosts (e.g., Cohen 1976; Never and Saedler 1977; Syvanen 1984). As mentioned above, a TE may carry a gene for drug or metal resistance and may thus confer a selective advantage to its host bacterium under certain adverse conditions. Moreover, in chemostat competition experiments Chao et al. (1983) found that transposon *Tn10* was able to increase the growth rate of *E. coli* cells by increasing the mutation rate of its host genome; the mutagenic effect was apparently due to transposition of *Tn10* or its flanking insertion sequence *IS10.* The same advantage was also found for the insertion sequence *IS50;* since *IS50* improved the growth rate of its host without transposition, the advantage was probably not mutational in origin but might have been derived from the effect of the transposase of *IS50* on one or more cellular processes such as DNA replication or repair (Hartl et al. 1983). In addition to the above observations, it has also often been argued that TEs can generate rearrangements that increase evolutionary versatility. However, such a long-term evolutionary advantage cannot explain the initial spread of TEs within a population, though it may be able to explain the persistence of the phenomenon in a population (see Doolittle 1982). Although the examples of drug and metal resistance and the chemostat experiments do provide support for the selectionist view, there is much reservation that this view can be a general explanation for the spread of TEs in natural populations, particularly in higher organisms.

The other view is that TEs are intragenomic parasites or "selfish DNA" (e.g., Doolittle and Sapienza 1980; Orgel and Crick 1980). The argument is that insertion of a TE into a chromosomal site is more likely to cause deleterious rather than beneficial effects because it may interfere with the expression or may disrupt the coding frame of a gene (see above). Despite such disadvantages, a TE can spread through a population because it replicates faster than the host genome. Indeed, Hickey (1982) showed that a high rate of transposition can overcome even extremely severe deleterious effects of transposition. Furthermore, Ginzburg et al. (1984) and Uyenoyama (1985) showed that a hybrid dysgenesis factor can rapidly spread through a population, despite its selective disadvantages.

The preceding models are essentially single-locus models. Sawyer and Hartl (1986) studied models with elements dispersed over multiple genomic sites but with limited possibilities of sexual transmission between individuals (and recombination between elements at different sites); such models are suitable for bacterial transposable elements. Models of randomly mating, diploid populations with elements dispersed over multiple genomic sites and with relatively high frequencies of recombination between different sites have been studied by Charlesworth and Charlesworth (1983), Langley et al. (1983), Charlesworth (1985), Brookfield (1986), and others (see reviews by Charlesworth 1988, and Charlesworth and Langley 1989).

Let us consider a randomly mating population with enough recombination so that linkage disequilibrium can be ignored. Let *u* be the probability that a transposable element produces a new genomic copy, i.e., the probability of replicative transposition. Let *v* be the probability that the element is excised. The values of *u* and *v* have been determined experimentally for several TEs in *Drosophila melanogaster* (see Charlesworth and Langley 1989). Transposition rates were found to vary among the TEs studied but on the average were in the order of 10^{-4} per element per generation. Excision rates were about one order of magnitude lower. Therefore, in the absence of selection against the TE, the number of copies in the genome is expected to increase indefinitely. A simple way to take selection into account is to assume that the fitness of an individual, *w*, decreases with the copy number, *n* (Charlesworth 1985). The justification for this assumption is that TE insertion frequently alters the expression of adjacent genes and that, with increasing numbers of TEs, the probability of a deleterious alteration of gene expression increases. Charlesworth (1985) showed that as long as w decreases with n, then regardless of the exact relationship between n and w, the mean fitness ($\overline{W}$) of a population at equilibrium relative to an individual lacking TE is

$$\overline{W} = e^{-n(u-v)} \tag{12.1}$$

In the case of *Drosophila melanogaster* there are approximately 50 families of TEs, each appearing in the genome on the average 10 times (Finnegan and Fawcett 1986). Thus, $n \approx 500$. Since v is smaller than u by at least an order of magnitude, $u - v \approx u = 10^{-4}$. Solving equation (12.1), we obtain $\overline{W} = 0.95$. Therefore, the reduction in fitness is $s = 1 - 0.95 = 0.05$. The stability of the equilibrium given by Equation 12.1 requires that the logarithm of fitness declines more steeply than linearly with increasing n (Charlesworth 1985). For simplicity, however, if we assume linearity, then the reduction in fitness with each additional copy is approximately $0.05/500 = 10^{-4}$. Such a small selective coefficient suggests that the number of copies of transposable elements within an organism is strongly affected by random genetic drift. Equation 12.1 is deterministic. For stochastic treatments, see Charlesworth (1985) and Kaplan et al. (1985).

The above calculation shows that a very weak intensity of selection is sufficient to check the spread of elements. On the other hand, the studies of Mukai and colleagues (summarized in Simmons and Crow 1977) on viability mutations indicate a mean selection coefficient of the order of 2% against homozygous detrimental mutations and 0.7% against heterozygotes. If all insertions result in a mutation with this magnitude of reduction in fitness, elements would never be able to spread in a population, unless transposition rates are higher than seem realistic for most elements (Charlesworth and Langley 1989). Of course, there probably exist many different classes of insertion sites. For example, sites that are sufficiently remote from coding regions may be selectively neutral or subject to only very weak selection, whereas sites in coding regions are likely to be subject to strong selection. Obviously, elements are more likely to be found in sites where insertions have little or no negative effects on fitness than in sites where insertions have deleterious effects. This uneven distribution of elements will be conspicuous in large populations because selection is effective when the population size is large. This distribution is in fact observed in restriction-map surveys of segments of the *Drosophila* genome (see the review by Charlesworth and Langley 1989).

Noting that the selection intensity required to check the spread of elements is very weak, Charlesworth (1985, 1988) and Charlesworth and Langley (1989) suggested that selection is not the only factor involved in stabilizing element frequencies in natural populations. Another possible factor is the regulation of transposition rates; if the rate of transposition is high when the copy number is low, then an element may be able to spread at least to some extent. For example, in the case of *P* element transposition, events proceed rapidly when a genome has only one or a small number of copies (see Engels 1986). However, the evidence for transposition regulation is equivocal for other elements in *Drosophila* (see Charlesworth and Langley 1989). Another possible factor is crossing-over between homologous elements located at different chromosomal sites (ectopic exchange) (Langley et al. 1988). This type of crossing-over can lead to deleterious chromosomal rearrangements and, as its rate increases with the number of elements in the genome, it can retard the spread of an element.

Reviewing theoretical models and data from *Drosophila*, bacteria, and yeast, Charlesworth (1988) and Charlesworth and Langley (1989) concluded that there was nothing strikingly inconsistent with the view that TEs are maintained in populations as a result of transpositional increase in copy number in spite of negative selection. In fact, this view can provide detailed explanations for a range of phenomena associated with TEs. In particular, the fact that elements in bacteria, yeast, and *Drosophila* are usually present at low frequencies at individual chromosomal sites into which they can insert can be convincingly explained by this view, but is almost impossible to be accommodated under the hypothesis that elements persist as a result of favorable mutations associated with their transpositional activities. Under the latter hypothesis, the frequencies would have been considerably higher than observed.

HORIZONTAL GENE TRANSFER: MECHANISMS

Horizontal or **lateral gene transfer** is defined as the transfer of genetic material from one genome to another, specifically between two species. It is distinct from the normal mode of transmission from parents to offspring, which is commonly known as vertical transfer. As is well known, horizontal transfer can occur between an organelle and a nuclear genome or between different organelles in the same species (see, e.g., Timmis and Scott 1984; Baldauf and Palmer 1990; Nugent and Palmer 1991; Collura and Stewart 1995). However, we shall not discuss this latter situation but shall focus on between-species transfers. In this section we shall discuss possible mechanisms of horizontal transfer, and in the next two sections we shall discuss how to detect gene transfer, and some plausible examples of horizontal transfer.

In bacteria, a commonly known mechanism of horizontal transfer is **transformation,** the uptake of DNA by competent cells (see Mazodier and Davies 1991). A substantial number of bacterial species are naturally competent for transformation and recalcitrant species (e.g., *E. coli*) can be rendered competent by physical or chemical treatments. Transformation consists of the following steps (Mazodier and Davies 1991):

1. Release or appearance of DNA in environment.
2. Induction of competent state in recipient host cells.
3. Interaction of cells and DNA.
4. Entry of DNA and processing in cell.
5. Functional integration and expression of entering DNA into cell operations.

Note that, except for the first step, these steps are common to all types of gene transfer.

The other two known mechanisms of gene transfer in bacteria are **conjugation** and **transduction,** which, unlike transformation, may require a vehicle (vector) to transport the genetic information between organisms. In bacteria, two types of vectors are known: plasmids and bacteriophages (i.e., bacterial viruses). Conjugation requires the help of a plasmid. For example, the F fertility factor (plasmid) in *E. coli* can initiate conjugation between a male (F^+) and a female (F^-) bacterium that leads to the replication of the F^+ chromosome and the transfer of a progeny F^+

chromosome to the F⁻ cell. In this process a piece of host DNA carried by a plasmid can be transferred from one bacterial strain to another. In contrast, transduction occurs via bacteriophages. A phage particle may encapsulate a section of the host DNA, and when the particle attaches to another host cell, the fragment of the bacterial chromosome is injected into the cell. It can then engage in crossing over with the host chromosome, being integrated into the host genome.

Horizontal gene transfers from eukaryotes to prokaryotes may occur by transformation, as is commonly practiced in molecular biology laboratories. Gene transfers from prokaryotes to eukaryotes may occur by conjugation. The best studied case is the transfer of the **T-DNA** (transferred DNA), from the tumor inducing (Ti) plasmid within the bacterium *Agrobacterium tumefaciens* to a plant cell genome (see Zambryski 1988, 1989). The T-DNA, which contains oncogenic functions for tumor initiation and maintenance, is the cause of crown gall tumors. Another path of transfer is via symbiosis, as is well known in the transfer of organelle genes to the nuclear genome (see, e.g., Timmis and Scott 1984; Baldauf and Palmer 1990). It is not clear how horizontal gene transfers between eukaryotes can occur except via retroviruses as vectors. Retroviruses are capable of both incorporating chromosomal DNA into their genomes and crossing species boundaries (Benveniste and Todaro 1976; Bishop 1981). In fact, there are several examples of endogeneous retroviral sequences being transferred between vertebrate species (reviewed in Benveniste 1985). In the case of the transfer of the *P* transposable element between *Drosophila* species, mites were suggested to have been the vector (Houck et al. 1991).

HORIZONTAL GENE TRANSFER: DETECTION METHODS

One may be able to detect a horizontal-transfer event through the discovery of an outstanding discontinuity in the phylogenetic distribution of the gene in question. For example, the bacterium *Salmonella typhimurium* contains a histone-like gene which, as far as we know, has no counterpart in other bacteria (Higgins and Hillyard 1988). Also, inferring horizontal transfers that have occurred in the recent past is not difficult, especially if the donor species and the selective advantage of the transfer, such as in the case of the transfer of an antibiotic-resistant gene (e.g., Maynard Smith et al. 1991; Spratt et al. 1992), are known. In other situations, however, horizontal gene transfers are difficult to prove, and a case can be made only if it is supported by a rigorous test. Three tests are described below.

First, we consider the situation where the phylogeny of the species from which the gene sequences were obtained is known. For example, suppose that the true phylogenetic relationships among species A, B, and C is as shown in Figure 12.17a; D is a known outgroup. However, a phylogenetic analysis of the sequences of a gene gives the tree in Figure 12.17b. This disparity does not necessarily mean that a horizontal transfer has occurred, because factors other than horizontal transfer can cause a discrepancy between the species tree and a gene tree (Chapter 5). Therefore, the case is plausible only if the tree in Figure 12.17b is statistically better than that in Figure 12.17a. Such a test can be conducted using the methods described in Chapter 5. For example, using the bootstrap technique, one may examine whether there is statistical support for a horizontal transfer, i.e., whether the bootstrap proportion for the B–C clade is 95% or higher.

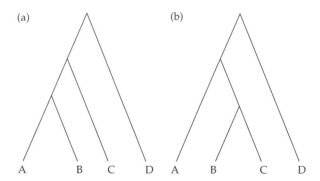

Figure 12.17   (a) True tree. (b) Inferred tree.

Second, the phylogeny of the taxa from which the sequence data were obtained is not known. In this situation, the problem of detecting a horizontal-transfer event becomes more complex. In the past, a claim of horizontal transfer was often made when the phylogenies inferred from different genes were different. To make such a claim statistical, Lawrence and Hartl (1992) proposed a method for testing whether the inconsistencies between the taxonomic relationships inferred from two data sets are statistically significant. Suppose that two sets of nucleotide sequences are available from two different genes; the two sets are arbitrarily called the reference data set and the test data set, respectively (Figure 12.18). First, a matrix of identity coefficients is generated for each data set, in which the i, j-th entry is the proportion of sites that are identical between taxa i and j. To make the matrices commensurate, the magnitudes of the relationships are ranked within each row to create corresponding matrices of ranks. The two matrices of ranks are compared row by row using the Spearman rank-correlation statistic. This statistic ranges from +1.0, indicating a perfect correlation between ranks, to −1.0, indicating a perfect negative correlation between the ranks. The Spearman statistics for all rows in the matrices of ranks are averaged to yield an overall similarity coefficient. A similarity coefficient less than 1.0 indicates some discrepancy in the taxonomic relationships inferred from the two genes. Second, to assess the statistical significance of a discrepancy, the bootstrap resampling technique is used. Positions from the aligned nucleotide sequences of the reference data set are chosen at random with replacement to create a simulated data set with the same sample size, that is, to create a new test data set (Figure 12.18). A matrix of identity coefficients and a matrix of ranks are generated, and a similarity coefficient between the bootstrapped matrix of ranks and the original matrix of ranks is computed. This process is repeated to compile a distribution of similarity coefficients, which is used to determine if the similarity coefficient of the two original data sets is significantly smaller than the distribution of similarity coefficients generated from the bootstrapped data sets. A case of horizontal transfer is plausible only if the test is significant. This procedure may also be used to identify the taxon in which the inferred transfer event had occurred, because exclusion of this taxon should eliminate the discrepancy, whereas exclusion of any other taxon would not. In some cases it might make a difference which of the two original data sets is chosen as the reference data set. So, significance should be corroborated by retesting after interchange of the reference and test data sets.

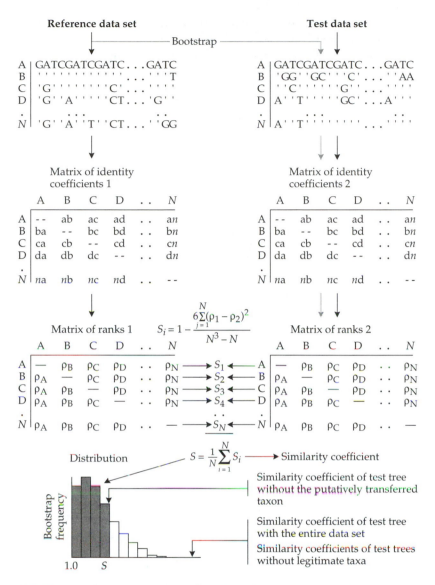

Figure 12.18   Flow chart describing the similarity coefficient and generation of the test distribution by resampling. One of the two original data sets is used as the reference data set and the other as the test data set. If a discrepancy between the two data sets is found, the bootstrap technique is used to test whether the discrepancy is significant. In the bootstrap procedure, the data set resampled from the reference data set is used as the test data set, and this process is repeated to generate a large number of test data sets and the test distribution. From Lawrence and Hartl (1992).

    The above two tests are based on phylogenetic analysis and several cautions should be made (Kidwell 1993; Smith et al. 1992; Syvanen 1994). First, as unequal rates of evolution among lineages may positively mislead phylogenetic analysis (Chapter 5), one should make sure that an "out-of-place" lineage in the inferred tree is not due to the unequal-rate effect. Second, one should try to use an extensive data set and also a suitable outgroup to root the tree. Third, one should make sure that no paralogous genes are mistaken as orthologous genes. The importance of such cautions can be seen from the following two examples. One is the case of ponyfish Cu-Zn superoxide dismutase (SOD). Initially, it was claimed that the symbiotic bacterium *Photobacterium leiognathi* acquired the gene from its ponyfish host (Barnister and Parker 1985), but the availability of additional sequence data suggested that the *P. leiognathi* sequence was actually of prokaryotic origin (Leunnissen and de Jong 1986; Smith and Doolittle 1992). The second example is that the glyceraldehyde-3-phosphate dehydrogenase gene in *E. coli* was originally suggested to be a eukaryote-to-prokaryote transfer because it was the sole prokaryote in a eukaryotic clade. However, the finding of an orthologous sequence in a distantly related bacterial species, *Anabaena*, suggested that the direction of transfer might have been from an ancestral bacterium to the eukaryotes, rather than the other way around.

Third, Stephens (1985), Sawyer (1989), and Maynard Smith (1992) have developed methods for testing whether a gene consisting of a mosaic of regions was derived from different ancestral genes by either recombination or gene conversion. We discuss the method by Maynard Smith because it allows an (approximate) determination of the crossover points. The method considers three situations of the availability of sequence data: (1) two parental sequences and a derived sequence, (2) a set of sequences with no information about ancestry, and (3) just two sequences.

The method, called the maximum chi-squared (MCS) method, will be explained by reference to Figure 12.19. Suppose that the DNA region under study contains n polymorphic sites, sites that show variation among sequences. Let us compare sequences 1 and 3, which differ at s sites. Suppose that for an arbitrary cut after the kth site, the two sequences differ at r sites. Then the proportion of site differences between the two sequences is r/k before the cut and $(s - r)/(n - k)$ after the cut. The value, c, of the 2×2 χ^2 is calculated. Among all the possible cuts, find the site, k_{max}, for which the value of c is a maximum, c_{max}. If the difference between sequences 1 and 3 has a mosaic structure with only two blocks, then the optimal boundary between the blocks occurs after site k_{max}.

To test whether this division is significant, construct T sequence pairs of length n, differing at s randomly distributed sites, and for each trial pair find c_{max}. If the value of c_{max} for the real data is greater than the value for every trial pair, then the observed mosaic structure is significant at the level $P < 1/T$.

One difficulty in using this method is in making an appropriate choice of polymorphic sites. Suppose that we are comparing sequences 1 and 3, but we do not have sequence 5. If we treat as polymorphic any site that varies in the set of sequences 1, 2, 3, and 4, we will be in danger of identifying a false division at site b, because in this case the divergence at polymorphic sites between sequences 1 and 3 is only 33% in the a—b block but 100% in the b–c block (see the table in Figure 12.19). This danger may be avoided if we consider all sites, polymorphic or not. There are, however, two drawbacks to this: the computations take much longer and the test becomes less powerful. The most satisfactory procedure is to consider only the two parental and the derived sequences (1, 2, and 3) and to treat as polymorphic only those sites that vary within this set of three. However, if we do not know the ancestry, we should consider all sequences, and if only two sequences are available, we should consider all sites.

Note also that in determining the block structure of sequence 3 in Figure 12.19, it can be compared either with parent 1 or parent 2. These two comparisons

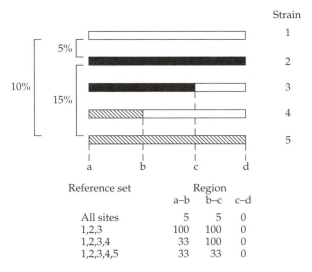

Reference set	Region		
	a–b	b–c	c–d
All sites	5	5	0
1,2,3	100	100	0
1,2,3,4	33	100	0
1,2,3,4,5	33	33	0

Figure 12.19 Schematic presentation of sequence comparison. It is assumed that sequences 1 and 2 differ by 5%, sequences 1 and 5 differ by 10%, and sequences 2 and 5 differ by 15%. Sequence 3 is derived from a recombination at point c between sequences 1 and 2, and sequence 4 is derived from a recombination at point b between sequences 1 and 5. In the table, "all sites" means all nucleotide sites, including sites that show no variation among the sequences; "1,2,3" means the sites that show variation among sequences 1, 2, and 3, with similar definitions for "1,2,3,4" and "1,2,3,4,5". The percent divergences between sequences 1 and 3 are 5, 5, and 0 for blocks ab, bc, and cd, repsectively, when all nucleotide sites are compared, but are 100, 100, and 0, when only polymorphic sites among sequences 1, 2, and 3 are compared, i.e., when a site is defined as polymorphic if it varies among sequences 1, 2, and 3. From Maynard Smith (1992).

need not give the same answer. If only one of them is statistically significant, one may accept that division as being optimal. If both comparisons are significant but the cuts identified are different, then an ambiguity occurs, but one may assume that the true cut is somewhere in between.

In the above discussion we assumed that the region consists of only two blocks. In principle, the method can be extended to more than two blocks, but the computation time may be prohibitive. The following procedure (Figure 12.20) is

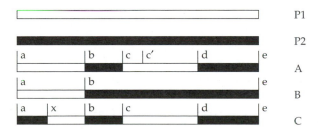

Figure 12.20 Successive stages in the analysis of mosaic structure. From Maynard Smith (1992).

heuristic but faster. (1) On the basis of a visual comparison of the sequence with the potential parental sequences (P1 and P2), propose a starting block structure, A. (2) Check, using the MCS method, that none of the individual blocks, a–b, b–c, c–d, and d–e, contain significant mosaic structure. (3) Check whether all proposed cuts (b, c, and d) are optimal and significant. To do this, first consider region a–c, and check, using the MCS method, whether b is optimal and significant. If it is, check whether cut c is optimal and significant for region b–d. If it is, check whether cut d is optimal and significant for region c–e. Suppose that one of the cuts proves not to be optimal. For example, suppose that cut c' instead of c is optimal for region b–d. If so, replace c by c' and start the procedure again at the beginning. This process of successive adjustment can be continued until a starting structure is found for which all the proposed cuts are both optimal and significant. (4) Suppose that one of the proposed cuts proves not to be statistically significant. For example, suppose that b is optimal and significant, but that no significant cut exists in region b–d. Then the proposed starting structure A must be abandoned, and replaced by B.

Note that step 2, checking that the proposed individual blocks do not themselves contain mosaic structure, is important. Thus, suppose that C is the true structure, and that cuts x, b, c, and d are optimal and significant. If the existence of cut x was overlooked, and A was taken as the starting structure, then calculations on region a–c might fail to find a significant cut. This failure in turn can lead to the structure of the rest of the sequence being wrongly identified or missed altogether.

HORIZONTAL GENE TRANSFER: EXAMPLES

Many cases of horizontal transfer have been proposed (see the reviews by Smith et al. 1992; Kidwell 1993; Syvanen 1994). We present two examples below.

Evolution of Penicillin Resistance in Neisseria Species

Penicillin-resistant strains of *Neisseria gonorrhoeae* have become increasingly prevalent during the last two decades. In many of these isolates, resistance is due to the acquisition of a β-lactamase, which inactivates the antibiotic. In other isolates, resistance is due to alterations of the chromosomal genes encoding two high-molecular-weight penicillin-binding proteins (PBPs 1 and 2), combined with reductions in the permeability of the outer membrane (Spratt 1989). These PBPs are enzymes that catalyze the final stages of peptidoglycan (cell wall) biosynthesis: in penicillin-susceptible strains, penicillin binds to these PBPs and inactivates them, leading to impairment of cell wall synthesis and cell death. In resistant strains, the affinities of these PBPs for penicillin are reduced, so that higher concentrations of the antibiotic are required for their inactivation (see Spratt 1989).

Isolates of the other pathogenic member of the genus *Neisseria*, *N. meningitidis*, that have increased levels of resistance to penicillin have been reported in the last few years. These low-level penicillin-resistant isolates have reductions in the affinity of PBP 2, but not of PBP 1. Penicillin resistance due to alterations of PBP 2 has also emerged recently in some of the commensal *Neisseria* species (see references in Spratt et al. 1992).

Using restriction-enzyme analysis, Spratt et al. (1989) found that the PBP 2 genes (*penA*) of penicillin-susceptible strains of *N. meningitidis* were rather uniform, whereas the *penA* genes of 42 penicillin-resistant isolates were heterogeneous and all had very different patterns of restriction fragments from those of the susceptible strains. These results suggested that the *penA* genes of penicillin-resistant isolates of *N. meningitidis* are very different in nucleotide sequence from those of susceptible isolates. The authors obtained the sequence of *penA* from a resistant strain of *N. meningitidis* and compared it with those from susceptible isolates. They found that the *penA* gene of the resistant isolate has acquired blocks of DNA from the *penA* genes of a closely related commensal *Nesseria* species, *N. flavescens*. *Nesseria* is naturally competent for transformation, so it is plausible that these blocks of DNA have been acquired by this recombinational mechanism. Similar mosaic *penA* genes have also been found in penicillin-resistant isolates of *N. gonorrhoeae* (Spratt 1988) and *N. lactamica* (Lujan et al. 1991).

N. flavescens was identified as the donor of the blocks of DNA in the *penA* genes of the above resistant isolates. *N. flavescens* isolates, including those from the preantibiotic era, fortuitously produce a PBP 2 that has a lower affinity for penicillin than PBP 2 of *N. meningitidis*, *N. gonorrhoeae*, and *N. lactamica* (Zhang 1991). Apparently, genetic transformation has provided a mechanism by which the latter *Neisseria* species can obtain increased levels of resistance to penicillin by acquiring the PBP 2 genes (or the relevant parts of them) from *N. flavescens*.

In a more detailed study, Spratt et al. (1992) obtained the sequences of the *penA* genes from 23 *Neisseria* isolates and found complex mosaic structures. Part of their results is shown in Figure 12.21. The *penA* genes of susceptible strains of *N. meningitidis* were uniform: the two most dissimilar strains differed by only 0.5%. These DNA sequences are not distinguished in Figure 12.21 and are denoted as NmS1; the *N. flavescens* sequence (strain *U*) is denoted as Nf1.

Using the maximum chi-squared method described in the preceding section with a very stringent criterion, i.e., $P = 0.001$, Spratt et al. (1992) found that the *penA* genes of the resistant *N. meningitidis* strains *B*, *C*, *D*, *E*, and *F* have acquired blocks of DNA from isolates that are very similar to the type strain of the commensal species *N. flavescens* (Figure 12.21). For example, strain *B* has acquired two blocks from *N. flavescens*. Interestingly, strains *C* and *D*, although isolated from different countries, have mosaic *penA* genes that are identical in sequence.

The *penA* gene of the susceptible *N. lactamica* strain (*Q*) has a mosaic structure. The region between sites 870 and 1254 differs from susceptible *N. meningitidis* by 58/385 (15.1 %) nucleotides, whereas the two species differ by only 26/992 (2.6%) in the regions before and after this central block. The presence of this block in the *penA* gene of all three of the *N. lactamica* isolates that were sequenced (*Q*, *R*, and *S*), and in the closely related species *N. polysaccharea* (strain *T*), suggests that it represents an ancient recombinational event that occurred (unrelated to the evolution of penicillin resistance) in the common ancestor of these strains.

The two penicillin-reistant *N. lactamica* strains (*R* and *S*) have acquired a block of DNA from *N. flavescens*. Their *penA* genes have identical block structures, but they differ in sequence at 26 sites. The *penA* gene of the type strain of *N. polysaccharea* (T) closely resembles that of the susceptible *N. lactamica* (*Q*), but has acquired blocks of DNA from *N. flavescens*. In recent years, isolates of *N. polysaccharea* with increased levels of resistance to penicillin have become common in some countries.

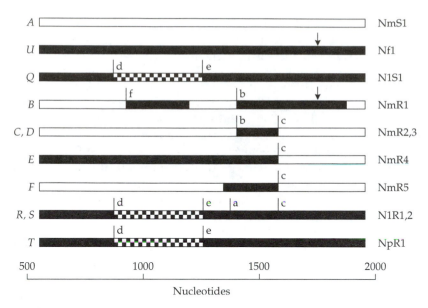

Figure 12.21 Mosaic *penA* genes containing blocks of *N. flavescens* DNA. Each line represents the *penA* (nucleotides 571–1947) of the *Neisseria* isolate indicated. The arrowheads indicate the positions of codon insertions; a–f mark the positions of common crossover points. The different shading in the mosaic *penA* genes of the *N. meningitidis, N. lactamica,* and *N. polysaccharea* isolates indicate the proposed origins of the different blocks. Black bar indicates *N. flavescens* DNA; unshaded bar indicates *N. lactamica* DNA; checkerboard bar indicates a more diverged region of *N. lactamica* DNA (see text). The unshaded regions in the genes from penicillin-resistance *N. meningitidis* isolates are similar in sequence to the corresponding regions in penicillin-susceptible isolates of *N. meningitidis.* All of the unshaded regions, except those downstream of crossover point c, differ from the corresponding regions in the penicillin-susceptible *meningitidis* (strain *A*) at < 0.6% of nucleotide sites. The unshaded regions downstream of crossover point c differed ≤ 4.2% from that of strain *A*. From Spratt et al. (1992).

Spratt et al. (1992) also gave examples of mosaic *penA* gene structures in *N. gonorrhoeae* and *N. cinerea.* In conclusion, it is clear that horizontal gene transfer involving the *penA* gene has occurred many times in *Neisseria* species.

Horizontal Transfer of P Transposable Elements

Another example of horizontal gene transfer involves the *P* elements in *Drosophila melanogaster.* As mentioned previously, *P* elements have rapidly spread through natural populations of *D. melanogaster* within the last four decades. *P* elements do not exist in closely related species of the *melanogaster* group, such as *D. mauritiana, D. sechellia, D. simulans,* and *D. yakuba.* Where then did these elements come from? An extensive survey of hundreds of *Drosophila* species by Daniels et al. (1990) has shown that, with the exception of *D. melanogaster, P* sequences are not found in any other species of the *melanogaster* subgroup. In contrast, all species of the distantly related *willistoni* and *saltans* groups contain *P* elements. In particular, the *P* element from *D. willistoni* was found to be identical to the one in *D. melanogaster* with the exception of a single base substitution, indicating that *D.*

willistoni may have served as the donor species in the horizontal gene transfer of *P* elements to *D. melanogaster*. Houck et al. (1991) have identified a possible vector for such a transfer: the parasitic mite, *Proctolaelaps regalis*, which share with these two species the same geographic range and ecology.

There are several reasons to suspect that this horizontal gene transfer occurred quite recently. First, the near identity between the *P* sequences from *D.*

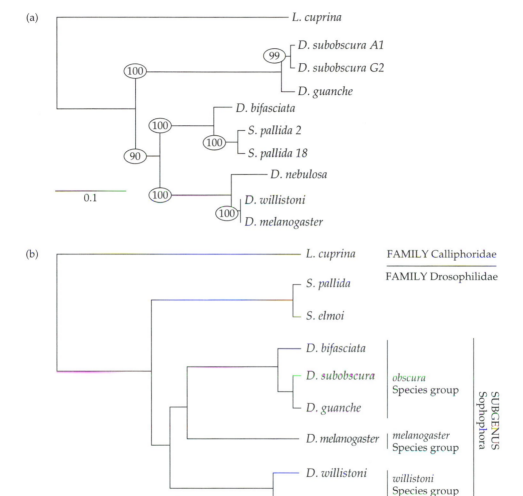

Figure 12.22 (a) Phylogeny of *P*-element nucleotide sequences inferred by the neighbor-joining method. The distance matrix was calculated by Kimura's (1980) two-parameter method. The branch lengths are given in terms of nucleotide substitutions per site. The numbers on the nodes denote bootstrap proportions. (b) Phylogeny of the nine Diptera species established from molecular, morphological and fossil data. The genus names are D = Drosophila, S = Scaptomyza and L = Lucilia. From Clark et al. (1994).

melanogaster and *D. willistoni* suggests a very short time of divergence. Second, the near absence of genetic variability in *P* sequences from *D. melanogaster* from even very distant geographic locations indicates that the time since the introduction of *P* elements into *D. melanogaster* is too short for genetic variability to accumulate. And finally, the geographical pattern of appearance of *P* elements in *D. melanogaster*, with populations in the American continents acquiring it first, seems to indicate that a very recent invasion (probably within the last 50) years, was involved.

Clark et al. (1994) have recently conducted a phylogenetic analysis of the *P*-element nucleotide sequences from nine species of *Drosophila* and other Diptera. The neighbor-joining tree obtained (Figure 12.22a) is well supported by bootstrap resampling and the tree is also supported by both parsimony and maximum-likelihood analyses (Clark et al. 1994). However, it is incongruent with the phylogeny of the nine species reliably inferred from molecular, morphological, and fossil data (Figure 12.22b). A minimum of three horizontal transfer events would be required to explain the discrepancy between the two trees.

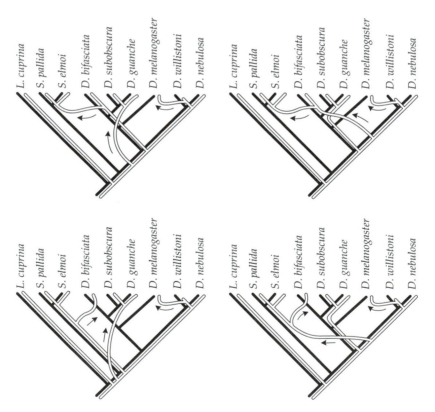

Figure 12.23 Four possible scenarios, involving horizontal transfer, to explain the incongruence between the *P*-element tree and the species tree (Figures 12.22a and b). The black trees underneath represent the species phylogeny; and the white trees above show the hypothesized descent of the *P* elements. In each scenario, there are three independent horizontal-transfer events (highlighted by arrows) that are consistent with the observed sequence data. From Clark et al. (1994).

Four possible alternative scenarios are outlined in Figure 12.23. In each, there is both a transfer between *D. willistoni* and *D. melanogaster* and an independent transfer between the ancestors of *Scaptomyza* and *D. bifasciata*. A third transfer, which occurs deeper in the phylogeny, is of ambiguous position. With the exception of the *D. willistoni–D. melanogaster* transfer, which has apparently occurred relatively recently, it is difficult to unequivocally identify when, and in what direction, any of the other horizontal transfers occurred. For example, solely on the basis of the comparisons presented here, it is not possible to determine whether there was a transfer from *D. bifasciata* to *Scaptomyza* or, as proposed by Hagemann et al. (1992), from *Scaptomyza* to *D. bifasciata*. Note that other explanations of the discrepancy between Figures 12.22a and b are possible but less plausible (Clark et al. 1994).

CHAPTER 13

Genome Organization and Evolution

T HERE ARE SEVERAL ASPECTS OF GENOME ORGANIZATION AND EVOLUTION. The first is genome size, which varies enormously among organisms. How did this variation occur and what mechanisms can increase or decrease genome size to produce such variation? The second topic is the genetic information included within genomes. Do genomes mostly consist of genic DNA? Or, is the genome made of mostly nongenic sequences? Are there many repetitive sequences in the genome, and if so, what is the pattern of chromosomal distribution of the repeated sequences? Does the nongenic fraction have a function, or is it merely "junk"? The third topic concerns the variation in nucleotide composition over different parts of the genome. Is there heterogeneity in composition among different regions of the genome? What mechanisms can give rise to localized differences in nucleotide composition? Finally, why are there numerous spliceosomal introns in the genome of higher eukaryotes, but no such introns have ever been found in prokaryotic genes? What is the origin of introns and what is their evolutionary significance?

GENOME SIZE VARIATION AND THE C-VALUE PARADOX

The genome size of an organism is defined as the amount of DNA in the haploid genomic set, such as that in the sperm nucleus. It is also commonly called the C value, where C stands for "constant" or "characteristic," to denote the fact that the size of the haploid genome is fairly constant within any one species. In contrast, C values vary widely among species in both prokaryotes and eukaryotes. Large genomes such as the nuclear genomes of eukaryotes are usually measured in picograms (pg) of DNA (1 pg = 10^{-12}g), whereas small genomes such as the genomes of bacteria and viruses are more commonly measured in base pairs (bp) or kilobase pairs (Kb) or in daltons. For convenience, we shall use mainly the unit of one million base pairs (Mb) because most genome sizes are larger than 1 Mb and because this unit is now commonly used in the human genome project. The conversion factors are 1 pg $\approx 0.98 \times 10^9$ bp = 980 Mb and 1 dalton $\approx 1.62 \times 10^{-3}$ bp.

Genome Sizes of Bacteria

Bacterial genome sizes vary over a 20-fold range, from about 0.6 Mb in some obligatory intracellular parasites, to more than 10 Mb in several cyanobacterial species (Table 13.1). The smallest prokaryotes, the mycoplasmas, have a genome of ~0.6 Mb, which consists of about 470 protein-coding genes including some 50 ribosomal proteins, one or two sets of rRNA genes (5S, 16S, and 23S), and about 33 tRNA genes (Fraser et al. 1995). Therefore, the genome of a mycoplasma contains about 500 genes, which is still smaller but probably close to the minimum number for supporting a free-living cell or life organism (see Fraser et al. 1995). The number of genes in *Escherichia coli* is probably between 3000 and 6000. The numbers of genes in other bacteria range very roughly from about 500 to 8000. In other words, among characterized bacterial species, the variation in the number of genes is the same as the variation in C values. Therefore, bacteria do not seem to contain large quantities of nongenic DNA.

The genomes of bacteria can be divided into three fractions: (1) chromosomal DNA, (2) DNA that originated in plasmids, and (3) transposable elements (Hartl et al. 1986). The chromosomal fraction contains protein-coding genes required for growth and metabolic functions (70–80%), spacers and various signals (20–30%), RNA-specifying genes (~1%), and a number of short repeated sequences, usually on the order of a few dozens of base pairs in length. Bacteria carry many plasmids as extrachromosomal genetic elements. In some instances, however, genes derived from plasmids are found in the bacterial chromosome. Transposable elements are common components of some bacterial genomes. For example, the chromosome of *E. coli* contains 1–10 copies of at least six different types of insertion sequences. The nonchromosomal fraction of the genome (including insertion sequences and plasmid-derived genes in the chromosome) seems to be one order of magnitude smaller than the chromosomal fraction.

The distribution of genome sizes in bacteria is discontinuous, showing major peaks with modal values of about 0.8, 1.6, 4.0 Mb, and several minor peaks at 7.2 and 8.0 Mb (Herdman 1985). This distribution has led to the suggestion that the larger genomes evolved from smaller genomes by successive genome duplications. Since there seems to be no notable relationship between genome size and bacterial phylogeny, it has been suggested that genome dupli-

TABLE 13.1   Range of C values in bacteria

	Range in genome size (kb)	Ratio (highest/lowest)
Eubacteria	580–13,200	20
Mycoplasmas	580–1,800	3
Gram negative	650–7,800	12
Gram positive	1,600–11,600	7
Cyanobacteria	3,100–13,200	4
Archaebacteria	1,600–4,100	3

From Cavalier-Smith (1985), updated by knowledge of the *Mycoplasma genitalium* (Fraser et al. 1995).

cations have occurred frequently in the evolution of bacterial lineages (Wallace and Morowitz 1973).

Using a tentative phylogeny of bacteria based on comparison of rRNA sequences, Herdman (1985) was able to relate changes in genome size to phylogenetic history. The results indicate that genome duplication occurred independently in different bacterial lineages. Interestingly, many of the duplications seem to have occurred coincidentally in several bacterial lineages at a rather specific time in evolution, i.e., soon after the appearance of oxygen in the atmosphere, approximately 1.8 billion years ago.

In summary, the distribution of genome sizes in bacteria can be explained by a combination of several processes: (1) genome duplication during the independent evolution of a respiratory metabolism in several lines, (2) subsequent whole genome duplications occurring independently in many lineages, (3) small-scale deletions and insertions, (4) duplicative transposition, (5) horizontal transfer of genes derived mainly from plasmids, and (6) loss of massive chunks of DNA in many parasitic lines.

Genome Sizes of Eukaryotes

C values in eukaryotes are usually much larger than in prokaryotes (Tables 13.2 and 13.3); for example, the genome size of humans (~3400 Mb) is about 1000 times that of *E. coli* (~4.7 Mb). However, there are exceptions. For instance, the yeast *Saccharomyces cerevisiae* has a smaller genome than many Gram-positive bacteria and most cyanobacteria. Since the eukaryotic genome has multiple origins of replication, while most prokaryotes seem to have only one, the eukaryotes are able to replicate much larger amounts of DNA in an amount of time not much larger than that required for a much smaller prokaryotic genome.

The variation in C values in eukaryotes is much larger than that in bacteria, from ~9 Mb to ~690,000 Mb, approximately a 80,000-fold range (Table 13.2). Unicellular protists, particularly sarcodine amoebae, show the greatest variation in C values. The three amniote classes (mammals, birds, and reptiles), on the contrary, are exceptional among eukaryotes in their small variation in genome size (up to fourfold). Other classes, for which a substantial body of C-value data exists, show variation of at least 100-fold.

Interestingly, the huge interspecific variation in genome sizes among eukaryotes seems to bear no clear relationship to either organismic complexity or the likely number of genes encoded by the organism. For example, several unicellular protozoans possess much more DNA than mammals (Table 13.3), which are presumably more complex. Moreover, organisms that seem similar in complexity (e.g., flies and locusts, onion and lily, *Paramecium aurelia* and *P. caudatum*) possess vastly different C values (Table 13.3). This lack of correspondence between C values and the presumed amount of genetic information contained within genomes has become known in the literature as the **C-value paradox**. The C-value paradox is also evident in comparisons of sibling species (i.e., species that are so similar to each other morphologically as to be indistinguishable phenotypically). In protists, bony fishes, amphibia, and flowering plants, many sibling species differ greatly in their C values, in spite of the fact that, by the definition of sibling species, no difference in organismic complexity exists. Since we cannot assume that a species possesses less DNA than the amount required for specify-

---

TABLE 13.2   **Range of C values in various eukaryotic groups of organisms**

	Genome size range (kb)	*Ratio (highest/lowest)*
Protists	23,500–686,000,000	29,191
Euglenozoa	98,000–2,350,000	24
Ciliophora	23,500–8,620,000	367
Sarcodina	35,300–686,000,000	19,433
Fungi	8,800–1,470,000	167
Animals	49,000–139,000,000	2,837
Sponges	49,000–53,900	1
Annelids	882,000–5,190,000	6
Molluscs	421,000–5,290,000	13
Crustaceans	686,000–22,100,000	32
Insects	98,000–7,350,000	75
Echinoderms	529,000–3,230,000	6
Agnathes	637,000–2,790,000	4
Sharks and rays	1,470,000–15,800,000	11
Bony fishes	382,000–139,000,000	364
Amphibians	931,000–84,300,000	91
Reptiles	1,230,000–5,340,000	4
Birds	1,670,000–2,250,000	1
Mammals	1,420,000–5,680,000	4
Plants	50,000–307,000,000	6,140
Algae	80,000–30,000,000	375
Pteridophytes	98,000–307,000,000	3,133
Gymnosperms	4,120,000–76,900,000	17
Angiosperms	50,000–125,000,000	2,500

Compiled by Li and Graur (1991) from Cavalier-Smith (1985) and other sources.

---

ing its vital functions, we have to explain why so many species contain vast excesses of DNA.

The first question to be clarified is whether or not a correlation exists between genome size and the number of genes. In other words, are the differences in genome sizes attributable to genic DNA or to nongenic DNA? Eukaryotes show approximately 20-fold variation in the number of protein-coding genes, from about 6000 in yeast (Coffeau et al. 1996) to roughly 60,000–80,000 in mammals (Antequera and Bird 1993; Fields et al. 1994). This 20-fold variation is insufficient to explain the 80,000-fold variation in nuclear DNA content. One way to compare the gene numbers in two genomes is to analyze the polysomal RNA complexity in the same tissue of the two organisms; the RNA complexity refers to the total length of different species of RNA. Such data are available in several vertebrates (Table 13.4). In *Xenopus laevis* and *Triturus cristatus* egg RNA complexities are of the order of 30-40 million nucleotides (Mnt). Thus, the two

TABLE 13.3 **C values from eukaryotic organisms ranked by size**

Species	C value (kb)
Navicola pelliculosa (diatom)	35,000
Drosophila melanogaster (fruitfly)	180,000
Paramecium aurelia (ciliate)	190,000
Gallus domesticus (chicken)	1,200,000
Erysiphe cichoracearum (fungus)	1,500,000
Cyprinus carpio (carp)	1,700,000
Lampreta planeri (lamprey)	1,900,000
Boa constrictor (snake)	2,100,000
Parascaris equorum (roundworm)	2,500,000
Carcarias obscurus (shark)	2,700,000
Rattus norvegicus (rat)	2,900,000
Xenopus laevis (toad)	3,100,000
Homo sapiens (human)	**3,400,000**
Nicotiana tabaccum (tobacco)	3,800,000
Paramecium caudatum (ciliate)	8,600,000
Schistocerca gregaria (locust)	9,300,000
Allium cepa (onion)	18,000,000
Coscinodiscus asteromphalus (diatom)	25,000,000
Lilium formosanum (lily)	36,000,000
Amphiuma means (newt)	84,000,000
Pinus resinosa (pine)	68,000,000
Protopterus aethiopicus (lungfish)	140,000,000
Ophioglossum petiolatum (fern)	160,000,000
Amoeba proteus (amoeba)	290,000,000
Amoeba dubia (amoeba)	670,000,000

Compiled by Li and Graur (1991) from Cavalier-Smith (1985), Sparrow et al. (1972), and other references. The C value for humans is highlighted for reference.

amphibians have an equivalent RNA transcribing capacity, despite the eightfold difference in genome size. As another example, the chicken has a genome size one-third of that of the mouse, yet the polysomal RNA complexities in chicken and mouse livers are virtually the same. Therefore, the large interspecific variation in genome size among organisms appears to be mostly due to nongenic DNA rather than genic DNA.

A positive correlation between the degree of repetition of several RNA-specifying genes and genome size has been found (Chapter 10). Similarly, a correlation exists between genome size and the number of copies of such regulatory sequences as telomeres, centromeres, and replicator genes, which are required for chromosome replication, segregation, and recombination during meiosis and mitosis. However, all these genes constitute only a minute fraction of the genome,

---

TABLE 13.4   RNA sequence complexity in vertebrates

Species	Genome size (Mb)	Percent scDNA	Tissue/ stage	Poly(A)+ mRNA (Mnt)
Xenopus laevis	3,000	75	Egg	27–40
			Tadpole	30
Triturus cristatus	23,000	47	Egg	40
Gallus domesticus	1,150	87	Liver	20
			Oviduct	30
			Myofibril	32
Mus musculus	3,360	70	Embryo	20
			Kidney	20
			Liver	10
			Brain	110–140

From John and Miklos (1988).

Notation: Mb, million base pairs; scDNA, single-copy DNA; Mnt, million nucleotides.

---

such that the variation in the number of RNA-specifying genes and regulatory sequences cannot explain the variation in genome size.

In summary, we are left with the nongenic DNA fraction as the sole culprit for the C-value paradox. In other words, a substantial portion of the eukaryotic genome consists of DNA that does not contain genetic information. It has been estimated that the amount of nongenic DNA per genome varies in eukaryotes from about 3 Mb to over 100,000 Mb (a 100,000-fold range) and constitutes anything from less than 30% to almost 100% of the genome (Cavalier-Smith 1985).

THE REPETITIVE STRUCTURE OF THE EUKARYOTIC GENOME

The eukaryotic genome is characterized by the presence of repetitive DNA consisting of nucleotide sequences of various lengths and compositions that occur from a few times to millions of times in the genome either in tandem or in a dispersed fashion. Segments of DNA that are not repeated are referred to as **single-copy DNA (scDNA)** or *unique DNA*. The proportion of the genome taken up by repetitive sequences varies widely among taxa. In yeast, this proportion amounts to about 20%. In animals, the proportion ranges from a few percent as in *Chironomus tentans*, an arthropod, to close to 90% (i.e., only 10% scDNA) in *Necturus masculosus,* an amphibian (Table 13.5). In plants, the proportion of repetitive DNA can exceed 80% (Flavell 1986).

Classical studies of the kinetics of DNA reassociation by Britten and Kohne (1968) showed that the genome of eukaryotes can roughly be divided into four fractions: **foldback DNA, highly repetitive DNA, middle-repetitive DNA,** and single-copy DNA (Figure 13.1). Foldback DNA consists of palindromic DNA sequences that can form hairpin double-stranded structures as soon as the denatured DNA is allowed to renature. The highly repetitive fraction is made up of short sequences, from a few to hundreds of nucleotides long, which are repeated thousands or even millions of times. The middle-repetitive fraction consists of much longer sequences, hundreds or thousands of base pairs on the average, which appear in the genome from dozens to thousands of times.

On the basis of the pattern of dispersion of repeats, the repetitive fractions can be classified into two types of repeated families: localized and dispersed.

Localized Repetitive Sequences

Localized repetitive sequences usually occur as tandem arrays and, when this is the case, they are called **tandem repetitive DNA (TR-DNA)**. One type of TR-DNA consists of multigene families with members arranged in tandem; however, some multigene families have members dispersed to various chromosomal locations. The

TABLE 13.5   C values and percent single-copy DNA (scDNA) in animal genomes

Invertebrates	C value (× 10⁹ bp)	% scDNA	Vertebrates	C value (× 10⁹ bp)	% scDNA
Protozoa			Protochordata		
Tetrahymena pyriformis	0.2	90	*Ciona intestinalis*	0.2	70
Coelenterata			Pisces		
Aurelia aurita	0.7	70	*Scyliorhinus stellatus*	6.0	39
			Leuascus cephalus	5.4	44
Nemertini (Rhyncocoela)			*Raja montagui*	3.3	47
Cerebratulus	1.4	60	Amphibia		
			Necturus masculosus	81.3	12
Mollusca			*Bufo bufo*	6.9	20
Aplysia californica	1.8	55	*Triturus cristatus*	20.6	47
Crassostrea virginica	0.7	60	*Xenopus laevis*	3.0	75
Spisula solidissima	1.2	75			
Loligo loligo	2.7	75	Reptilia		
			Natrix natrix	2.5	47
Arthropoda			*Terrapene carolina*	4.0	54
Prosimulium multidentatum	0.2	56	*Caiman crocodylus*	2.6	66
Drosophila melanogaster	0.2	60	*Python reticulatus*	1.7	71
Limulus polyphemus	2.7	70	Aves		
Musca domestica	0.9	90	*Gallus domesticus*	1.2	80
Chironomus tentans	0.2	95			
			Mammalia		
Echinodermata			*Homo sapiens*	3.4	64
Strongylocentrotus purpuratus	0.9	75	*Mus musculus*	3.4	70

From John and Miklos (1988).

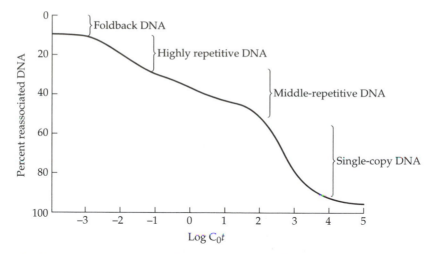

Figure 13.1 A reassociation profile of mammalian DNA. DNA is purified, sheared, thermally melted into single strands, and then allowed to reassociate through gradual cooling. The percentage of reassociated double-stranded DNA on the vertical axis is shown as a function of the product of DNA concentration and time (C_0t) on the horizontal axis. Modified from Schmid and Deininger (1975).

repeat unit in a gene family generally includes the entire sequence of the gene (exons, introns, and untranscribed flanking sequences) and spacer regions, and thus can be several kilobases or even larger in length. Well-known examples of tandemly arrayed multigene families are rRNA and tRNA genes in mammals. For example, the human genome contains approximately 350 rRNA genes and 1300 tRNA genes; the rRNA genes are located in the nucleolar organizers of the five acrocentric chromosomes (13, 14, 15, 21, and 22) (Wellauer and Dawid 1979). The β-globin cluster in humans represents an example of a low copy number multigene family (Figure 10.9).

The major type of TR-DNA consists of tandemly arrayed, simple repetitive sequences, mostly noncoding sequences. In some species this type of repetitive DNA can account for the majority of the DNA in the genome. For example, in the kangaroo rat *Dipodomys ordii*, more than 50% of the genome consists of three families of repeated sequences: AAG (2.4 billion times), TTAGGG (2.2 billion times), and ACACAGCGGG (1.2 billion times) (see Widegren et al. 1985). Of course, these families are not completely homogeneous but contain many variants that differ from the consensus sequence in one or two nucleotides. For example, some sequences in the TTAGGG family are actually TTAGAG. Notwithstanding, many of the localized highly repeated sequences have such a uniform nucleotide composition that upon fractionalization of the genomic DNA and separation by density gradient they form one or more thick bands that are clearly distinguishable from the main band of DNA and from the smear created by the other fragments of more heterogeneous composition. These bands are called **satellite DNA**.

In some species members of a satellite DNA are found on more than one chromosome, while in others they are restricted to a particular chromosomal location. For example, alpha satellite or alphoid DNA, the most abundant family of satellite DNA in the human genome, is found at the centromeres of all chromosomes

(Willard and Waye 1987). In contrast, 60% of the genome of *Drosophila nasutoides* consists of satellite DNA, and all of it is localized on one of the four chromosomes (Figure 13.2), which seems to contain little else (Miklos 1985). In general, satellite DNA is located inside heterochromatic or centromeric regions. Other types of simple tandem repetitive sequences will be discussed in detail in the next section.

Dispersed Repetitive Sequences

The second class of repetitive DNA consists of sequences that are dispersed throughout the genome. Single copies of dispersed repetitive sequences are found in introns, flanking regions of genes, intergenic regions, and nongenic DNA. There are two major kinds of dispersed highly repetitive sequences: short interspersed repeats (SINEs) and long interspersed repeats (LINEs) (Chapter 12). In addition, there are moderately repetitive sequences, which are usually transposable elements or processed pseudogenes (retropseudogenes) and are usually dispersed over the genome (see Chapter 12).

SINEs are retropseudogenes, i.e., pseudogenes derived from POL III RNA transcripts. As has been discussed in Chapter 12, they are abundant in mammalian genomes. For example, there are more than 500,000 *Alu* repeats in the human genome.

LINEs were originally described as DNA sequences longer than 5 kb and present in 10^4 or more copies per genome (Singer 1982). The human genome apparently contains only one LINE family, L1 (Hutchison et al. 1989). The consensus L1 sequence is about 6 kb in length, has a poly-A tail at one end, and is flanked by short direct repeats generally less than 20 bp long. The copy number is about 100,000 per haploid genome. Most L1 sequences are truncated at their 5' end. Intact L1 sequences contain two large open reading frames, *ORF*-1 and *ORF*-2, with about 375 and 1,300 codons, respectively. *ORF*-2 contains amino acid sequence motifs characteristic of reverse transcriptase enzymes. The above

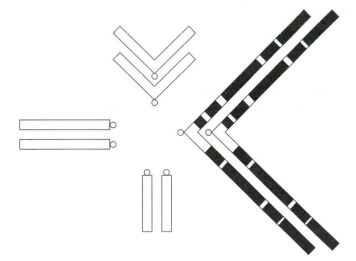

Figure 13.2   Highly repetitive DNA sequences (black areas) in *Drosophila nasutoides* are localized on a single chromosome. Modified from Miklos (1985) by Li and Graur (1991).

features suggest that L1 elements are produced by reverse transcription of polyadenylated mRNAs and the subsequent insertion of the reverse transcripts into the genome. Since L1 sequences do not possess long terminal repeats, they are classified as retroposons (Chapter 12). However, most L1 sequences are nonfunctional and such LINEs should be classified as retrosequences. LINEs homologous to the human L1 family have been found in all mammals, including marsupials (see Hutchison et al. 1989). Nonmammalian elements related to L1 include the *I*, *F*, *G*, and *D* factors in *Drosophila* species, *Ingi* in *Trypanosoma brucei*, and *R2* in *Bombyx mori*.

Most L1 sequences are truncated, containing no intact reading frames. Such defective elements may not be able to transpose and were found to evolve much more rapidly than intact elements, probably because they are no longer subjected to functional constraints. The fact that most L1 sequences are defective implies that the propagation of L1 elements within the genome depends on only a small number of source elements (Deininger et al. 1992). For these reasons, the degree of L1 sequence divergence among species is much greater than the degree of divergence among conspecific L1 copies. For example, L1 sequences from mice and humans differ from each other on average by about 30%, whereas the average sequence divergence among L1 elements in mice is only 4% (Hutchison et al. 1989). Hardies et al. (1986) estimated that in rodents more than half of the L1 elements are no more than 3 million years old.

The Genomic Location of Transcribed Genes

RNA–DNA hybridization experiments have shown that only a very small fraction of the transcribed genes are located within the repeated fractions of the genome. Most transcribed genes reside within the unique DNA fraction, most of which is in euchromatic regions. Even in the unique fraction, most sequences are not transcribed. Indeed, it has been estimated that only about 3% of the nonrepetitive DNA sequences in humans are transcribed (see Lewin 1994). These studies constitute further support for the view that most of the eukaryotic genome is devoid of genetic information.

TANDEM REPETITIVE SEQUENCES

Tandem repetitive DNA (TR-DNA) has attracted much attention because many TR-DNA loci are highly variable among individuals in a population and so are very useful genetic markers in biology and medicine. This line of research started with Wyman and White's (1980) fortuitous observation of a locus highly polymorphic for restriction length variation and the finding by Bell et al. (1982) that the hypervariable locus near the insulin gene consisted of tandem repeats, with multiallelic variation arising through variation in repeat copy number. Subsequently, probes constructed from a short tandem-repeat region in an intron of the human myoglobin gene (Weller et al. 1984) were shown by Southern blot hybridization to detect an astonishingly complex, variable array of DNA fragments, not just from humans but also from several other animal species that happened to be on the Southern blot (see Jeffreys 1993). This was the beginning of the so-called DNA fingerprinting study (Jeffreys et al. 1985a, b).

Classification of Tandem Repetitive DNA

Table 13.6 shows a classification of simple tandem repetitive sequences, according to the size of the repeat unit, the number of repeat units per array, and the genomic location of the tandem arrays. The first class, satellite DNA, has already been discussed in the preceding section; satellite DNA has rarely been used in fingerprinting because it is too large for PCR amplification. The rest of this section concerns the other four classes. It should be noted that there is presently no consensus in the classification of simple tandem repeats and the terminology used varies among authors.

The term **minisatellite** was originally coined and is still used by Jeffreys and colleagues (e.g., Jeffreys et al. 1991) to denote any tandem repeat region that is much shorter than but has a structural similarity to classical satellite DNA. However, it is used here and by other authors (e.g., Shriver 1993) in a more restricted sense, i.e., as a tandem array of repeats with a unit length 9 bp or longer, while tandem arrays of repeats with a shorter unit length are classified into three other classes (Table 12.6). The advantage of having more classes of TR-DNA in terms of repeat length is that the primary mechanism for generating repeated sequences may be different for different length classes (see later). The term *short tandem repeats (STRs)* refers to tandem repeats with repeat units 3–5 bp long. It has been estimated that there are 300,000 trinucleotide and tetranucleotide STRs in the human genome or, on average, one array in every 10 kb of genomic DNA (Edwards et al. 1991; Beckmann and Weber 1992). *Microsatellites* are TR-DNA arrays that have repeat units of 1–2 bp; some authors (e.g., Beckmann and Weber 1992) also include STRs in the class of microsatellites. The most common human microsatellites are dinucleotide arrays of $(CA)_n$, which means n repeats of CA (Beckmann and Weber 1992). There are approximately 50,000 $(CA)_n$ arrays in the human genome (Tautz and Renz 1984; Hamada et al. 1982) or about one array every 30 kb (Stallings et al. 1991; Hudson et al. 1992). *Alu* tails, the last class, are

---

TABLE 13.6   **Five classes of simple tandem repetitive sequences**

Class	Examples	Repeat size (base pairs)	Array size (repeat units)	Genomic distribution
Satellites	Alpha[a], Beta[b]	2–100s	≥1000s	Centromeric, heterochromatic
Minisatellites	Apo B[c], D1S7[d]	9–100	10–100	Dispersed, subtelomeric
Short tandem repeats	FMR-1[e], MYCN[f]	3–5	10–100	Dispersed
Microsatellites	Apo CII[g], PLAT[h]	1–2	10–100	Dispersed
Alu tail arrays	IN20-Rep[i], HMG-CoA[j]	1–5	10–40	Dispersed

Modified from Shriver (1993).

References: [a](Willard and Waye 1987), [b](Waye and Willard 1989), [c](Knott et al. 1986, Huang and Breslow 1987), [d](Jeffreys et al. 1988), [e](Aslanidis et al. 1992), [f](Fougerousse et al. 1992), [g](Fornage et al. 1992), [h](Sadler et al. 1991), [i](Shriver et al. 1992), [j](Zuliani and Hobbs 1990).

tandem arrays of repeat units 1–5 bp in size that are present immediately 3' of *Alu* elements. The most common *Alu* tail arrays are $(A)_n$ and $(AAAN)_n$, where $N = G$, C, or T (Beckmann and Weber 1992). Interestingly, tandem arrays with repeat unit length 6–8 bp are rarely observed and so are not included in Table 13.6. See Shriver (1993) for a review of TR-DNA characteristics.

A tandem repeat region that shows variation in number of repeats among chromosomes in a population is commonly referred to as a **VNTR** locus, where VNTR stands for **variable number of tandem repeat** (Nakamura et al. 1987). Since most tandem repeat regions are variable, VNTRs include virtually all the last four classes of loci in Table 13.6; satellite DNA loci are not included because variation in size between individuals is difficult to type. The most prominant feature that distinquishes VNTR loci from classical gene loci is that they are usually highly polymorphic, having a large number of alleles per locus and a high heterozygosity. Indeed, the heterozygosity can be close to 1 (Table 13.7). The high polymorphism appears to be largely due to exceptionally high rates of mutation; e.g., the mutation rate per gamete at a minisatellite locus can be a few percent per generation (Table 13.7), in contrast to 10^{-5} per generation or lower at classical gene loci.

Currently VNTRs are heavily used in biology, medicine, and forensic science. Since a large number of alleles usually exist at a VNTR locus, a collection of several VNTR loci can lead to a huge number of possible combinations or DNA profiles, making VNTR loci extremely powerful for identity analysis and forensic use. For example, in the investigation of a crime where a biological specimen has been obtained from the crime scene, DNA from the specimen and from the suspect are typed for size at several unlinked VNTR loci. If the profiles do not match, the suspect is excluded. If they match, the probability of this match occurring in the population by chance alone can be calculated from the frequencies of alleles in the population. This probability is usually extremely low because of high polymorphism at VNTR loci. In the United States, however, the application of VNTRs in crime courts is presently still controversial, largely because of disagreement on how to compute the frequency of a specific profile, e.g., whether it should be computed on the basis of the ethnic group the crime suspect belongs to or on the basis of the general population (see Lewontin and Hartl 1991; Chakraborty and Kidd 1991; for a review, see Weir 1992, 1995). The high polymorphism and the ease of typing make VNTR loci also very useful for genetic mapping (e.g., Nakamura et

TABLE 13.7   Summary of heterozygosities and mutation rates at five human minisatellite loci

Locus	Repeat length (bp)	Heterozygosity (%)	Mutation rate per gamete
D5S43	29, 30	85.1	0.000
D12S11	45	95.9	0.000
D7S22	37	97.4	0.003
D7S21	20	98.0	0.007
D1S7	9	99.4	0.052

From Jeffreys et al. (1988).

al. 1987; Edwards et al. 1991; Silver 1992), studies of population structure (e.g., Flint et al. 1989; Chakraborty and Jin 1992), detecting loss of heterozygosity in tumor initiation and progression (e.g., Kraemer et al. 1989), ethological studies (e.g., Packer et al. 1991), and disease association studies (e.g., Friedl et al. 1990).

The interest in VNTR loci has been further stimulated by the recent finding that mutation at some VNTR loci can cause a severe disease. For example, the following four genetic diseases have been associted with unstable VNTR loci: Kennedy's disease (spinobulbar muscular atrophy, La Spada et al. 1991), fragile X mental retardation syndrome (Fu et al. 1991), myotonic dystrophy (a progressive neuromuscular disorder, Fu et al. 1992), and Huntington's disease (McDonald et al. 1993). All four diseases and at least six others (see Hummerich and Lehrach 1995) are caused by amplification of trinucleotide tandem arrays, i.e., an allele becomes disease-causing when it has more repeats than some threshold number. For example, in the case of fragile X the threshold number is about 200; the normal gene contains 6 to 54 copies of the CGG repeat and a **premutation** allele has 50 to 200 repeat units (Fu et al. 1991). Surprisingly, this VNTR locus is in the 5′ untranslated region of the fragile X gene and in the case of myotonic muscular dystrophy the VNTR locus is in the 3′ untranslated region rather than in the coding region of a protein kinase gene (Caskey et al. 1992; Fu et al. 1992).

Mutational Mechanisms

Current models of size change at VNTR loci all involve mispairing of repeats but differ in the timing of the DNA mispairing event: replication, mitosis, and meiosis. **Slipped-strand mispairing (SSM)** or **replication slippage** occurs during DNA replication; **unequal sister-chromatid exchange (USCE)** occurs during mitosis and, less frequently, during meiosis; and unequal genetic recombination or crossing-over occurs during meiosis (Chapter 1). SSM and USCE are intrachromosomal phenomena because they involve exchange between two copies of the same chromosome, whereas meiotic unequal crossing-over is an interchromosomal phenomenon because it involves two different chromosomes. Meiotic unequal crossing-over causes an exchange of flanking markers, whereas SSM and USCE do not. This property can be used to detect meiotic unequal crossing-over events. Wolff et al. (1989) found that only 1 in 12 size-change events at the *D1S7* minisatellite locus had a recombination between two flanking markers, and Vergnaud et al. (1991) observed no exchange of flanking markers in 52 mutation events at the *CEB1* locus. Another way to study the mechanism of mutation is to determine from which parental allele(s) a mutant allele is derived. Jeffreys et al. (1990, 1991) developed the minisatellite variant repeat (MVR) mapping technique, which utilizes sequence differences between repeat units to assay the interdispersion pattern of different repeat units along an allele. Using this technique, Jeffreys et al. (1990) determined that each of the 38 mutant alleles isolated from sperm and the 37 mutant alleles from blood at the *D1S8* minisatellite locus could be traced back to a single parental allele; thus none of these mutant alleles had arisen by recombination between homologous chromosomes. In a subsequent study, Jeffreys et al. (1991) found that 2 of the 7 additional mutant alleles at the *D1S8* locus had arisen by recombination events between parental alleles. The above data taken together suggest that meiotic unequal crossing-over is a minor mechanism of mutation at minisatellite loci or, in other words, size change at minisatellite loci occurs mainly at mitosis. This conclusion is also supported by the observation that only 1 of the 52 mutations at the *CEB1* locus was of maternal origin (Vergnaud et al. 1991). (As explained in Chapter 8, in the formation of mammalian germ cells, sperm undergo many more rounds of mitotic division than do eggs but the same number of meiotic divisions. Thus, a high proportion of mutants of paternal origin suggests that mutation occurs mainly at mitosis.) Unfortunately, the above techniques cannot distinguish between USCE and SSM. However, USCE events tend to be larger in size than SSM events because USCE is less constrained in misalignment of repeats than is SSM (see Kornberg et al. 1964; Wells et al. 1967a,b; Schlötterer and Tautz 1992; Worton 1992). The average size of change for the 52 mutations at the *CEB1* locus was 250 bp or 9% of the size of the parental allele. Moreover, the average size change for the 32 mutations at the *D1S7* locus studied by Jeffreys et al. (1988) was 34 repeat units or 5% of the parental allele. The relatively large size of these mutation events suggest that USCE is the primary mechanism of mutation at these minisatellite loci. However, recent data (Jeffreys et al. 1995) suggest that minisatellites do not primarily mutate by processes such as slippage and unequal crossover intrinsic to the tandem repeat array. Instead, germline repeat instability is largely regulated by *cis*-acting elements near the arrays and involves unexpectedly complex processes of gene conversion.

Limited data on the size change and parental origin of mutation events at microsatellite and STR loci are also available (Table 13.8). Of the 7 mutation events where the parental origin was determined, 6 were of paternal origin and only 1 was maternal. The prevalence of paternal origin suggests that size mutation at these loci occurs mainly at mitosis. Of the 9 mutations 6 were changes of 1 repeat unit, 2 were changes of 2 repeat units and 1 was a change of 3 repeat units. The small size of the mutation events suggests SSM as the primary mechanism of mutation for microsatellite and STR loci. Thus, while USCE seems to be the primary mechanism of mutation at minisatellite loci, SSM seems to be the primary mechanism of mutation at microsatellite and STR loci. However, since this conclusion is based on limited data, it needs to be substantiated by further studies.

TABLE 13.8 Characterized microsatellite and STR mutations

Locus type	Parental origin[a] and number of repeats					
	−3	−2	−1	+1	+2	+3
$(CA)_n$				P		
$(CA)_n$			P		P	
$(CA)_n$						U
$(CA)_n$			U		P	
Tetra-STR				M	P	
Total	0	2	2	4	0	1

Taken from Shriver (1993).

[a]Parental origin of alleles are indicated by characters: P, paternal; M, maternal; and U, undetermined.

Population Genetics of VNTRs

Another way to study the mutational mechanisms is to use the theory of population genetics (e.g., Harding et al. 1993; Shriver et al. 1993; Valdes et al. 1993). The change in the number of repeats in an allele can be assumed to follow the stepwise mutation model of Ohta and Kimura (1973). The simplest model is the one-step model, which assumes that each mutation increases or decreases the number of repeats in the allele by one unit. The distribution of allele frequencies under this model can be compared with data. One interesting observation from data is that the distribution of allele frequencies versus the allele size (number of repeats) is often multimodal; some examples are shown in Figure 13.3a. One might wonder whether the multimodality arises from population subdivision, migration, or selection. However, Valdes et al. (1993) showed that a multimodal distribution often arises in a random mating population under the one-step model without migration or selection (Figure 13.3b). Actually, it is not difficult to imagine in terms of gene genealogy how a multimodal distribution can arise from stepwise mutation. For example, if the genealogy of the sequences in the sample is as that in Figure 13.4, where the sequences are clustered into two groups of closely related sequences, then it is likely to have a bimodal distribution

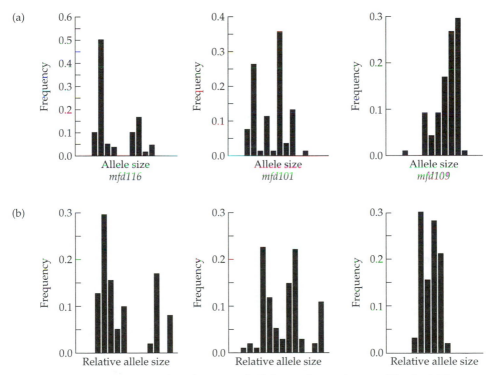

Figure 13.3 (a) Frequency distributions of alleles at three microsatellite loci. These are all dinucleotide repeat loci. (b) Frequency distributions of alleles for three independent replicates in a computer simulation. The one-step model with $4Nu = 10$ and a sample size of 100 were used. The relative allele sizes are all centered on 0. In both (a) and (b), adjacent bars in the histograms indicate frequencies of alleles that differ by one repeat unit. From Valdes et al. (1993).

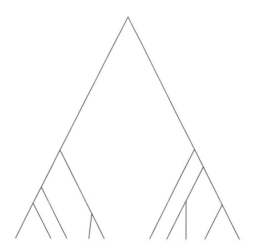

Figure 13.4 A genealogical tree for a sample of DNA sequences.

because the sequences on each side of the root would have similar numbers of repeats but would differ from those on the other side in repeat number. Valdes et al. (1993) also compared the observed variances of the distribution of allele size at 108 microsatellite loci (arrays of dinucleotide repeats) with simulation results, and concluded that the observed distributions are consistent with the results of the one-step model.

Shriver et al. (1993) conducted a more detailed study by investigating four summary measures of allele frequency distributions—number of alleles, number of modes, range in allele size, and heterozygosity—by using computer simulations of the one-step model. The VNTR loci analyzed were classified into microsatellites (1–2-bp repeat unit), STRs (3–5-bp repeat unit) and minisatellites (15–70-bp repeat unit). They chose data from the literature that were obtained from large homogeneous populations to avoid the effects of population substructure and recent decreases in population size. For the 12 STR loci examined there were no significant deviations for any summary measure predictions, and so it appeared that these loci have a mutational mechanism very much like the one-step model. Of the hypothesized mechanisms of VNTR mutation, replication slippage is most like the stepwise mutation model and so is the most likely mechanism of mutation at STR loci. By comparison, 35% of the 31 microsatellite loci examined had significant deviations from the simulation expectations for the number of alleles or their size range and 19% of the loci were borderline significant for deviation from the expected number of modes. Thus, it appeared that the mutational mechanism of microsatellite loci is close to but is not exactly a one-step type mechanism. A plausible explanation for the small, but significant, excess in number of alleles and allelic range for this class of loci is that the mutational mechanism is stepwise with a low frequency of multistep mutations in addition to single-step mutations. Finally, only 3 of the 11 minisatellite loci showed no significant differences between the observed summary measures and their simulation results. Multiple mechanisms may be involved in the mutation of minisatellite loci (see Deka et al. 1991).

The infinite-allele model has also been examined by comparing its predictions with data or with simulation results (e.g., Chakraborty and Daiger 1991; Harding et al. 1993; Shriver et al. 1993). In contrast to the stepwise model, this model assumes that each mutation produces a new allele. The general conclusion is that this model fits data less well than does the stepwise mutation model, particularly for STR and microsatellite data. Harding et al.'s (1993) simulation results based on unequal sister-chromatin exchange also show the superiority of the stepwise model over the infinite-allele model.

Many questions concerning the population genetics of VNTRs remain unsolved. For example, the effects of migration, change in population size, and natural selection on the distribution of allele frequencies have not been well studied. A better understanding of the population genetics of VNTRs will strengthen their applications, e.g., in forensic science and in making biological inferences such as estimation of effective population size and the relationships among populations.

MECHANISMS FOR INCREASING GENOME SIZE

Before attempting to explain the maintenance of the vast amounts of nongenic DNA in the genome of eukaryotes (next section), we consider possible mechanisms for increasing the size of a genome. The putative mechanisms should be able to explain not only the existence of nongenic DNA *per se*, but also its repetitiveness and particular genomic distribution.

We distinguish between two types of genome increase: (1) global increases, in which the entire genome or a substantial part of it, such as a chromosome, is duplicated, and (2) regional increases, in which a particular sequence is multiplied to generate repetitive DNA. The latter mechanisms include transposition, replication slippage, unequal sister-chromatid exchange, interchromosomal unequal crossing-over, and DNA amplification. Transposition produces dispersed repetitive sequences, while the other mechanisms produce localized repetitive sequences.

Genome Duplication

Genome duplication occurs as a consequence of lack of disjunction between all the daughter chromosomes following DNA replication. Since it immediately doubles the size of a genome, it would be the most effective mechanism for increasing genome size, provided that it does not cause deleterious effects. Indeed, if genome duplication were the sole means for genome enlargement, it would require only 10 rounds of genome duplication to explain the thousandfold increase in genome size from *E. coli* to mammals (Nei 1969). Of course, the assumption of no deleterious effects is often not true, and so the question is: How much has genome duplication really contributed to the increase in genome size in the evolution from bacteria to mammals or during the evolution of a group of organisms?

In plants, polyploidy is a widespread phenomenon (Müntzing 1936; Stebbins 1971, 1980; Lewis 1980). For example, in monocotyledons genome sizes exhibit a polymodal distribution with peaks at 0.60, 1.18, 2.16, 4.51, and 8.53×10^9 bp, a distribution suggesting a series of genome duplication (Sparrow and Nauman

1976). Numerous studies have been made to understand the prevalence of poly-ploidy in plants. Artificially produced autopolyploids are generally inferior to their diploid progenitors. The inferiority is expressed in lower fertility and also often in lowered ability to compete with diploids in artificially controlled exper-iments and under more or less natural conditions (Ellerström and Hagberg 1954; Sakai 1955; Stebbins 1971). The inferiority seems to be mainly due to physiolog-ical effects such as genetic imbalances and prolongation of cell-division time and partly due to irregularities of chromosomal segregation (Müntzing 1936; Steb-bins 1971; White 1978). These experimental results and the fact that the perfor-mance of induced polyploids as agricultural crops has consistently fallen short of expectations (Dewey 1980) have led the majority of workers to believe that genome multiplication *per se* is selectively disadvantageous. The widespread success of polyploidy in plants has therefore been attributed to hybridization between individuals from different populations such as ecotypes, races, or species; hybridization increases genetic diversity and may therefore enable the polyploids to compete with their progenitors or to colonize new habitats.

Polyploidy is extremely rare in bisexually reproducing animal groups. Muller (1925) proposed that this is because in bisexual animals the two sexes are differ-entiated by means of a process involving the diploid mechanism of segregation and combination, and polyploidy invariably disturbs this process. [Muller's expla-nation has been well accepted for organisms with the XY/XX (or WZ/ZZ) scheme of chromosomal sex-determining mechanism, but has been questioned for organ-isms with the dominant Y (or W) principle of sex determination (White 1978). However, even in the latter case, Muller's argument that when a tetraploid arises it will usually have to breed with diploids and will thus obliterate itself should also apply. Of course, Muller's principle cannot explain the rarity of tetraploids in animals such as insects and reptiles that are capable of producing parthenogenet-ic polyploids (Bogart 1980; Lokki and Saura 1980); the main reason here is proba-bly because in animals asexual reproduction is generally less successful than sex-ual reproduction.] In amphibians and fish the chromosomal determiners of the opposite sexes are still in a rather initial state of differentiation, and the X and the Y or the Z and the W can substitute for each other (Ohno 1970). In these animals genome duplication would not result in sexual imbalance, and many tetraploid species have been found (Ohno 1970; Bogart 1980; Schultz 1980). Furthermore, Bogart and Wasserman (1972) propose that in amphibians tetraploids can evolve through an intermediate triploid state. Such a situation cannot occur in animals with a well-developed chromosomal sex-determining mechanism (Muller 1925).

Complete and Partial Chromosomal Duplication

The duplication of a chromosome or part of a chromosome is likely to cause a severe imbalance in gene product, and its chance of incorporation into the popu-lation is small. For example, in *Drosophila* the viability of terminal triploids declines with the amount of the triplicated material, becoming generally lethal when more than one-half of an autosomal arm is present in three doses (Lindsley et al. 1972). In humans, trisomies of larger chromosomes are lethal conditions, and even those of smaller ones cause sterility (see, e.g., Hamerton 1971). Well-known examples include such frequent abnormalities as Down's syndrome (tri-somy 21), and trisomy of chromosome 18. Similar deleterious manifestations are associated with duplications of parts of chromosomes. Thus, polysomy and partial polysomy have apparently contributed little to evolution, except in some plants such as *Datura stramonium* (Blakeslee 1930).

Transposition

As described in Chapter 12, transposition can be duplicative or conservative. In the former case, the original copy is retained at the original site while a new copy is inserted elsewhere in the genome. Thus, duplicative transposition increases the copy number of the transposable element. In conservative transposition, the sequence is excised and reinserted elsewhere, and so there is no increase in copy number. However, a circular intermediate may form from the excised sequence and replicate autonomously. In this case, the number of copies of the element can increase. Note also that a transcribed sequence may undergo reverse transcription and reinsertion, i.e., retroposition (Chapter 12). If this occurs in the germline, there will be an increase in the copy number of the sequence. As noted in Chapter 12, the *Alu* sequences and many other processed pseudogenes have been produced by retroposition.

It has been suggested that all the middle-repetitive DNA fraction in eukaryotes has originated in transposable elements. Most of the elements are, however, no longer mobile, because their ability to transpose has been destroyed by mutations or the insertion of other elements. A large part of the heterochromatin in *Drosophila*, for instance, may in fact be a graveyard of such dead elements.

Comparison of dispersed repeats in related species suggests that transposition and retroposition can rapidly increase the copy number of transposable elements. For example, the genome fraction of dispersed repeats is only about 3% in *Drosophila simulans* but about 21% (24 million bp) in *D. melanogaster* (Dowsett and Young 1982). This large difference in the number of dispersed repeats between the two sibling species is likely to have been mainly due to a rapid increase by transposition rather than due to loss of elements, because excision or loss of transposable elements usually does not occur rapidly. It has also been noted that the number of *Alu* sequences has increased rapidly by retroposition during the evolution of higher primates (Chapter 12).

Replication Slippage and Unequal Crossing-Over

As noted in the preceding section, unequal sister-chromatid exchange seems to be the primary mechanism of mutation at minisatellite loci, while replication slippage seems to be the primary mechanism of mutation at microsatellite and STR loci. Interchromosomal unequal crossing-over is commonly believed to be an important mechanism for the creation of multigene families; however, there is evidence that unequal sister-chromatid exchange is also an effective means for recombination between genes in a family (Hu and Worton 1992).

There is evidence that the copy number at a VNTR locus can increase rapidly. For example, in humans an allele at the *MS32* minisatellite locus may contain more than 600 repeats, whereas the locus is monomorphic with 3–4 repeats in great apes, Old World monkeys, and New World monkeys (Gray and Jeffreys 1991). The latter organization presumably represents the relatively stable, ancestral precursor state and the high number in some alleles in humans represents a large increase in recent time.

DNA Amplification

It has been noted that replication slippage and unequal crossing-over tend to remove tandem arrays more often than to increase their size and copy number, and so these mechanisms cannot explain the existence of all the localized repeated DNA (Walsh 1987). In order to explain the existence of localized, highly repetitive sequences such as satellite DNA, DNA amplification has been proposed. **DNA amplification** refers to any event that increases the number of copies of a gene or a DNA sequence far above the level characteristic for an organism. In particular, it refers to events that occur within the lifespan of an organism and cause a sudden increase in the copy number of a DNA sequence.

One of the most powerful methods of amplification is the **rolling-circle model** of DNA replication (Figure 13.5; Hourcade et al. 1973; Bostock 1986). This type of replication is used in the amplification of rRNA genes in amphibian oocytes. Amplification in this case involves the formation of an extrachromosomal circular copy of a DNA sequence, which can then produce many additional extrachromosomal units containing tandem repeats of the original sequence. If

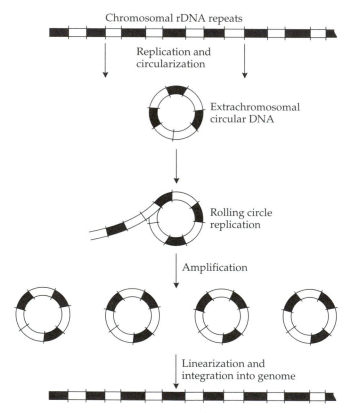

Figure 13.5 The rolling-circle model of gene amplification in amphibian oocytes. The chromosomal rRNA is arranged in a tandem array containing transcribed (black) and nontranscribed (white) parts. Amplification involves the formation of an extrachromosomal circular copy containing a variable number of repeats, which is then amplified by multiple rounds of rolling-circle replication. Modified from Bostock (1986) by Li and Graur (1991).

such units become integrated back into the genome, there will be an addition to the genome consisting of identical repeated sequences.

MAINTENANCE OF NONGENIC DNA

Solving the C-value paradox and accounting for the structure of the eukaryotic genome require finding evolutionary mechanisms for the long-term maintenance of vast quantities of nongenic, seemingly superfluous DNA. This, in turn, is intimately linked with the question of what function this DNA might have, if any. Numerous attempts have been made to provide evolutionary explanations for this phenomenon. The following are four such hypotheses:

1. The nongenic DNA performs essential functions, such as global regulation of gene expression (Zuckerkandl 1976). According to this hypothesis, the excess of DNA is only apparent, and the DNA is wholly functional. Consequently, deletions or removal of such DNA will have a deleterious effect on fitness.
2. The nongenic DNA is useless "junk" (Ohno 1972), carried passively by the chromosome merely because of its physical linkage to functional genes. According to this view, the excess DNA does not affect the fitness of the organism and thus will be carried indefinitely from generation to generation.
3. The nongenic DNA is a functionless "parasite" (Östergren 1945) or "selfish DNA" (Orgel and Crick 1980; Doolittle and Sapienza 1980) that accumulates and is actively maintained by intragenomic selection.
4. DNA has a structural or nucleoskeletal function, i.e., a function related to the determination of nuclear volume but unrelated to the task of carrying genetic information (Cavalier-Smith 1978).

There is very little evidence for the first hypothesis. In fact, most indications are that most nongenic DNA is indeed devoid of function and can be deleted without discernable phenotypic effects (see the review by John and Miklos 1988). It also seems that excess DNA in eukaryotes does not greatly tax the metabolic system and that the cost in energy and nutrients of maintaining and replicating large amounts of nongenic DNA is not excessive. It is thus possible that most nongenic DNA is indeed junk or selfish DNA.

However, there may be some drawbacks in maintaining a large amount of nongenic DNA. First, large genomes have been found to exhibit greater sensitivity to mutagens than small genomes (Heddle and Athanasiou 1975). Second, maintaining and replicating a large amount of DNA may impose a certain burden on the organism. Indeed, the fact that the genome of yeast contains much fewer introns than do those of animals and plants suggests that it is advantageous for yeast to have a streamlined genome. Also, some degree of streamlining in genome size seems to be advantageous for *Drosophila* and other invertebrates, for their intron lengths are, on average, considerably shorter than those in mammals (Hawkins 1988; Fields 1990); for example, more than half of the 209 *Drosophila* introns surveyed in one study were shorter than 80 bp, whereas such short introns are rare in mammals (Mount et al. 1992).

As mentioned in the preceding section, most middle-repetitive sequences are believed to be transposable elements or their degenerated variants. This class of repetitive sequences would have their origin as selfish DNA. Note that the sequence of a transposable element can become so degenerated by mutation that it is no longer recognizable as a transposable element. It is quite possible that a substantial fraction of the genome of higher eukaryotes is due to such sequences, which would indeed be junk in the sense of the word.

All satellite DNA families appear to be transient in existence, for no satellite family has been found to have persisted for a long evolutionary time. For example, no satellite family is shared by all members of the *Drosophila melanogaster* species group (Barnes et al. 1978), suggesting that every satellite that existed in the common ancestor of this *Drosophila* group has become lost in one or more species of the group. The transient nature of satellite DNA argues against the existence of any significant functional role for satellite DNA. This view is also consistent with the observation that most satellite DNAs are found in chromosomal regions with a low rate of recombination (e.g., heterochromatin). This observation suggests that satellite DNAs are dispensable and that a satellite is likely to persist longer if it is in a region with reduced recombination, for in such a region the chance of being deleted by recombination is lower (Charlesworth et al. 1986; Stephan 1986, 1987).

In addition to satellite DNA, families of localized, moderately repetitive DNA are also good candidates for junk DNA. It is possible that most families of localized simple-repetitive sequences are devoid of function and they neither lower nor increase the fitness of their carriers. Consequently, the evolution of such sequences is not affected by natural selection but is determined mainly by gene conversion and unequal crossing-over (Chapter 11). These mechanisms will result in two outcomes: (1) sequence homogeneity and (2) wide fluctuations in numbers over evolutionary times (Charlesworth et al. 1986). It has also been suggested that the rate of turnover of localized arrays of repetitive sequences is quite high; that is, existing arrays may be removed by unequal crossing-over, while new arrays are continuously created by processes of DNA duplication (Walsh 1987).

Of course, not all tandemly repeated simple sequences are junk, for there are some families of such sequences that do have a function. As mentioned in Chapter 1, telomeres and centromeres are such examples. Telomeres are short repetitive sequences at the end of eukaryotic chromosomes and provide protection for the end of a chromosome against exonucleotic degradation (see Zakian 1989). Centromeres, which consist of short repetitive sequences similar to those of telomeres, provide specific sites for attachment of the chromosome to segregation machinery during meiosis and mitosis (Allshire et al. 1988; Meyne 1990).

Another interesting example of functional tandem repeats is the *Responder (Rsp)* locus in natural populations of *Drosophila melanogaster* (Wu et al. 1988). This locus consists of 20–2500 copies of an AT-rich 120-bp-long sequence. In a competition experiment involving a mixed population consisting of flies containing 700 copies of the repeat and flies containing 20 copies, it was observed that the frequency of the flies containing 20 repeats decreased with time (Wu et al. 1989). Therefore, it has been concluded that flies containing 700 copies have a higher fitness than flies containing only 20 copies. At the present time, the function of the *Rsp* locus is not known, but it is apparently not junk DNA since its length affects the fitness of the flies.

Most functional repetitive sequences require some sequence specificities and may be classified as "genic" DNA in the broad sense. The skeletal-DNA hypothesis (Cavalier-Smith 1978, 1985) proposes a function for nongenic DNA that requires no sequence specificity but depends only on the amount of DNA. It postulates that DNA acts directly as a nucleoskeleton that is the primary determinant of interphase nuclear volume and that the changes in nuclear volume occurring in different diploid somatic cells of multicellular organisms are brought about primarily by controlling the degree of folding or unfolding of the DNA. Since larger cells require large nuclei, selection for a particular cell volume will secondarily result in selection for a particular genome size. According to this scheme, excess DNA is maintained by selection, but its nucleotide composition can be changed at random. Whether nongenic DNA has the proposed versatility of controlling nuclear volume for different types of cells remains to be tested by experiments.

No single explanation is likely to solve the C-value paradox. All the above mechanisms, and some additional ones, may contribute to the maintenance of "excess" genomic size, and our task in the future will be to determine the relative contribution of each.

GC CONTENT IN BACTERIA

Among eubacteria, the mean percentage of guanine and cytosine in genomic sequences, or the **(GC) content** varies from approximately 25% to 75%. Bacterial GC contents are phylogenetically correlated, with closely related bacteria having similar GC contents (Figure 13.6).

There are essentially two types of hypotheses to explain the variation in GC content in bacteria. The selectionist view regards the GC content as a form of adaptation to environmental conditions. For example, in thermophilic bacteria, which inhabit very hot environments, strong preferential usage of thermally stable amino acids encoded by GC-rich codons (e.g., alanine and arginine) and strong avoidance of thermally unstable amino acids encoded by GC-poor codons (e.g., serine and lysine) have been reported (e.g., Argos et al. 1979; Kagawa et al. 1984; Kushiro et al. 1987). Therefore, GC content may be a trait that is determined by selection. Another selectionist scenario invokes UV radiation as the selective force. Since T–T dimers are produced by radiation, microorganisms in the upper layers of the soil, which are exposed to sunlight should have a higher GC content than bacteria that are not exposed, for example, intestinal bacteria such as *Escherichia coli* (Singer and Ames 1970).

The mutationalist view invokes biases in the mutation patterns to explain the variation in GC contents (Sueoka 1964; Muto and Osawa 1987). According to this view, the GC content of a given bacterial species is determined by the balance between (1) the rate of substitution from G or C to T or A, denoted as u, and (2) the rate of substitution from A or T to G or C, denoted as v. At equilibrium, the GC content is expected to be

$$P_{GC} = v / (v + u) \tag{13.1}$$

The ratio u/v is also called the **GC mutational pressure**. When u/v is 3.0, the GC content at equilibrium will be 25%. Such is the situation in *Mycoplasma capricolum*. When the ratio is 1, the GC content will be 50%, as in *Escherichia coli*. When it is 0.33, the GC content will be 75%, as in *Micrococcus luteus*.

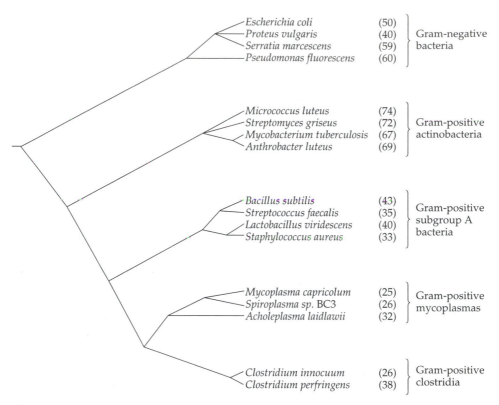

Figure 13.6 Phylogenetic tree of several eubacteria based on 5S rRNA sequences. Genomic GC contents are shown in parentheses. From Muto et al. (1986).

In addition to the GC mutational pressure, mutational changes are subject to selective constraint. The weaker the constraint is in a particular region, the stronger the effect the GC mutational pressure will be on the GC composition. Figure 13.7 shows the correlation between the total GC content and the GC content at the three codon positions for 11 bacterial species covering a broad range of GC-content values. The correlation at the third position resembles the expectation for the case of no selection. On the other hand, the correlations at the first and second positions, while positive, show a more moderate slope. This result is easily explained by the fact that selection at the mostly degenerate third position of codons is expected to be much less stringent than at the first and second positions (Chapter 7), so that the GC level at the third position is largely determined by mutation pressure. Whether mutation or selection is more important or both are important remains to be determined by further studies.

COMPOSITIONAL ORGANIZATION OF THE VERTEBRATE GENOME

Figure 13.8 shows the GC content in different groups of organisms. While the genome sizes of multicellular eukaryotes are generally larger than those of prokaryotes, their GC content exhibits a much smaller variation. In particular, vertebrate genomes show quite a uniform GC content, ranging from about 35% to

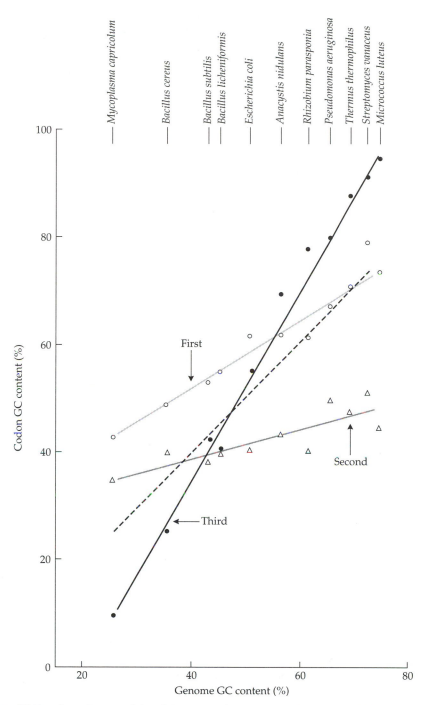

Figure 13.7 Correlation of the GC content between total genomic DNA and the first (open circles), second (open triangles), and third (solid circles) codon positions. The dashed line represents the theoretical expectation of a perfect correspondence between the GC content in the genome and that in the codons. From Muto and Osawa (1987).

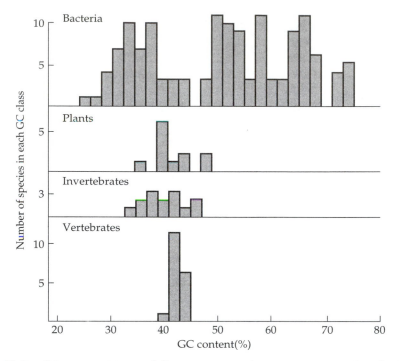

Figure 13.8 GC composition in different groups of organisms. From Sueoka (1964).

45% (Sueoka 1964). Part of the reason for the small range in GC content in vertebrates might be because vertebrates, unlike bacteria, have not diverged long enough from one another to allow for considerable differences in GC content to accumulate.

The uniformity of GC content notwithstanding, vertebrate genomes have a much more complex compositional organization than prokaryotic genomes. When vertebrate genomic DNA is randomly sheared into fragments 30–100 kb in size and the fragments are separated by their base composition, the fragments cluster into a small number of classes distinguished from each other by their GC content (GC-rich fragments being heavier than AT-rich fragments). Each class is characterized by bands of similar, although not identical, base compositions (Bernardi et al. 1985; Bernardi 1995). Such compositionally homogeneous stretches of DNA have been termed *isochores*, which are 100–300 kb or even larger in size (Bernardi et al. 1985; Bernardi 1995).

There are conspicuous differences in compositional organization between the genomes of warm- and cold-blooded vertebrates (Bernardi et al. 1985, 1988). Figure 13.9a shows the relative amounts and buoyant densities of the major DNA components from the carp *Cyprinus carpio* and the amphibian *Xenopus laevis* (left panel) and from three warm-blooded vertebrates: chicken, mouse, and human (right panel). In the latter genomes, there are two light components (L_1 and L_2), and two or three heavy components (H_1, H_2, and H_3). In the human genome, the GC-poor L_1 and L_2 isochore families make up about 62% of the genome, the GC-rich H_1 and H_2 isochore families represent about 22% and 9%, and the GC-richest

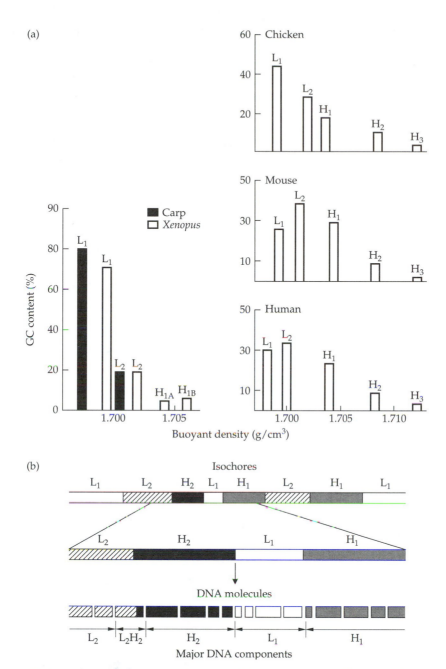

Figure 13.9　(a) Histograms showing the relative amounts and buoyant densities of the major DNA components from *Cyprinus carpio* and *Xenopus laevis* (left panel) and from chickens, mice, and humans (right panel). (b) Scheme depicting the mosaic organization of nuclear DNA from warm-blooded vertebrates. When the isochores undergo breakage during DNA preparation, four major families of molecules with different GC contents are generated. From Bernardi et al. (1985).

isochore family H_3 represents about 3–4%; the remaining part of the genome is formed by satellite DNAs and other minor components such as ribosomal DNA (see Bernardi 1995). In contrast, genomic DNA from most cold blooded vertebrates comprises mostly light components (Figure 13.9a). For example, in *Xenopus*, DNA fragments with a density higher than 1.704 g/cm³, which is the density for H_1, represent less than 10% of the genome, as compared with 30–40% for warm-blooded vertebrates.

Figure 13.9b shows the *mosaic* organization of nuclear DNA from warm-blooded vertebrates (i.e., the alternation of light and heavy isochores), illustrating how the four major GC-content classes are generated when DNA is sheared.

The finding that the genome of warm-blooded vertebrates is mosaic is consistent with chromosome-banding studies. Metaphase chromosomes of warm-blooded vertebrates show distinct **Giemsa dark bands (G bands)** and **light bands (R bands)** when treated with fluorescent dyes, proteolytic digestion, or differential denaturing conditions. A subset of the R bands, the **T bands**, are the most resistant to heat. In contrast, metaphase chromosomes of cold-blooded vertebrates show either little banding or no banding at all. According to Saccone et al. (1993), T bands are formed mainly by the H_2 and H_3 families, the R′ bands (i.e., R bands exclusive of T bands) are formed, with almost equal proportions, by the H_1 family (with a minor contribution of the H_2 and H_3 families) and by the L_1 and L_2 families; and G bands essentially consist of the L_1 and L_2 families, with a minor contribution from the H_1 family. The relative amounts of different isochore families are schematically shown in Figure 13.10. Studies on the replication timing of genes show that genes localized in GC-rich isochores (T bands) replicate early in the cell cycle, whereas genes localized in GC-poor isochores (G bands) replicate late (Goldman et al. 1984; Bernardi et al. 1985; Bernardi 1989).

The first few genes localized in compositional DNA fractions suggested that genes are not distributed randomly in the human genome but are concentrated in GC-rich isochores (Bernardi et al. 1985). Subsequent investigations indeed indi-

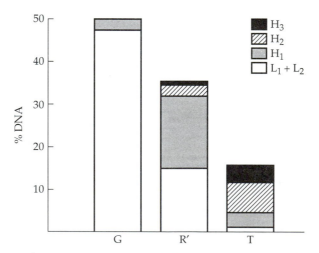

Figure 13.10   A schematic presentation of the relative amounts of isohcore families $L_1 + L_2$, H_1, H_2, and H_3 in the chromosomal bands of the human karyotype. R′ bands are R bands exclusive of T bands. From Saccone et al. (1993).

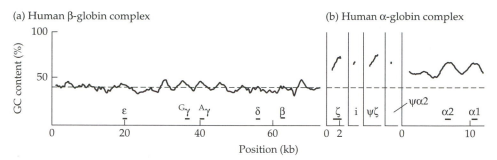

Figure 13.11   Distribution of GC content along human globin DNA sequences. (a) The β-globin gene cluster; (b) the α-globin gene cluster (incomplete). The genes are arranged in the same order as in Figure 10.7. The gene names are shown at the bottom of the figure; region i is the intergenic region between ζ and ψζ. In the β-globin cluster and the region covering the α_1- and α_2-globin genes each point represents the average of the GC composition of the 2001 nucleotides surrounding the point, while in the other regions each point represents the average of 1401 nucleotides. The horizontal broken line represents the overall GC content of the human genome (40%). From Ikemura and Aota (1988).

cated that gene concentration is low in GC-poor isochores, increases with increasing GC in isochores families H_1 and H_2, and reaches a maximum in isochore family H_3, which exhibits up to a 20-fold higher gene concentration compared to GC-poor isochores (see Bernardi 1995).

Analyses of DNA sequence data have revealed a positive correlation between the GC levels of genes and the GC level in the large DNA regions in which the genes are embedded (Bernardi et al. 1985; Ikemura 1985; Aota and Ikemura 1986; Bernardi 1995). Figure 13.11 contrasts the α- and β-globin clusters in humans. The β and β-like globin genes are low in GC content and are embedded in a low-GC region. The α and α-like globin genes, on the other hand, are GC rich and are embedded in a GC-rich region. The same situation is found in rabbits, goats, and mice. In chickens, both β- and α-globin genes are GC rich, and both are embedded in GC-rich regions. In contrast, the α- and β-globin genes in *Xenopus* are GC-poor, and both are embedded in a GC-poor region.

In the vast majority of cases, the GC content in coding regions tends to be higher than that in the flanking regions (Figure 13.12). We also see that the GC level at the third-codon position is on average higher than that in introns, which in turn is higher than that in the 5' and 3' flanking regions. The GC level in the 5' flanking region tends to be higher than that in the 3' flanking region, probably because the promoter and its surrounding regions tend to be GC rich.

ORIGINS OF ISOCHORES

The origin of GC-rich isochores is mysterious and controversial. Note that what is at issue is the general tendency of long DNA segments (100 kb or longer) to be homogeneous in GC content, not the localized variation in GC content, such as that observed among the various regions of a gene. Bernardi et al. (1985, 1988)

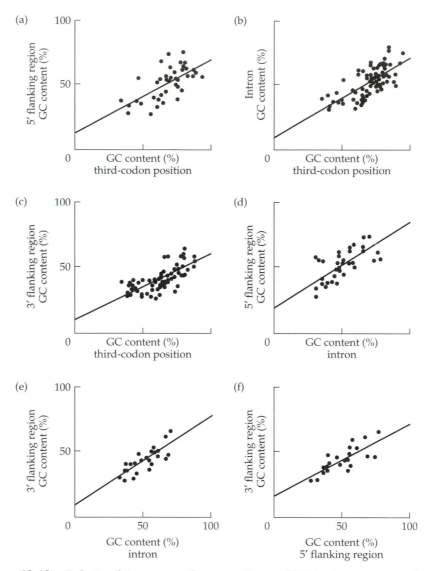

Figure 13.12 Relationships among the percentages of GC in the various regions of the gene. (a) The third-codon position and the 5′ flanking region; (b) The third-codon position and introns; (c) The third-codon position and the 3′ flanking region; (d) Introns and the 5′ flanking region; (e) Intron and the 3′ flanking region; and (f) 5′ and 3′ flanking regions. From Aota and Ikemura (1986).

proposed that isochores arose because of a functional (i.e., selective) advantage. Their main argument is that in warm-blooded organisms an increase in GC content can protect DNA, RNA, and proteins from degradation by heat (see below), because G–C bonds are stronger than A–T bonds (Chapter 1). We call this view the selectionist hypothesis.

The opposing view is that GC-rich isochores arose because of mutational pressure (Filipski 1987; Sueoka 1988, 1992; Wolfe et al. 1989a; Holmquist and Filipski

1994). In particular, the hypothesis proposed by Wolfe et al. (1989a) stipulates that isochores arose from mutational biases due to compositional changes in the precursor nucleotide pool during the replication of germline DNA. The GC-rich isochores are carried on DNA regions that replicate early in the germline cell cycle during which the precursor pool has a high GC content and, thus, a propensity to mutate to G or C. The AT-rich isochores, on the other hand, are replicated late in the cell cycle, when the precursor pool has a high AT content and a propensity to mutate to AT. We call this the mutationalist hypothesis. This hypothesis is based on the observations that the composition of the nucleotide precursor pool changes during the cell cycle and that such changes can in fact lead to altered base ratios in the newly synthesized DNA (Leeds et al. 1985). Note that the replication of the mammalian genome is quite a lengthy process, taking eight hours or more (Holmquist 1987). Gu and Li (1994) recently showed that, under suitable conditions, mutation pressure can indeed lead to isochore formation.

An argument for the selectionist hypothesis is that increases in GC content at the first- and second-codon positions may confer thermal stability to proteins, while increases in GC content in introns, third-codon positions, and untranslated regions can both increase the thermal stability of primary mRNA transcripts and stabilize chromosomal structures, possibly through effects on DNA–protein interactions. Indeed, in thermophilic bacteria, strong preferential usage of GC-rich codons has been reported (see earlier discussion). However, the body temperature of warm-blooded vertebrates is much lower than the temperatures experienced by thermophilic bacteria, so temperature may not be a very important factor in the evolution of proteins and DNA sequences in vertebrates.

One difficulty with the selectionist hypothesis is the fact that a substantial proportion of the mammalian and avian genes are low in GC content. This hypothesis also cannot explain why some duplicate genes have opposite GC contents. For example, in mammals, the β-globin cluster is low in GC, while the α-globin cluster is GC-rich (Figure 13.11), although the two types of globin genes are expressed in the same cells and serve the same function. Another example is the insulin-like growth factors I and II (IGF-I and IGF-II) genes; in humans IGF-II is closely liked to the insulin gene and located on chromosome 11 whereas IGF-I is located on chromosome 12. In cold-blooded vertebrates, insulin and the IGFs are similar in GC content. In contrast, insulin and IGF-II demonstrate dramatic increases in GC content in mammals, though no such trend occurs in IGF-I (Ellsworth et al. 1994); for example, the GC content in introns is only 40–43% in human IGF-I but higher than 60% in the IGF-II and insulin genes (Table 13.9). Similarly, some immunoglobulin genes are located in GC-rich regions while others are located in AT-rich regions (see Aota and Ikemura 1986). Bernardi et al.'s (1985, 1988) explanation is that isochores represent the unit of selection and the α-globin cluster in mammals has been translocated to a high-GC isochore, whereas the β cluster remained in a low-GC isochore. According to this argument, the α and β clusters in chickens should have both been translocated to GC-rich isochores because both of them have a high-GC content. This argument, however, raises the issue of the functional advantage of a GC-rich isochore. If the advantage does not come from the genes it contains, where does it come from?

The mutationalist hypothesis can explain the large difference in GC content between the α- and β-globin clusters in the mammalian genome by assuming that

TABLE 13.9 Percentage of GC in the coding regions, introns, and flanking regions of the insulin and insulin-like growth factor (IGF) genes in humans

Gene	Coding region	Introns			Flanking regions	
		B	C	D	5'	3'
IGF-I	54.8(456)[a]	40.1(781)	40.0(2,365)	42.5(1,505)	–	34.5(6,027)
IGF-II	65.4(537)	64.2(2,635)	61.5(1,702)	63.8(293)	66.0(1,393)	62.6(1,630)
Insulin	65.1(327)	68.7(179)	64.4(786)	–	67.1(2,185)	66.0(1,393)

From Ellsworth et al. (1994).
[a]Number in parentheses denotes the sequence length (bp).

they are located in an early- and a late-replicating region (replicon), respectively. However, this hypothesis also has difficulties (Bernardi et al. 1988). First, it is difficult to explain why GC-rich isochores rarely occur in cold-blooded vertebrates. Second, it is not clear why in both the avian and mammalian lineages the increase in GC content occurred preferentially in regions with higher gene concentrations (see Bernardi 1995). Of course, it is possible that the majority of such regions happened to be on early replicons. Third, constitutive heterochromatin, such as satellite DNAs, which are mostly GC-rich, and facultative heterochromatin, such as the inactive X chromosome, replicate at the end of the cell cycle, and in these cases, no connection between changes in nucleotide pools and DNA composition is observed.

In conclusion, the presently available data seem to be insufficient to distinguish between the two hypotheses. It is also possible that both mutation pressure and natural selection have played roles in shaping the compositional organization of the genome of warm-blooded vertebrates. One line of future research for resolving the controversy is to examine the degree of discontinuity in GC content at isochore borders. The isochore border between an L (GC-poor) and an H_2 (GC-rich) isochore in the MHC region was recently sequenced and found to have a sharp compositional discontinuity (Fukagawa et al. 1995). Limited available data (see Bernardi 1995) suggest that this represents the general case. Such discontinuities can be explained by the mutationist hypothesis by assuming that a discontinuity occurs because the isochores on the two sides of the border are located on replicons that are replicated at very different times during germline DNA replication. Incidentally, it is interesting that the boundary sequence in the MHC region is highly similar to the sequence at the pseudoautosomal boundary on the human X and Y chromosomes (Fukagawa et al. 1995).

ORIGINS OF INTRONS

The discovery of intervening sequences (introns) in eukaryotic genes was a great surprise and soon raised questions about their origin and significance in evolution. Gilbert (1978) suggested that this gene organization could speed up evolution by providing mechanisms for the generation of novel proteins from old ones (i.e., exon shuffling), Doolittle (1978) and Darnell (1978) speculated that the "genes-in-pieces" structure is a primitive form that was present in the genome of the progenote (see definition below), and Blake (1978) suggested that "genes-in-pieces" might imply "proteins-in-pieces," that is, exons originally corresponded to structural units of proteins. This view has been known as the introns-early hypothesis. Later, Gilbert (1987) modified it to the exon theory of genes, proposing that exons are the descendants of ancient minigenes and introns are the descendants of spacers between them. The opposing view, i.e., the introns-late view, assumes that early genes had no introns and that the addition of introns and splicing mechanism occurred after the emergence of the eukaryotic cell or the emergence of mitochondria (Cavalier-Smith 1978, 1985, 1991; Rogers 1985, 1989; Palmer and Logsdon 1991). The debate between the two schools has been lively and fascinating—it has taken several turns in the past two decades, with the weight of evidence favoring sometimes one view and sometimes the other.

Before discussing the debate, we define **cenancestor** and **progenote**. *Cenancestor* refers to "the most recent common ancestor to all organisms that are alive today" (Fitch and Upper 1987). In the view that the eukaryotes are derived from archaebacteria, the cenancestor refers to the common ancestor of the eubacteria and archaebacteria (Figure 13.13). *Progenote* is Woese and Fox's (1977b) descriptor for early biological entities that had only "a rudimentary, imprecise linkage between its [their] genotype and phenotype" and were much simpler than contemporary prokaryotic cells. Originally, they thought that the eukaryotic lineage (cytoplasm) was very ancient and that it was at the progenote stage, not the prokaryote stage, that the line of descent leading to the eukaryotic cytoplasm diverged from the bacterial lines of descent. In this sense, the progenote would be the cenancestor. Woese and Fox's view is not supported by current data, and the term progenote is now commonly used to refer to the most primitive organism (probably with an RNA genome) from which all living cells have descended; it is probably considerably more ancient than the cenancestor (Figure 13.13). See Doolittle and Brown (1994) for more discussion.

The debate is on the origin of the classical nuclear **spliceosomal introns,** often called **nuclear introns**. They are different from group I and group II introns, which have self-splicing activities (see Lewin 1994). In the introns-late view, spliceosomal introns were derived either from transposable elements (Cavalier-Smith 1978, 1985; Orgel and Crick 1980) or from group II introns (e.g., Rogers 1989; Cavalier-Smith 1991; Nilsen 1994; Michel and Ferat 1995). In the latter proposal, some group II introns from mitochondria or chloroplasts invaded the nuclear genes of the eukaryotic cell by retroposition and evolved into the nuclear introns we see today. This may be called the mitochondrial (or organelle) origin of introns (Figure 13.13b). On the other hand, the introns-early view states that spliceosomal introns are descendants of self-splicing introns (ribozymes) that were present in the progenote. After the divergence of the eubacterial, archeabacterial, and eukaryotic lineages, most or all of these ancient introns were lost in the Eubacteria and Archaebacteria, whereas in the eukaryotes these introns evolved to the spliceosomal introns (see the review by Long et al. 1995a). Note that the exon theory of genes and the introns-early hypothesis have commonly been mistaken as equivalent, but actually they are not, because introns might be present in the progenote but they were not derived from spacers as Gilbert (1987) proposed.

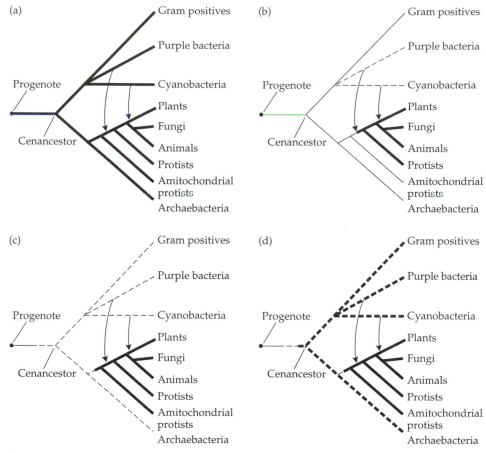

Figure 13.13 Schematic representations of four possible senarios for the evolution of introns. Amitochondrial protists are protists without mitochondria. The figures are not drawn to the evolutionary time scale. The first arrow indicates the origin of the mitochondria and the second arrow indicates the origin of the chloroplasts. Dashed lines indicate the presence of group II introns and heavy lines the presence of spliceosomal introns. (a) The progenote origin of introns (the introns-early hypothesis); self-splicing introns were present in the progenote and some of them evolved into spliceosomal introns early in evolution. (b) The mitochondrial (organelle) origin of introns; group II introns from the mitochondrial endosymbiont spread to nuclear DNA, where they evolved into spliceosomal introns. (c) The nuclear origin; group II introns are inherited by the nucleus and degenerate into spliceosomal introns. (d) This prokaryotic origin model has the degeneration of group II introns in a prokaryote before the divergence of eukaryotes. It implies the coexistence of group II and spliceosomal introns in prokaryotes (heavy dashed lines). For more details of the last two scenarios, see Roger et al. (1994).

Note also that in addition to these two scenarios, which are depicted in Figure 13.13a and b, there are other possible alternatives. One of them is that group II introns were present in all or most prokaryotes before the emergence of eukaryotes and were inherited by the nucleus, and degenerate into spliceosomal introns in eukaryotes (the nuclear origin hypothesis; Figure 13.13c) (Roger et al. 1994).

Another scenario is that some group II introns evolved into spliceosomal introns in a prokaryote before the divergence of the eukaryotes (the prokaryote origin hypothesis; Figure 13.13d).

The presence or absence of introns in the progenote or early primitive organisms has profound implications for the evolution of early genes. In the introns-early view, the primordial introns played an important role in the early assembly of proteins and accelerated the early evolution of proteins by exon-shuffling. In contrast, in the introns-late view, there was no spliceosomal intron in the genome of the progenote and so exon shuffling played no role in the assembly and evolution of early genes; self-splicing introns might have emerged early in evolution, but such introns have structural requirements that make them less suitable for exon-shuffling than are spliceosomal introns (see Patthy 1991b).

Relationships Between Exons and Units of Protein Structure

One line of evidence that has been used to support the introns-early view is the correlation between modules and exons (Gō 1981; Gilbert et al. 1986). For example, Gō (1981) identified four modules, F1–F4, in the α and β globins, showed that F1, F2+F3, and F4 correspond to exons 1, 2, and 3 in vertebrate α- and β-globin genes, and predicted the existence of an intron at a position between F2 and F3 in an ancient globin gene. The finding of a third intron exactly at the predicted region in the gene of leghemoglobin from soybean (Jensen et al. 1981) was taken as strong evidence for the exon theory of genes. A similar prediction of the existence of an intron at a particular region in the triose phosphate isomerase (TPI) gene was made (Gilbert et al. 1986) and found to be true (Tittiger et al. 1993). However, as more sequences of a gene from diverse organisms become available, the number of introns increases, and the correlation between exons and protein structure units tends to decrease or even disappear. For instance, 12 rather than 3 introns have later been found at different locations of globin-related genes, and when these 12 intron positions were analyzed, no correlation between exons and units of protein structure was found (Stoltzfus et al. 1994). The same conclusion was reached for alcohol dehydrogenase, pyruvate kinase, and TPI (Stoltzfus et al. 1994; Logsdon et al. 1995). In response, Long et al. (1995a) argued that there are two problems with this type of analysis. The first, and major, one is the definition of ancestral introns, because today's intron/exon structures are the final products of intron loss, intron sliding, and possibly intron gain. The second problem is that sequence alignment may be problematic—in comparing divergent genes, there can be regions that lack enough sequence similarity to make a reliable alignment. Thus, what appear to be different intron positions in such regions may only reflect alignment artifacts. Obviously, the two schools hold very different views on this issue.

The introns-early/introns-late controversy is unlikely to be resolved by studying the relationships between exons and structural units of proteins for several reasons. First, it involves the issue of intron gain and loss. Obviously, a change in intron number either by loss or gain can reduce the correlation between exons and structural units. Intron gain is a controversial but important issue and will be discussed again later. Second, it involves the issue of **intron-sliding,** a shift in the position of an intron in the gene, a situation that may arise from mutations that create a new splice junction and block the old one (see also Cerff 1995). If sliding had occurred, the resultant intron might be mistaken for a new intron and the correlation between exons and structural units would be reduced; however, Stoltzfus et al.'s (1994, 1995) test may still be valid, if the sliding is small. Intron-sliding has been invoked to infer homology of introns located at slightly different positions in different organisms (e.g., Gilbert et al. 1986), but this approach has been challenged by Logsdon et al. (1995). A good candidate for intron-sliding has been proposed for two duplicated genes in *Volvox* (Muller and Schmitt 1988). Third, an ancient structural unit in a gene may degenerate if it becomes no longer useful to the gene. In this case, the correlation between exons and structural units is reduced. Fourth, it is difficult to know whether a structural unit such as an α helix or a β sheet in a gene is ancient or derived (see Long et al. 1995a). Finally, although absence of a correlation between exons and protein structural units may be taken as evidence against the exon theory of genes, it can not refute the introns-early view, because, as noted above, introns might be present in the progenote but had nothing or little to do with protein structure. Note also that this type of study cannot distinguish among the three scenarios depicted in Figure 13.13b, c, and d.

Phylogenetic Distribution of Introns

Another line of evidence used to support the introns-early view is the correspondence of intron positions between plant and animal genes (e.g., Shah et al. 1983; Marchionni and Gilbert 1986). For example, four TPI introns were shared by the plants and animals studied (Gilbert et al. 1986). However, when more diverse organisms including protists were studied, the phylogenetic distribution of introns became difficult to interpret. For example, only a few of the dozens of introns in genes for actin and α and β tubulins were found to be widely shared by plants, animals, fungi, and late-arising protists, and it was argued that such distributions are difficult to explain under the introns-early view, whereas each of them could be explained by a single recent gain without subsequent loss (Dibb and Newman 1989). Later, several genes, e.g., glyceraldehyde-3-phosphate dehydrogenase (*GAPDH*, Michels et al. 1991) and RNA polymerase II large subunit (Nawrath et al. 1990), were found to lack introns in some, if not all protists, but contain many introns in animals, plants, and fungi. A summary of extensive data is shown in Figure 13.14, which reveals a complete absence of nuclear introns from all examined genes of the earliest protistan lineages. Palmer and Logsdon (1991) took this as strong evidence that nuclear introns are relatively recently derived features of eukaryotes. Figure 13.14 also shows that introns have increased markedly in number in plant, animal, and fungal lineages relative to those (late) protistan lineages that do contain introns. To explain the observed intron distribution, the introns-early view would have to postulate the parallel loss of tens of thousands of introns from many different protistan lineages, including complete intron extinction from the several earliest lineages.

Although the observed phylogenetic distribution of introns poses a serious difficulty for the introns-early hypothesis, it cannot refute the hypothesis for two reasons. First, if intron loss is selectively advantageous in prokaryotes and protists because of selection for an increased rate of genome replication (i.e., selection for genome streamlining), then most introns may indeed become lost in these organisms, given their long history of evolution. There is strong evidence that

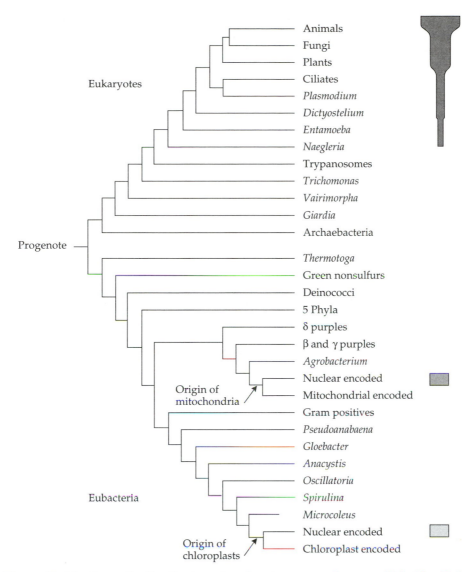

Figure 13.14 Sporadic distribution of nuclear introns on the tree of life. The clado-gram is based primarily on rRNA gene sequences of eukaryotes and eubacteria; the exact relationships among some lineages are not certain but this does not affect the general conclusion to be drawn in the text. The width of the shaded boxes in the right column indicates the approximate number of nuclear-type pre-mRNA introns per kilobase (kb) of coding sequence, with animals containing on average 6 introns per kb of coding sequence and *Dictyostelium* 1 intron per kb. The animal estimate is for vertebrates only, and the fungal estimate excludes yeast. *Physarum* contains introns, but has been excluded from the tree because its phylogenetic position is unclear. The absence of a shaded box means that no spliceosomal-dependent introns have been found in that organism. In some cases the data for protists are based on relatively few genes. Modified from Palmer and Logsdon (1991).

intron loss can indeed occur; for example, the apolipoprotein A-IV gene has lost the first of its three introns (see Li et al. 1988) and the mouse insulin I gene has lost the second of its two introns (Chapter 12; Lomedico 1979). However, just how easily an intron can become lost is a subject of dispute (see, e.g., Cavalier-Smith 1991). Second, intron gain may be selectively disadvantageous in lower organisms (for the same reason as above), but not so, or less so, in higher organisms. Note also that intron gain can spread more easily in sexual organisms than in asexual organisms, for one of two reasons (Hurst 1994). If introns are junk or self-ish DNA, they may spread in a population even if deleterious, but only if out-breeding events are regular enough (Hickey and Benkel 1986). Alternatively, the existence of introns may reduce the chance of mispairing and thus the rate of recombination between paralogous sequences located at nonhomologous chromosomal sites. Of course, it is difficult to know whether the above factors together can produce the phylogenetic distribution in Figure 13.14. There is, however, another possible factor that may increase the number of introns in higher organisms (see later).

Another issue concerns the interpretation of the existence of introns at the same positions in nuclear genes and homologous mitochondria- or chloroplast-derived genes now present in the nuclear genome (see Figure 13.14). Some such examples are: (1) five introns at identical positions between nuclear and chloroplast *GAPDH* (Shih et al. 1988; Kersanach et al. 1994), (2) two identical introns between cytosolic and mitochondrial malate dehydrogenase (Obaru et al. 1988), and (3) five identical introns between cytosolic and mitochondrial aspartate aminotransferase (Setoyama et al. 1988; Iwabe et al. 1990). In the introns-early view (e.g., Kersanach et al. 1994; Long et al. 1995a), these examples represent ancient introns that existed before the divergence between eukaryotes and prokaryotes and are not due to coincidences resulting from nonrandom insertions into so-called protosplice sites, as suggested by the introns-late school (Dibb and Newman 1989; Logsdon and Palmer 1994; Stoltzfus 1994). However, Palmer and Logsdon (1991) have challenged the claim that all of these examples represent introns at identical positions and have argued that because chloroplasts (with a similar argument for mitochondria) emerged so late in evolution (~1 billion years ago), the introns-early hypothesis would force one to postulate retention of these introns for two-thirds of eubacterial evolution (before the derivation of chloroplasts from cyanobacteria), as well as their parallel and independent loss in many separate eubacterial lineages. They therefore favored the interpretation of parallel insertion of introns at the same or similar positions.

Long et al. (1995b) have recently studied the distribution of intron phases in ancient conserved regions (ACRs), which are complete genes or portions of genes that are shared by both eukaryotes and prokaryotes. These regions have no introns in the prokaryotes but may have introns in the eukaryotic counterparts. On such introns-late models as those in Figure 13.13b and c, these introns must have been inserted. For such models, they could not participate in exon shuffling because the ACR regions are colinear with the ancestral molecules, which on the introns-late hypothesis came into existence before there were introns. Long et al. examined these ACR introns for phase bias and for phase correlation in the same genes. (As defined in chapter 10, intron phase refers to the position of the intron within a codon.) They found that about 55% of the introns studied were of phase zero and that intron phases in the same genes are correlated; in contrast, under the assumption of random insertion, the expected frequency of each intron phase should be 1/3 and there should be no correlation between intron phases in a gene. They found an excess of symmetric exons (exons flanked by the same phase introns) or symmetric sets of exons, significant at the 1% level. They interpreted this excess as evidence of exon shuffling (duplication), arguing that the shuffling would have had to have occurred in the progenote or early organisms, and hence these introns would have existed in the progenote. They therefore took their finding as evidence for the introns-early hypothesis. Note, however, that the putative shuffling actually only had to have occurred before the cenancestor, not necessarily in the progenote. Therefore, Long et al.'s observation is also compatible with the scenario depicted in Figure 13.13d, i.e., the prokaryotic origin of introns.

Mechanisms of Intron Insertion

In the current version of the introns-late hypothesis, nuclear introns are originally derived from group II introns (e.g., Hickey et al. 1989; Cavalier-Smith 1991; Palmer and Logsdon 1991). Recent demonstrations of reverse self-splicing of group II introns *in vitro* (Mörl and Schmelzer 1990; see also Nilsen 1994) suggest a ready mechanism for the insertion of group II introns into nuclear genomes (see Palmer and Logsdon 1991). An intron that was spliced out is self-spliced into an RNA. That intron containing RNA molecule is then later copied by reverse transcriptase into cDNA, and the intron is inserted by recombination into the nuclear gene; this process is similar to the creation of processed pseudogenes or retrosequences (Chapter 12). The intron thus created would be a "perfect intron" because it could be precisely excised from the pre-mRNA by self-splicing. Palmer and Logsdon (1991) suggest that nuclear genes are continually invaded by group II introns by this reverse self-splicing mechanism. Rapid degeneration of a group II intron into a spliceosome-dependent form could then occur, perhaps stemming from a single nucleotide substitution that would establish a nuclear-intron consensus splice site (Rogers 1989; Cavalier- Smith 1991). One question is, Where do these group II introns come from? Noting that many chloroplast and mitochondrial genomes do contain group II introns and that organelle sequences are frequently transferred into nuclear genomes (Timmis and Scott 1984), Palmer and Logsdon (1991) suggest that the organelles have been seeding the nucleus with group II introns. The failure to find functional group II introns in nuclear genomes is assumed to be due to their relatively rapid degeneration into standard nuclear introns.

However, with respect to more recent intron gains it is not clear that organelles have been seeding the nucleus with enough introns, especially in view of the fact that group II introns appear to be absent from animal mitochondrial genomes (see Gray 1989) and that animals do not have chloroplasts. A more plausible process of intron gain may be the reverse splicing of existing introns (by a process similar to that discussed above), as has recently been proposed for the addition of spliceosomal introns to the *U6* small nuclear RNA genes in yeasts, one in *S. pombe*, and five in different species of *Rhodotorula* (Tani and Ohshima 1989, 1991; Takahashi et al. 1993). An intron thus created is already a perfect spliceosomal intron: it requires no further evolution, and there is no need to find an intermediate form as in the scheme proposed by Palmer and Logsdon (1991). Long et al. (1995a) suggest that this is a general mechanism for the movement or insertion of spliceosomal introns into RNA. They also note from the yeast examples of insertion that there is no evidence of protosplice sites as proposed by Dibb and Newman (1989). Another possible source of recent intron gain is from transposable elements—there is evidence that some transposable elements in maize can be spliced out like an intron (Wessler 1988; Purugganan 1993; Giroux et al. 1994).

Note that the preceding process may be able to explain why higher organisms have numerous introns, because the more introns an organism has, the higher the chance for this process to occur. If addition of introns creates no or little selective disadvantage to their carriers, as is commonly thought for higher organisms, the added introns can be retained and spread to the species. Of course, although this may explain why there are far more introns in higher organisms than in lower ones, it does not explain the origin of introns.

In conclusion, although the introns-early view was once the prevailing view and has even been cited as an established theory in textbooks on molecular and cell biology, it is facing serious challenges. In fact, current evidence seems to be more in favor of the introns-late hypothesis rather than the introns-early hypothesis. However, the controversy is still unsettled and may continue for a long time.

CHAPTER 14

Roles of Mutation and Selection in Molecular Evolution

*I*N THIS FINAL CHAPTER WE COME BACK TO THE LONG-STANDING CONTROVERSY on the relative roles of mutation and selection in molecular evolution. Mutation here is taken in the broad sense, including not just point mutation but also genetic changes caused by other mechanisms such as insertion, deletion, duplication, and transposition. Also, it should be emphasized that we are concerned with molecular evolution, not the evolution of form, function, or fitness, where presumably selection reigns. The presentation will no doubt be biased by my own view of the issue, but I shall provide data to support my argument. For a different view of the issue, readers may refer to Gillespie (1991).

As described in the Introduction and Chapter 2, the essence of the neutralist-selectionist controversy is whether mutation or selection is the major driving force of molecular evolution. One approach to this question is to see whether the upper limit of the rate of evolution is determined by mutation or selection. Considerable insight might also be gained by considering the relative roles of mutation and selection in the emergence of new functions, in the evolution of genetic code, and in the switch in the most preferred synonymous codons between species. It will also be interesting to see whether directional mutation pressure changes the amino acid composition of proteins.

At the end of the chapter we shall also examine whether neo-Darwinism is adequate for explaining adaptive evolution at the molecular level. That is, we shall examine whether the rate of molecular evolution is determined by the rate of environmental change as postulated by neo-Darwinism, even if we consider only the class of advantageous evolutionary changes.

IS SEQUENCE EVOLUTION MUTATION OR SELECTION DRIVEN?

The selectionist view of evolution, which is essentially the same as the neo-Darwinism, contains the following two key elements (see King 1972; Nei 1987): First, natural populations contain a sufficient amount of genetic variability to respond to almost any kind of selection. Second, evolution is determined mainly by environmental changes and natural selection. Since there is enough genetic variability, no new mutations are required for a population to respond to an environmental change. Therefore, there is no relationship between the rate of mutation and the rate of evolutionary change. In this view, natural selection is the driving factor of evolution. In contrast, neutralists believe that mutation plays a more important role in molecular evolution than does selection, so that the rate of molecular evolution is mainly determined by the rate of mutation (Kimura 1968a, 1983; King and Jukes 1969; Ohta 1974; Nei 1975, 1987). In other words, mutation is the driving factor in the rate of molecular evolution (see Nei 1987).

As the two views give very different predictions for the relationship between the rate of mutation and the rate of evolution, we can test the predictions using molecular data. Under the neutralist view, the rate of nucleotide substitution should be higher for an organism with a higher mutation rate than for an organism with a lower mutation rate. In contrast, no such relationship exists under the selectionist view. Note that the controversy is concerned mainly with nucleotide changes that cause amino acid changes (i.e., nonsynonymous substitutions), not with synonymous or silent nucleotide changes. Also, the issue here is the average rate of nonsynonymous substitution in the genome of an organism, not the rate in a single gene or certain codons.

Table 14.1 shows the rates of synonymous and nonsynonymous substitution in animals and in the human immunodeficiency virus, a retrovirus. The effect of natural selection on the nonsynonymous rate can be inferred by the variation in nonsynonymous rate among genes in the same genome. This variation is likely to be largely due to the effect of natural selection because the differences in mutation rate among regions in a genome should be relatively small. In Table 14.1, the highest nonsynonymous rate is less than 1000 times the lowest one in the same genome. As explained in Chapter 7, the variation in nonsynonymous rate among genes in a genome is mainly due to differences in functional constraints among genes, that is, mainly due to differences in the intensity of negative selection among genes. Therefore, the variation in nonsynonymous rate that can be attrib-

TABLE 14.1 Rates of synonymous and nonsynonymous substitution in various organisms[a]

Gene	Nonsynonymous	Synonymous	Ratio
Mammals[b]			
Lowest (actin α)	0.01×10^{-9}	2.92×10^{-9}	0.003
Highest (interferons γ)	3.06×10^{-9}	5.50×10^{-9}	0.556
Average	0.74×10^{-9}	3.51×10^{-9}	0.210
Drosophila[c]			
Lowest (arrestin B)	0.23×10^{-9}	11.62×10^{-9}	0.020
Highest (Cp18 Chorion prot. 4)	5.63×10^{-9}	18.38×10^{-9}	0.306
Average	1.91×10^{-9}	15.60×10^{-9}	0.122
Human immunodeficiency virus (HIV-1)[d]			
Lowest (pol gene)	1.6×10^{-3}	11.0×10^{-3}	0.145
Highest (env hypervariable regions)	14.0×10^{-3}	17.2×10^{-3}	0.814
Average	3.9×10^{-3}	10.3×10^{-3}	0.379

[a]All rates are in units of substitutions per site per year.
[b]From Table 7.1, excluding histone genes because their synonymous rates are difficult to estimate.
[c]From Table 7.6.
[d]From Li et al. (1988a).

uted to differences in the frequency or intensity of positive selection among genes is likely to be much smaller than 1000 times. On the other hand, the effect of mutation on the nonsynonymous rate can be inferred from comparisons between different organisms. Here we use the synonymous rate as a rough estimate of the mutation rate in an organism, though the synonymous rate may be significantly reduced by codon usage constraints. It is seen that the synonymous rate and the nonsynonymous rate increase approximately in parallel, by about one million times, from animals to retroviruses. One might argue that the nonsynonymous rate in retroviruses is very high because natural selection is very effective in an organism with rapid generations. However, the increase in the nonsynonymous rate from animals to retroviruses is only of the same order of magnitude as the increase in the synonymous rate, and so both increases can be largely attributed to the increase in mutation rate. At any rate, the data clearly show a strong positive correlation between the rate of mutation and the rate of nonsynonymous substitution. This correlation is predicted by the neutralist view but not by the selectionist view.

Since the nonsynonymous rate in an organism is low when the mutation rate is low and is high when the mutation rate is high, the upper limit of the nonsynonymous rate is determined by the rate of mutation. Because negative selection is much more prevalent than positive selection, the presence of selection actually reduces rather than elevates the upper limit of the rate of evolution. In fact, the highest nonsynonymous rate in a genome is still lower than that of the synonymous rate (Table 14.1), which is likely to be lower than the mutation rate. Therefore, it appears that mutation is the major driving force in molecular evolution, including protein sequence evolution.

MUTATION PRESSURE VERSUS AMINO ACID COMPOSITION

As mentioned above, selectionists believe that evolution occurs by environmental changes and natural selection. Under this view, the amino acid composition of a protein is determined by natural selection and is not or is little affected by the pattern of mutation. In contrast, under the neutralist view the amino acid composition of a protein may be affected by the pattern of mutation because amino acid changes in evolution are mainly caused by mutation. It is well known that many bacterial species have extreme nucleotide compositions (GC contents, i.e., G+C%), which are thought to have arisen from biased mutation pressures (Chapter 13). These observations can be used to test the predictions of the two views. The neutralist view predicts that directional mutation pressure will affect the amino acid composition of proteins, whereas the selectionist view predicts no such effect.

Support for the neutralist prediction actually already existed in Sueoka's 1961 finding that the levels of alanine and glycine tend to increase but those of isoleucine and tyrosine tend to decrease with the genomic GC content of the bacterial species studied. Because alanine (codons GCU, GCC, GCA, GCG) and glycine (GGU, GGC, GGA, GGG) are encoded by GC-rich codons, whereas isoleucine (AUU, AUC, AUA) and tyrosine (UAU, UAC) are encoded by GC-poor codons, Sueoka's finding suggested that the amino acid composition of proteins in a bacterium is affected by its genomic GC content. This suggestion was supported by Jukes and Bhushan's (1986) study of the *trp*A and *trp*B proteins and Osawa et al.'s (1992) study of eight ribosomal proteins in bacteria.

X. Gu, D. Hewett-Emmett, and W.-H. Li (unpublished) have used extensive DNA sequence data from bacteria to test the neutralist prediction. The 20 amino acids are classified into three groups according to the GC content of their codons. Group A consists of proline, alanine, glycine, and tryptophan, which are encoded by codons with a high GC content; e.g., alanine is encoded by GCU, GCC, GCA, or GCG. Group B consists of valine, threonine, histidine, glutamic acid, aspartic acid, glutamine, cysteine, and serine, which are encoded by codons with an intermediate GC content; e.g., aspartic acid is encoded by either GAU or GAC. Group C consists of phenylalanine, tyrosine, asparagine, lysine, isoleucine, and methionine, which are encoded by codons with a high AT (AU) content; e.g., phenylalanine is encoded by either UUU or UUC. Note that arginine and leucine are not included in these groups because each of them is encoded by two families of codons that belong to two different groups. Denote the frequencies of groups A, B, and C in a protein by f_A, f_B, and f_C, respectively. Under the neutralist hypothesis, f_A should increase whereas f_C should decrease with increasing GC content of the genome, while f_B would not be much affected by changes in GC content.

Figure 14.1 shows the result for the *dna*A gene. In this figure GC_4 denotes the GC content at the fourfold degenerate sites of the gene and is used as a measure

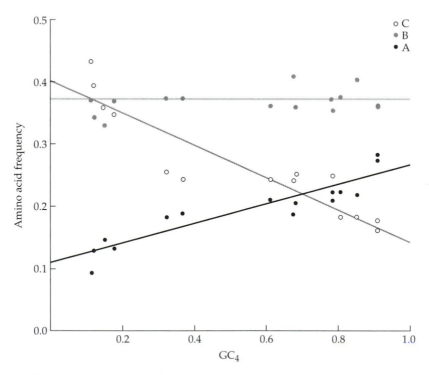

Figure 14.1 Correlation between amino acid frequencies and the GC content (GC_4) at the fourfold degenerate sites of the gene (*dna*A) in bacteria. Lines *A*, *B*, and *C* refer to the frequencies of groups A, B, and C of amino acids, respectively. From X. Gu, D. Hewett-Emmet, and W.-H. Li (unpublished).

TABLE 14.2   Correlations between amino acid frequencies and genomic GC content

Gene	Number of species	Correlations Group A	Group B	Group C
dnaN	12	0.90**	0.46	−0.84**
dnaA	15	0.91**	0.37	−0.92**
aroA	17	0.83**	0.00	−0.85**
trpC	14	0.93**	0.25	−0.88**
grpE	12	0.92**	0.00	−0.79**
gyrB	16	0.80**	0.61*	−0.83**
trpE	17	0.84**	0.07	−0.82**
trpD	14	0.80**	0.38	−0.83**
dnaJ	14	0.77**	0.03	−0.76**
trpG	12	0.79**	0.61*	−0.97**
gyrA	14	0.91**	0.06	−0.83**
ftsZ	14	0.76**	0.01	−0.79**
nusG	11	0.88**	0.18	−0.81**
rpoB	15	0.73**	0.64*	−0.79**
hsp70	15	0.66**	0.54*	−0.79**

* significant at the 5% level.
** significant at the 1% level.

of the intensity of the GC mutational pressure. Clearly, f_A increases whereas f_C decreases with increasing GC_4, though f_B is little affected by change in GC_4. Table 14.2 shows the correlations between amino acid frequencies and GC_4 for 15 genes. The results are: (1) f_A is positively correlated with GC_4; the correlation coefficient r ranges from 0.66 to 0.93, (2) f_C is negatively correlated with GC_4; r ranges from −0.76 to −0.97, and (3) the correlation between f_B and GC_4 is weak—11 of the 15 genes show no significant correlation. All these results are as predicted by the neutralist hypothesis.

The observed effects of mutation pressure on the amino acid composition of proteins are not predicted by the selectionist view. Under this view one will have to assume that amino acids encoded by GC-rich codons are advantageous in species with a high GC genome, whereas amino acids encoded by GC-poor codons are favored in species with a GC-poor genome. It is doubtful that such a nice correlation occurs in so many genes in so many species. In contrast, these observations can be easily explained under the neutralist view by assuming that either most of these amino acid changes are nearly neutral or their selective disadvantages are so small that substitutions can occur under strong mutation pressure. The effect of mutation pressure on the amino acid composition is expected to be stronger on proteins that are not subject to strong functional constraints (and thus can evolve relatively fast) than on proteins that are subject to strong functional constraints. Indeed, the correlation coefficient between f_A (or f_C) and GC_4 is, on average, higher for the faster evolving proteins than for the more slowly evolving proteins in Table 14.2 (data not shown). For example, dnaN has evolved at a relatively high rate and shows a strong correlation between f_A and GC_4 ($r = 0.90$), whereas hsp70 is highly conservative and shows a much weaker correlation between f_A and GC_4 ($r = 0.66$). The reason for a weak correlation in highly conservative proteins is that in such proteins few amino acid substitutions would be allowed because of functional requirements. Viewed from this perspective, it is remarkable that all the 15 proteins studied show a significant correlation between f_A (f_C) and GC_4.

SWITCHES IN SPECIES-SPECIFIC CODON PREFERENCES

In Chapter 7 we noted that the usage of synonymous codons is highly nonrandom in most of the species studied to date. We further noted that the preferred (i.e., the most frequently used) codon for an amino acid may differ among species (Table 7.10). For example, the preferred codon for leucine (Leu) is CUG in *Escherichia coli* but UUG in the yeast *Saccharomyces cerevisiae*. Therefore, a switch in codon preference must have occurred either in the *E. coli* lineage or in the *S. cerevisiae* lineage since their separation. Many such switches appear to have occurred in evolution (e.g., see Table 7.10; Shields 1990). The question is then, What causes such switches?

One possible explanation is as follows. We note that in an organism the preferred codon for an amino acid is usually recognized by the tRNA species that is the most abundant among the cognate tRNAs for that amino acid. For example, the preferred codon and the most abundant cognate tRNA species for Leu are CUG and $\text{tRNA}_1^{\text{Leu}}$ in *E. coli*, but are UUG and $\text{tRNA}_3^{\text{Leu}}$ in *S. cerevisiae* (Figure 7.6). Thus, it seems that species-specific codon preferences are determined by tRNA abundances, as a result of selection for translational efficiency and accuracy (Chapter 7). So, intuitively a simple explanation for a switch in codon preference is that it is caused by a switch in the abundances of tRNAs (see Ikemura 1981). For example, one may assume that the switch in the preferred Leu codon from CUG in *E. coli* to UUG in *S. cerevisiae* was caused by the switch in the most abundant cognate tRNA species from $\text{tRNA}_1^{\text{Leu}}$ in *E. coli* to $\text{tRNA}_3^{\text{Leu}}$ in *S. cerevisiae*. Such a switch in tRNA abundance, however, has the following problems: Since the codon UUG was originally rare, a large increase in the expression level of $\text{tRNA}_3^{\text{Leu}}$ would be a waste of energy and disadvantageous, and since the codon CUG was originally abundant, a large reduction in the expression level of $\text{tRNA}_1^{\text{Leu}}$ would be highly disadvantageous. Of course, a gradual switch may be possible. For example, a small elevation in the expression level of $\text{tRNA}_3^{\text{Leu}}$ may be nearly neutral. This elevation would allow an increase in the frequency of UUG, a reduction in the frequency of CUG and a reduction in the expression level of $\text{tRNA}_1^{\text{Leu}}$. Repetition of such a gradual shift could eventually lead to a complete switch. Note, however, until the codon UUG became abundant, there would be no selective advantage for the proposed shifts to occur. So, it is doubtful that a switch in tRNA abundance is the major cause of switches in codon preference.

A more likely cause for a codon preference switch is a change in mutation pressure that exerts a strong pressure for change in codon frequencies (Shields 1990). We note that changes in codon frequencies may exert selection pressure on tRNA abundance (Kimura 1983; Bulmer 1987). So, suppose that there was greater mutation pressure toward a higher AT content in the *S. cerevisiae* lineage than in the *E. coli* lineage. Then the codon CUG would have a tendency to change to the codon UUG and there would be a selective advantage for tRNA$_3^{Leu}$ to become abundant and a selective advantage for tRNA$_3^{Leu}$ to become less abundant. (There would have been a pressure for UUG to change to UUU, but UUU codes for phenylalanine, not leucine.) This explanation provides a simple scenario for the switch in the preferred leucine codon from CUG in *E. coli* to UUG in *S. cerevisiae*. In this respect, it is interesting to note that the AT content of the *S. cerevisiae* genome (62%) is indeed higher than that of the *E. coli* genome (50%). The following observations (see Osawa et al. 1992) also support the importance of mutation pressure as a cause of switches in codon preference. *Micrococcus luteus*, which has a genomic GC content of 74%, uses almost exclusively CUC and CUG for leucine, whereas *Mycoplasma capricolm*, which has a genomic GC content of only 25%, prefers UUA. For arginine, *Micrococcus luteus* prefers CGC and CGG, whereas *Mycoplasma capricolm* prefers AGA.

In both of the above schemes the switch in codon preference is initiated by mutation, not by selection, though selection may indeed play a role at a later stage. Note that in the process of a switch numerous synonymous substitutions would have occurred because there are many genes in a genome. Indeed, a change in mutation pressure may cause switches in many codon families and, as a consequence, a very large number of synonynous substitutions.

EMERGENCE OF NONUNIVERSAL GENETIC CODES

Almost all living organisms, from bacteria to man, use the universal genetic code (Chapter 1). A plausible explanation for this striking phenomenon is the frozen accident theory (Crick 1968), which states that the code is universal because any change would alter the amino acid sequences of many highly evolved proteins and would therefore be highly detrimental to the organism. The persistence of the universal code in millions of lines of life is a good example of the view that natural selection often inhibits further evolution of a system or the emergence of novelty. However, the code is not really frozen and a number of exceptions to the universal code have been found (Chapter 1). The question here is how a change in the code may evolve. The best answer to this question seems to be the codon capture hypothesis, which postulates that a codon may disappear during evolution from the coding sequences of the genome of an organism or organelle, and then may later reappear with a different assignment (see Chapter 1; Jukes 1985; Osawa and Jukes 1989; Osawa et al. 1992). We consider two examples below to show how this hypothesis may explain the emergence of a new code.

The first example is that *Mycoplasma capricolum* uses UGA (a stop codon) as well as UGG for tryptophan. *Mycoplasma* DNA is very high in AT (about 75%), indicating the existence of directional mutation pressure from GC to AT. Under this pressure, all UGA stop codons mutated to UAA and thus disappeared from the genome; a rare codon may also become lost by random drift. UGA then became an unassigned codon. (This situation caused no harm because UGA was absent from the genome.) The gene for tryptophan tRNA with anticodon CCA duplicated, and one of the duplicates became UCA, which pairs with regular tryptophan codon UGG by wobble pairing between G and U. Then some UGG tryptophan codons mutated to UGA, which pairs with anticodon UCA. The result is that UGA became "captured" by tryptophan, no longer being used as a stop codon in *M. capricolum*. Various experimental findings substantiate this evolutionary scenario. First, the genes in *M. capricolum* for two tryptophan tRNAs, one with anticodon CCA and one with UCA, are arranged in tandem, close together on the chromosome. This suggests gene duplication followed by a mutation of anticodon CCA to UCA (Yamao et al. 1988). Next, tRNA with anticodon CCA in *Mycoplasma*, which is no longer needed, appears to be disappearing. Yamao et al. (1988) reported failure to find tRNA (CCA) appreciably charged in vivo with tryptophan, and Inamine et al. (1990) found that the gene for tRNA Trp (CCA) has disappeared in some of the mycoplasma species such as *M. pneumoniae* and *M. genitalium*. It seems probable that the change from tryptophan coding from UGG to UGA in *Mycoplasma* is an example of "evolution in action."

The second example is that AAA codes for asparagine instead of lysine in mitochondrial DNA of some animals such as platyhelminth (*Fasciola hepatica*) and sea urchin (*Strongylocentrotus purpuratus*) (see Ohama et al. 1990). Normally, lysine is coded by AAA with anticodon UUU and by AAG with antocodon CUU. GC pressure would change all AAA codons to AAG. Anticodon UUU, which pairs with AAA, is thiolated, so that it does not pair strongly with AAG, and it would tend to disappear under GC pressure, which was assumed to be exerted. Anticodon GUU could be converted to IUU, just as in the case of other GNN anticodons that have been converted by evolution to INN in eukaryotes. I in anticodon INN pairs with U, C, and A in third positions of codons. AAA codons, reappeared by mutations of AAU and AAC (asparagine), would pair with IUU and code for Asn. The changes are shown schematically in Table 14.3. Note that although some codons have changed, the amino acid sequence remains unchanged.

In the above two examples we assumed that the disappearance of a codon is due to mutation pressure. Note, however, that a codon may also disappear because of selection against the codon (with or without the assistance of random drift). So, loss of a codon can be initiated either by mutation or selection. However, almost all the other steps in the evolution of a new code such as the loss of a tRNA gene from the genome, the duplication of a tRNA gene, mutation of an anticodon, and the reappearance of a codon are all due to mutation. Moreover, since the amino aicd sequence remained unchanged during the process of the reasignment of a codon, no natural selection was involved in the evolution of the DNA sequence. Therefore, one may conclude that mutation plays a more important role in the emergence of nonuniversal genetic codes than does selection.

TABLE 14.3 Changes in codons (underlined) while amino acid sequence remains constant

	Asn ...	Lys ...	Lys ...	Asn ...	Lys ...	Asn ...	Lys ...	Asn
Original sequence	AAC ...	AAG ...	AAA ...	AAC ...	AAG ...	AAU ...	AAA ...	AAU
Anticodon	GUU	CUU	UUU	GUU	CUU	GUU	UUU	GUU
GC pressure, AAA and UUU disappear	AAC ...	AAG ...	AAG ...	AAC ...	AAG ...	AAC ...	AAG ...	AAC
	GUU	CUU	CUU	GUU	CUU	GUU	CUU	GUU
New sequence	AAA ...	AAG ...	AAG ...	AAC ...	AAG ...	AAA ...	AAG ...	AAU
Anticodon change	IUU	CUU	CUU	IUU	CUU	IUU	CUU	IUU
Amino acid sequence	Asn ...	Lys ...	Lys ...	Asn ...	Lys ...	Asn ...	Lys ...	Asn

From Jukes and Osawa (1991).

EMERGENCE OF NEW FUNCTIONS

The emergence of a new function in a DNA or protein sequence is supposedly advantageous and is commonly believed to have occurred by advantageous mutations. However, acquiring a new function may require many mutational steps, and a point that needs emphasis is that the early steps might have been selectively neutral because the new function might not be manifested until a certain number of steps had already occurred.

Let us consider the emergence of the regulation of hemoglobin function by 2,3-diphosphoglycerate (DPG or 2,3-DPG) (see also the last section of Chapter 7). This is a relatively simple case, so that it would be easier to infer the likely steps and mechanisms involved. DPG is the most abundant organic phosphate in the red cells of most mammals. It can assist hemoglobin in unloading its oxygen to the tissues, because when it binds to hemoglobin, oxygen is released from hemoglobin (see Dickerson and Geis 1983). DPG binds between the ends of the two β chains of hemoglobin: The negative charges of DPG interact with the positive charges on the side chains of the histidine at position 2 (His2), the lysine at position 82 (Lys82), and the histidine at position 143 (His143), as well as the amino termini of the β chains. Therefore, the critical residues for binding DPG are His2, Lys82, and His143 of the β chain; the protonated N-terminus is also important but less critical (Table 14.4). For example, cat hemoglobin B does not respond to DPG because the second residue of the β chain is phenylalanine rather than histidine and because its N-terminal serine is blocked by an acetyl group (see Table 14.4; Bunn and Forget 1986). In contrast, cat hemoglobin A responds to DPG, though only weakly, because the free N-terminal glycine of the β chain compensates to some extent for the absence of histidine at position 2. The failure of ruminant hemoglobins to respond to DPG is due to the deletion of the second residue of the a chain; because of this deletion, the N termini of the β chains in the native deoxygenated hemoglobin apparently become too far apart to bind DPG. Human fetal

TABLE 14.4   Changes in primary structure of the β chain at the DPG binding site

Peptide	Residue position in β chain					Reactivity with 2,3-DPG
	1	2	3	82	143	
Human β globin	H_3N^+—Val	His	Leu	Lys	His	+++
Human γ globin	H_3N^+—Gly	His	Leu	Lys	Ser	+
Lemur β globin	H_3N^+—Thr	*Leu*	Leu	Lys	His	+
Horse β globin	H_3N^+—Val	Gln	Leu	Lys	His	+++
Llama β globin	H_3N^+—Val	*Asn*	Leu	Lys	His	+
Ruminant β globin	H_3N^+—Met	–	Leu	Lys	His	0
Cat β globin A	H_3N^+—Gly	*Phe*	Leu	Lys	His	+
Cat β globin B	*Acetyl*-N—Ser	*Phe*	Leu	Lys	His	0
Human α globin	H_3N^+—Val	–	Leu	Pro	Ser	0
Sperm whale myoglobin	H_3N^+—Val	–	Leu	Glu	Ala	0

Modified from Bunn and Forget (1986).

globin ($\alpha_2\gamma_2$) has only a weak response to DPG because His143 is replaced by serine in the γ chains.

DPG binding of hemoglobin apparently evolved after the divergence between the α and β globins because neither the α globin chain nor myoglobin binds to DPG. The second residue of the β globin is evidently an insertion that occurred after the divergence between the α- and β-globin genes because both the α chain and myoglobin lack this residue (Table 14.4). A comparison of the β chain with the α chain and myoglobin suggests that Lys82 and His143 of the β chain also evolved after the α-β globin divergence. One possible scenario for the evolution of these three critical residues is that Lys82 and His143 evolved before His2. In this case the first two changes would have no selective advantage because before the insertion of His2 the β chain would be equivalent to the situation in ruminant β globin, which has no DPG binding affinity (Table 14.4). (Of course, one cannot rule out the possibility that either Lys82 or His143 or both have evolved because of a selective advantage other than DPG binding.) Another possible scenario is that the insertion at the second position occurred before Lys82 and His143 evolved. This insertion alone would have no DPG binding affinity and thus would have no selective advantage, even if the amino acid inserted was histidine. In short, regardless of the path via which the evolution had occurred, some early steps might have been selectively neutral.

In the above example, the evolution of the new function involved only three critical sites. In a more complex case, more critical sites would be involved and the new function might not be manifested until many of these sites had the right amino acids. In such a case, many early mutational steps might have been selectively neutral. In summary, neutral mutations may serve as the initial steps toward the evolution of a new function.

CONCLUSIONS

It is now well recognized that the evolution of noncoding regions, especially nonfunctional sequences, is mainly determined by mutation pressure. For example, mutation pressure provides a simple explanation for why the genome of vertebrates is AT-rich—it is due to the facts that in vertebrates G and C nucleotides are, on average, more mutable than A and T nucleotides (see Chapter 1) and that the vertebrate genome is made up mostly of noncoding sequences (see Chapter 13). Although the origin of GC-rich isochores in the genome of warm-blooded vertebrates remains controversial, a simple explanation is that it arose from biased local mutation pressure (Chapter 13).

Evidence was given above that mutation pressure is the major cause for switches in species-specific codon preference. The conclusion was that when directional mutation pressure is strong, it can force not only a switch in codon preference but also a switch in tRNA abundance, so that a new correspondence between tRNA abundance and codon usage is achieved.

The genetic code was thought to be frozen and invariable, but new codes have been found in some organisms and organelles (Chapter 1). Yet, what was also surprising is that these seemingly impossible changes were apparently mainly caused by directional mutation pressure. The sigificance of mutation is difficult to appreciate because the rate of mutation is low, so that mutation cannot produce any significant change in a short time period. However, because it is ever-present, mutation pressure can produce unexpected evolutionary changes when the time scale is millions or billions of years.

The neutralist-selectionist controversy, however, pertains mainly to protein evolution, and a highly contentious point is whether mutation or selection is the driving factor of the rate of protein evolution. As discussed above, a contrast between the substitution rates in animals and viruses strongly suggested that mutation is the major determinant. It was concluded that the higher the mutation rate is, the higher the nonsynonymous rate will be. This conclusion and the fact that directional mutation pressure causes biased amino acid composition in proteins suggest that mutation rather than selection is the major driving force in protein evolution. This conclusion does not deny the existence of advantageous substitutions but implies that such substitutions are less frequent than neutral or nearly neutral substitutions in protein evolution.

Given the above conclusion, one might wonder, What role does selection play in evolution? There are two types of selection, negative (purifying) and positive selection. Negative selection acts like a policeman (Ohno 1970). It eliminates disadvantageous mutants or genotypes from the population. As the vast majority of nonneutral mutations are deleterious, negative selection is the prevailing type of selection. For this reason, the presence of selection tends to slow down the rate of evolution, and this explains why in almost all genes the nonsynonymous rate is lower than the synonymous rate and the rates in nonfunctional sequences such as pseudogenes (Chapter 7). The power of purifying selection is best seen in highly conserved proteins such as histones and ubiquitins, which have undergone virtually no amino acid substitution in the past billion years of evolution (Chapter 7; Dayhoff 1972). In such proteins, negative selection has obviously greatly reduced the rate of evolution.

Positive selection is powerful in comparison with random drift. It can greatly increase the probability for an advantageous mutation to increase its frequency and become fixed in the population. For example, the probability of fixation of a mutation is $1/(2N) = 0.00005$ for a neutral mutation in a population of $N = 10,000$, whereas it is approximately $2s = 0.002$ for an advantageous mutation with $s = 0.001$ (see Chapter 2). In this case, positive selection increases the probability of fixation by fortyfold! Positive selection also occurs when a disadvantageous or effectively neutral allele becomes advantageous because of environmental change. Such switches in selective advantage as well as advantageous mutations can help a population cope with environmental changes.

Positive selection is very important for the evolution of a new function. Although it was stressed in the previous discussion that some early mutational steps in the evolution of a new function may be selectively neutral, sooner or later an advantageous mutation must occur, otherwise some already-established steps may become lost by mutation and random drift, and the process would be disrupted. When an advantageous mutation occurs, positive selection helps the new mutation to become established in the population. The emerging system is much less likely to be disrupted once an advantageous mutation is incorporated into it. The system will become even more advantageous, and positive selection becomes even stronger if another favorable mutational step occurs. Thus, positive selection may play a key role at a certain stage of the evolution of a new function.

Positive selection may have played an especially important role in the evolution of duplicate genes. As discussed in Chapter 10, there are two evolutionary paths for a redundant duplicate—either it fixes a null mutation and becomes a pseudogene (path a in Figure 14.2), or it fixes an advantageous mutation (path b), so that it has a better chance of evolving a new function. The ratio (ρ) of the rate of advantageous mutation to that of null mutation is expected to be very low, so that the second path is unlikely to occur without the help of positive selection (see Ohta 1987). Indeed, Walsh (1995) showed that the probability for the second path is close to ρ if the selective advantage (s) for the mutation is small, so that $S = 4N_e s$ is considerably smaller than 1. In contrast, this probability is increased to ρS if $S \gg 1$ (but $\rho S \ll 1$). For example, if ρ is 0.001, then the probability of the fixation of an advantageous mutation before the fixation of a null mutation is approximately only 0.001 if $S < 1$ but is approximately 0.04 if $S = 40$; the increase is 40 times. Note further that if $S < 1$, the fixation of an advantageous mutation is of little help in retaining the duplicate gene, because the duplicate gene can still readily fix a null mutation and become a pseudogene (i.e., path c in Figure 14.2). Only if $S \gg 1$, will there be sufficient selective advantage for retaining the duplicate gene, allowing time for it to accumulate further advantageous mutations (i.e., path d in Figure 14.2). Clearly, positive selection is required in the early stage of the evolution of a new function from a duplicate gene.

Phase I: Initial advantageous fixation

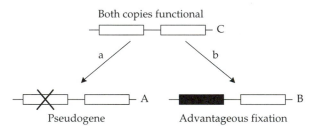

Phase II: Continuing differentiation

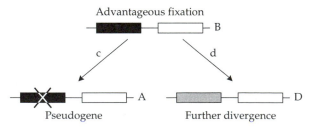

Figure 14.2   Phase I: A duplicate gene (state C) may first fix a null mutation and become a pseudogene (path a, state A) or may first fix an advantageous mutation (path b, state B). Phase II: Following the fixation of the first advantageous mutation, the duplicate locus may still fix a null mutation (path c), or it may fix another advantageous mutation (path d). Modified from Walsh (1995).

Having described the relative roles of mutation and selection in molecular evolution in general, we now consider *adaptive* molecular evolution in particular. That is, we now consider only the class of *advantageous evolutionary changes*. It appears that even for this class of evolutionary changes, the rate of evolution is still mainly determined by the rate of mutation, not by the rate of environmental change as proposed by selectionists. As noted above, neo-Darwinism assumes that whenever the environment changes, the population has sufficient genetic variation, i.e., it has a "right" mutant variant, to respond to the change. Although this variant was originally disadvantageous, it has been maintained in the population by recurrent mutation; it becomes advantageous, and its frequency starts to increase when the environment changes. This view is unlikely to hold at the molecular level. Most molecular changes are unique rather than recurrent, and most of them will soon become lost if there is no selective advantage (see Chapter 2). Thus, it is likely that in the majority of cases when the environment changes, no "right" variant is there in the population and it will take sometime for a right mutation to occur. A more plausible view is that occasionally a mutation happens to be advantageous in the existing environment. As the mutation has an immediate selective advantage, it has a definite, though small, probability of becoming fixed in the population (see Chapter 2). In this view the rate of evolution is mainly determined by how often advantageous mutations occur. Positive selection also affects the rate of evolution in the sense that the probability of fixation of an advantageous mutation and the rate of substitution increase with the magnitude of the selective advantage. This view is similar to the view proposed by Morgan (1925, 1932) and elaborated in the framework of modern molecular genetics by Nei (1987); Nei has pointed out that Morgan's view is equivalent to the mutation-selection theory or classical theory in Dobzhansky's (1955) sense. To compare this view and neo-Darwinism, let us consider a few examples.

First, let us consider the emergence of trichromatic vision in higher primates. Trichromatic vision is helpful for detecting colored fruits and has evolved independently in New World monkeys (NWMs) and in the common ancestor of the Old World monkeys (OWMs), apes, and humans (Chapter 10; Shyue et al. 1995). The two systems differ in that OWMs, apes and humans possess two X-linked opsin genes, which code for the red and green pigments, respectively, whereas most NWMs possess only one X-linked opsin locus, but the locus has three alleles, two of which have spectral sensitivities similar to those of the human red and green pigments, respectively (see Chapter 10). In the evolution of either system there would have been at most only one relevant environmental change in monkeys, that is, a change from a diet without fruits to one with fruits. However, the emergence of each system required multiple mutational steps; for example, at least 5 of the 15 amino acid differences between the human red and green pigments are responsible for the differences in spectral sensitivity between the two pigments (Nathans et al. 1986; Neitz et al. 1991; Shyue 1994). Obviously, most, if not all, of these steps required new mutations, and the rate at which the system evolved would have depended on how fast these mutations had occurred, not on how fast the environment had changed.

Second, the emergence of regulation of hemoglobin function by organic phosphates such as DPG requires at least three mutational steps (see the preceding section). However, the evolution of this regulation would require at most only one environmental change, that is, the physiology of vertebrates became ready to accommodate this regulation, i.e., to synthesize DPG (in mammals) or inositol pentaphasphate (in birds). Therefore, how fast the regulation had evolved would have depended on the rate of occurrence of the three critical residues, not on the rate of environmental change. Interestingly, this regulatory system is preadaptive to life at high altitudes. For example, it has enabled Bolivians to adapt to life at 15,000 ft in the Andes; these Bolivians have 20% higher levels of DPG in their red blood cells than do people who live at sea level (see Dickerson and Geis 1983; Bunn and Forget 1986). This example suggests that a new function can be preadaptive, rather than being a response to an environmental change, and may enable an organism to explore a new niche.

Third, as explained in Chapter 10, a domain duplication or a domain shuffling may have an immediate selective advantage in the existing environment because it may immediately improve the function of the protein. Here, no environmental change is necessary. As emphasized in Chapter 10, domain duplication and domain shuffling have played an important role in the evolution of proteins.

Fourth, whether a duplicate gene can be retained in the population depends on how often advantageous mutations occur and how large the selective advantages are (see above), not on how often the environment changes. Subsequently, how fast the duplicate gene evolves into a new gene depends on how often advantageous novel mutations occur, not on how often the environment changes. For example, as mentioned above, during the evolution of a pair of duplicate genes into the red and green pigment genes in OWMs, apes and humans there would have been at most only one environmental change, but there were multiple amino acid substitutions. So, the rate of evolution of the two genes depended on the rate of occurrence of these mutations, not on the rate of environmental change.

The above examples revealed the inadequacy of neo-Darwinism even for explaining *adaptive* molecular evolution. This is not surprising because neo-Darwinism was developed long before the development of modern genetics and molecular biology. A good understanding of an evolutionary process or the emergence of a new system requires not only the theory of evolution and population genetics but also knowledge of the molecular steps involved.

Since the time of Darwin it has been apparent that the great shaping force in phenotypic evolution is natural selection. Selection, acting on the genetic variability in natural populations, is responsible for the wondrous diversity of animal and plant life that is so apparent as we look around us. Yet, molecular biology has revealed a deeper and more extensive level of genetic change among individuals or organisms. Most of these changes have no visible effect and the neo-Darwinian paradigm is inadequate. Such mutation-driven molecular changes far exceed selection-driven phenotypic changes. Thus, inside the "forest" of selectively determined phenotypic changes are an enormously greater number of "trees" mostly marching to a different drummer—the tempo of mutation. Moreover, a new paradigm is needed even for adaptive evolution at the molecular level. In conclusion, molecular studies have greatly deepened our understanding of how evolution works, and the rapid progress in the past two decades is only the beginning.

Answers to Problems

CHAPTER 1

Problem 3

(a) About 33.3%; (b) about 26%; and (c) about 44.4%.

Problem 4

(a) About 11%; (b) about 13%; and (c) about 46.5%.

CHAPTER 2

Problem 1

$$\frac{p^2 qs}{(p+q)(p+q)+qs(2p+q)} = \frac{p^2 qs}{1+qs(p+1)}$$

Problem 3

From (2.6), (1) $3.215/s$; (2) $0.847/s$; and (3) $3.208/s$.

Corollary: The change in allele frequency is faster or, in other words, natural selection is more effective if the frequency is closer to 0.5 than to 0 or 1.

Problem 5

This is a straightforward binomial probability computation,

$$P_i = \frac{(2N)!}{i!(2N-i)!} = p^i q^{2N-i}$$

which is performed for $p = 0.3$, $N = 10$ and $i = 0, 1, \dots 20$. We will compute, for example, for $i = 0$, $i = 5$, and $i = 10$:

for $i = 0$, $P_i = 0.0008$;
for $i = 5$, $P_i = 0.179$;
for $i = 10$, $P_i = 0.031$

Problem 7

8/9.

Problem 8

N_e is approximately 60.

Problem 9

About 0.00038.

CHAPTER 3

Problem 3

From (3.17), $t > 9.24 \times 10^7$. Since $\beta < \alpha$ in this particular case, $Y(t) > Z(t)$ for every positive t.

Problem 8

From (3.27) and (3.28):

	t_1 = 100 million years (Myr)	t_2 = 200 Myr
$P(t)$	0.281	0.255
$Q(t)$	0.432	0.491

Problem 9

From (3.32),

For $a = 0.5$, $\bar{I} = 0.585$;
For $a = 1.0$, $\bar{I} = 0.5$;
For $a = 2.0$, $\bar{I} = 0.438$.

CHAPTER 4

Problem 2

JC distance: 15 substitutions out of 175 sites (excluding deletions and insertions). Therefore, $K = 0.09$.

Kimura distance: 10 transitions and 5 transversions, $a = 1.17$, and $b = 1.07$. Therefore, $K = 0.106$.

Formula (4.9): The determinant is 1,751,801; $F_1 = 2,503,872$; $F_2 = 2,710,950$. Therefore, the answer is 0.099.

Problem 3

(a) $K = 0.824$

(b) For $a = 0.5$, $K = 3$; for $a = 1.0$, $K = 1.5$; for $a = 2.0$, $K = 1.098$.

Problem 4

Using Li et al.'s (1985b) method, $K_S = 1.628$, $K_A = 0.298$.

Using Li's (1993) method, $K_S = 1.634$, $K_A = 0.318$.

Problem 5

Consider the following two alignments:

```
AATGCTTGCATGGGGC....TAGTT
ATTGCT.GCATGAGGCGCGCTAGT.
```

and

```
AATGCTTGCATGGGGCTAGTT...
ATTGCT.GCATGAGGCGCGCTAGT
```

Both look attractive; however, if we assume a constant gap penalty of 2, the second alignment is preferable.

For a constant gap penalty of 10, the following alignment "wins":

```
AATGCTTGCATGGGGCTAGTT..
ATTGCTGCATGAGGCGCGCTAGT
```

Problem 6

$a = 1$ and $b = 0$.

CHAPTER 5

Problem 3

Tree [(12)(34)] is supported by 11 sites; tree [(13)(24)] is supported by 7 sites; tree [(14)(23)] is supported by 3 sites. Therefore, the first tree is MP.

Problem 4

If we assume additivity, then, following (5.6),

$$x = \frac{d_{AC} + d_{BD} - d_{AB} - d_{CD}}{2}$$

Alternatively, we can use (5.23) to obtain a least-squares estimate:

$$\hat{x} = \frac{d_{AC} + d_{BD} + d_{BC} + d_{AD} - 2d_{AB} - 2d_{CD}}{4}$$

Problem 5

	1	2	3
2	0.14		
3	0.17	0.20	
4	0.24	0.19	0.16

And, by four-point condition, we infer the tree [(12)(34)], where x, following (5.23), is estimated to be 0.05.

Problem 8

Substituting (5.11) and (5.12), we have

$$P = p_{AA} \times p_{AG} \times p_{CG} \times p_{GT} \times p_{TT} \times p_{TT},$$

where

$$p_{AA}(t_1 + t_2 + t_3) = 1/4 + (3/4)e^{-4h(t_1+t_2+t_3)/3}$$

$$p_{AG}(t_1) = 1/4 - (1/4)e^{-4h(t_1)/3}$$

$$p_{GC}(t_2 + t_3) = 1/4 - (1/4)e^{-4h(t_2+t_3)/3}$$

$$p_{GT}(t_2) = 1/4 - (1/4)e^{-4h(t_2)/3}$$

$$p_{TT}(t_3) = 1/4 + (3/4)e^{-4h(t_3)/3}$$

Problem 9

$$L(A, A, G, T) = \sum_X (1/4) P_{XT}(t_1 + t_2 + t_3)$$

$$\times \sum_Y \{P_{XY}(t_1) \times P_{YG}(t_1 + t_2)\}$$

$$\times \sum_Z \{P_{YZ}(t_2) \times P_{ZA}(t_3) \times P_{ZA}(t_3)\}$$

Problem 10

$$L = C \prod_{i=1}^{i=15} U_i^{N_i}$$

where

$$C = \frac{N!}{N_1! \dots N_{15}!}$$

Also we need to assume equilibrium, the one-parameter model, etc.

Configuration table:

$$U_1 = 4f(iiii)$$
$$U_2 = 12f(iiij)(i \neq j)$$
$$U_3 = 12f(iiji)(i \neq j)$$
$$\vdots$$
$$U_{15} = 24f(ijkl)(i \neq j \neq k \neq l)$$

Since comb. coefficients are already in U's,

$$L = C \prod_{i=1}^{i=15} U_i^{M_i}$$

where M_i is an observed number of the ith configuration.

Computing M_i's,

$M_1 = 103$
$M_2 = 6$
$M_3 = 2$
$M_4 = 2$
$M_5 = 2$
$M_6 = 11$
$M_7 = 7$
$M_8 = 3$
$M_{13} = 1$
$M_{14} = 2$
$M_{15} = 1$

Other M_i's are zero.

Now we are ready to write down the likelihood function. We start with

$$P_{xx}(t) = 1/4 + (3/4)e^{-4\gamma t/3}$$

$$P_{xy}(t) = 1/4 - (1/4)e^{-4\gamma t/3}, \quad (x \neq y)$$

Then, we substitute these P's into

$$f(ijkl) = \sum_X \left[(1/4) P_{Xl}(t_1 + t_2 + t_3) \right]$$

$$\times \sum_Y \left[P_{XY}(t_1) \times P_{Yk}(t_2 + t_3) \right]$$

$$\times \sum_Z \left[P_{YZ}(t_2) \times P_{Zi}(t_3) \times P_{Zj}(t_3) \right]$$

Using these f's, we obtain

$$L = C \left[4f(iiii) \right]^{103}$$

$$\times \left[12f(iiij) \right]^6$$

$$\vdots$$

$$\times 24 f(ijkl)$$

As an additional exercise, one could use v's instead of γt's and try to maximize the likelihood with respect to v. (This would require either Matlab-type software, separate optimization software, or good working knowledge of computer programming.)

Literature Cited

Acher, R. and J. Chauvet. 1953. La structure de la vasopressine de boeuf. Biochim. Biophys. Acta. 12: 487–488.

Adachi, J. and M. Hasegawa. 1995. Phylogeny of whales: Dependence of the inference on species sampling. Mol. Biol. Evol. 12: 177–179.

Aguadé, M. and C. Langley. 1994. Polymorphism and divergence in regions of low recombination in *Drosophila*. pp. 67–76. In: Golding, B. (ed.) *Non-Neutral Evolution: Theories and Molecular Data*. Chapman & Hall, New York.

Aguadé, M., N. Miyashita, and C. H. Langley. 1989. Reduced variation on the *yellow-achaete-scute* region in natural populations of *Drosophila melanogaster*. Genetics 122: 607–615.

Akashi, H. 1994. Synonymous codon usage in *Drosophila melanogaster*: Natural selection and translational accuracy. Genetics 136: 927–935.

Akashi, H. 1995. Inferring weak selection from patterns of polymorphism and divergence at "silent" sites in *Drosophila* DNA. Genetics 139: 1067–1076.

Albert, V. A., A. Backlund, K. Bremer, M. W. Chase, J. R. Manhart, B. D. Mishler, and K. C. Nixon. 1994. Functional constraints and *rbc*L evidence for land plant phylogeny. Ann. Missouri Bot. Gard. 81: 534–567.

Allaby, R. G., M. K. Jones, and T. A. Brown. 1994. DNA in charred wheat grains from the Iron Age hillfort at Danebury, England. Antiquity 68: 126–132.

Allendorf, F. W., F. M. Utter, and B. P. May. 1975. Gene duplication within the family Salmonidae. II. Detection and determination of the genetic control of duplicate loci through inheritance studies and the examination of populations. pp. 415–432. In: Markert, C. L. (ed.) *Isozymes IV: Genetics and Evolution*. Academic Press, New York.

Allshire, R. C., J. R. Gosden, S. H. Cross, G. Cranston, D. Rout, N. Sugawara, J. W. Szostak, P. A. Fantes, and N. D. Hastie. 1988. Telomeric repeat from *T. thermophila* cross hybridizes with human telomeres. Nature 332: 656–659.

Altschul, S. F. and D. J. Lipman. 1989. Trees, stars, and multiple biological sequence alignment. SIAM J. Appl. Math. 49: 197–209.

Anderson, S. and 11 others. 1981. Sequence and organization of the human mitochondrial genome. Nature 290: 457–464.

Andrews, P. 1987. Aspects of hominoid phylogeny. pp. 23–53. In: Patterson, C. (ed.) *Molecules and Morphology in Evolution: Conflict or Compromise?* Cambridge University Press, Cambridge.

Anfinsen, C. B. 1959. *The Molecular Basis of Evolution*. Wiley, New York.

Antequera, F. and A. Bird. 1993. Number of CpG islands and genes in human and mouse. Proc. Natl. Acad. Sci. USA 90: 11995–11999.

Antoine, M., C. Erbil, E. Munch, S. Schnell, and J. Niessing. 1987. Genomic organization and primary structure of five homologous pairs of intron-less genes encoding secretory globins from the insect *Chironomus thummi thummi*. Gene 56: 41–51.

Anxolabéhère, D., M. G. Kidwell, and G. Periquet. 1988. Molecular characteristics of diverse populations are consistent with the hypothesis of recent invasion of *Drosophila melanogaster* by mobile *P* elements. Mol. Biol. Evol. 5: 252–269.

Aota, S. and T. Ikemura. 1986. Diversity in G+C content at the third position of codons in vertebrate genes and its cause. Nucleic Acids Res. 14: 6345–6355.

Aquadro, C. F. and B. D. Greenberg. 1983. Human mitochondrial DNA variation and evolution: Analysis of nucleotide sequences from seven individuals. Genetics 103: 287–312.

Aquadro, C. F., D. J. Begun, and E. C. Kindahl. 1994. Selection, recombination, and DNA polymorphism in *Drosophila*. pp. 46–56. In: Golding, B. (ed.) *Non-Neutral Evolution: Theories and Molecular Data*. Chapman & Hall, New York.

Argos, P., M. G. Rossmann, U. M. Grau, A. Zuber, G. Franck, and J. D. Tratschin. 1979. Thermal stability and protein structure. Biochemistry 18: 5698–5703.

Arnason, U. and A. Guillberg. 1994. Relationship of baleen whales established by cytochrome *b* gene sequence comparison. Nature 367: 726–728.

Arnason, U., S. Gretarsdottir, and A. Gullberg. 1993. Comparison between the 12S rRNA, 16S rRNA, NADH1 and COI genes of sperm and fin whale mitochondrial DNA. Biochem. Syst. Ecol. 21: 115–122.

Arnheim, N. 1983. Concerted evolution of multigene families. pp. 38–61. In: Nei, M. and R. K. Koehn (eds.) *Evolution of Genes and Proteins*. Sinauer Associates, Sunderland, MA.

Arnheim, N., M. Krystal, R. Schmickel, G. Wilson, O. Ryder, and E. Zimmer. 1980. Molecular evidence for genetic exchange among ribosomal genes on nonhomologous chromosomes in man and apes. Proc. Natl. Acad. Sci. USA 77: 7323–7327.

Aslanidis, C. and 15 others. 1992. Cloning of the essential myotonic dystrophy region and mapping of the putative defect. Nature 355: 548–551.

Atchley, W. R. and W. M. Fitch. 1991. Gene trees and the origins of inbred strains of mice. Science 254: 554–558.

Avise, J. C. 1992. Molecular population structure and the biogeographic history of a regional fauna: A case history with lessons for conservation biology. Oikos 63: 62–76.

Avise, J. C. 1994. *Molecular Markers, Natural History and Evolution*. Chapman & Hall, New York.

Axelrod, D. I. 1952. A theory of angiosperm evolution. Evolution 6: 29–60.

Axelrod, D. I. 1970. Mesozoic paleogeography and early angiosperm history. Bot. Rev. 36: 277–319.

Ayala, F. J. (ed.). 1976. *Molecular Evolution*. Sinauer Associates, Sunderland, MA.

Bailey, G. S., R. T. M. Poulter, and P. A. Stockwell. 1978. Gene duplication in tetraploid fish: Model for gene silencing at unlinked duplicate loci. Proc. Natl. Acad. Sci. USA 75: 5575–5579.

Bailey, W. J., D. H. A. Fitch, D. A. Tagle, J. Czelusniak, J. L. Slightom, and M. Goodman. 1991. Molecular evolution of the $\psi\eta$-globin gene locus: Gibbon phylogeny and the hominoid slowdown. Mol. Biol. Evol. 8: 155–184.

Bailliet, G., F. Rothhammer, F. R. Carnese, C. M. Bravi, and N. O. Bianchi. 1994. Founder mitochondrial haplotypes in Amerindian populations. Am. J. Hum. Genet. 54: 27–33.

Bains, W. 1986. The multiple origins of human *Alu* sequences. J. Mol. Evol. 23: 189–199.

Baker, B. S. 1989. Sex in flies: The splice of life. Nature 340: 521–524.

Baker, C. S. and S. R. Palumbi. 1996. Population structure, molecular systematics forensic identification of whales and dolphins. In: Avise, J. C. and J. L. Hamrick (eds.) *Conservation Genetics: Case Histories from Nature*. Chapman & Hall, London.

Baker, C. S. and 13 others. 1993. Abundant mitochondrial DNA variation and world-wide population structure in humpback whales. Proc. Natl. Acad. Sci. USA 90: 8239–8243.

Baldauf, S. L. and J. D. Palmer. 1990. Evolutionary transfer of the chloroplast *Tuf* A gene to the nucleus. Nature 344: 262–265.

Baldauf, S. L., J. D. Palmer, and W. F. Doolittle. 1996. The root of the universal tree and the origin of eukaryotes based on elongation factor phylogeny. Proc. Natl. Acad. Sci. USA 93: 7749–7754.

Baldwin, J. M. 1980. The structure of human carbonmonoxy haemoglobin at 2.7 Å resolution. J. Mol. Biol. 136: 103–128.

Baltimore, D. 1981. Gene conversion: Some implications for immunoglobulin genes. Cell 24: 592–594.

Bannister, J. V. and M. Parker. 1985. The presence of a copper/zinc superoxidase dismutase in the bacterium *photobacterium leiognathi*: A likely case of gene transfer from eukaryotes to prokaryotes. Proc. Natl. Acad. Sci. USA 82: 149–152.

Banyai, L., A. Varadi, and L. Patthy. 1983. Common evolutionary origin of the fibrin-binding structures of fibronectin and tissue-type plasminogen activator. FEBS Letters. 163: 37–41.

Barker, W. C. and M. O. Dayhoff. 1977. Evolution of lipoproteins deduced from protein sequence data. Comp. Biochem. Physiol. 576: 309–315.

Barker, W. C. and M. O. Dayhoff. 1980. Evolutionary and functional relationships of homologous physiological mechanisms. BioScience 30: 593–600.

Barker, W. C., L. K. Ketcham, and M. O. Dayhoff. 1978. Duplication in protein sequences. pp. 359–362. In: Dayhoff, M. O. (ed.) *Atlas of Protein Sequence and Structure*. Vol. 5, supplement 3., Natl. Biomed. Res. Found., Silver Spring, MD.

Barnes, L. G., D. P. Doming, and C. E. Ray. 1985. Status of studies on fossil marine mammals. Mar. Mammal. Sci. 1: 15–53.

Barnes, S. R., D. A. Webb, and G. A. Dover. 1978. The distribution of satellite and main-band DNA components in the melanogaster species subgroup of *Drosophila*. I. Fractionation of DNA in actinomycin D and distamycin A density gradients. Chromosoma 67: 341–363.

Barry, D. and J. A. Hartigan. 1987. Statistical analysis of hominoid molecular evolution. Stat. Sci. 2: 191–210.

Barsh, G. S., P. H. Seeburg, and R. E. Gelinas. 1983. The human growth hormone gene family: Structure and evolution of the chromosomal locus. Nucleic Acids Res. 11: 3939–3958.

Bartels, J. L., M. T. Murtha, and F. H. Ruddle. 1993. Multiple Hox/HOM-class homeoboxes in platyhelminthes. Mol. Phylogenet. Evol. 2: 143–151.

Basten, C. J. and B. S. Weir. 1990. Effect of gene conversion on variances of digenic identity measures. Theor. Pop. Biol. 38: 125–148.

Beale, D. and H. Lehmann. 1965. Abnormal hemoglobin and the genetic code. Nature 207: 259–261.

Becak, M. L., W. Becak, and M. N. Rabello. 1966. Cytological evidence of constant tetraploidy in the bisexual South American frog, *Odontophrynus americanus*. Chromosoma 19: 188–193.

Beckmann, J. S. and J. L. Weber. 1992. Survey of human and rat microsatellites. Genomics 12: 627–631.

Beech, R. N. and A. J. Leigh-Brown. 1989. Insertion-deletion variation at the *yellow-achaete-scute* region in two natural populations of *Drosophila melanogaster*. Genet. Res. 53: 7–15.

Begun, D. J. and C. F. Aquadro. 1992. Levels of naturally occurring DNA polymorphism correlate with recombination rates in *D. melangaster*. Nature 356: 519–520.

Bell, G. I., M. J. Selby, and W. J. Rutter. 1982. The highly polymorphic region near the human insulin gene is composed of simple tandemly repeating sequences. Nature 295: 31–35.

Bender, W., P. Spierer, and D. S. Hogness. 1983. Chromosomal walking and jumping to isolate DNA from the *Ace* and *rosy* loci and the *Bithorax* complex in *Drosophila melanogaster*. J. Mol. Biol. 168: 17–33.

Benne, R., J. van den Burg, J. P. Brakenhoff, P. Sloof, J. H. Van Boom, and M. C. Tromp. 1986. Major transcript of the frameshifted *coxII* gene from trypanosome mitochondria contains four nucleotides that are not encoded in the DNA. Cell 46: 819–826.

Benveniste, R. E. 1985. The contributions of retroviruses to the study of mammalian evolution. pp. 359–417. In: MacIntyre, R. I. (ed.) *Molecular Evolutionary Genetics*. Plenum Press, New York.

Benveniste, R. E. and G. J. Todaro. 1976. Evolution of type C viral genes: Evidence for an Asian origin of man. Nature 261: 101–108.

Berg, D. E. and M. M. Howe. 1989. *Mobile DNA*. American Society for Microbiology Press, Washington, DC.

Bernardi, G. 1989. The isochore organization of the human genome. Annu. Rev. Genet. 23: 637–661.

Bernardi, G. 1995. The human genome: Organization and evolutionary history. Annu. Rev. Genet. 29: 445–476.

Bernardi, G. and G. Bernardi. 1985. Codon usage and genome composition. J. Mol. Evol. 22: 363–365.

Bernardi, G., B. Olofsson, J. Filipski, M. Zerial, J. Salinas, G. Cuny, M. Meunier-Rotival, and F. Rodier. 1985. The mosaic genome of warm-blooded vertebrates. Science 228: 953–958.

Bernardi, G., D. Mouchiroud, C. Gautier, and G. Bernardi. 1988. Compositional patterns in vertebrate genomes: Conservation and change in evolution. J. Mol. Evol. 28: 7–18.

Bernardi, G., D. Mouchiroud, and C. Gautier. 1993. Silent substitutions in mammalian genomes and their evolutionary implications. J. Mol. Evol. 37: 583–589.

Berry, A. J., J. W. Ajioka, and M. Kreitman. 1991. Lack of polymorphism on the *Drosophila* fourth chromosome resulting from selection. Genetics 129: 1111–1117.

Bingham, P. M., M. G. Kidwell, and G. M. Rubin. 1982. The molecular basis of PM hybrid dysgenesis: The role of the P element, a P-strain-specific transposon family. Cell 29: 995–1004.

Birky Jr., C. W. and R. V. Skavaril. 1976. Maintenance of genetic homogeneity in systems with multiple genomes. Genet. Res. 27: 249–265.

Birky Jr., C. W. and J. B. Walsh. 1988. Effects of linkage on rates of molecular evolution. Proc. Natl. Acad. Sci. USA 85: 6414–6418.

Bishop, J. M. 1981. Enemies within: The genesis of retrovirus oncogenes. Cell 23: 5–6.

Black, J. A. and G. H. Dixon. 1968. Amino acid sequence of α chains of human haptoglobins. Nature 218: 736–741.

Black, J. A. and D. Gibson. 1974. Neutral evolution and immunoglobulin diversity. Nature 250: 327–328.

Blaisdell, B. E. 1985. A method of estimating from two aligned present-day DNA sequences their ancestral composition and subsequent rates of substitution, possibly different in the two lineages, corrected for multiple and parallel substitutions at the same site. J. Mol. Evol. 22: 69–81.

Blake, C. C. F. 1978. Do genes-in-pieces imply protein in pieces? Nature 273: 267.

Blakeslee, A. F. 1930. Extra chromosomes: A source of variation in the Jimson weed. Smithsonian Reports 431–450.

Bodmer, W. F. 1972. Evolutionary significance of the HL-A system. Nature 237: 139–145.

Bodmer, W. F. and L. L. Cavalli-Sforza. 1976. *Genetics, Evolution, and Man*. Freeman, San Francisco.

Boedtker, H., M. Finer, and S. Aho. 1985. The structure of the chicken α2 collagen gene. Ann. N.Y. Acad. Sci. 460: 85–116.

Bogart, J. P. 1980. Evolutionary implications of polyploidy in amphibians and reptiles. pp. 341–378. In: Lewis, W. H. (ed.) *Polyploidy: Biological Relevance*. Plenum Press, New York.

Bogart, J. P. and A. Wasserman. 1972. Diploid-polyploid cryptic pairs: A possible clue to evolution by polyploidization in anuran amphibians. Cytogenetics 11: 7–24.

Boguski, M. S., N. Elshourbagy, J. M. Taylor, and J. I. Gordon. 1984. Rat apolipoprotein A-IV contains 13 tandem repetitions of a 22-amino acid segment with amphipathic helical potential. Proc. Natl. Acad. Sci. USA 81: 5021–5025.

Borts, R. H. and J. E. Haber. 1987. Meiotic recombination in yeast: Alteration by multiple heterozygosities. Science 237: 1459–1465.

Bostock, C. J. 1986. Mechanisms of DNA sequence amplification and their evolutionary consequences. Phil. Trans. R. Soc. London 312B: 261–273.

Boyes, D. C., C.-H. Chen, T. Tantikanjana, J. J. Esch, and J. B. Nasrallah. 1991. Isolation of a second S-locus-related cDNA from *Brassica oleracea*: Genetic relationships between the S locus and two related loci. Genetics 127: 221–228.

Braunitzer, G., R. Gehring-Muller, N. Hilschmann, K. Hilse, G. Hobom, V. Rudloff, and B. Wittmann-Liebold. 1961. Die konstitution des normalen adulten human hämoglobins. Z. Physiol. Chem. Hoppe Seyler 325: 283–286.

Braunitzer, G., A. Stangl, B. Schrank, C. Krombach, and H. Wiesner. 1984. The primary structure of the haemoglobin of the African elephant (*Loxodonta africana*, Proboscidea): Asparagine in position 2 of the β-chain. Z. Physiol. Chem. Hoppe-Seyler 365: S743–S749.

Braverman, J. M., R. R. Hudson, N. L. Kaplan, C. H. Langley, and W. Stephan. 1995. The hitchhiking effect on the site frequency spectrum of DNA polymorphism. Genetics 140: 783–796.

Bregliano, J.-C. and M. G. Kidwell. 1983. Hybrid dysgenesis determinants. pp. 363–410. In: Shapiro, J. A. (ed.) *Mobile Genetic Elements*. Academic Press, New York.

Bridges, C. and T. H. Morgan. 1923. The third chromosome group of mutant characters in *Drosophila melanogaster*. Carnegie Inst. Wash. Publ. 327: 1–251.

Britten, R. J. 1986. Rates of DNA sequence evolution differ between taxonomic groups. Science 231: 1393–1398.

Britten, R. J. 1994. Evidence that most human *Alu* sequences were inserted in a process that ceased about 30 million years ago. Proc. Natl. Acad. Sci. USA 91: 6148–6150.

Britten, R. J. 1996. Cases of ancient mobile element DNA insertions that now affect gene regulation. Mol. Phylogenet. Evol. 5: 13–17.

Britten, R. J. and D. E. Kohne. 1968. Repeated sequences in DNA. Science 161: 529–540.

Britten, R. J., W. F. Baron, D. B. Stout, and E. H. Davidson. 1988. Sources and evolution of human *Alu* repeated sequences. Proc. Natl. Acad. Sci. USA 85: 4770–4774.

Brookfield, J. F. 1986. Population biology of transposable elements. Philos. Trans. R. Soc. Lond. [B] 312: 217–226.

Brookfield, J. F. Y. and P. M. Sharp. 1994. Neutralism and selection face up to DNA data. Trends Genet. 10: 109–111.

Brown, A. J. L. and D. Ish-Horowicz. 1981. Evolution of the 87A and 87C heat-shock loci in *Drosophila*. Nature 290: 677–682.

Brown, D. D. and K. Sugimoto. 1973. 5S DNAs of *Xenopus laevis* and *Xenopus mulleri*: Evolution of a gene family. J. Mol. Biol. 78: 397–415.

Brown, D. D., P. C. Wensink, and E. Jordan. 1972. A comparison of the ribosomal DNAs of *Xenopus laevis* and *Xenopus mulleri*: Evolution of tandem genes. J. Mol. Biol. 63: 57–73.

Brown, H., F. Sanger, and R. Kitai. 1955. The structure of pig and sheep insulins. Biochem. J. 60: 556–565.

Brown, J. R. and W. F. Doolittle. 1995. Root of the universal tree of life based on ancient aminoacyl-tRNA synthetase gene duplications. Proc. Natl. Acad. Sci. USA 92: 2441–2445.

Brown, T. A. and K. A. Brown. 1994. Ancient DNA: Using molecular biology to explore the past. BioEssays 16: 719–726.

Brown, W. M., M. George Jr., and A. C. Wilson. 1979. Rapid evolution of animal mitochondrial DNA. Proc. Natl. Acad. Sci. USA 76: 1967–1971.

Brown, W. M., E. M. Prager, A. Wang, and A. C. Wilson. 1982. Mitochondrial DNA sequences of primates: Tempo and mode of evolution. J. Mol. Evol 18: 225–239.

Brownell, E. 1983. DNA/DNA hybridization studies of muroid rodents: Symmetry and rates of molecular evolution. Evolution 37: 1034–1051.

Brues, A. M. 1969. Genetic load and its varieties. Science 164: 1130–1136.

Bruns, T. D., R. Fogel, and J. N. Taylor. 1990. Amplification and sequencing of DNA from fungal herbarium specimens. Mycologia 82: 175–184.

Brutlag, D. L. 1980. Molecular arrangement and evolution of heterochromatic DNA. Annu. Rev. Genet. 14: 121–144.

Bulmer, M. 1986. Neighboring base effects on substitution rates in pseudogenes. Mol. Biol. Evol. 3: 322–329.

Bulmer, M. 1987. Coevolution of codon usage and transfer RNA abundance. Nature 325: 728–730.

Bulmer, M. 1988. Are codon usage patterns in unicellular organisms determined by selection-mutation balance? J. Evol. Biol. 1: 15–26.

Bulmer, M. 1989. Estimating the variability of substitution rates. Genetics 123: 615–619.

Bulmer, M. 1991. Strand symmetry of mutation rates in the β-globin region. J. Mol. Evol. 33: 305–310.

Bulmer, M., K. H. Wolfe, and P. M. Sharp. 1991. Synonymous nucleotide substitution rates in mammalian genes: implications for the molecular clock and the relationship of mammalian orders. Proc. Natl. Acad. Sci. USA 88: 5974–5978.

Buneman, P. 1971. The recovery of trees from measurements of dissimilarity. pp. 387–395. In: Hodson, F. R., D. G. Kendall, and P. Tautu (eds.) *Mathematics in the Archeological and Historical Sciences*. Edinburgh University Press, Edinburgh.

Bunn, H. F. and B. G. Forget. 1986. *Hemoglobin: Molecular, Genetic and Clinical Aspects*. W. B. Saunders, Philadelphia, PA.

Buongiorno-Nardelli, M., F. Amaldi, and P. A. Lava-Sanchez. 1972. Amplification as a rectification mechanism for the redundant rRNA genes. Nature 238: 134–137.

Bürglin, T. R. 1994. A comprehensive classification of homeobox genes. pp. 27–71. In: Duboule, D. (ed.) *Guidebook to the Homeobox Genes*. Oxford University Press, New York.

Bürglin, T. R. and G. Ruvkun. 1993. The *Caenorhabditis elegans* homeobox gene cluster. Curr. Opin. Genes Dev. 3: 615–620.

Bürglin, T. R., G. Ruvkun, A. Coulson, N. C. Hawkins, J. D. McGhee, D. Schaller, C. Wittmann, F. Müller, and R. H. Waterston. 1991. Nematode homeobox cluster. Nature 351: 703.

Buroker, N. E. 1983. Population genetics of the American oyster *Crassostrea virginica* along the Atlantic coast and the Gulf of Mexico. Mar. Biol. 75: 99–112.

Buth, D. G. 1979. Duplicate gene expression in tetraploid fishes of the tribe Moxostomatini (Cypriniformes, Catastomidae). Comp. Biochem. Physiol. 63B: 7–12.

Caccone, A. and J. R. Powell. 1989. DNA divergence among hominoids. Evolution 43: 925–942.

Cairns-Smith, A. G. 1982. *Genetics Takeover and Mineral Origins of Life*. Cambridge University Press, Cambridge.

Calabretta, B., D. L. Robberson, H. A. Barrera-Saldana, T. P. Lambrou, and G. F. Saunders. 1982. Genome instability in a region of human DNA enriched in *Alu* repeat sequences. Nature 296: 219–225.

Callan, H. G. 1967. The organization of genetic units in chromosomes. J. Cell Sci. 2: 1–7.

Cann, R. L., M. Stoneking, and A. C. Wilson. 1987. Mitochondrial DNA and human evolution. Nature 325: 31–36.

Cano, R. J. and M. K. Borucki. 1995. Revival and identification of bacterial spores in 25- to 40-million-year-old Dominican amber. Science 268: 1060–1064.

Cano, R. J., H. N. Poinar, N. J. Pieniazek, A. Acra, and G. O. Poinar. 1993. Amplification and sequencing of DNA from a 120–135-million-year-old weevil. Nature 363: 536–538.

Cao, Y., J. Adachi, A. Janke, S. Pääbo, and M. Hasegawa. 1994. Phylogenetic relationships among eutherian orders estimated from inferred sequences of mitochondrial proteins: instability of a tree based on a single gene. J. Mol. Evol. 39: 519–527.

Carrillo, H. and D. Lipman. 1988. The multiple sequence alignment problem in biology. SIAM J. Appl. Math. 48: 1073–1082.

Carroll, S. B. 1995. Homeotic genes and the evolution of arthropods and chordates. Nature 376: 479–485.

Cartwright, P., M. Dick, and L. M. Buss. 1993. HOM/Hox type homeoboxes in the chelicerate *Limulus polyphemus*. Mol. Phylogenet. Evol. 2: 185–192.

Carulli, J. P., D. E. Krane, D. L. Hartl, and H. Ochman. 1993. Compositional heterogeneity and patterns of molecular evolution in the *Drosophila* genome. Genetics 134: 837–845.

Caskey, C. T., A. Pizzuti, Y.-H. Fu, R. G. Fenwick Jr., and D. L. Nelson. 1992. Triplet repeat mutations in human disease. Science 256: 784–789.

Catzeflis, F. M., F. H. Sheldon, J. A. Ahlquist, and C. G. Sibley. 1987. DNA-DNA hybridization evidence of the rapid rate of muroid rodent DNA evolution. Mol. Biol. Evol. 4: 242–253.

Cavalier-Smith, T. 1978. Nuclear volume control by nucleoskeletal DNA, selection for cell volume and cell growth rate and the solution to the DNA C-value paradox. J. Cell. Sci. 34: 247–278.

Cavalier-Smith, T. 1985. *The Evolution of Genome Size*. John Wiley, New York.

Cavalier-Smith, T. 1991. Intron phylogeny: A new hypothesis. Trends Genet. 7: 145–148.

Cavalli-Sforza, L. L. and A. W. F. Edwards. 1967. Phylogenetic analysis: models and estimation procedures. Am. J. Hum. Genet. 19: 233–257.

Cavender, J. A. 1978. Taxonomy with confidence. Math. Biosci. 40: 271–280.

Cavender, J. A. 1989. Mechanized derivation of linear invariants. Mol. Biol. Evol. 6: 301–316.

Cavender, J. A. 1991. Necessary conditions for the method of inferring phylogeny by linear invariants. Math. Biosci. 103: 69–75.

Cavender, J. A. and J. Felsenstein. 1987. Invariants of phylogenies: simple case with discrete states. J. Classif. 4: 57–71.

Cedergren, R., M. W. Gray, Y. Abel, and D. Sankoff. 1988. The evolutionary relationships among known life forms. J. Mol. Evol. 28: 98–112.

Cerff, R. 1995. The chimeric nature of nuclear genomes and the antiquity of introns as demonstrated by the GAPDH gene system. pp. 205–227. In: Gō, M. and P. Schimmel (eds.) *Tracing Biological Evolution in Protein and Gene Structures*. Elsevier Science, Amsterdam.

Chakraborty, R. and S. P. Daiger. 1991. Polymorphisms at VNTR loci suggest homogeneity of the white population of Utah. Hum. Biol. 63: 571–587.

Chakraborty, R. and L. Jin. 1992. Heterozygote deficiency, population substructure and their implications in DNA finger-printing. Hum. Genet. 88: 267–271.

Chakraborty, R. and K. K. Kidd. 1991. The utility of DNA typing in forensic work. Science 254: 1735–1739.

Chakraborty, R., P. A. Fuerst, and M. Nei. 1978. Statistical studies on protein polymorphism in natural populations. II. Gene differentiation between populations. Genetics 88: 367–390.

Chan, L. 1993. RNA editing: Exploring one mode with apolipoprotein B mRNA. BioEssays 15: 33–41.

Chan, L., E. Boerwinkle, and W.-H. Li. 1990. Molecular genetics of the plasma apolipoproteins. pp. 183–219. In: Chien, S. (ed.) *Molecular Biology of the Cardiovascular System*. Lea & Febiger, Philadelphia, PA.

Chang, B. H.-J. and W.-H. Li. 1995. Estimating the intensity of male-driven evolution in rodents by using X-linked and Y-linked *Ube* 1 genes and pseudogenes. J. Mol. Evol. 40: 70–77.

Chang, B. H.-J., L. C. Shimmin, S.-K. Shyue, and D. Hewett-Emmett. 1994. Weak male-driven molecular evolution in rodents. Proc. Natl. Acad. Sci. USA 91: 827–831.

Chao, L., C. Vargas, B. B. Spaear, and E. C. Cox. 1983. Transposable elements as mutator genes in evolution. Nature 303: 633–635.

Chao, S., R. Sederoff, and C. S. Levings III. 1984. Nucleotide sequence and evolution of the 18S ribosomal RNA gene in maize mitochondria. Nucleic Acids Res. 12: 6629–6645.

Charlesworth, B. 1985. The population genetics of transposable elements. pp. 213–232. In: Ohta, T. and K. Aoki (eds.) *Population Genetics and Molecular Evolution*. Springer-Verlag, Berlin.

Charlesworth, B. 1988. The maintenance of transposable elements in natural populations. pp. 189–212. In: Nelson, O. J. (ed.) *Plant Transposable Elements*. Plenum Press, New York.

Charlesworth, B. 1994. The effect of background selection against deleterious mutations on weakly selected, linked variants. Genet. Res. 63: 213–227.

Charlesworth, B. 1994. The effect of background selection against deleterious alleles on weakly selected, linked variants. Genet. Res. 63: 213–228.

Charlesworth, B. and D. Charlesworth. 1983. The population dynamics of transposable elements. Genet. Res. 42: 1–27.

Charlesworth, B. and C. H. Langley. 1989. The population genetics of *Drosophila* transposable elements. Annu. Rev. Genet. 23: 251–287.

Charlesworth, B. and C. H. Langley. 1991. Population genetics of transposable elements in *Drosophila*. pp. 150–176. In: Selander, R. K., A. G. Clark, and T. S. Whittam (eds.) *Evolution at the Molecular Level*. Sinauer Associates, Sunderland, MA.

Charlesworth, B., C. Langley, and W. Stephan. 1986. The evolution of restricted recombination and the accumulation of repeated DNA sequences. Genetics 112: 947–962.

Charlesworth, B., J. A. Coyne, and N. H. Barton. 1987. The relative rates of evolution of sex chromosomes and autosomes. Am. Nat. 130: 113–146.

Charlesworth, B., M. T. Morgan, and D. Charlesworth. 1993. The effect of deleterious mutations on neutral molecular variation. Genetics 134: 1289–1303.

Charlesworth, D., B. Charlesworth, and M. T. Morgan. 1995. The pattern of neutral molecular variation under the background selection model. Genetics 141: 1605–1617.

Chase, M. W. and 41 others. 1993. DNA sequence phylogenetics of seed plants: An analysis of nucleotide sequences from the plastid gene *rbc*L. Ann. Missouri Bot. Gard. 80: 528–580.

Chaw, S.-M., H. Long, B.-S. Wang, A. Zharkikh, and W.-H. Li. 1993. The phylogenetic position of Taxaceae based on 18S rRNA sequences. J. Mol. Evol. 37: 624–630.

Chaw, S.-M., H.-M. Sung, H. Long, A. Zharkikh, and W.-H. Li. 1995. The phylogenetic positions of the conifer genera *Amentotaxus*, *Phyllocladus*, and *Nageia* inferred from 18S rRNA sequences. J. Mol. Evol. 41: 224–230.

Chen, S.-H. and 12 others. 1987. Apolipoprotein B–48 is the product of a messenger RNA with an organ-specific in-frame stop codon. Science. 238: 363–366.

Ciochon, R. L. 1985. Hominoid cladistics and the ancestry of modern apes and humans. pp. 345–362. In: Ciochon, R. L. and J. C. Fleagle (eds.) *Primate Evolution and Human Origin*. Benjamin/Cummings, Menlo Park, California.

Clark, A. G. 1993. Evolutionary inferences from molecular characterization of self-incompatibility alleles. pp. 79–108. In: Takahata, N. and A. G. Clark (eds.) *Mechanisms of Molecular Evolution*. Sinauer Associates, Sunderland, MA.

Clark, J. B., W. P. Maddison, and M. G. Kidwell. 1994. Phylogenetic analysis supports horizontal transfer of *P* transposable elements. Mol. Biol. Evol. 11: 40–50.

Clarke, B. 1970a. Selective constraints on amino acid substitutions during the evolution of proteins. Nature 228: 159–160.

Clarke, B. 1970b. Darwinian evolution of proteins. Science 168: 1009–1011.

Clarke, B. 1971. Natural selection and the evolution of proteins. Nature 232: 487.

Cleary, M. L., E. A. Schon, and J. B. Lingrel. 1981. Two related pseudogenes are the result of a gene duplication in the goat β-globin locus. Cell 26: 181–190.

Clifford, H. T. and J. Constantine. 1980. *Fern, Fern Allies, and Conifers of Australia*. University Queensland Press, Brisbane.

Coen, E. S. and G. A. Dover. 1983. Unequal exchanges and the coevolution of X and Y rDNA arrays in *Drosophila melanogaster*. Cell 33: 849–855.

Cohen, S. N. 1976. Transposable genetic elements and plasmid evolution. Nature 263: 731–735.

Cohen, S. N. and J. A. Shapiro. 1980. Transposable genetic elements. Sci. Am. 242: 40–49.

Collura, R. V. and C.-B. Stewart. 1995. Insertions and duplications of mtDNA in the nuclear genomes of Old World monkeys and hominoids. Nature 378: 485–489.

Comeron, J. M. 1995. A method for estimating the numbers of synonymous and nonsynonymous substitutions per site. J. Mol. Evol. 41: 1152–1159.

Cooper, G. M. 1997. *The Cell: A Molecular Approach.* American Society for Microbiology Press/Sinauer Associates, Sunderland, MA

Coulondre, C., J. H. Miller, P. J. Farabaugh, and W. Gilbert. 1978. Molecular basis of base substitution hotspots in *Escherichia coli.* Nature 274: 775–780.

Covello, P. S. and M. W. Gray. 1989. RNA editing in plant mitochondria. Nature 341: 662–666.

Covello, P. S. and M. W. Gray. 1990. Differences in editing at homologous sites in messenger RNAs from angiosperm mitochondria. Nucleic Acids Res. 18: 5189–5196.

Crane, P. R., E. M. Friis, and K. R. Pedersen. 1995. The origin and early diversification of angiosperms. Nature 374: 27–33.

Crick, F. H. C. 1968. The origin of the genetic code. J. Mol. Biol. 38: 367–379.

Crow, J. F. 1985. The neutrality-selection controversy in the history of evolution and population genetics. pp. 1–18. In: Ohta, T. and K. Aoki (eds.) *Population Genetics and Molecular Evolution.* Springer-Verlag, New York.

Crow, J. F. and M. Kimura. 1970. *An Introduction to Population Genetics Theory.* Harper & Row, New York.

Curtis, S. E. and M. T. Clegg. 1984. Molecular evolution of chloroplast DNA sequences. Mol. Biol. Evol. 1: 291–301.

Czelusniak, J., M. Goodman, D. Hewett-Emmett, M. L. Weiss, P. J. Venta, and R. E. Tashian. 1982. Phylogenetic origins and adaptive evolution of avian and mammalian haemoglobin genes. Nature 298: 297–300.

Daniels, G. R. and P. L. Deininger. 1985. Repeat sequence families derived from mammalian tRNA genes. Nature 317: 819–822.

Daniels, S. B., S. H. Clark, M. G. Kidwell, and A. Chovnick. 1987. Genetic transformation of *Drosophila melanogaster* with an autonomous *P* element: Phenotypic and molecular analyses of long-established transformed lines. Genetics 115: 711–723.

Daniels, S. B., K. R. Peterson, L. D. Strausbaugh, M. G. Kidwell, and A. Chovnick. 1990. Evidence for horizontal transmission of the *P* transposable element between *Drosophila* species. Genetics 124: 339–355.

Darnell, J. E. 1978. Implications of RNA-RNA splicing in evolution of eukaryotic cells. Science 202: 1257–1260.

Darwin, C. 1871. *The Descent of Man and Selection in Relation to Sex.* Appleton, New York.

Dayhoff, M. O. 1972. *Atlas of Protein Sequence and Structure.* Vol. 5, Natl. Biomed. Res. Found., Washington, DC.

Dayhoff, M. O. 1978. *Atlas of Protein Sequence and Structure.* Vol. 5, Suppl. 3, Natl. Biomed. Res. Found., Washington, DC.

De Nettancourt, D. 1977. *Incompatibility in Angiosperms.* Springer-Verlag, New York.

DeBry, R. W. 1992. The consistency of several phylogeny-inference methods under varying evolutionary rates. Mol. Biol. Evol. 9: 537–551.

Deininger, P. L. 1989. SINEs: Short interspersed repeated DNA elements in higher eukaryotes. pp. 619–636. In: Berg, D. E. and M. M. Howe (eds.) *Mobile DNA.* American Society for Microbiology Press, Washington, DC.

Deininger, P. L. and M. A. Batzer. 1994. Evolution of retroposons. Evol. Biol. 27: 157–196.

Deininger, P. L. and G. R. Daniels. 1986. The recent evolution of mammalian repetitive elements. Trends Genet. 2: 76–80.

Deininger, P. L., M. A. Batzer, C. A. Hutchison, and M. H. Edgell. 1992. Master genes in mammalian repetitive DNA amplification. Trends Genet. 8: 307–311.

Deka, R., R. Chakraborty, and R. E. Ferrell. 1991. A population genetic study of six VNTR loci in three ethnically defined populations. Genomics 11: 83–92.

DeSalle, R., J. Gatesy, W. Wheeler, and D. Grimaldi. 1992. DNA sequences from a fossil termite in Oligo-Miocene amber and their phylogenetic implications. Science 257: 1933–1936.

DeSalle, R., M. Barcia, and C. Wray. 1993. PCR jumping in clones of 30-million-year-old DNA fragments from amber preserved termite (*Mastotermes electrodominicus*). Experientia 49: 906–909.

Devos, R., R. Contreras, J. Van Emmela, and W. Fiers. 1979. Identification of the translocatable element *IS1* in a molecular chimera constructed with pBR 322 into which MS2 DNA copy was inserted by poly(dA.dT) linked method. J. Mol. Biol. 128: 621–632.

Dewey, D. R. 1980. Some applications and misapplications of induced polyploidy to plant breeding. pp. 445–470. In: Lewis, W. H. (ed.) *Polyploidy: Biological Relevance.* Plenum Press, New York.

Dibb, N. J. and A. J. Newman. 1989. Evidence that introns arose at protosplice sites. EMBO J. 8: 2015–2021.

DiCaprio, L. and P. J. Hastings. 1976. Gene conversion and intragenic recombination at the *SUP6* locus and the surrounding region in *Saccharomyces cerevisiae.* Genetics 84: 697–721.

Dick, M. H. and L. M. Buss. 1994. A PCR-based survey of homeobox genes in *Ctenodrilus serratus* (Annelida: Polychaeta). Mol. Phylogenet. Evol. 3: 146–158.

Dickerson, R. E. 1971. The structure of cytochrome *c* and the rates of molecular evolution. J. Mol. Evol. 1: 26–45.

Dickerson, R. E. and I. Geis. 1969. *The Structure and Action of Proteins*. Harper & Row,New York.

Dickerson, R. E. and I. Geis. 1983. *Hemoglobin*. Benjamin/Cummings, Menlo Park, CA.

Djian, P. and H. Green. 1989. Vectorial expansion of the involucrin gene and the relatedness of the hominoids. Proc. Natl. Acad. Sci. USA 86: 8447–8451.

Dobzhansky, T. 1955. A review of some fundamental concepts and problems of population genetics. Cold Spring Harbor Symp. Quant. Biol. 20: 1–15.

Doherty, P. C. and R. M. Zinkernagel. 1975. Enhanced immunological surveillance in mice heterozygous at the H-2 gene complex. Nature 256: 50–52.

Doolittle, R. F. 1981. Similar amino acid sequences: chance or common ancestry? Science 214: 149–159.

Doolittle, R. F. 1985. The genealogy of some recently evolved vertebrate proteins. Trends Biochem. Sci. 10: 233–237.

Doolittle, R. F. 1987. The evolution of the vertebrate plasma proteins. Biol. Bull. 172: 269–283.

Doolittle, R. F. 1990. Searching through sequence databases. pp. 99–110. In: Doolittle, R. F. (ed.) *Molecular Evolution: Computer Analysis of Protein and Nucleic Acid Sequences. Methods in Enzymology*. Academic Press, New York.

Doolittle, R. F. 1994. Convergent evolution—the need to be explicit. Trends Biochem. Sci. 19: 15–18.

Doolittle, R. F. 1995. The multiplicity of domains in proteins. Annu. Rev. Biochem. 64: 287–314.

Doolittle, R. F., D.-F. Feng, M. S. Johnson, and M. A. McClure. 1989. Origins and evolutionary relationships of retroviruses. Q. Rev. Biol. 64: 1–31.

Doolittle, W. F. 1978. Genes in pieces: Were they ever together? Nature 272: 581–582.

Doolittle, W. F. 1982. Selfish DNA after fourteen months. pp. 3–28. In: Dover, G. A. and R. B. Flavell (eds.) *Genome Evolution*. Academic Press, New York.

Doolittle, W. F. and J. R. Brown. 1994. Tempo, mode, the progenote, and the universal root. Proc. Natl. Acad. Sci. USA 91: 6721–6728.

Doolittle, W. F. and C. Sapienza. 1980. Selfish genes, the phenotype paradigm and genome evolution. Nature 284: 601–603.

Dorit, R. L. and W. Gilbert. 1991. The limited universe of exons. Curr. Opin. Genet. Dev. 1: 464–469.

Dorit, R. L., L. Schoenbach, and W. Gilbert. 1990. How big is the universe of exons? Science 250: 1377–1382.

Dorit, R. L., H. Akashi, and W. Gilbert. 1995. Absence of polymorphism at the ZFY locus on the human Y chromosome. Science 268: 1183–1185.

Dover, G. A. 1982. Molecular drive: A cohesive mode of species evolution. Nature 299: 111–117.

Dover, G. A. 1986. Molecular drive in multigene families: How biological novelties arise, spread and are assimilated. Trends Genet. 2: 159–165.

Dover, G. A. 1987. DNA turnover and the molecular clock. J. Mol. Evol. 26: 47–58.

Dover, G. A. and R. B. Flavell. 1984. Molecular coevolution: DNA divergence and the maintenance of function. Cell 38: 622–623.

Dowsett, A. P. and M. W. Young. 1982. Differing levels of dispersed repetitive DNA among closely related species of *Drosophila*. Proc. Natl. Acad. Sci. USA 79: 4570–4574.

Doyle, J. A. 1978. Origin of angiosperms. Annu. Rev. Ecol. Syst. 9: 365–392.

Doyle, J. A. and M. J. Donoghue. 1986. Seed plant phylogeny and the origin of angiosperms: An experimental cladistic approach. Bot. Rev. 52: 321–431.

Doyle, J. A., M. J. Donoghue, and E. A. Zimmer. 1994. Integration of morphological and ribosomal RNA data on the origin of the angiosperms. Ann. Missouri Bot. Gard. 81: 419–450.

Drake, J. W. 1991. A constant rate of spontaneous mutation in DNA-based microbes. Proc. Natl. Acad. Sci. USA 88: 7160–7164.

Drake, J. W. 1993. Rates of spontaneous mutation among RNA viruses. Proc. Natl. Acad. Sci. USA 90: 4171–4175.

Drolet, S. and D. Sankoff. 1990. Quadratic invariants for multivalued characters. J. Theor. Biol. 144: 117–129.

Du Vigneaud, V., C. Ressler, and S. Trippett. 1953. The sequence of amino acids in oxytocin, with a proposal for the structure of oxytocin. J. Biol. Chem. 205: 949–957.

Dynan, W. S. 1986. Promoters for houskeeping genes. Trends Genet. 2: 196–197.

Dyson, F. 1985. *Origins of Life*. Cambridge University Press, Cambridge.

Eanes, W. F. 1994. Patterns of polymorphism and between species divergence in the enzymes of central metabolism. pp. 18–28. In: Golding, B. (ed.) *Non-Neutral Evolution: Theories and Molecular Data*. Chapman & Hall, New York.

Eanes, W. F., J. W. Ajioka, J. Hey, and C. Wesley. 1989. Restriction-map variation associated with G6PD polymorphism in natural populations of *Drosophila melanogaster*. Mol. Biol. Evol. 6: 384–397.

Eanes, W. F., M. Kirchner, and J. Yoon. 1993. Evidence for adaptive evolution of the *G6pd* gene in the *Drosophila melanogaster* and *Drosophila simulans* lineages. Proc. Natl. Acad. Sci. USA 90: 7475–7479.

Easteal, S. 1985. Generation time and the rate of molecular evolution. Mol. Biol. Evol. 2: 450–453.

Easteal, S. 1991. The relative rate of cDNA evolution in primates. Mol. Biol. Evol. 8: 115–127.

Easteal, S. and C. Collet. 1994. Consistent variation in amino acid substitution rate, despite uniformity of mutation rate: Protein evolution in mammals is not neutral. Mol. Biol. Evol. 11: 643–647.

Eck, R. V. and M. O. Dayhoff. 1966. *Atlas of Protein Sequence and Structure*. Natl. Biomed. Res. Found., Washington, DC.

Edelman, G. M. and J. A. Gally. 1970. Arrangement and evolution of eukaryotic genes. pp. 962–972. In: Schmitt, F. O. (ed.) *The Neurosciences: Second Study Program*. Rockefeller University Press, New York.

Edgell, M. H., S. C. Hardies, B. Brown, C. Voliva, A. Hill, S. Phillips, M. Comer, F. Burton, S. Weaver, and C. A. Hutchison III. 1983. Evolution of the mouse β-globin complex locus. pp. 1–13. In: Nei, M. and R. K. Koehn (eds.) *Evolution of Genes and Proteins*. Sinauer Associates, Sunderland, MA.

Edwards, A. W. F. and L. L. Cavalli-Sforza. 1964. Reconstruction of evolutionary trees. pp. 67–76. In: Heywood, V. H. and J. McNeill (eds.) *Phenetic and Phylogenetic Classification*. Systematics Association Publ. No. 6.

Edwards, A., A. Civitello, H. A. Hammond, and C. T. Caskey. 1991. DNA typing and genetic mapping with trimeric and tetrameric tandem repeats. Am. J. Hum. Genet. 49: 746–756.

Edwards, Y. H. and D. A. Hopkinson. 1977. Developmental changes in the electrophoretic patterns of human enzymes and other proteins. pp. 19–78. In: Rattazzi, M. C., J. G. Scandalios, and G. S. Whitt (eds.) *Isozymes: Current Topics in Biological and Medical Research*. Vol. 1, Alan R. Liss, New York.

Efron, B. 1982. *The Jackknife, the Bootstrap and Other Resampling Plans*. Soc. Ind. Appl. Math. CBMS/Natl. Sci. Found. Monogr. 38.

Efstratiadis, A. and 14 others. 1980. The structure and evolution of the human β-globin gene family. Cell 21: 653–668.

Eickbush, T. E. 1994. Origin and evolutionary relationships of retroelements. pp. 121–157. In: Morse, S. S. (ed.) *The Evolutionary Biology of Viruses*. Raven Press, New York.

Eigen, M. 1992. *Steps Towards Life*. Oxford University Press, Oxford.

Eigen, M. and P. Schuster. 1979. *Hypercycle: A Principle of Natural Self-Organisation*. Springer-Verlag, Heidelberg, Germany.

Elgin, S. C. R. and H. Weintraub. 1975. Chromosomal proteins and chromatin structure. Annu. Rev. Biochem. 44: 725–774.

Ellegren, H. 1991. DNA typing of museum birds. Nature 354: 113.

Ellerström, S. and A. Hagberg. 1954. Competition between diploids and tetraploids in mixed rye populations. Hereditas 40: 535–537.

Ellsworth, D. L., D. Hewett-Emmett, and W.-H. Li. 1993. Insulin-like growth factor II intron sequences support the hominoid rate-slowdown hypothesis. Mol. Phylogenet. Evol. 2: 315–321.

Ellsworth, D. L., D. Hewett-Emmett, and W.-H. Li. 1994. Evolution of base composition in the insulin and insulin-like growth factor genes. Mol. Biol. Evol. 11: 875–885.

Emerson, S. 1939. A preliminary survey of the *Oenothera organensis* population. Genetics 24: 535–537.

Engels, W. R. 1981. Hybrid dysgenesis in *Drosophila* and the stochastic loss hypothesis. Cold Spring Harbor Symp. Quant. Biol. 45: 561–565.

Engels, W. R. 1986. On the evolution and population genetics of hybrid dysgenesis causing transposable elements in *Drosophila*. Philos. Trans. R. Soc. London [B] 312: 205–215.

Epstein, C. J. 1967. Non-randomness of amino-acid changes in the evolution of homologous proteins. Nature 215: 355–359.

Esposito, M. S. and J. E. Wagstaff. 1981. Mechanisms of mitotic recombination. pp. 341–370. In: Strathern, J. N., E. W. Jones, and J. R. Broach (eds.) *The Molecular Biology of the Yeast* Saccharomyces*: Life Cycle and Inheritance*. Cold Spring Harbor Laboratory Press, Cold Spring Harbor, NY.

Everse, J. and N. Kaplan. 1975. Mechanisms of action and biological functions of various dehydrogenase isozymes. pp. 29–43. In: Markert, C. L. (ed.) *Isozymes II: Physiological Function*. Academic Press, New York.

Ewens, W. J. 1964. The pseudo-transient distribution and its uses in genetics. J. Appl. Prob. 1: 141–156.

Ewens, W. J. 1972. The sampling theory of selectively neutral alleles. Theor. Pop. Biol. 3: 87–112.

Ewens, W. J. 1979. *Mathematical Population Genetics*. Springer-Verlag, Berlin.

Fan, W., M. Kasahara, J. Gutknecht, D. Klein, W. E. Mayer, M. Jonker, and J. Klein. 1989. Shared class II MHC polymorphisms between humans and chimpanzees. Hum. Immunol. 26: 107–121.

Farris, J. S. 1977. On the phenetic approach to vertebrate classification. pp. 823–850. In: Hecht, M. K., P. C. Goody, and B. M. Hecht (eds.) *Major Patterns in Vertebrate Evolution*. Plenum, New York.

Feller, W. 1968. *An Introduction to Probability Theory and its Applications*, 3rd Ed. John Wiley, New York.

Felsenstein, J. 1973. Maximum-likelihood and minimum-steps methods for estimating evolutionary trees from data on discrete characters. Syst. Zool. 22: 240–249.

Felsenstein, J. 1978. The number of evolutionary trees. Syst. Zool. 27: 27–33.

Felsenstein, J. 1981. Evolutionary trees from DNA sequences: A maximum likelihood approach. J. Mol. Evol. 17: 368–376.

Felsenstein, J. 1985a. Confidence limits on phylogenies: an approach using the bootstrap. Evolution 39: 783–791.

Felsenstein, J. 1985b. Confidence limits on phylogenies with a molecular clock. Syst. Zool. 34: 152–161.

Felsenstein, J. 1988. Phylogenies from molecular sequences: Inference and reliability. Annu. Rev. Genet. 22: 521–565.

Felsenstein, J. 1992a. Estimating effective population size from samples of sequences: inefficiency of pairwise and segregation sites as compared to phylogenetic estimates. Genet. Res. 56: 139–147.

Felsenstein, J. 1992b. Estimating effective population size from samples of sequences: A bootstrap Monte Carlo approach. Genet. Res. 60: 209–220.

Felsenstein, J. and H. Kishino. 1993. Is there something wrong with the bootstrap on phylogenies? A reply to Hillis and Bull. Syst. Biol. 42: 193–200.

Feng, D.-F. and R. F. Doolittle. 1987. Progressive sequence alignment as a prerequisite to correct phylogenetic trees. J. Mol. Evol. 25: 351–360.

Feng, D.-F. and R. F. Doolittle. 1990. Progressive alignment and phylogenetic tree construction of protein sequences. pp. 375–387. In: Doolittle, R. F. (ed.) *Molecular Evolution: Computer Analysis of Protein and Nucleic Acid Sequences. Methods in Enzymology.* Academic Press, New York.

Fermi, G. and M. F. Perutz. 1981. *Haemoglobin and Myoglobin. Atlas of Biological Structure.* Vol. 2, Clarendon Press, Oxford.

Ferris, S. D. and G. S. Whitt. 1977. Loss of duplicate gene expression after polyploidization. Nature 265: 258–260.

Ferris, S. D. and G. S. Whitt. 1979. Evolution of the differential regulation of duplicate genes after polyploidization. J. Mol. Evol. 12: 267–317.

Ferris, S. D., S. L. Portnoy, and G. S. Whitt. 1979. The roles of speciation and divergence time in the loss of duplicate gene expression. Theor. Pop. Biol. 15: 114–139.

Ferris, S. D., A. C. Wilson, and W. M. Brown. 1981. Evolutionary tree for apes and humans based on cleavage maps of mitochondrial DNA. Proc. Natl. Acad. Sci. USA 78: 2432–2436.

Fields, C. 1990. Information content of *Caenorhabditis elegans* splice site sequences varies with intron length. Nucleic Acids Res. 18: 1509–1512.

Fields, C., M. D. Adams, O. White, and J. C. Venter. 1994. How many genes in the human genome? Nature Genetics 7: 345–346.

Figueroa, F., E. Günther, and J. Klein. 1988. MHC polymorphisms pre-dating speciation. Nature 335: 265–271.

Filipski, J. 1987. Correlation between molecular clock ticking, codon usage, fidelity of DNA repair, chromosome banding and chromatin compactness in germline cells. FEBS Lett. 217: 184–186.

Finnegan, D. J. and D. H. Fawcett. 1986. Transposable elements in *Drosophila*. Oxford Surv. Eukaryot. Genet. 3: 1–62.

Fisher, R. A. 1930. *The Genetical Theory of Natural Selection.* Clarendon Press, Oxford.

Fisher, R. A. 1935. The sheltering of lethals. Am. Nat. 69: 446–455.

Fitch, W. M. 1971. Toward defining the course of evolution: minimum change for a specific tree topology. Syst. Zool. 20: 406–416.

Fitch, W. M. 1977a. Phylogenies constrained by the crossover process as illustrated by human hemoglobins and a 13-cycle, 11-amino-acid repeat in human apolipoprotein A-I. Genetics 86: 623–644.

Fitch, W. M. 1977b. On the problem of discovering the most parsimonious tree. Am. Nat. 111: 223–257.

Fitch, W. M. 1981. A non-sequential method for constructing trees and hierarchical classifications. J. Mol. Evol. 18: 30–37.

Fitch, W. M. and W. R. Atchley. 1985. Evolution in inbred strains of mice appears rapid. Science 228: 1169–1175.

Fitch, W. M. and E. Margoliash. 1967. Construction of phylogenetic trees. A method based on mutation distances as estimated from cytochrome *c* sequences is of general applicability. Science 155: 279–284.

Fitch, W. M. and T. F. Smith. 1983. Optimal sequence alignments. Proc. Natl. Acad. Sci. USA 80: 1382–1386.

Fitch, W. M. and K. Upper. 1987. The phylogeny of tRNA sequences provides evidence for ambiguity reduction in the origin of the genetic code. Cold Spring Harbor Symp. Quant. Biol. 52: 759–767.

Flavell, R. B. 1986. Repetitive DNA and chromosome evolution in plants. Philos. Trans. R. Soc. London [B] 312: 227–242.

Fleagle, J. G., T. M. Bown, J. D. Obradovich, and E. L. Simons. 1986. Age of the earliest African anthropoids. Science 234: 1247–1249.

Flemington, E., H. D. Bradshaw, V. Traina-Dorge Jr., V. Slagel, and P. L. Deininger. 1987. Sequence, structure and promoter characterization of the human thymidine kinase gene. Gene 52: 267–277.

Flint, J., A. J. Boyce, J. J. Martinson, and J. B. Clegg. 1989. Population bottleneck in Polynesia revealed by minisatellites. Hum. Genet. 83: 257–263.

Florin, R. 1948. On the morphology and relationships of Taxaceae. Bot. Gaz. 110: 31–39.

Florin, R. 1954. The female reproductive organs of conifers and taxads. Biol. Rev. 29: 367–389.

Flower, W. H. 1883. On whales, present and past and their probable origin. Proc. Zool. Soc. Lond. 1883: 466–513.

Fogel, S. and R. K. Mortimer. 1969. Informational transfer in meiotic gene conversion. Proc. Natl. Acad. Sci. USA 62: 96–103.

Fogel, S., R. K. Mortimer, K. Lusnak, and F. Tavares. 1978. Meiotic gene conversion: A signal of the basic recombination event in yeast. Cold Spring Harbor Symp. Quant. Biol. 43: 1325–1341.

Fogel, S., R. K. Mortimer, and K. Lusnak. 1981. Mechanisms of meiotic gene conversion, or "Wanderings on a Foreign Strand." pp. 289–339. In: Strathern, J. N., E. W. Jones, and J. R. Broach (eds.) *The Molecular Biology of the Yeast Saccharomyces.* Cold Spring Harbor Laboratory Press, Cold Spring Harbor, NY.

Fornage, M., L. Chan, G. Siest, and B. Boerwinkle. 1992. Allele frequency distribution of the $(TG)_n \cdot (AG)_n$ microsatellite in the apolipoprotein C-II gene. Genomics 12: 63–68.

Fougerousse, F., R. Meloni, R. Roudaut, and R. S. Beckman. 1992. Tetranucleotide repeat polymorphism at the human N-MYC gene (MYCN). Nucleic Acids Res. 20: 1165.

Fox, G. E. and 18 others. 1980. The phylogeny of prokaryotes. Science 209: 457–463.

Fox, T. D. and C. J. Leaver. 1981. The *Zea mays* mitochondrial gene coding cytochrome oxidase subunit II has an intervening sequence and does not contain TGA codons. Cell 26: 315–323.

Fraser, C. M. and 28 others. 1995. The minimal gene complement of *Mycoplasma genitalium*. Science 270: 397–403.

Friedl, W., E. H. Ludwig, B. Paulweber, F. Sandhofer, and B. McCarthy. 1990. Hypervariability in a minisatellite 3′ of the apolipoprotein B gene in patients with coronary heart disease compared with normal controls. J. Lipid Res. 31: 659–665.

Friedmann, W. 1990. Double fertilization in *Ephedra*, a non-flowering seed plant: Its bearing on the origin of angiosperms. Science 247: 951–954.

Friedmann, W. 1994. The evolution of embryogeny in seed plants and the developmental origin and early history of endosperm. Am. J. Bot. 81: 1468–1486.

Fu, Y.-H. and 12 others. 1991. Variation of the CGG repeat at the fragile X site results in genetic instability: Resolution of the Sherman paradox. Cell 67: 1047–1058.

Fu, Y.-H. and 14 others. 1992. An unstable triplet repeat in a gene related to myotonic muscular dystrophy. Science 255: 1256–1258.

Fu, Y.-X. 1994a. A phylogenetic estimator of effective population size or mutation rate. Genetics 136: 685–692.

Fu, Y.-X. 1994b. Estimating effective population size or mutation rate using the frequencies of mutations of various classes in a sample of DNA sequences. Genetics 138: 1375–1386.

Fu, Y.-X. and W.-H. Li. 1991. Necessary and sufficient conditions for the existence of certain quadratic invariants under a phylogenetic tree. Math. Biosci. 105: 229–238.

Fu, Y.-X. and W.-H. Li. 1992a. Necessary and sufficient conditions for the existence of linear invariants in phylogenetic inference. Math. Biosci. 108: 203–218.

Fu, Y.-X. and W.-H. Li. 1992b. Construction of linear invariants in phylogenetic inference. Math. Biosci. 109: 201–228.

Fu, Y.-X. and W.-H. Li. 1993a. Statistical tests of neutrality of mutations. Genetics 133: 693–709.

Fu, Y.-X. and W.-H. Li. 1993b. Maximum likelihood estimation of population parameters. Genetics 134: 1261–1270.

Fuetterer, J. and T. Hohn. 1987. Involvement of nucleocapsids in reverse transcription: A general phenomenon? Trends Biochem. Sci. 12: 92–95.

Fukagawa, T., K. Sugaya, K.-I. Matsumoto, K. Okumura, A. Ando, H. Inoko, and T. Ikemura. 1995. A boundary of long-range G+C% mosaic domains in the human MHC locus: Pseudoautosomal boundary-like sequence exists near the boundary. Genomics 25: 184–191.

Galas, D. J. and M. Chandler. 1989. Bacterial insertion sequences. pp. 109–162. In: Berg, D. E. and M. M. Howe (eds.) *Mobile DNA*. American Society for Microbiology Press, Washington, DC.

Garcia-Fernández, J. and P. W. H. Holland. 1994. Archetypal organization of the *Amphioxus Hox* gene cluster. Nature 37: 563–566.

Gascuel, O. 1994. A note on Sattath and Tversky's, Saitou and Nei's, and Studier and Keppler's algorithms for inferring phylogenies from evolutionary distances. Mol. Biol. Evol. 11: 961–963.

Gaut, B. S., S. V. Muse, W. D. Clark, and M. T. Clegg. 1992. Relative rates of nucleotide substitution at the *rbc*L locus of monocotyledonous plants. J. Mol. Evol. 35: 292–303.

Gehring, W. J. 1994. A history of the homeobox. pp. 1–10. In: Duboule, D. (ed.) *Guidebook to the Homeobox Genes*. Oxford University Press, New York.

Gehring, W. J., M. Affolter, and T. Bürglin. 1994. Homeodomain proteins. Annu. Rev. Biochem. 63: 487–526.

Gensel, P. G. and H. N. Andrews. 1984. *Plant Life in the Devonian*. Praeger, New York.

Gesteland, R. F. and J. F. Atkins. 1993. *The RNA World*. Cold Spring Harbor Laboratory Press, Cold Spring Harbor, NY.

Gibbs, A. J. and G. A. McIntyre. 1970. The diagram, a method for comparing sequences. Euro, J. Biochem. 16: 1–11.

Gifford, E. M. and A. S. Foster. 1988. *Morphology and Evolution of Vascular Plants*. W.H. Freeman, New York.

Gilbert, W. 1978. Why genes in pieces? Nature 271: 501.

Gilbert, W. 1987. The exon theory of genes. Cold Spring Harbor Symp. Quant. Biol. 52: 901–905.

Gilbert, W., M. Marchionni, and G. McKnight. 1986. On the antiquity of introns. Cell 46: 151–153.

Gillespie, J. H. 1986a. Variability of evolutionary rates of DNA. Genetics 113: 1077–1091.

Gillespie, J. H. 1986b. Rates of molecular evolution. Annu. Rev. Ecol. Syst. 17: 637–665.

Gillespie, J. H. 1987. Molecular evolution and neutral allele theory. Oxford Surveys Evol. Biol. 4: 10–37.

Gillespie, J. H. 1988. More on the overdispersed molecular clock. Genetics 118: 385–386.

Gillespie, J. H. 1989. Lineage effects and the index of dispersion of molecular evolution. Mol. Biol. Evol. 6: 636–647.

Gillespie, J. H. 1991. *The Causes of Molecular Evolution*. Oxford University Press, New York.

Gillespie, J. H. 1994. Alternatives to the neutral theory. pp. 1–17. In: Golding, B. (ed.) *Non-Neutral Evolution: Theories and Molecular Data*. Chapman & Hall, New York.

Gillespie, J. H. and C. H. Langley. 1979. Are evolutionary rates really variable? J. Mol. Evol. 13: 27–34.

Gingerich, P. D. 1984. Primate evolution: Evidence from the fossil record, comparative morphology, and molecular biology. Yrbk. Phys. Anthrop. 27: 57–72.

Gingerich, P. D., B. H. Smith, and E. L. Simons. 1990. Hind limbs of eocene *Basilosaurus*: Evidence of feet in whales. Science 249: 154–157.

Gingerich, P. D., S. M. Raza, M. Arif, M. Anwar, and X. Zhou. 1994. New whale from the Eocene of Pakistan and the origin of cetacean swimming. Nature 368: 844–847.

Ginzburg, L. R., P. M. Bingham, and S. Yoo. 1984. On the theory of speciation induced by transposable elements. Genetics 107: 331–341.

Giovannoni, S. J., S. Turner, G. J. Olsen, S. Barns, D. J. Lane, and N. R. Pace. 1988. Evolutionary relationships among cyanobacteria and green chloroplasts. J. Bacteriol. 170: 3584–3592.

Giroux, M. J., M. Clancy, J. Baier, L. Ingham, D. McCarty, and L. C. Hannah. 1994. *De novo* synthesis of an intron by the maize transposable element *Dissociation*. Proc. Natl. Acad. Sci. USA 91: 12150–12154.

Gō, M. 1981. Correlation of DNA exonic regions with protein structural units in haemoglobin. Nature 291: 90–92.

Gō, M. and M. Nosaka. 1987. Protein architecture and the origin of introns. Cold Spring Harbor Symp. Quant. Biol. 52: 915–924.

Goad, W. B. and M. I. Kanehisa. 1982. Pattern recognition in nucleic acid sequences. I. A general method for finding local homologies and symmetries. Nucleic Acids Res. 10: 247–263.

Goffeau, A. and 15 others. 1996. Life with 6000 genes. Science 274: 546–567.

Gogarten, J. P. and 12 others. 1989. Evolution of the vacuolar H^+-ATPase: Implications for the origin of eukaryotes. Proc. Natl. Acad. Sci. USA 86: 6661–6665.

Gojobori, T. and M. Nei. 1984. Concerted evolution of the immunoglobulin V_H gene family. Mol. Biol. Evol. 1: 195–211.

Gojobori, T. and S. Yokoyama. 1985. Rates of evolution of the retroviral oncogene of moloney murine sarcoma virus and of its cellular homologues. Proc. Natl. Acad. Sci. USA 82: 4198–4201.

Gojobori, T., K. Ishii, and M. Nei. 1982a. Estimation of average number of nucleotide substitutions when the rate of substitution varies with nucleotide. J. Mol. Evol. 18: 414–423.

Gojobori, T., W.-H. Li, and D. Graur. 1982b. Patterns of nucleotide substitution in pseudogenes and functional genes. J. Mol. Evol. 18: 360–369.

Gojobori, T., E. N. Moriyama, and M. Kimura. 1990. Molecular clock of viral evolution, and the neutral theory. Proc. Natl. Acad. Sci. USA 87: 10015–10018.

Golding, B. 1994. Using maximum-likelihood to infer selection from phylogenies. pp. 126–139. In: Golding, B. (ed.) *Non-Neutral Evolution: Theories and Molecular Data*. Chapman & Hall, New York.

Golding, B. and J. Felsenstein. 1990. A maximum likelihood approach to the detection of selection from a phylogeny. J. Mol. Evol. 31: 511–523.

Golding, G. B. and R. S. Gupta. 1995. Protein-based phylogenies support a chimeric origin for the eukaryotic genome. Mol. Biol. Evol. 12: 1–6.

Golding, G. B., C. F. Aquadro, and C. H. Langley. 1986. Sequence evolution within populations under multiple types of mutation. Proc. Natl. Acad. Sci. USA 83: 427–431.

Goldman, D., P. R. Giri, and S. J. O'Brien. 1987. A molecular phylogeny of the hominoid primates as indicated by two-dimensional protein electrophoresis. Proc. Natl. Acad. Sci. USA 84: 3307–3311.

Goldman, M. A., G. P. Holmquist, M. C. Gray, L. A. Caston, and A. Nag. 1984. Replication timing of genes and middle repetitive sequences. Science 224: 686–692.

Goldman, N. 1993. Statistical tests of models of DNA substitution. J. Mol. Evol. 36: 182–198.

Goldman, N. 1994. Variance to mean ratio, $R(t)$ for Poisson processes on phylogenetic trees. Mol. Phylogenet. Evol. 3: 230–239.

Golenberg, E. M., D. E. Giannasi, M. T. Clegg, C. J. Smiley, M. Durbin, D. Henderson, and G. Zurawski. 1990. Chloroplast DNA sequence from a Miocene *Magnolia* species. Nature 344: 656–658.

Goloubinoff, P., S. Pääbo, and A. C. Wilson. 1993. Evolution of maize inferred from sequence diversity of an *adh2* gene segment from archaeological specimens. Proc. Natl. Acad. Sci. USA 90: 1997–2001.

Good, A. G., G. Meister, H. Brock, T. Grigliatti, and D. Hickey. 1989. Rapid spread of transposable elements in experimental populations of *Drosophila melanogaster*. Genetics 122: 387–396.

Goodman, M. 1961. The role of immunochemical differences in the phyletic development of human behavior. Hum. Biol. 33: 131–162.

Goodman, M. 1962. Immunochemistry of the primates and primate evolution. Ann. N.Y. Acad. Sci. 102: 219–234.

Goodman, M. 1963. Serological analysis of the systematics of recent hominoids. Hum. Biol. 35: 377–424.

Goodman, M. 1976. Protein sequences in phylogeny. pp. 141–159. In: Ayala, F. J. (ed.) *Molecular Evolution*. Sinauer Associates, Sunderland, MA.

Goodman, M. 1981. Decoding the pattern of protein evolution. Prog. Biophys. Mol. Biol. 38: 105–164.

Goodman, M., E. Poulik, and M. D. Poulik. 1960. Variations in the serum specificities of higher primates detected by two-dimensional starch-gel electrophoresis. Nature 188: 78–79.

Goodman, M., J. Barnabas, G. Matsuda, and G. W. Moore. 1971. Molecular evolution in the descent of man. Nature 233: 604–613.

Goodman, M., G. W. Moore, and G. Matsuda. 1975. Darwinian evolution in the genealogy of haemoglobin. Nature 253: 603–608.

Goodman, M., A. E. Romero-Herrera, H. Dene, J. Czelusniak, and R. E. Tashian. 1982. Amino acid sequence evidence on the phylogeny of primates and other eutherians. pp. 115–191. In: Goodman, M. (ed.) *Macromolecular Sequences in Systematic and Evolutionary Biology*. Plenum Press, New York.

Goodman, M., B. F. Koop, J. Czelusniak, M. L. Weiss, and J. L. Slightom. 1984. The η-globin gene: Its long evolutionary history in the β-globin gene family of mammals. J. Mol. Biol. 180: 803–823.

Goremykin, V., V. Bobrova, J. Pahnke, A. Troitsky, A. Antonov, and W. Martin. 1996. Noncoding sequences from the slowly evolving chloroplast inverted repeat in addition to *rbc*L data do not support gnetalean affinities of angiosperms. Mol. Biol. Evol. 13:383–396.

Gouy, M. and W.-H. Li. 1989. Phylogenetic analysis based on rRNA sequences supports the archaebacterial rather than the eocyte tree. Nature 339: 145–147.

Grantham, R. 1974. Amino acid difference formula to help explain protein evolution. Science 185: 862–864.

Grantham, R., C. Gautier, M. Gouy, R. Mercier, and A. Pave. 1980. Codon catalog usage and the genome hypothesis. Nucleic Acids Res. 8: r49-r62.

Graur, D. 1985. Amino acid composition and the evolutionary rates of protein-coding genes. J. Mol. Evol. 22: 53–63.

Graur, D. and D. G. Higgins. 1994. Molecular evidence for the inclusion of cetaceans within the order artiodactyla. Mol. Biol. Evol. 11: 357–364.

Graur, D. and W.-H. Li. 1991. Neutral mutation hypothesis test. Nature 354: 114–115.

Graur, D., Y. Shuali, and W.-H. Li. 1989. Deletions in processed pseudogenes accumulate faster in rodents than in humans. J. Mol. Evol. 28: 279–285.

Gray, I. C. and A. J. Jeffreys. 1991. Evolutionary transience of hypervariable minisatellites in man and the primates. Proc. R. Soc. Lond. [B] 243: 241–253.

Gray, M. W. 1989. Origin and evolution of mitochondrial DNA. Annu. Rev. Cell Biol. 5: 25–50.

Green, S. and P. Chambon. 1986. A superfamily of potentially oncogenic hormone receptors. Nature 324: 615–617.

Gregory, W. K. 1910. The orders of mammals. Bull. Am. Mus. Nat. Hist. 27: 1–524.

Griffith, F. 1928. The significance of pneumococcal types. J. Hyg. 27: 113–159.

Griffiths, R. C. and S. Tavaré. 1994. Sampling theory for neutral alleles in a varying environment. Philos. Trans. R. Soc. [B] 344: 403–410.

Grosjean, H. and W. Fiers. 1982. Preferential codon usage in prokaryotic genes: The optimal codon-anticodon interaction energy and the selective codon usage in efficiently expressed genes. Gene 18: 199–209.

Gruskin, K. D., T. F. Smith, and M. Goodman. 1987. Possible origin of a calmodulin gene that lacks intervening sequences. Proc. Natl. Acad. Sci. USA 84: 1605–1608.

Gu, X. and W.-H. Li. 1992. Higher rates of amino acid substitution in rodents than in humans. Mol. Phylogenet. Evol. 1: 211–214.

Gu, X. and W.-H. Li. 1994. A model for the correlation of mutation rate with GC content and the origin of GC-rich isochores. J. Mol. Evol. 38: 468–475.

Gu, X. and W.-H. Li. 1995. The size distribution of insertions and deletions in human and rodent pseudogenes suggests the logarithmic gap penalty for sequence alignment. J. Mol. Evol. 40: 464–473.

Gu, X. and W.-H. Li. 1996a. A general additive distance with time-reversibility and rate variation among nucleotide sites. Proc. Natl. Acad. Sci. USA. 93:4671–4676.

Gu, X. and W.-H. Li. 1996b. Bias-corrected paralinear and LogDet distances and tests of molecular clocks and phylogenies under nonstationary nucleotide. Mol. Biol. Evol. 13: 1375–1383.

Gupta, R. and G. B. Golding. 1993. Evolution of *HSP70* gene and its implications regarding relationships between Archaebacteria, Eubacteria, and Eukaryotes. J. Mol. Evol. 37: 573–582.

Gutz, H. 1971. Site specific induction of gene conversion in *Schizosaccharomyces pombe*. Genetics 69: 317–337.

Haeckel, E. 1894. *Systematische Phylogenie der Protisten und Pflanzen*. Verlag von Gerog Reimer, Berlin.

Hagelberg, E. and J. B. Clegg. 1993. Genetic polymorphisms in prehistoric Pacific islanders determined by analysis of ancient bone DNA. Proc. R. Soc. Lond. [B] 252: 163–170.

Hagelberg, E., B. Sykes, and R. Hedges. 1989. Ancient bone DNA amplified. Nature 342: 485.

Hagelberg, E., I. C. Gray, and A. J. Jeffreys. 1991. Identification of the skeletal remains of a murder victim by DNA analysis. Nature 352: 427–429.

Hagelberg, E., S. Quevedo, D. Turbon, and J. B. Clegg. 1994. DNA from ancient Easter Islanders. Nature 369: 25–26.

Hagemann, S., W. J. Miller, and W. Pinsker. 1992. Identification of a complete *P* element in the genome of *Drosophila bifasciata*. Nucleic Acids Res. 20: 409–413.

Haldane, J. B. S. 1932. *The Causes of Evolution*. Longmans and Green, London.

Haldane, J. B. S. 1933. The part played by recurrent mutation in evolution. Am. Nat. 67: 5–19.

Haldane, J. B. S. 1935. The rate of spontaneous mutation of a human gene. J. Genet. 33: 317–326.

Haldane, J. B. S. 1947. The mutation rate of the gene for hemophilia, and its segregation ratios in males and females. Ann. Eugen. 13: 262–271.

Haldane, J. B. S. 1957. The cost of natural selection. J. Genet. 55: 511–524.

Hall, P. and M. A. Martin. 1988. On bootstrap resampling and iteration. Biometrika 75: 661–671.

Hamada, H., M. G. Petrino, and T. Kakunaga. 1982. A novel repeated element with Z-DNA-forming potential is widely found in evolutionary diverse eukaryotic genomes. Proc. Natl. Acad. Sci. USA 79: 6465–6469.

Hambor, J. E., J. Mennone, M. E. Coon, J. H. Hanke, and P. Kavathas. 1993. Identification and characterization of an *Alu*-containing, T-cell-specific enhancer located in the last intron of the human CD8 gene. Mol. Cell. Biol. 13: 7056–7070.

Hamby, R. K. and E. A. Zimmer. 1992. Ribosomal RNA as a phylogenetic tool in plant systematics. pp. 50–91. In: Soltis, D. E. and J. J. Doyle (eds.) *Molecular Systematics in Plants*. Chapman & Hall, New York.

Hamerton, J. L. 1971. *Human Cytogenetics*. Academic Press, New York.

Hammer, M. F. and S. Horai. 1995. Y chromosomal DNA variation and the peopling of Japan. Am. J. Hum. Genet. 56: 951–962.

Hancock, J. M. and G. A. Dover. 1988. Molecular coevolution among cryptically simple expansion segments of eukaryotic 26S/28S rRNAs. Mol. Biol. Evol. 5: 377–391.

Hänni, C., V. Laudet, M. Sakka, A. Begue, and D. Stehelin. 1990. Amplification de fragments d'ADN mitochondrial a partir de dents de d'os humains anciens. Comptes rendus de l'Academie des Sciences, Paris, series III 310: 365–370.

Hardies, S. C., S. L. Martin, C. F. Voliva, C. A. Hutchison III, and M. H. Edgell. 1986. An analysis of replacement and synonymous changes in the rodent L1 repeat family. Mol. Biol. Evol. 3: 109–125.

Harding, R. M., A. J. Boyce, J. J. Martinson, J. Flint, and J. B. Clegg. 1993. A computer simulation study of VNTR population genetics: Constrained recombination rules out the infinite alleles model. Genetics 135: 911–922.

Hardison, R. C. and J. B. Margot. 1984. Rabbit globin pseudogene ψβ2 is a hybrid of δ- and β-globin gene sequences. Mol. Biol. Evol. 1: 302–316.

Harris, H. 1966. Enzyme polymorphisms in man. Proc. R. Soc. Lond. [B] 164: 298–310.

Harris, H., D. A. Hopkinson, J. E. Luffman, and S. Rapley. 1967. Electrophoretic variation in erythrocyte enzymes. pp. 1–20. In: Beutler, E. (ed.) *Hereditary Disorders of Erythrocyte Metabolism. City of Hope Symposium Series*. Vol. 1, Grune & Stratton, New York.

Harris, J. I., F. Sanger, and M. A. Naughton. 1956. Species differences in insulin. Arch. Biochem. Biophys. 65: 427–438.

Harris, T. M. 1976. The mesozoic gymnosperms. Rev. Palaebot. Palynol. 21: 119–134.

Hart, J. A. 1987. A cladistic analysis of conifers: preliminary results. J. Arnold Arb. 68: 296–307.

Hartl, D. L. and A. G. Clark. 1989. *Principles of Population Genetics*, 2nd Ed. Sinauer Associates, Sunderland, MA.

Hartl, D. L., D. E. Dykhuizen, R. D. Miller, L. Green, and J. de Framond. 1983. Transposable element IS50 improves growth rate of *E. coli* cells without transposition. Cell 35: 503–510.

Hartl, D. L., M. Medhora, L. Green, and D. E. Dykhuizen. 1986. The evolution of DNA sequences in *Escherichia coli*. Philos. Trans. R. Soc. Lond. [B] 312: 191–204.

Hartl, D. L., E. N. Moriyama, and S. Sawyer. 1994. Selection intensity for codon bias. Genetics 138: 227–234.

Hasebe, M., R. Kofuji, M. Ito, K. Iwatsuki, and K. Ueda. 1992. Phylogeny of gymnosperms inferred from *rbc*L gene sequences. Bot. Mag. Tokyo 105: 673–679.

Hasegawa, M., H. Kishino, and T. A. Yano. 1985. Dating of the human-ape splitting by a molecular clock of mitochondrial DNA. J. Mol. Evol. 22: 160–174.

Hasegawa, M., H. Kishino, and N. Saitou. 1991. On the maximum likelihood method in molecular phylogenetics. J. Mol. Evol. 32: 443–445.

Hashimoto, T. and M. Hasegawa. 1996. Origin and early evolution of eukaryotes inferred from the amino acid sequences of translation elongation factors 1α/Tu and 2/G. Adv. Biophys. 32: 73–120.

Hawkins, J. D. 1988. A survey on intron and exon lengths. Nucleic Acids Res. 16: 9893–9908.

Hayashida, H., K. Kuma, and T. Miyata. 1992. Interchromosomal gene conversion as a possible mechanism for explaining divergence patterns of ZFY-related genes. J. Mol. Evol. 35: 181–183.

Heddle, J. A. and K. Athanasiou. 1975. Mutation rate, genome size and their relation to the *rec*. concept. Nature 258: 359–361.

Hedges, R. W. and A. E. Jacob. 1974. Transposition of ampicillin resistance from RP4 to other replicons. Mol. Gen. Genet. 132: 31–40.

Hedges, S. B. 1992. The number of replications needed for accurate estimation of the bootstrap P value in phylogenetic studies. Mol. Biol. Evol. 9: 366–369.

Hedges, S. B. and M. H. Schweitzer. 1995. Detecting dinosaur DNA. Science 268: 1191–1192.

Hedges, S. B., S. Kumar, K. Tamura, and M. Stoneking. 1992. Human origins and analysis of mitochondrial DNA sequences. Science 255: 737–739.

Hedrick, P. W. and G. Thomson. 1983. Evidence for balancing selection at HLA. Genetics 104: 449–456.

Hedrick, P. W., W. Klitz, W. P. Robinson, M. K. Kuhner, and G. Thomson. 1991. Population genetics of HLA. pp. 248–271. In: Selander, R. K., A. G. Clark, and T. S. Whittam (eds.) *Evolution at the Molecular Level*. Sinauer Associates, Sunderland, MA.

Hein, J. 1990. Unified approach to alignment and phylogenies. pp. 626–644. In: Doolittle, R. F. (ed.) *Molecular Evolution: Computer Analysis of Protein and Nucleic Acid Sequences. Methods in Enzymology*. Academic Press, New York.

Hendriks, W., J. W. M. Mulders, M. A. Bibby, C. Slingsby, H. Bloemendal, and W. W. De Jong. 1988. Duck lens F-crystallin and lactate dehydrogenase B4 are identical: A single-copy gene product with two distinct functions. Proc. Natl. Acad. Sci. USA 85: 7114–7118.

Hendy, M. D. and D. Penny. 1982. Branch and bound algorithms to determine minimal evolutionary trees. Math. Biosci. 59: 277–290.

Hendy, M. D. and D. Penny. 1989. A framework for the quantitative study of evolutionary trees. Syst. Zool. 38: 297–309.

Herdman, M. 1985. The evolution of bacterial genomes. pp. 37–68. In: Cavalier-Smith. T. (ed.) *The Evolution of Genome Size*. John Wiley, London.

Hess, J. F., M. Fox, C. Schmid, and C.-K. J. Shen. 1983. Molecular evolution of the human adult α-globin-like gene region: Insertion and deletion of *Alu* family repeats and non-*Alu* DNA sequences. Proc. Natl. Acad. Sci. USA 80: 5970–5974.

Hey, J. and R. M. Kliman. 1993. Population genetics and phylogenetics of DNA sequence variation at multiple loci within the *Drosophila melanogaster* species complex. Mol. Biol. Evol. 10: 804–822.

Hibner, B. L., W. D. Burke, and T. H. Eickbush. 1991. Sequence identity in an early chorion multigene family is the result of localized gene conversion. Genetics 128: 595–606.

Hickey, D. A. 1982. Selfish DNA: A sexually-transmitted nuclear parasite. Genetics 101: 519–531.

Hickey, D. A. and B. Benkel. 1986. Introns as relict retrotransposons: Implications for the evolutionary origin of eukaryotic mRNA splicing mechanisms. J. Theor. Biol. 121: 283–291.

Hickey, D. A., B. F. Benkel, and S. M. Abukashawa. 1989. A general model for the evolution of nuclear pre-mRNA introns. J. Theor. Biol. 137: 41–53.

Hickey, L. J. and J. A. Doyle. 1977. Early Cretaceous fossil evidence for angiosperm evolution. Bot. Review 43: 3–104.

Hiebl, I., G. Braunitzer, and D. Schneeganss. 1987. The primary structures of the major and minor hemoglobin-components of adult Andean goose (*Chloephaga melanoptera*, Anatidae): The mutation Leu → Ser in position 55 of the β-chains. Z. Biol. Chem. Hoppe-Seyler 368: 1559–1569.

Hiebl, I., R. E. Weber, D. Schneeganss, J. Kösters, and G. Braunitzer. 1988. Structural adaptations in the major and minor hemoglobin components of adult Rüppell's griffon (*Gyps rueppellii*, Aegypiinae): A new molecular pattern for hypoxic tolerance. Z. Biol. Chem. Hoppe-Seyler 369: 217–232.

Hiebl, I., G. Braunitzer, R. E. Weber, and J. Kösters. 1989. Structural adaptations for high-altitude respiration in bird hemoglobins: Barheaded goose (*Anser indicus, Gorni gus*), Andean goose (*Chloephaga melanoptera*) and Rüppell's griffon (*Gyps rueppellii*). pp. 299–313. In: Koenig, W. A. and W. Voelter (eds.) *Proc. of VIth USSR-FRG Symposium "Chemistry of Peptides and Proteins."* Vol. 4, Walter de Gruyter & Co., Berlin.

Hiesel, R., B. Combettes, and A. Brennicke. 1994. Evidence for RNA editing in mitochondria of all major groups of land plants except the Bryophyta. Proc. Natl. Acad. Sci. USA 91: 629–633.

Higgins, D. G., A. J. Bleasby, and R. Fuchs. 1992. CLUSTAL V: Improved software for multiple sequence alignment. CABIOS 8: 189–191.

Higgins, N. P. and D. Hillyard. 1988. Primary structure and mapping of the *hupA* gene of *Salmonella typhimurium*. J. Bacteriol. 170: 5751–5758.

Higuchi, R., B. Bowman, M. Freiberger, O. A. Ryder, and A. C. Wilson. 1984. DNA sequences from the quagga, an extinct member of the horse family. Nature 312: 282–284.

Higuchi, R. G., C. H. Von Beroldingen, G. F. Sensabaugh, and H. A. Erlich. 1988. DNA typing from single hairs. Nature 332: 543–546.

Hill, A. V. S., C. E. M. Allsopp, D. Kwiatkowski, N. M. Anstey, P. Twumasi, P. A. Rowe, S. Bennett, D. Brewster, A. J. McMichael, and B. M. Greenwood. 1991. Common West African HLA antigens are associated with protection from severe malaria. Nature 352: 595–600.

Hill, A. V. S., D. Kwiatkowski, A. J. McMichael, B. M. Greenwood, and S. Bennett. 1992. Maintenance of MHC polymorphism. Nature 355: 403.

Hilliker, A. J. and A. Chovnick. 1981. Further observations on intragenic recombination in *Drosophila melanogaster*. Genet. Res. 38: 281–296.

Hillis, D. M. and J. J. Bull. 1993. An empirical test of bootstrapping as a method for assessing confidence on phylogenetic analysis. Syst. Biol. 42: 182–192.

Hillis, D. M., J. J. Bull, M. E. White, M. R. Badgett, and I. J. Molineux. 1992. Experimental phylogenetics: generation of a known phylogeny. Science 255: 589–592.

Hillis, D. M., J. P. Huelsenbeck, and C. W. Cunningham. 1994. Application and accuracy of molecular phylogenies. Science 264: 671–677.

Hixson, J. E. and W. M. Brown. 1986. A comparison of the small ribosomal RNA genes from the mitochondrial DNA of the great apes and humans: Sequence, structure, evolution and phylogenetic implications. Mol. Biol. Evol. 3: 1–18.

Hobbs, H. H., M. S. Brown, J. L. Goldstein, and D. W. Russell. 1986. Deletion of exon encoding cysteine rich repeat of low density lipoprotein receptor alters its binding specificity in a subject with familial hypercholesterolemia. J. Biol. Chem. 261: 13114–13120.

Hogeweg, P. and B. Hesper. 1984. The alignment of sets of sequences and the construction of phyletic trees: An integrated method. J. Mol. Evol. 20: 175–186.

Holland, J., K. Spindler, F. Horodyski, E. Grabau, S. Nichol, and S. VandePol. 1982. Rapid evolution of RNA genomes. Science 215: 1577–1585.

Holland, M. M., D. L. Fisher, L. G. Mitchell, W. C. Rodriquez, J. J. Canik, C. R. Merril, and V. W. Weedn. 1993. Mitochondrial DNA sequence analysis of human skeletal remains: identification of remains from the Vietman war. J. Forensic Science 38: 542–553.

Holliday, R. 1964. A mechanism for gene conversion in fungi. Genet. Res. 5: 282–304.

Holmquist, G. P. 1987. Role of replication time in the control of tissue-specific gene expression. Am. J. Hum. Genet. 40: 151–173.

Holmquist, G. P. and J. Filipski. 1994. Organization of mutations along the genome: A prime determinant of genome evolution. Trends Ecol. Evol. 9: 65–69.

Holmquist, R. 1972. Theoretical foundations for a quantitative approach to paleogenetics. I: DNA. J. Mol. Evol. 1: 115–133.

Holmquist, R. and D. Pearl. 1980. Theoretical foundations for quantitative paleogenetics III. The molecular divergence of nucleic acids and proteins for the case of genetic events of unequal probability. J. Mol. Evol. 16: 211–267.

Holmquist, R., M. M. Miyamoto, and M. Goodman. 1988. Analysis of higher-primate phylogeny from transversion differences in nuclear and mitochondrial DNA by Lake's methods of evolutionary parsimony and operator metrics. Mol. Biol. Evol. 5: 217–236.

Holum, J. R. 1978. *Organic and Biological Chemistry*. John Wiley, New York.

Hood, L., J. H. Campbell, and S. C. R. Elgin. 1975. The organization, expression and evolution of antibody genes and other multigene families. Annu. Rev. Genet. 9: 305–353.

Horai, S. and K. Hayasaka. 1990. Intraspecific nucleotide sequence differences in the major noncoding region of human mitochondrial DNA. Am. J. Hum. Genet. 46: 828–842.

Horai, S., K. Hayasaka, K. Murayama, N. Wate, H. Koike, and N. Nakai. 1989. DNA amplification from ancient human skeletal remains and their sequence analysis. Proc. Jap. Acad. Series B 65: 229–233.

Horai, S., Y. Satta, K. Hayasaka, R. Kondo, T. Inoue, T. Ishida, S. Hayashi, and N. Takahata. 1992. Man's place in hominoidea revealed by mitochondrial DNA genealogy. J. Mol. Evol. 35: 32–43.

Horai, S., R. Kondo, Y. Nakagawa-Hattorri, S. Hayashi, S. Sonoda, and K. Tajima. 1993. Peopling of the Americas, founded by four major lineages of mitochondrial DNA. Mol. Biol. Evol. 10: 23–47.

Horai, S., K. Hayasaka, R. Kondo, K. Tsugane, and N. Takahata. 1995. Recent African origin of modern humans revealed by complete sequences of hominoid mitochondrial DNAs. Proc. Natl. Acad. Sci. USA 92: 532–536.

Hori, H., B.-L. Lim, and S. Osawa. 1985. Evolution of green plants as deduced from 5S rRNA sequences. Proc. Natl. Acad. Sci. USA 82: 820–823.

Horowitz, M., S. Luria, G. Rechavi, and D. Givol. 1984. Mechanism of activation of the mouse c-*mos* oncogene by the LTR of an intracisternal A-particle gene. EMBO J. 3: 2937–2941.

Höss, M. and S. Pääbo. 1993. DNA extraction from pleistocene bones by a silica-based purification method. Nucleic Acids Res. 21: 3913–3914.

Houck, C. M., F. P. Rinehart, and C. W. Schmid. 1979. Ubiquitous family of repeated DNA sequences in the human genome. J. Mol. Biol. 132: 289–306.

Houck, M. A., J. B. Clark, K. R. Peterson, and M. G. Kidwell. 1991. Possible horizontal transfer of *Drosophila* genes by the mite *Proctolaelaps regalis*. Science 253: 1125–1129.

Hourcade, D., D. Dressler, and J. Wolfson. 1973. The amplification of ribosomal RNA genes involves a rolling circle intermediate. Proc. Natl. Acad. Sci. USA 70: 2926–2930.

Hu, X. and R. G. Worton. 1992. Partial gene duplication as a cause of human disease. Hum. Mutat. 1: 3–12.

Huang, H., S. Crews, and L. Hood. 1981. An immunoglobulin V_H pseudogene. J. Mol. Appl. Genet. 1: 93–101.

Huang, L.-S. and J. L. Breslow. 1987. A unique AT-rich hypervariable minisatellite 3′ to the *apo B* gene defines a high information restriction fragment length polymorphism. J. Biol. Chem. 262: 8952–8955.

Hudson, R. R. 1982. Estimating genetic variability with restriction endonucleases. Genetics 100: 711–719.

Hudson, R. R. 1990. Gene genealogies and the coalescent process. Oxford Surveys Evol. Biol. 7: 1–44.

Hudson, R. R. 1993. The how and why of generating gene genealogies. pp. 23–36. In: Takahata, N. and A. G. Clark (eds.) *Mechanisms of Molecular Evolution: Introduction to Molecular Paleopopulation Biology*. Sinauer Associates, Sunderland, MA.

Hudson, R. R., M. Kreitman, and M. Aguadé. 1987. A test of neutral molecular evolution based on nucleotide data. Genetics 116: 153–159.

Hudson, T. J., M. Engelstein, M. K. Lee, E. C. Ho, M. J. Rubenfield, C. P. Adams, D. E. Housman, and N. C. Dracopoli. 1992. Isolation and chromosomal assignment of 100 highly informative human simple sequence repeat polymorphisms. Genomics 13: 622–629.

Huelsenbeck, J. P. and D. M. Hillis. 1993. Success of phylogenetic methods in the four-taxon case. Syst. Biol. 42: 247–264.

Hughes, A. L. and M. Nei. 1988. Pattern of nucleotide substitution at major histocompatibility complex class I loci reveals overdominant selection. Nature 335: 167–170.

Hughes, A. L. and M. Nei. 1989. Nucleotide substitution at major histocompatibility complex class II loci: Evidence for overdominant selection. Proc. Natl. Acad. Sci. USA 86: 958–962.

Hughes, A. L. and M. Nei. 1992. Maintenance of MHC polymorphism. Nature 355: 402–403.

Hummerich, H. and H. Lehrach. 1995. Trinucleotide repeat expansion and human disease. Electrophoresis 16: 1698–1704.

Hurst, L. D. 1994. The uncertain origin of introns. Nature 371: 381–382.

Hutchison, C. A. III, S. C. Hardies, D. D. Loeb, W. R. Shehee, and M. H. Edgell. 1989. LINEs and related retroposons: Long interspersed repeated sequences in the eukaryotic genome. pp. 593–617. In: Berg, D. E. and M. M. Howe (eds.) *Mobile DNA*. American Society for Microbiology Press, Washington, DC.

Hutchinson, G. B., S. E. Andrew, H. McDonald, Y. P. Goldberg, R. Graham, J. M. Rommens, and M. R. Hayden. 1993. An *Alu* element retroposition in two families with Huntington disease defines a new active *Alu* subfamily. Nucleic Acids Res. 21: 3379–3383.

Ibbotson, R., D. M. Hunt, J. K. Bowmaker, and J. D. Mollon. 1992. Sequence divergence and copy number of the middle- and long-wave photopigment genes in Old World monkeys. Proc. R. Soc. Lond. [B] 247: 145–154.

Ikemura, T. 1981. Correlation between the abundance of *Escheriachia coli*: Transfer RNAs and the occurrence of the respective codons in its protein genes: A proposal for a synonymous codon choice that is optimal for the *E. coli* translational system. J. Mol. Biol. 151: 389–409.

Ikemura, T. 1982. Correlation between the abundance of yeast transfer RNAs and the occurrence of the respective codons in protein genes: Differences in synonymous codon choice patterns of yeast and *Escherichia coli* with reference to the abundance of isoaccepting transfer RNAs. J. Mol. Biol. 158: 573–597.

Ikemura, T. 1985. Codon usage and tRNA content in unicellular and multicellular organisms. Mol. Biol. Evol. 2: 13–34.

Ikemura, T. and S.-I. Aota. 1988. Global variation in G+C content along vertebrate genome DNA: Possible correlation with chromosome band structures. J. Mol. Biol. 203: 1–13.

Ina, Y. 1995. New methods for estimating the numbers of synonymous and nonsynonymous substitutions. J. Mol. Evol. 40: 190–226.

Ina, Y., M. Mizokami, K. Ohba, and T. Gojobori. 1994. Reduction of synonymous substitutions in the core protein gene of hepatitis C virus. J. Mol. Evol. 38: 50–56d.

Inamine, J. M., K. Ho, S. Loechel, and P. Hu. 1990. Evidence that UGA is read as tryptophan rather than stop by *Mycoplasma pneumoniae, Mycoplasma genitalium* and *Mycoplasma gallisepticum*. J. Bacteriol. 172: 504–506.

Inouye, S., M.-Y. Hsu, S. Eagle, and M. Inouye. 1989. Reverse transcriptase associated with the biosynthesis of the branched RNA-linked msDNA in *Myxococcus xanthus*. Cell 56: 709–717.

Ioerger, T. R., A. G. Clark, and T.-H. Kao. 1990. Polymorphism at the self-incompatibility locus in Solanaceae predates speciation. Proc. Natl. Acad. Sci. USA 87: 9732–9735.

Irwin, D. M. and A. C. Wilson. 1990. Concerted evolution of ruminant stomach lysozymes. Characterization of lysozyme cDNA clones from sheep and deer. J. Biol. Chem. 265: 4944–4952.

Irwin, D. M., T. D. Kocher, and A. C. Wilson. 1991. Evolution of the cytochrome *b* gene of mammals. J. Mol. Evol. 32: 128–144.

Iwabe, N., K. I. Kuma, M. Hasegawa, S. Osawa, and T. Miyata. 1989. Evolutionary relationship of archaebacteria, eubacteria and eukaryotes inferred from phylogenetic trees of duplicated genes. Proc. Natl. Acad. Sci. USA 86: 9355–9359.

Iwabe, N., K. Kuma, H. Kishino, M. Hasegawa, and T. Miyata. 1990. Compartmentalized isoenzymes and the origin of intron. J. Mol. Evol. 31: 205–210.

Jackson, J. A. and G. R. Fink. 1985. Meiotic recombination between duplicated genetic elements in *Saccharomyces cerevisiae*. Genetics 109: 303–332.

Jacobs, G. H. and J. Neitz. 1986. Spectral mechanisms and color vision in the tree shrew (*Tupaia belangeri*). Vision Res. 26: 291–298.

Jacobs, L. L. and D. Pilbeam. 1980. Of mice and men: Fossil-based divergence dates and molecular "clocks." J. Hum. Evol. 9: 551–555.

Jacq, C., J. R. Miller, and G. G. Brownlee. 1977. A pseudogene structure in 5S DNA of *Xenopus laevis*. Cell 12: 109–120.

Jagadeeswaran, P., B. G. Forget, and S. M. Weissman. 1981. Short, interspersed repetitive DNA elements in eukaryotes: Transposable DNA elements generated by reverse transcription of RNA pol III transcripts? Cell 26: 141–142.

Janczewski, D. N., N. Yukhi, D. A. Gilbert, G. T. Jefferson, and S. J. O'Brien. 1992. Molecular phylogenetic inference from saber-toothed cat fossils of Rancho La Brea. Proc. Natl. Acad. Sci. USA 89: 9769–9773.

Jeffreys, A. 1979. DNA sequence variants in $^G\gamma$-, $^A\gamma$-, δ- and β-globin genes of man. Cell 18: 1–10.

Jeffreys, A. J. 1993. 1992 William Allan Award Address. Am. J. Hum. Genet. 53: 1–5.

Jeffreys, A. J., V. Wilson, and S. L. Thein. 1985a. Individual-specific 'fingerprints' of human DNA. Nature 316: 76–79.

Jeffreys, A. J., V. Wilson, and S. L. Thein. 1985b. Hypervariable 'minisatellite' regions in human DNA. Nature 314: 67–73.

Jeffreys, A. J., N. J. Royle, V. Wilson, and Z. Wong. 1988. Spontaneous mutation rates to new length alleles at tandem-repetitive hypervariable loci in human DNA. Nature 332: 278–281.

Jeffreys, A. J., R. Neumann, and V. Wilson. 1990. Repeat unit sequence variation in minisatellites: A novel source of DNA polymorphism for studying variation and mutation by single molecule analysis. Cell 60: 473–485.

Jeffreys, A. J., A. MacLeod, K. Tamaki, D. L. Neil, and D. G. Monckton. 1991. Minisatellite repeat coding as a digital approach to DNA typing. Nature 354: 204–209.

Jeffreys, A. J., M. J. Allen, J. A. L. Armour, A. Collick, Y. Dubrova, N. Fretwell, T. Guram, M. Jobling, C. A. May, D. L. Neil, and R. Neumann. 1995. Mutation processes at human minisatellites. Electrophoresis 16: 1577–1585.

Jensen, E. Q., K. Paludan, J. J. Hyldig-Nielsen, P. Jorgensen, and K. A. Markere. 1981. The structure of a chromosomal leghaemoglobin gene from soybean. Nature 291: 677–679.

Jessen, T.-H., R. E. Weber, G. Fermi, J. Tame, and G. Braunitzer. 1991. Adaptation of bird hemoglobins to high altitudes: Demonstration of molecular mechanism by protein engineering. Proc. Natl. Acad. Sci. USA 88: 6519–6522.

Jin, L. and M. Nei. 1990. Limitations of the evolutionary parsimony method of phylogenetic analysis. Mol. Biol. Evol. 7: 82–102.

Jinks-Robertson, S. and T. D. Petes. 1985. High-frequency meiotic gene conversion between repeated genes on non-homologous chromosomes in yeast. Proc. Natl. Acad. Sci. USA 82: 3350–3354.

Jinks-Robertson, S. and T. D. Petes. 1986. Chromosomal translocations generated by high-frequency meiotic recombination between repeated yeast genes. Genetics 114: 731–752.

John, B. and G. L. G. Miklos, 1988. *The Eukaryote Genome in Development and Evolution.* Allen and Unwin, London.

Johnson, W. E. and R. K. Selander. 1971. Protein variation and systematics in kangaroo rats (genus *Dipodomys*). Syst. Zool. 20: 377–405.

Jollès, J., E. M. Prager, E. S. Alnemri, P. Jollès, I. M. Ibrahimi, and A. C. Wilson. 1990. Amino acid sequences of stomach and nonstomach lysozymes of ruminants. J. Mol. Evol. 30: 370–382.

Judd, S. R. and T. D. Petes. 1988. Physical lengths of meiotic and mitotic gene conversion tracts in *Saccharomyces cerevisiae*. Genetics 118: 401–410.

Jukes, T. H. 1985. A change in the genetic code in *Mycoplasma capricolum*. J. Mol. Evol. 22: 361–362.

Jukes, T. H. and V. Bhushan. 1986. Silent nucleotide substitutions and G+C content of some mitochondrial and bacterial genes. J. Mol. Evol. 24: 39–44.

Jukes, T. H. and C. R. Cantor. 1969. Evolution of protein molecules. pp. 21–123. In: Munro, H. N. (ed.) *Mammalian Protein Metabolism.* Academic Press, New York.

Jukes, T. H. and J. L. King. 1971. Deleterious mutations and neutral substitutions. Nature 231: 114–115.

Jukes, T. H. and S. Osawa. 1991. Recent evidence for evolution of the genetic code. pp. 79–95. In: Osawa, S. and T. Honjo (eds.) *Evolution of Life: Fossils, Molecules, and Culture.* Springer-Verlag, Tokyo.

Jurka, J. 1995. Origin and evolution of *Alu* repetitive elements. pp. 25–41. In: Maraia, R. J. (ed.) *The Impact of Short Interspersed Elements (SINEs) on the Host Genome.* R. G. Landes Company, Austin, TX.

Jurka, J. and A. Milosavljevic. 1991. Reconstruction and analysis of human *Alu* genes. J. Mol. Evol. 32: 105–121.

Jurka, J. and T. Smith. 1988. A fundamental division in the *Alu* family of repeated sequences. Proc. Natl. Acad. Sci. USA 85: 4775–4778.

Kafatos, F. C., A. Efstratiadis, B. G. Forget, and S. M. Weissman. 1977. Molecular evolution of human and rabbit β-globin mRNAs. Proc. Natl. Acad. Sci. USA 74: 5618–5622.

Kagawa, Y., H. Nojima, N. Nukiwa, M. Ishizuka, T. Nakajima, T. Yasuhara, T. Tanaka, and T. Oshima. 1984. High guanine plus cytosine content in the third letter of codons of an extreme thermophile. J. Biol. Chem. 259: 2956–2960.

Kan, Y. W., J. P. Holland, A. M. Dozy, S. Charache, and H. H. Kazazian. 1975. Deletion of the β-globin structure gene in hereditary persistence of foetal haemoglobin. Nature 258: 162–163.

Kaplan, N. L. and R. R. Hudson. 1987. On the divergence of genes in a multigene family. Theor. Pop. Biol. 31: 178–194.

Kaplan, N. L. and K. Risko. 1982. A method for estimating rates of nucleotide substitution using DNA sequence data. Theor. Pop. Biol. 21: 318–328.

Kaplan, N., T. Darden, and C. H. Langley. 1985. Evolution and extinction of transposable elements in Mendelian populations. Genetics 109: 459–480.

Kaplan, N. L., T. Darden, and R. R. Hudson. 1988. The coalescent process in models with selection. Genetics 120: 819–829.

Kaplan, N. L., R. R. Hudson, and C. H. Langley. 1989. The "hitchhiking effect" revisited. Genetics 123: 887–899.

Kaplan, N., R. R. Hudson, and M. Lizuka. 1991. The coalescent process in models with selection, recombination and geographic subdivision. Genet. Res. 57: 83–91.

Kappen, C. and F. H. Ruddle. 1993. Evolution of a regulatory gene family: HOM/HOX genes. Curr. Opin. Gen. Dev. 3: 931–938.

Kappen, C., K. Schughart, and F. H. Ruddle. 1989. Two steps in the evolution of Antennapedia-class vertebrate homeobox genes. Proc. Natl. Acad. Sci. USA 86: 5459–5463.

Kappen, C., K. Schughart, and F. H. Ruddle. 1993. Early evolutionary origin of major homeodomain sequence classes. Genomics 18: 54–70.

Karathanasis, S. K., V. I. Zannis, and J. L. Breslow. 1983. Isolation and characterization of the human apolipoprotein A-I gene. Proc. Natl. Acad. Sci. USA 80: 6147–6151.

Karin, M. and R. I. Richards. 1984. The human metallothionein gene family: Structure and expression. Environ. Health Perspect. 54: 111–115.

Karl, S. A. and J. C. Avise. 1992. Balancing selection at allozyme loci in oysters: Implications from nuclear RFLPs. Science 256: 100–102.

Keng, H. 1974. The phylloclade of *phyllocladus* and its possible bearing on the branch systems of progymnosperms. Ann. Bot. 38: 757–764.

Keng, H. 1978. The genus *Phyllocladus* (Phyllocladaceae). J. Arnold Arb. 59: 249–273.

Kersanach, R., H. Brinkmann, M.-F. Liaud, D.-X. Zhang, W. Martin, and R. Cerff. 1994. Five identical intron positions in ancient duplicated genes of eubacterial origin. Nature 367: 387–389.

Kessler, L. G. and J. C. Avise. 1985. A comparative description of mitochondrial DNA differentiation in selected avian and other vertebrate genera. Mol. Biol. Evol. 2: 109–125.

Ketterling, R. P., E. Vielhaber, C. D. K. Bottema, D. J. Schaid, M. P. Cohen, C. L. Sexauer, and S. S. Sommer. 1993. Germ-line origins of mutation in families with Hemophilia B: The sex ratio varies with the type of mutation. Am. J. Hum. Genet. 52: 152–166.

Kidwell, M. G. 1979. Hybrid dysgenesis in *Drosophila melanogaster*: The relationship between the *P-M* and *I–R* interaction systems. Genet. Res. 33: 205–217.

Kidwell, M. G. 1983. Evolution of hybrid dysgenesis determinants in *Drosophila melanogaster*. Proc. Natl. Acad. Sci. USA 80: 1655–1659.

Kidwell, M. G. 1993. Lateral transfer in natural populations of eukaryotes. Annu. Rev. Genet. 27: 235–256.

Kidwell, M. G. and J. F. Kidwell. 1976. Selection for male recombination in *Drosophila melanogaster*. Genetics 84: 333–351.

Kim, A., C. Terzian, P. Santamaria, A. Pelisson, N. Prud'homme, and A. Bucheton. 1994. Retroviruses in invertebrates: The *gypsy* retrotransposon is apparently an infectious retrovirus of *Drosophila melanogaster*. Proc. Natl. Acad. Sci. USA 91: 1285–1289.

Kim, J. H., C. Y. Yu, A. Bailey, R. Hardison, and C. K. Shen. 1989. Unique sequence organization and erythroid cell-specific nuclear factor-binding of mammalian θ1 globin promoters. Nucleic Acids Res. 17: 5687–5700.

Kimura, M. 1962. On the probability of fixation of mutant genes in populations. Genetics. 47: 713–719.

Kimura, M. 1968a. Evolutionary rate at the molecular level. Nature 217: 624–626.

Kimura, M. 1968b. Genetic variability maintained in a finite population due to mutational production of neutral and nearly neutral isoalleles. Genet. Res. 11: 247–269.

Kimura, M. 1969. The rate of molecular evolution considered from the standpoint of population genetics. Proc. Natl. Acad. Sci. USA 63: 1181–1188.

Kimura, M. 1977. Preponderance of synonymous changes as evidence for the neutral theory of molecular evolution. Nature 267: 275–276.

Kimura, M. 1979. The neutral theory of molecular evolution. Sci. Am. 241: 94–104.

Kimura, M. 1980. A simple method for estimating evolutionary rates of base substitutions through comparative studies of nucleotide sequences. J. Mol. Evol. 16: 111–120.

Kimura, M. 1981. Estimation of evolutionary distances between homologous nucleotide sequences. Proc. Natl. Acad. Sci. USA 78: 454–458.

Kimura, M. 1983. *The Neutral Theory of Molecular Evolution*. Cambridge University Press, Cambridge.

Kimura, M. and J. F. Crow. 1964. The number of alleles that can be maintained in a finite population. Genetics 49: 725–738.

Kimura, M. and J. L. King. 1979. Fixation of a deleterious allele at one of two "duplicate" loci by mutation pressure and random drift. Proc. Natl. Acad. Sci. USA 76: 2858–2861.

Kimura, M. and T. Ohta. 1969. The average number of generations until fixation of a mutant gene in a finite population. Genetics 61: 763–771.

Kimura, M. and T. Ohta. 1971a. Protein polymorphism as a phase of molecular evolution. Nature 229: 467–469.

Kimura, M. and T. Ohta. 1971b. *Theoretical Aspects of Population Genetics*. Princeton University Press, Princeton, New Jersey.

Kimura, M. and T. Ohta. 1972. On the stochastic model for estimation of mutational distance between homologous proteins. J. Mol. Evol. 2: 87–90.

Kimura, M. and T. Ohta. 1973. Mutation and evolution at the molecular level. Genet. Suppl. 73: 19–35.

Kimura, M. and T. Ohta. 1974. On some principles governing molecular evolution. Proc. Natl. Acad. Sci. USA 71: 2848–2852.

King, J. L. 1967. Continuously distributed factors affecting fitness. Genetics 55: 483–492.

King, J. L. 1972. The role of mutation in evolution. Proc. 6th Berkeley Symp. Math. Statist. Probab. 5: 69–100.

King, J. L. and T. H. Jukes. 1969. Non-Darwinian evolution. Science 164: 788–798.

Kingman, J. F. C. 1982. On the genealogy of large populations. J. Appl. Probab. 19A: 27–43.

Kishino, H. and M. Hasegawa. 1989. Evaluation of the maximum likelihood estimate of the evolutionary tree topologies from DNA sequence data, and the branching order in Hominoidea. J. Mol. Evol. 29: 170–179.

Kishino, H., T. Miyata, and M. Hasegawa. 1990. Maximum likelihood inference of protein phylogeny and the origin of chloroplasts. J. Mol. Evol. 31: 151–160.

Kiss, T., C. Marshallsay, and W. Filipowicz. 1991. Alteration of the RNA polymerase specificity of U3 snRNA genes during evolution and in vitro. Cell 65: 517–526.

Kiyasu, P. K. and M. G. Kidwell. 1984. Hybrid dysgenesis in *Drosophila melanogaster*: The evolution of mixed P and M populations maintained at high temperature. Genet. Res. 44: 251–259.

Klein, H. L. 1984. Lack of association between intrachromosomal gene conversion and reciprocal exchange. Nature 310: 748–753.

Klein, H. L. and T. D. Petes. 1981. Intrachromosomal gene conversion in yeast. Nature 289: 144–148.

Klein, J. 1986. *Natural History of the Major Histocompatibility Complex*. John Wiley, New York.

Klein, J. and F. Figueroa. 1986. Evolution of the major histocompatibility complex. CRC Crit. Rev. Immunol. 6: 295–386.

Klein, J., N. Takahata, and F. J. Ayala. 1993. MHC polymorphism and human origins. Sci. Am. 269: 78–83.

Kleinschmidt, T., J. März, K. D. Jürgens, and G. Braunitzer. 1986. Interaction of allosteric effectors with α-globin chains and high altitude respiration of mammals. The primary structure of two tylopoda hemoglobins with high oxygen affinity: Vicuna (*Lama vicugna*) and Alpaca (*Lama pacos*). Z. Biol. Chem. Hoppe-Seyler 367: 153–160.

Kliman, R. M. and J. Hey. 1994. The effects of mutation and natural selection on codon bias in the genes of *Drosophila*. Genetics 137: 1049–1056.

Klotz, L. C., N. Komar, R. L. Blanken, and R. M. Mitchell. 1979. Calculation of evolutionary trees from sequence data. Proc. Natl. Acad. Sci. USA 76: 4516–4520.

Knott, T. J., S. C. Wallis, R. J. Pease, L. M. Powell, and J. Scott. 1986. A hypervariable region 3' to the human apolipoprotein B gene. Nucleic Acids Res. 14: 9215–9217.

Kocher, T. D and A. C. Wilson. 1991. Sequence evolution of mitochondrial DNA in human and chimpanzees: Control region and a protein-coding region. pp. 391–413. In: Osawa, S. and T. Honjo (eds.) *Evolution of Life: Fossils, Molecules, and Culture*. Springer-Verlag, Tokyo.

Kohne, D. E. 1970. Evolution of higher-organism DNA. Q. Rev. Biophys. 33: 1–48.

Kohne, D. E., J. A. Chiscon, and B. H. Hoyer. 1972. Evolution of primate DNA sequences. J. Hum. Evol. 1: 627–644.

Konigsberg, W. and G. N. Godson. 1983. Evidence for use of rare codons in the *dnaG* gene and other regulatory genes of *Escherichia coli*. Proc. Natl. Acad. Sci. USA 80: 687–691.

Konigsberg, W., G. Guidotti, and R. J. Hill. 1961. The amino acid sequence of the α-chain of human hemoglobin. J. Biol. Chem. 236: PC 55–56.

Konings, D. A. M., P. Hogeweg, and B. Hesper. 1987. Evolution of the primary and secondary structure of the E1a mRNAs of the adenovirus. Mol. Biol. Evol. 4: 300–314.

Koop, B. F., M. Goodman, P. Xu, K. Chan, and J. L. Slightom. 1986. Primate η-globin DNA sequences and man's place among the great apes. Nature 319: 234–238.

Kornberg, A. 1982. *Supplement to DNA Replication*. W.H. Freeman, San Francisco.

Kornberg, A., L. L. Bertsch, J. E. Jackson, and H. G. Khorana. 1964. Enzymatic synthesis of deoxyribonucleic acid XVL: Oligonucleotides as templates and the mechanism of their replication. Biochemistry 51: 315–323.

Kornegay, J. R., J. W. Schilling, and A. C. Wilson. 1994. Molecular adaptation of a leaf-eating bird: Stomach lysozyme of the hoatzin. Mol. Biol. Evol. 11: 921–928.

Kraemer, P. M., R. L. Ratliff, M. F. Bartholdi, N. C. Brown, and J. L. Longmire. 1989. Use of variable number of tandem repeat sequences for monitoring chromosomal instability. Prog. in Nucl. Acids Res. and Mol. Bio. 36: 187–204.

Krajewski, C., A. C. Driskell, P. R. Baverstock, and M. J. Braun. 1992. Phylogenetic relationships of the thylacine (Mammalia, Thylacinidae) among dasyuroid marsupials: Evidence from cytochrome *b* DNA sequences. Proc. R. Soc. Lond. [B] 250: 19–27.

Krane, D. E. and R. C. Hardison. 1990. Short interspersed repeats in rabbit DNA can provide functional polyadenylation signals. Mol. Biol. Evol. 7: 1–8.

Kreitman, M. 1983. Nucleotide polymorphism at the alcohol dehydrogenase locus of *Drosophila melanogaster*. Nature 304: 412–417.

Kreitman, M. and M. Aguadé. 1986. Excess polymorphism at the *Adh* locus in *Drosophila melanogaster*. Genetics 114: 93–110.

Kreitman, M. and R. R. Hudson. 1991. Inferring the evolutionary histories of the *Adh* and *Adh-dup* loci in *Drosophila melanogaster* from patterns of polymorphism and divergence. Genetics 127: 565–582.

Krystal, M., P. D'Eustachio, F. H. Ruddle, and N. Arnheim. 1981. Human nucleolus organizers on non-homologous chromosomes can share the same ribosomal gene variants. Proc. Natl. Acad. Sci. USA 78: 5744–5748.

Kuhner, M. K. and J. Felsenstein. 1994. A simulation comparison of phylogeny algorithms under equal and unequal evolutionary rates. Mol. Biol. Evol. 11: 459–468.

Kuhner, M. K., J. Yamato, and J. Felsenstein. 1995. Estimation of effective population size and mutation rate from sequence data using Metropolis-Hastings sampling. Genetics 140: 1421–1430.

Kumar, S., K. Tamura, and M. Nei. 1993. *MEGA: Molecular Evolutionary Genetic Analysis, version 1. 0.* Pennsylvania State University, University Park, PA.

Kushiro, A., M. Shimizu, and K.-I. Tomita. 1987. Molecular cloning and sequence determination of the *tuf* gene coding for the elongation factor Tu of *Thermus thermophilus* HB8. Eur. J. Biochem. 170: 93–98.

La Spada, A. R., E. M. Wilson, D. B. Lubahn, A. E. Harding, and H. Fishbeck. 1991. Androgen receptor gene mutation in X-linked spinal and bulbar muscular atrophy. Nature 352: 77–79.

Laird, C. D., B. L. McConaughy, and B. J. McCarthy. 1969. Rate of fixation of nucleotide substitutions in evolution. Nature 224: 149–154.

Lake, J. A. 1987. A rate-independent technique for analysis of nucleic acid sequences: Evolutionary parsimony. Mol. Biol. Evol. 4: 167–191.

Lake, J. A. 1988. Origin of the eukaryotic nucleus determined by rate-invariant analysis of rRNA sequences. Nature 331: 184–186.

Lake, J. A. 1994. Reconstructing evolutionary trees from DNA and protein sequences: Paralinear distances. Proc. Natl. Acad. Sci. USA 91: 1455–1459.

Lake, J. A., E. Henderson, M. Oakes, and M. W. Clark. 1984. Eocytes: A new ribosome structure indicates a kingdom with a close relationship to eukaryotes. Proc. Natl. Acad. Sci. USA 81: 3786–3790.

Lakovaara, S., A. Saura, and C. T. Falk. 1972. Genetic distance and evolutionary relationships in the *Drosophila obscura* group. Evolution 26: 177–184.

Lamb, B. C. and S. Helmi. 1982. The extent to which gene conversion can change allele frequencies in populations. Genet. Res. 39: 199–217.

Lampson, B. C., J. Sun, M.-Y. Hsu, J. Vallejo-Ramirez, S. Inouye, and M. Inouye. 1989. Reverse transcriptase in a clinical strain of *Escherichia coli*: Production of branched RNA-linked msDNA. Science 243: 1033–1038.

Lanave, C., G. Preparata, C. Saccone, and G. Serio. 1984. A new method for calculating evolutionary substitution rates. J. Mol. Evol. 20: 86–93.

Lanfear, J. and P. W. H. Holland. 1991. The molecular evolution of *ZFY*-related genes in birds and mammals. J. Mol. Evol. 32: 310–315.

Langley, C. H. and W. M. Fitch. 1974. An examination of the constancy of the rate of molecular evolution. J. Mol. Evol. 3: 161–177.

Langley, C. H., J. F. Y. Brookfield, and N. Kaplan. 1983. Transposable elements in Mendelian populations. I. A theory. Genetics 104: 457–471.

Langley, C. H., E. A. Montgomery, R. H. Hudson, N. L. Kaplan, and B. Charlesworth. 1988. On the role of unequal exchange in the containment of transposable element copy number. Genet. Res. 52: 223–235.

Langley, C. H., J. MacDonald, N. Miyashita, and M. Aguadé. 1993. Lack of correlation between interspecific divergence and intraspecific polymorphism at the suppressor of forked region in *Drosophila melanogaster* and *Drosophila simulans*. Proc. Natl. Acad. Sci. USA 90: 1800–1803.

Laroche, J., P. Li, and J. Bousquet. 1995. Mitochondrial DNA and monocot-dicot divergence time. Mol. Biol. Evol. 12: 1151–1156.

Larson, A. 1991. Evolutionary analysis of length-variable sequences: divergent domains of ribosomal RNA. pp. 221–248. In: Miyamoto, M. M. and J. Cracraft (eds.) *Phylogenetic Analysis of DNA Sequences*. Oxford University Press, New York.

Lawlor, D. A., F. E. Ward, P. D. Ennis, A. P. Jackson, and P. Parham. 1988. HLA-A and B polymorphisms predate the divergence of humans and chimpanzees. Nature 335: 268–271.

Lawlor, D. A., C. D. Dickel, W. W. Hauswirth, and P. Parham. 1991. Ancient HLA genes from 7,500-year-old archaeological remains. Nature 349: 785–788.

Lawrence, J. G. and D. L. Hartl. 1992. Inference of horizontal genetic transfer from molecular data: An approach using the bootstrap. Genetics 131: 753–760.

Lawson, F. S., R. L. Charlebois, and J.-A. R. Dillon. 1996. Phylogenetic analysis of carbamoylphosphate synthetase genes: Complex evolutionary history includes an internal duplication within a gene which can root the tree of life. Mol. Biol. Evol. 13: 970–977.

Leder, P. 1982. The genetics of antibody diversity. Sci. Am. 246: 102–115.

Leeds, J. M., M. B. Slabourgh, and C. K. Mathews. 1985. DNA precursor pools and ribonucleotide reductase activity: Distribution between the nucleus and cytoplasm of mammalian cells. Mol. Cell. Biol. 5: 3443–3450.

Leffers, H., J. Kjems, Østergaard L, N. Larsen, and R. A. Garrett. 1987. Evolutionary relationships amongst archaebacteria: A comparative study of 23S ribosomal RNAs of a sulphur-dependent extreme thermophile, an extreme halophile and a thermophilic methanogen. J. Mol. Biol. 195: 43–61.

Lehninger, A. L. 1982. *Principles of Biochemistry.* Worth Publishers, New York.

Lehrman, M. A., D. W. Russell, J. L. Goldsmith, and M. S. Brown. 1986. Exon-*Alu* recombination deletes 5 kilobases from the low density lipoprotein receptor gene, producing a null phenotype in familial hypercholesterolemia. Proc. Natl. Acad. Sci. USA 83: 3679–3683.

Leunissen, J. A. M. and W. W. de Jong. 1986. Copper/Zinc superoxide dismutase: How likely is gene transfer from ponyfish to *Photobacterium leiognathi*?. J. Mol. Evol. 23: 250–258.

Levinson, G. and G. A. Gutman. 1987. Slipped-strand mispairing: A major mechanism for DNA sequence evolution. Mol. Biol. Evol. 4: 203–221.

Levy, J. A., H. Fraenkel-Conrat, and R. A. Owens. 1994. *Virology,* 3rd Ed. Prentice Hall, Englewood Cliffs, NJ.

Lewin, B. 1994. *Genes V*. Oxford University Press, Oxford.

Lewin, R. 1981. Evolutionary history written in globin genes. Science 214: 426–429.

Lewis, D. 1949. Incompatibility in flowering plants. Biol. Rev. Camb. Philos. Soc. 24: 472–496.

Lewis, E. B. 1978. A gene complex controlling segmentation in *Drosophila*. Nature 276: 565–570.

Lewis, W. H. 1980. *Polyploidy: Biological Relevance*. Plenum Press, New York.

Lewontin, R. C. 1974. *The Genetic Basis of Evolutionary Change*. Columbia University Press, New York.

Lewontin, R. C. and D. L. Hartl. 1991. Population genetics in forensic DNA typing. Science 254: 1745–1750.

Lewontin, R. C. and J. L. Hubby. 1966. A molecular approach to the study of genic heterozygosity in natural populations. II. Amount of variation and degree of heterozygosity in natural populations of *Drosophila pseudoobscura*. Genetics 54: 595–609.

Li, P. and J. Bousquet. 1992. Relative-rate test for nucleotide substitutions between two lineages. Mol. Biol. Evol. 9: 1185–1189.

Li, W.-H. 1976. Electrophoretic identity of proteins in a finite population and genetic distance between taxa. Genet. Res. 28: 119–127.

Li, W.-H. 1977a. Maintenance of genetic variability under mutation and selection pressures in a finite population. Proc. Natl. Acad. Sci. USA 74: 2509–2513.

Li, W.-H. 1977b. Distribution of nucleotide differences between two randomly chosen cistrons in a finite population. Genetics 85: 331–337.

Li, W.-H. 1980. Rate of gene silencing at duplicate loci: a theoretical study and interpretation of data from tetraploid fishes. Genetics 95: 237–258.

Li, W.-H. 1981. Simple method for constructing phylogenetic trees from distance matrices. Proc. Natl. Acad. Sci. USA 78: 1085–1089.

Li, W.-H. 1982. Evolutionary change of duplicate genes. pp. 55–92. In: Rattazzi, M. C., J. G. Scandalios, and G. S. Whitt (eds.), *Isozymes VI: Current Topics in Biological and Medical Research*. Alan R. Liss, New York.

Li, W.-H. 1983. Evolution of duplicate genes and pseudogenes. pp. 14–37. In: Nei, M. and R. K. Koehn (eds.) *Evolution of Genes and Proteins*. Sinauer Associates, Sunderland, MA.

Li, W.-H. 1986. Evolutionary change of restriction cleavage sites and phylogenetic inference. Genetics 113: 187–213.

Li, W.-H. 1987. Models of nearly neutral mutations with particular implications for nonrandom usage of synonymous codons. J. Mol. Evol. 24: 337–345.

Li, W.-H. 1989. A statistical test of phylogenies estimated from sequence data. Mol. Biol. Evol. 6: 424–435.

Li, W.-H. 1993. Unbiased estimation of the rates of synonymous and nonsynonymous substitution. J. Mol. Evol. 36: 96–99.

Li, W.-H. and M. Gouy. 1991. Statistical methods for testing molecular phylogenies. pp. 249–277. In: Miyamoto, M. M. and J. Cracraft (eds.) *Phylogenetic Analysis of DNA Sequences*. Oxford University Press, New York.

Li, W.-H. and D. Graur. 1991. *Fundamentals of Molecular Evolution*. Sinauer Associates, Sunderland, MA.

Li, W.-H. and M. Nei. 1977. Persistence of common alleles in two related populations or species. Genetics 86: 901–914.

Li, W.-H. and L. A. Sadler. 1991. Low nucleotide diversity in man. Genetics 129: 513–523.

Li, W.-H. and M. Tanimura. 1987. The molecular clock runs more slowly in man than in apes and monkeys. Nature 326: 93–96.

Li, W.-H. and A. Zharkikh. 1994. What is the bootstrap technique? Syst. Biol. 43: 424–430.

Li, W.-H. and A. Zharkikh. 1995. Statistical tests of DNA phylogenies. Syst. Biol. 44: 49–63.

Li, W.-H., C.-I. Wu, and C.-C. Luo. 1984. Nonrandomness of point mutation as reflected in nucleotide substitutions in pseudogenes and its evolutionary implications. J. Mol. Evol. 21: 58–71.

Li, W.-H., C.-C. Luo, and C.-I. Wu. 1985a. Evolution of DNA sequences. pp. 1–94. In: MacIntyre, R. J. (ed.) *Molecular Evolutionary Genetics*. Plenum Press, New York.

Li, W.-H., C.-I. Wu, and C.-C. Luo. 1985b. A new method for estimating synonymous and nonsynonymous rates of nucleotide substitution considering the relative likelihood of nucleotide and codon changes. Mol. Biol. Evol. 2: 150–174.

Li, W.-H., M. Tanimura, and P. M. Sharp. 1987a. An evaluation of the molecular clock hypothesis using mammalian DNA sequences. J. Mol. Evol. 25: 330–342.

Li, W.-H., K. H. Wolfe, J. Sourdis, and P. M. Sharp. 1987b. Reconstruction of phylogenetic trees and estimation of divergence times under nonconstant rates of evolution. Cold Spring Harbor Symp. Quant. Biol. 52: 847–856.

Li, W.-H., M. Tanimura, and P. M. Sharp. 1988a. Rates and dates of divergence between AIDS virus nucleotide sequences. Mol. Biol. Evol. 5: 313–330.

Li, W.-H., M. Tanimura, C.-C. Luo, S. Datta, and L. Chan. 1988b. The apolipoprotein multigene family: biosynthesis, structure, structure-function relationships, and evolution. J. Lipid Res. 29: 245–271.

Lichten, M., R. H. Borts, and J. E. Haber. 1987. Meiotic gene conversion and crossing over between dispersed homologous sequences occurs frequently in *Saccharomyces cerevisiae*. Genetics 115: 233–246.

Lidgard, S. and P. R. Crane. 1988. Quantitative analyses of early angiosperm radiation. Nature 331: 344–346.

Liebhaber, S. A., M. Goossens, and Y. W. Kan. 1981. Homology and concerted evolution at the α1 and α2 loci of human α-globin. Nature 290: 26–29.

Lindsley, D. L. and 15 others. 1972. Segmental aneuploidy and the genetic gross structure of the *Drosophila* genome. Genetics 11: 157–184.

Linial, M. 1987. Creation of a processed pseudogene by retroviral infection. Cell 49: 93–102.

Lipman, D. J. and W. R. Pearson. 1985. Rapid and sensitive protein similarity searches. Science 227: 1435–1441.

Lipman, D. J. and W. J. Wilbur. 1985. Interaction of silent and replacement changes in eukaryotic coding sequences. J. Mol. Evol. 21: 161–167.

Lipman, D. J., S. F. Altschul, and J. D. Kececioglu. 1989. A tool for multiple sequence alignment. Proc. Natl. Acad. Sci. USA 86: 4412–4415.

Little, P. F. R. 1982. Globin pseudogenes. Cell 28: 683–684.

Lockhart, P. J., M. A. Steel, M. D. Hendy, and D. Penny. 1994. Recovering evolutionary trees under a more realistic model of sequence evolution. Mol. Biol. Evol. 11: 605–612.

Loconte, H. and D. W. Stevenson. 1990. Cladistics of spermatophyta. Brittonia 42: 197–211.

Logsdon, J. M. Jr. and J. D. Palmer. 1994. Origin of introns—early or late? Nature 369: 526.

Logsdon, J. M. Jr., M. G. Tyshenko, C. Dixon, J. D. Jafari, V. K. Walker, and J. D. Palmer. 1995. Seven newly discovered intron positions in the triosephosphate isomerase gene: Evidence for the introns-late theory. Proc. Natl. Acad. Sci. USA 92: 8507–8511.

Lokki, J. and A. Saura. 1980. Polyploidy in insect evolution. pp. 277–312. In: Lewis, W. H. (ed.) *Polyploidy: Biological Relevance*. Plenum Press, New York.

Lomedico, P., N. Rosenthal, A. Efstratiadis, W. Gilbert, R. Kolodner, and R. Tizard. 1979. The structure and evolution of the two nonallelic rat preproinsulin genes. Cell 18: 545–558.

Long, M. and C. H. Langley. 1993. Natural selection and the origin of *jingwei*, a processed functional gene in *Drosophila*. Science 260: 91–95.

Long, M. and J. H. Gillespie. 1991. Codon usage divergence of homologous vertebrate genes and codon usage clock. J. Mol. Evol. 32: 6–15.

Long, M., S. J. de Souza, and W. Gilbert. 1995a. Evolution of the intron/exon structure of eukaryotic genes. Curr. Opin. Genet. Devel. 5: 774–778.

Long, M., C. Rosenberg, and W. Gilbert. 1995b. Intron phase correlations and the evolution of the intron/exon structure of genes. Proc. Natl. Acad. Sci. USA 92: 12495–12499.

Loomis, W. F. 1988. *Four Billion Years: An Essay on the Evolution of Genes and Organisms*. Sinauer Associates, Sunderland, MA.

Lujan, R., Q. Y. Zhang, J. A. Saez Nieto, D. M. Jones, and B. G. Spratt. 1991. Penicillin-resistant isolates of *Neisseria lactamica* produce altered forms of penicillin-binding protein 2 that arose by interspecies horizontal gene transfer. Antimicrob. Agents Chemother 35: 300–304.

Luo, C.-C., W.-H. Li, M. N. Moore, and L. Chan. 1986. Structure and evolution of the apolipoprotein multigene family. J. Mol. Biol. 187: 325–340.

Luo, C.-C., W.-H. Li, and L. Chan. 1989. Structure and expression of dog apolipoprotein A-I, E, and C-I mRNAs: Implications for the evolution and functional constraints of apolipoprotein structure. J. Lipid Res. 30: 1735–1746.

MacPherson, J. N., B. S. Weir, and A. J. Leigh-Brown. 1990. Extensive linkage disequilibrium in the *achaete-scute* complex of *Drosophila melanogaster*. Genetics 126: 121–129.

Maeda, N. and O. Smithies. 1986. The evolution of multigene families: Human haptoglobin genes. Annu. Rev. Genet. 20: 81–108.

Maeda, N., J. B. Bliska, and O. Smithies. 1983. Recombination and balanced chromosome polymorphism suggested by DNA sequence 5' to the human δ-globin gene. Proc. Natl. Acad. Sci. USA 80: 5012–5016.

Maeda, N., C.-I. Wu, J. Bliska, and J. Reneke. 1988. Molecular evolution of intergenic DNA in higher primates: Pattern of DNA changes, molecular clock and evolution of repetitive sequences. Mol. Biol. Evol. 5: 1–20.

Mahley, R. W. 1988. Apolipoprotein E: Cholesterol transport protein with expanding role in cell biology. Science 240: 622–630.

Maizel, J. and R. Lenk. 1981. Enhanced graphic matrix analysis of nucleic acid and protein sequences. Proc. Natl. Acad. Sci. USA 78: 7665–7669.

Maloney, D. and S. Fogel. 1987. Gene conversion, unequal crossing-over, and mispairing at a nontandem duplication during meiosis of *Saccharomyces cerevisiae*. Curr. Genet. 12: 1–7.

Marchionni, M. and W. Gilbert. 1986. The triosephosphate isomerase gene from maize: Introns antedate the plant/animal divergence. Cell 46: 133–141.

Margoliash, E. 1963. Primary structure and evolution of cytochrome *c*. Proc. Natl. Acad. Sci. USA 50: 672–679.

Markert, C. L. 1964. Cellular differentiation: An expression of differential gene function. pp. 163–174. In: *Congenital Malformations*. International Medical Congress, New York.

Markert, M. L., J. J. Hutton, D. A. Wiginton, J. C. States, and R. E. Kaufman. 1988. Adenosine deaminase (ADA) deficiency due to deletion of the ADA gene promoter and first exon by homologous recombination between two *Alu* elements. J. Clin. Invest. 81: 1323–1327.

Martin, A. P. and S. R. Palumbi. 1993. Body size, metabolic rate, generation time, and the molecular clock. Proc. Natl. Acad. Sci. USA 90: 4087–4091.

Martin, S. L., K. A. Vincent, and A. C. Wilson. 1983. Rise and fall of the δ-globin gene. J. Mol. Biol. 164: 513–528.

Martin, W., A. Gierl, and H. Saedler. 1989. Molecular evidence for pre-Cretaceous angiosperm origins. Nature 339: 46–48.

Martin-Campos, J. M., J. M. Comerón, N. Miyashita, and M. Aguadé. 1992. Intraspecific and interspecific variation at the *y-ac-sc* region of *Drosophila simulans* and *Drosophila melanogaster*. Genetics 130: 805–816.

Maruyama, T. and M. Kimura. 1974. A note on the speed of gene frequency changes in reverse directions in a finite population. Evolution 28: 162–163.

Matera, A. G., U. Hellmann, and C. W. Schmid. 1990. A transpositionally and transcriptionally competent *Alu* subfamily. Mol. Cell Biol. 10: 5424–5432.

Matheson, A. T., J. Auer, C. Ramírez, and A. Böck. 1990. Structure and evolution of archaebacterial ribosomal proteins. pp. 617–635. In: Hill, W. E., A. Dahlberg, R. A. Garrett, P. B. Moore, D. Schlessinger, and J. R. Warner (eds.) *The Ribosome: Structure, Function, and Evolution*. Am. Soc. Microbiol., Washington, D. C.

Maxson, L. R. and A. C. Wilson. 1975. Albumin evolution and organismal evolution in tree frogs (Hylidae). Syst. Zool. 24: 1–15.

Mayer, W. E., M. Jonker, D. Klein, P. Ivanyi, G. van Seventer, and J. Klein. 1988. Nucleotide sequences of chimpanzee MHC class I alleles: Evidence for trans-species mode of evolution. EMBO J. 7: 2765–2774.

Maynard Smith, J. 1968. 'Haldane's dilemma' and the rate of evolution. Nature 219: 1114–1116.

Maynard Smith, J. 1992. Analyzing the mosaic structure of genes. J. Mol. Evol. 34: 126–129.

Maynard Smith, J. and J. Haigh. 1974. The hitch-hiking effect of a favorable gene. Genet. Res. 23: 23–35.

Maynard Smith, J., C. G. Dowson, and B. G. Spratt. 1991. Localized sex in bacteria. Nature 349: 29–31.

Mazodier, P. and J. Davies. 1991. Gene transfer between distantly related bacteria. Annu. Rev. Genet. 25: 147–171.

McCarrey, J. R. and K. Thomas. 1987. Human testis-specific PGK gene lacks introns and possesses characteristics of a processed gene. Nature 326: 501–505.

McClintock, B. 1952. Chromosome organization and gene expression. Cold Spring Harbor Symp. Quant. Biol. 16: 13–47.

McClure, M. A., T. K. Vasi, and W. M. Fitch. 1994. Comparative analysis of multiple protein-sequence alignment methods. Mol. Biol. Evol. 11: 571–592.

McConnell, T. J., W. S. Talbot, R. A. McIndoe, and E. K. Wakeland. 1988. The origin of MHC class II gene polymorphism within the genus *Mus*. Nature 332: 651–654.

McDonald, J. F. 1993. *Transposable Elements and Evolution*. Kluwer Academic Publishers, Boston, MA.

McDonald, J. F. 1995. Transposable elements: Possible catalysts of organismic evolution. Trends Ecol. Evol. 10: 123–126.

McDonald, J. and M. Kreitman. 1991. Adaptive protein evolution at *adh* locus in *Drosophila*. Nature 351: 652–654.

McGinnis, W., R. L. Garber, J. Wirz, A. Kuroiwa, and W. J. Gehring. 1984. A homologous protein-coding sequence in *Drosophila* homeotic genes and its conservation in other metazoans. Cell 37: 403–408.

McKnight, G., T. Cardillo, and F. Sherman. 1981. An extensive deletion causing overproduction of yeast iso–2-cytochrome *c*. Cell 25: 409–419.

McLachlan, A. D. 1977. Growth and evolution of proteins by gene duplication. pp. 208–239. In: Kimura, M. (ed.) *Molecular Evolution and Polymorphism. Proc. 2nd Taniguchi International Symposium on Biophysics*. National Institute of Genetics, Mishima, Japan.

Meselson, M. and C. Radding. 1975. A general model for genetic recombination. Proc. Natl. Acad. Sci. USA 72: 358–361.

Metzenberg, A. B., G. Wurzer, T. H. Huisman, and O. Smithies. 1991. Homology requirements for equal crossing-over in humans. Genetics 128: 143–161.

Meyne, J., R. J. Baker, H. H. Hobart, T. C. Hsu, O. A. Ryder, O. G. Ward, J. E. Wiley, D. H. Wurster-Hill, T. L. Yates, and R. K. Moyzis. 1990. Distribution of non-telomeric sites of the (TTAGGG)n telomeric sequence in vertebrate chromosomes. Chromosoma 99: 3–10.

Michel, F. and J.-L. Ferat. 1995. Structure and activities of group II introns. Annu. Rev. Biochem. 64: 435–461.

Michels, P., M. Marchand, L. Kohl, S. Allert, R. K. Wierenga, and F. R. Opperdoew. 1991. The cytosolic and glycosomal isoenzymes of glyceraldehyde-3-phosphate dehydrogenase in *Trypanosoma brucei* have a distant evolutionary relationship. Eur. J. Biochem. 198: 421–428.

Miklos, G. L. G. 1985. Localized highly repetitive DNA sequences in vertebrate and invertebrate genomes. pp. 241–321. In: MacIntyre, R. J. (ed.) *Molecular Evolutionary Genetics*. Plenum Press, New York.

Milinkovitch, M. C., G. Ortí, and A. Meyer. 1993. Revised phylogeny of whales suggested by mitochondrial ribosomal DNA sequences. Nature 361: 346–348.

Milinkovitch, M. C., A. Meyer, and J. R. Powell. 1994. Phylogeny of all major groups of cetaceans based on DNA sequences from three mitochondrial genes. Mol. Biol. Evol. 11: 939–948.

Milinkovitch, M. C., G. Ortí, and A. Meyer. 1995. Novel phylogeny of whales revisited but not revised. Mol. Biol. Evol. 12: 518–520.

Milkman, R. D. 1967. Heterosis as a major cause of heterozygosity in nature. Genetics 55: 493–495.

Miyamoto, M. M. and J. Cracraft. 1991. *Phylogenetic Analysis of DNA Sequences*. Oxford University Press, New York.

Miyamoto, M. M., J. L. Slightom, and M. Goodman. 1987. Phylogenetic relationships of humans and African apes as ascertained from DNA sequences (7.1 kilobase pairs) of the $\psi\eta$-globin region. Science 238: 369–373.

Miyata, T. and T. Yasunaga. 1978. Evolution of overlapping genes. Nature 272: 532–535.

Miyata, T. and T. Yasunaga. 1980. Molecular evolution of mRNA: A method for estimating evolutionary rates of synonymous and amino acid substitution from homologous nucleotide sequences and its application. J. Mol. Evol. 16: 23–36.

Miyata, T., H. Hayashida, R. Kikuno, M. Hasegawa, M. Kobayashi, and K. Koike. 1982. Molecular clock of silent substitution: At least six-fold preponderance of silent changes in mitochondrial genes over those in nuclear genes. J. Mol. Evol. 19: 28–35.

Miyata, T., H. Toh, H. Hayashida, R. Kikuno, Y. Inokuchi, and K. Saigo. 1985. Sequence homology among reverse transcriptase-containing viruses and transposable genetic element: Functional and evolutionary implications. pp. 313–331. In: Ohta, T. and K. Aoki (eds.) *Population Genetics and Molecular Evolution*. Japan Sci. Soc. Press/Springer-Verlag, Tokyo.

Miyata, T., H. Hayashida, K. Kuma, K. Mitsuyasa, and T. Yasunaga. 1987. Male-driven molecular evolution: A model and nucleotide sequence analysis. Cold Spring Harbor Symp. Quant. Biol. 52: 863–867.

Miyata, T., K. Kuma, N. Iwabe, and N. Nikoh. 1994. A possible link between molecular evolution and tissue evolution demonstrated by tissue-specific genes. Jpn. J. Genet. 69: 473–480.

Mochizuki, K., M. Umeda, H. Ohtsubo, and E. Ohtsubo. 1992. Characterization of a plant SINE, p-SINE 1, in rice genome. Jpn. J. Genet. 67: 155–166.

Mooers, A. O. and P. H. Harvey. 1994. Metabolic rate, generation time, and the rate of molecular evolution in birds. Mol. Phylogenet. Evol. 3: 344–350.

Morgan, T. H. 1925. *Evolution and Genetics*. Princeton University Press, Princeton, NJ.

Morgan, T. H. 1932. *The Scientific Basis of Evolution*. Norton, New York.

Moriyama, E. N. and T. Gojobori. 1992. Rates of synonymous substitution and base composition of nuclear genes in *Drosophila*. Genetics 130: 855–864.

Moriyama, E. N. and D. L. Hartl. 1993. Codon usage bias and base composition of nuclear genes in *Drosophila*. Genetics 134: 847–858.

Mörl, M. and C. Schmelzer. 1990. Integration of group II intron *b11* into a foreign RNA by reversal of the self-splicing reaction in vitro. Cell 60: 629–636.

Mouchés, C., N. Pasteur, J. B. Bergé, O. Hyrien, M. Raymond, B. R. De Saint Vincent, M. De Silvestri, and G. P. Georghiou. 1986. Amplification of an esterase gene is responsible for insecticide resistance in a California *culex* mosquito. Science 233: 778–780.

Mount, S. M., C. Burks, G. Hertz, G. D. Stormo, O. White, and C. Fields. 1992. Splicing signals in *Drosophila*: Intron size, information content, and consensus sequences. Nucleic Acids Res. 20: 4255–4262.

Mourant, A. E., A. C. Kopec, and K. Domaniewska-Sobczak. 1976. *The Distribution of the Human Blood Groups and Other Polymorphisms*. Oxford University Press, Oxford.

Muller, H. J. 1925. Why polyploidy is rarer in animals than in plants. Am. Nat. 59: 346–353.

Muller, H. J. 1935. The origination of chromatin deficiencies as minute deletions subject to insertion elsewhere. Genetics 17: 237–252.

Muller, K. and R. Schmitt. 1988. Histone genes of *Volvox carteri*: DNA sequence and organization of two H3-H4 gene loci. Nucleic Acid Res. 16: 4121–4136.

Müntzing, A. 1936. The evolutionary significance of autopolyploidy. Hereditas 21: 263–378.

Murata, S., N. Takasaki, M. Saitoh, and N. Okada. 1993. Determination of the phylogenetic relationships among Pacific salmonids by using short interspersed elements (SINEs) as temporal landmarks of evolution. Proc. Natl. Acad. Sci. USA 90: 6995–6999.

Murtha, M. T., J. F. Leckman, and F. H. Ruddle. 1991. Detection of homeobox genes in development and evolution. Proc. Natl. Acad. USA 88: 10711–10715.

Muse, S. V. and B. S. Weir. 1992. Testing for equality of evolutionary rates. Genetics 132: 269–276.

Muto, A. and S. Osawa. 1987. The guanine and cytosine content of genomic DNA and bacterial evolution. Proc. Natl. Acad. Sci. USA 84: 166–169.

Muto, A., F. Yamao, H. Hori, and S. Osawa. 1986. Gene organization of *Mycoplasma capricolum*. Adv. Biophys. 21: 49–56.

Nadal-Ginard, B. and C. L. Markert. 1975. Use of affinity chromatography for purification of lactate dehydrogenase and for assessing the homology and function of the A and B subunits. pp. 45–67. In: Market, C. L. (ed.) *Isozymes II: Physiological Function*. Academic Press, New York.

Nag, D. K., M. A. White, and T. D. Petes. 1989. Palindromic sequences in heteroduplex DNA inhibit mismatch repair in yeast. Nature 340: 318–320.

Nagahashi, S., H. Endoh, Y. Suzuki, and N. Okada. 1991. Characterization of a tandemly repeated DNA sequence family originally derived by retroposition of tRNA$_{Glu}$ in the newt. J. Mol. Biol. 222: 391–404.

Nagylaki, T. 1974. The moments of stochastic integrals and the distribution of sojourn times. Proc. Natl. Acad. Sci. USA 71: 746–749.

Nagylaki, T. 1983. Evolution of a large population under gene conversion. Proc. Natl. Acad. Sci. USA 80: 5941–5945.

Nagylaki, T. 1984a. Evolution of multigene families under interchromosomal gene conversion. Proc. Natl. Acad. Sci. USA 81: 3796–3800.

Nagylaki, T. 1984b. The evolution of multigene families under intrachromosomal gene conversion. Genetics 106: 529–548.

Nagylaki, T. 1988. Gene conversion, linkage, and the evolution of multigene families. Genetics 120: 291–301.

Nagylaki, T. 1990. Gene conversion, linkage, and the evolution of repeated genes dispersed among multiple chromosomes. Genetics 126: 261–276.

Nagylaki, T. and T. D. Petes. 1982. Intrachromosomal gene conversion and the maintenance of sequence homogeneity among repeated genes. Genetics 100: 315–337.

Naito, M., H. Ishiguro, T. Fujisawa, and Y. Kurosawa. 1993. Presence of eight distinct homeobox-containing genes in cnidarians. FEBS Lett. 333: 271–274.

Nakamura, Y., M. Leppert, P. O'Connell, R. Wolff, T. Holm, M. Culver, E. Fujimoto, M. Hoff, E. Kumlin, and R. White. 1987. Variable number of tandem repeat (VNTR) markers for human genetic mapping. Science 235: 1616–1622.

Nasrallah, J. B. and M. E. Nasrallah. 1989. The molecular genetics of self-incompatibility in *Brassica*. Annu. Rev. Genet. 23: 121–139.

Nathans, J., D. Thomas, and D. S. Hogness. 1986. Molecular genetics of human color vision: The genes encoding blue, green, and red pigments. Science 232: 193–202.

Navidi, W. C., G. A. Churchill, and A. von Haeseler. 1991. Methods for inferring phylogenies from nucleic acid sequence data by using maximum likelihood and linear invariants. Mol. Biol. Evol. 8: 128–143.

Nawrath, C., J. Schell, and C. Koncz. 1990. Homologous domains of the largest subunit of eucaryotic RNA polymerase II are conserved in plants. Mol. Gen. Genet. 223: 65–75.

Needleman, S. B. and C. D. Wunsch. 1970. A general method applicable to the search of similarities in the amino acid sequence of two proteins. J. Mol. Biol. 48: 443–453.

Nei, M. 1969. Gene duplication and nucleotide substitution in evolution. Nature 221: 40–42.

Nei, M. 1971a. Fertility excess necessary for gene substitution in regulated populations. Genetics 68: 169–184.

Nei, M. 1971b. Interspecific gene differences and evolutionary time estimated from electrophoretic data on protein identity. Am. Nat. 105: 385–398.

Nei, M. 1972. Genetic distance between populations. Am. Nat. 106: 283–292.

Nei, M. 1975. *Molecular Population Genetics and Evolution*. North-Holland, Amsterdam.

Nei, M. 1976. Mathematical models of speciation and genetic distance. pp. 723–766. In: Karlin, S. and E. Nevo (eds.) *Population Genetics and Ecology*. Academic Press, New York.

Nei, M. 1987. *Molecular Evolutionary Genetics*. Columbia University Press, New York.

Nei, M. 1991. Relative efficiencies of different tree-making methods for molecular data. pp. 90–128. In: Miyamoto, M. M. and J. L. Cracraft (eds.) *Phylogenetic Analysis of DNA Sequences*. Oxford University Press, Oxford.

Nei, M. and T. Gojobori. 1986. Simple methods for estimating the numbers of synonymous and nonsynonymous nucleotide substitutions. Mol. Biol. Evol. 3: 418–426.

Nei, M. and D. Graur. 1984. Extent of protein polymorphism and the neutral mutation theory. pp. 73–118. In: Hecht, M. K., B. Wallace, and G. T. Prance (eds.) *Evolutionary Biology*. Vol. 17, Plenum, New York.

Nei, M. and A. L. Hughes. 1991. Polymorphism and evolution of the major histocompatibility complex loci in mammals. pp. 222–247. In: Selander, R. K., A. G. Clark, and T. S. Whittam (eds.) *Evolution at the Molecular Level*. Sinauer Associates, Sunderland, MA.

Nei, M. and Y. Imaizumi. 1966. Genetic structure of human populations. II. Differentiation of blood group gene frequencies among isolated populations. Heredity 21: 183–190.

Nei, M. and L. Jin. 1989. Variances of the average numbers of nucleotide substitutions within and between populations. Mol. Biol. Evol. 6: 290–300.

Nei, M. and W.-H. Li. 1979. Mathematical model for studying genetic variation in terms of restriction endonucleases. Proc. Natl. Acad. Sci. USA 76: 5269–5273.

Nei, M. and A. K. Roychoudhury. 1973. Probability of fixation of nonfunctional genes at duplicate loci. Am. Nat. 107: 362–372.

Nei, M. and A. K. Roychoudhury. 1974. Genetic variation within and between the three major races of man, Caucasoids, Negroids, and Mongoloids. Am. J. Hum. Genet. 26: 421–443.

Nei, M., J. C. Stephens, and N. Saitou. 1985. Methods for computing the standard errors of branching points in an evolutionary tree and their application to molecular data from humans and apes. Mol. Biol. Evol. 2: 66–85.

Neitz, M. and J. Neitz. 1995. Numbers and ratios of visual pigment genes for normal red-green color vision. Science 267: 1013–1016.

Neitz, M., J. Neitz, and G. H. Jacobs. 1991. Spectral tuning of pigments underlying red-green color vision. Science 252: 971–974.

Nelson, O. E. 1975. The waxy locus in Maize III. Effect of structural heterozygosity on intragenic recombination and flanking maker assortment. Genetics 79: 31–44.

Never, P. and H. Saedler. 1977. Transposable genetic elements as agents of gene instability and chromosome rearrangements. Nature 268: 109–115.

Nguyen, T. and T. P. Speed. 1992. A derivation of all linear invariants for a nonbalanced transversion model. J. Mol. Evol. 35: 60–76.

Nilsen, T. W. 1994. RNA-RNA interactions in the spliceosome: Unraveling the ties that bind. Cell 78: 1–4.

Nishioka, Y., A. Leder, and P. Leder. 1980. Unusual α-globin-like gene that has cleanly lost both globin intervening sequences. Proc. Natl. Acad. Sci. USA 77: 2806–2809.

Nixon, K. C., W. L. Crepet, D. Stevenson, and E. M. Friis. 1994. A reevaluation of seed plant phylogeny. Ann. Missouri Bot. Gard. 81: 484–533.

Novacek, M. J. 1994. Whales leave the beach. Nature 368: 807.

Nugent, J. M. and J. D. Palmer. 1991. RNA-mediated transfer of the gene coxII from the mitochondrion to the nucleus during flowering plant evolution. Cell 66: 473–481.

Nuttall, G. H. F. 1904. *Blood Immunity and Blood Relationship*. Cambridge University Press, Cambridge.

Obaru, K., T. Tsuzuki, C. Setoyama, and K. Shimada. 1988. Structural organization of the mouse aspartate aminotransferase isozyme genes: Introns antedate the divergence of cytosolic and mitochondrial isozymes genes. J. Mol. Biol. 200: 13–22.

Oberthür, W., W. Voelter, and G. Braunitzer. 1980. Die sequenz der Hämoglobine von streifengans (*Anser indicus*) und strauss (*Struthio camelus*). Inositpentaphosphat als modulator der evolutions-geschwindigkeit: Die überraschende sequenz α63 (E12) valin. Z. Physiol. Chem. Hoppe-Seyler 361: S969–S975.

Ochman, H. and A. C. Wilson. 1987. Evolution in bacteria: Evidence for a universal substitution rate in cellular genomes. J. Mol. Evol. 26: 74–86.

Ohama, T., S. Osawa, K. Watanabe, and T. H. Jukes. 1990. Evolution of the mitochondrial genetic code. IV. AAA as an asparagine codon in some animal mitochondria. J. Mol. Evol. 30: 329–332.

Ohno, S. 1970. *Evolution by Gene Duplication*. Springer-Verlag, Berlin.

Ohno, S. 1972. So much "junk" DNA in our genome. pp. 366–370. In *Evolution of Genetic Systems*. Vol. 23, Brookhaven Symp. Biol.

Ohta, T. 1973. Slightly deleterious substitutions in evolution. Nature 246: 96–98.

Ohta, T. 1974. Mutational pressure as the main cause of molecular evolution and polymorphism. Nature 252: 351–354.

Ohta, T. 1976. A simple model for treating the evolution of multigene families. Nature 263: 74–76.

Ohta, T. 1977. On the gene conversion model as a mechanism for maintenance of homogeneity in systems with multiple genomes. Genet. Res. 30: 89–91.

Ohta, T. 1978. Sequence variability of immunoglobulins considered from the standpoint of population genetics. Proc. Natl. Acad. Sci. USA 75: 5108–5112.

Ohta, T. 1980. *Evolution and Variation of Multigene Families*. Springer-Verlag., Berlin.

Ohta, T. 1982. Further study on the genetic correlation between gene numbers of a multigene family. Genetics 99: 555–571.

Ohta, T. 1983a. On the evolution of multigene families. Theor. Pop. Biol. 23: 216–240.

Ohta, T. 1983b. Time until fixation of a mutant belonging to a gene family. Genet. Res. 41: 47–55.

Ohta, T. 1984. Some models of gene conversion for treating the evolution of multigene families. Genetics 106: 517–528.

Ohta, T. 1985. A model of duplicative transposition and gene conversion for repetitive DNA families. Genetics 110: 513–524.

Ohta, T. 1986. Actual number of alleles contained in a multigene family. Genet. Res. 48: 119–123.

Ohta, T. 1987. Simulating evolution by gene duplication. Genetics 115: 207–213.

Ohta, T. 1993. An examination of the generation-time effect on molecular evolution. Proc. Natl. Acad. Sci. USA 90: 10676–10680.

Ohta, T. 1995. Synonymous and nonsynonymous substitutions in mammalian genes and the nearly neutral theory. J. Mol. Evol. 40: 56–63.

Ohta, T. and G. A. Dover. 1983. Population genetics of multigene families that are dispersed into two or more chromosomes. Proc. Natl. Acad. Sci. USA 80: 4079–4083.

Ohta, T. and M. Kimura. 1971. On the constancy of the evolutionary rate of cistrons. J. Mol. Evol. 1: 18–25.

Ohta, T. and M. Kimura. 1973. A model of mutation appropriate to estimate the number of electrophoretically detectable alleles in a finite population. Genet. Res. 22: 201–204.

O'hUigin, C. and W.-H. Li. 1992. The molecular clock ticks regularly in muroid rodents and hamsters. J. Mol. Evol. 35: 377–384.

Okada, N. and K. Ohshima. 1993. A model for the mechanism of initial generation of short interspersed elements (SINEs). J. Mol. Evol. 37: 167–170.

Olsen, G. J. 1987. The earliest phylogenetic branchings: comparing rRNA-based evolutionary trees inferred with various techniques. Cold Spring Harbor Symp. Quant. Biol. 52: 825–837.

Oparin, A. I. 1957. *The Origin of Life on Earth.* Academic Press, New York.

Orgel, L. E. and F. H. C. Crick. 1980. Selfish DNA: The ultimate parasite. Nature 284: 604–607.

Orito, E., M. Mizokami, Y. Ina, E. N. Moriyama, N. Kameshima, M. Yamamoto, and T. Gojobori. 1989. Host-independent evolution and a genetic classification of the hepadnavirus family based on nucleotide sequences. Proc. Natl. Acad. Sci. USA 86: 7059–7062.

Osawa, S. and T. H. Jukes. 1989. Codon reassignment (codon capture) in evolution. J. Mol. Evol. 28: 271–278.

Osawa, S., T. H. Jukes, K. Watanabe, and A. Muto. 1992. Recent evidence for evolution of the genetic code. Microbiological Reviews 56: 229–264.

Østergren, G. 1945. Parasitic nature of extra fragment chromosomes. Bot. Notiser 2: 157–163.

Ota, T. and M. Nei. 1994. Estimation of the number of amino acid substitutions per site when the substitution rate varies among sites. J. Mol. Evol. 38: 642–643.

Ouenzar, B., B. Agoutin, F. Reinisch, D. Weill, F. Perin, G. Keith, and T. Heyman. 1988. Distribution of isoaccepting tRNAs and codons for proline and glycine in collageneous and noncollageneous chicken tissues. Biochem. Biophys. Res. Commun. 150: 148–155.

Pääbo, S. 1985. Molecular cloning of ancient Egyptian mummy DNA. Nature 314: 644–645.

Pääbo, S. 1989. Ancient DNA: Extraction, characterization, molecular cloning, and enzymatic amplification. Proc. Natl. Acad. Sci. USA 86: 1939–1943.

Pääbo, S., J. A. Gifford, and A. C. Wilson. 1988. Mitochondrial DNA sequences from a 7000-year-old brain. Nucleic Acids Res. 16: 9775–9787.

Pääbo, S., R. G. Higuchi, and A. C. Wilson. 1989. Ancient DNA and the polymerase chain reaction. J. Biol. Chem. 264: 9709–9712.

Packer, C., D. A. Gilbert, A. E. Pusey, and S. J. O'Brien. 1991. A molecular genetic analysis of kinship and cooperation in African lions. Nature 351: 562–565.

Palmer, J. D. 1985a. Comparative organization of chloroplast genomes. Annu. Rev. Genet. 19: 325–354.

Palmer, J. D. 1985b. Evolution of chloroplast and mitochondrial DNA in plants and algae. pp. 131–240. In: MacIntyre, R. J. (ed.) *Molecular Evolutionary Genetics.* Plenum Press, New York.

Palmer, J. D. 1991. The molecular biology of plastid. pp. 5–53. In: Bogorad, L. and I. K. Vasil (eds.) *Cell Culture and Somatic Cell Genetics of Plants.* Vol. 7, Academic Press, San Diego.

Palmer, J. D. and L. A. Hebron. 1987. Unicircular structure of the *Brassica hirta* mitochondrial genome. Curr. Genet. 11: 565–570.

Palmer, J. D. and J. M. Logsdon. 1991. The recent origins of introns. Curr. Opin. Genet. Devel. 1: 470–477.

Palumbi, S. R. 1989. Rates of molecular evolution and the fraction of nucleotide positions free to vary. J. Mol. Evol. 29: 180–187.

Pamilo, P. and N. O. Bianchi. 1993. Evolution of the *ZFX* and *ZFY* genes: Rates and interdependence between the genes. Mol. Biol. Evol. 10: 271–281.

Patthy, L. 1985. Evolution of the proteases of blood coagulation and fibrinolysis by assembly from modules. Cell 41: 657–663.

Patthy, L. 1987. Intron-dependent evolution: preferred types of exons and introns. FEBS Letters. 214: 1–7.

Patthy, L. 1991a. Modular exchange principles in proteins. Curr. Opin. Struc. Biol. 1: 351–362.

Patthy, L. 1991b. Exons: Original building blocks of proteins? BioEssays 13: 187–192.

Patthy, P. 1994. Introns and exons. Curr. Opin. Struc. Biol. 4: 383–392.

Pendleton, J. W., B. K. Nagai, M. T. Murtha, and F. H. Ruddle. 1993. Expansion of the *Hox* gene family and the evolution of chordates. Proc. Natl. Acad. Sci. USA 90: 6300–6304.

Penny, D., M. D. Hendy, and I. M. Henderson. 1987. Reliability of evolutionary trees. Cold Spring Harbor Symp. Quant. Biol. 52: 857–862.

Penny, D., M. D. Hendy, and M. A. Steel. 1992. Progress with methods for constructing evolutionary trees. Trends Ecol. Evol. 7: 73–79.

Perelson, A. S. and G. I. Bell. 1977. Mathematical models for the evolution of multigene families by unequal crossing over. Nature 265: 304–310.

Perlman, P. S. and R. A. Butow. 1989. Mobile introns and intron-encoded proteins. Science 246: 1106–1109.

Perutz, M. F. 1983. Species adaptation in a protein molecule. Mol. Biol. Evol. 1: 1–28.

Perutz, M. F. and H. Lehman. 1968. Molecular pathology of human haemoglobin. Nature 219: 902–909.

Petes, T. D. and C. W. Hill. 1988. Recombination between repeated genes in microorganisms. Annu. Rev. Genet. 22: 147–168.

Petes, T. D., R. E. Malone, and L. S. Symington. 1991. Recombination in yeast. pp. 407–521. In: Broach, J., E. Jones, and J. Pringle (eds.) *The Molecular and Cellular Biology of the Yeast* Saccharomyces: *Genome Dynamics, Protein Synthesis and Energetics.* Vol. I, Cold Spring Harbor Laboratory Press, Cold Spring Harbor, NY.

Piatigorsky, J. and G. J. Wistow. 1989. Enzyme/crystallins: Gene sharing as an evolutionary strategy. Cell 57: 197–199.

Piatigorsky, J. and G. J. Wistow. 1991. The recruitment of crystallins: New functions precede gene duplication. Science 252: 1078–1079.

Piatigorsky, J., W. E. O'Brien, B. L. Norman, K. Kalumuck, G. J. Wistow, T. Borras, J. M. Nickerson, and E. F. Wawrousek. 1988. Gene sharing by δ-crystallin and argininosuccinate lyase. Proc. Natl. Acad. Sci. USA 85: 3479–3483.

Piccinini, M., T. Kleinschmidt, K. D. Jürgens, and G. Braunitzer. 1990. Primary structure and oxygen-binding properties of the hemoglobin from Guanaco (*Lama guanacoë*, Tylopoda). Z. Biol. Chem. Hoppe-Seyler 371: 641–648.

Pilbeam, D. 1984. The descent of hominoids and hominids. Sci. Am. 252: 84–96.

Pilger, R. 1926. Coniferae. pp. 121–407. In: Engler, A. and K. Prantl (eds.) *Die Natürlichen Pflänzenfamilien*. Vol. 13, 2nd Ed., W. Engelmann, Leipzig.

Poinar, H. N., M. Höss, J. L. Bada, and S. Pääbo. 1996. Amino acid racemization and the preservation of ancient DNA. Science 272: 864–866.

Poinar, H. N., R. J. Cano, and G. N. Poinar. 1993. DNA from an extinct plant. Nature 363: 677.

Ponticelli, A. S., E. P. Sena, and G. R. Smith. 1988. Genetic and Physical analysis of the M26 recombination hotspot of *Schizosaccharomyces pombe*. Genetics 119: 491–497.

Porter, C. A., I. Sampaio, H. Schneider, M. C. Schneider, J. Czelusniak, and M. Goodman. 1995. Evidence of primate phylogeny from ε-globin gene sequences and flanking regions. J. Mol. Evol. 40: 30–55.

Post, L. E., G. D. Strycharz, M. Nomura, H. Lewis, and P. P. Dennis. 1979. Nucleotide sequence of the ribosomal protein gene cluster adjacent to the gene for RNA polymerase subunit β in *Escherichia coli*. Proc. Natl. Acad. Sci. USA 76: 1697–1701.

Powell, L. M., S. C. Wallis, R. J. Pease, Y. H. Edwards, T. J. Knott, and J. Scott. 1987. A novel form of tissue-specific RNA processing produces apolipoprotein B-48 in intestine. Cell 50: 831–840.

Proudfoot, N. 1980. Pseudogenes. Nature 286: 840–841.

Pukkila, P. J., M. D. Stephens, D. M. Binninger, and B. Errede. 1986. Frequency and directionality of gene conversion events involving the *CYC7-H3* mutation in *Saccharomyces cerevisiae*. Genetics 114: 347–361.

Pulle, A. 1937. Remarks on the system of the spermatophytes. Med. Bot. Mus. Utrecht. 43: 1–17.

Purugganan, M. D. 1993. Transposable elements as introns: Evolutionary connections. Trends Ecol. Evol. 8: 239–243.

Purugganan, M. D. and S. Wessler. 1992. The splicing of transposable elements and its role in intron evolution. Genetica 86: 295–303.

Quentin, Y. 1988. The *Alu* family developed through successive waves of fixation closely connected with primate lineage history. J. Mol. Evol. 27: 194–202.

Quinn, C. J. 1987. The Phyllocladaceae Keng: A critique. Taxon 36: 559–565.

Rattazzi, M. C., J. G. Scandalios, and G. S. Whitt. 1982. *Isozymes: Current Topics in Biological and Medical Research*. Alan R. Liss, New York.

Raubeson, L. A. and R. K. Jansen. 1992. A rare chloroplast-DNA structural mutation is shared by all conifers. Biochem. Syst. Evol. 20: 17–24.

Razin, A. and A. D. Riggs. 1980. DNA methylation and gene function. Science 210: 604–610.

Reeb, C. A. and J. C. Avise. 1990. A genetic discontinuity in a continuously distributed species: Mitochondrial DNA in the American oyster, *Crassostrea virginica*. Genetics 124: 397–406.

Rhinesmith, H. S., W. A. Schroeder, and N. Martin. 1958. The N-terminal sequence of the β chains of normal adult human hemoglobin. J. Am. Chem. Soc. 80: 3358–3361.

Richards, E. J., H. M. Goodman, and F. M. Ausubel. 1991. The centromere region of *Arabidopsis thaliana* chromosome 1 contains telomere-similar sequences. Nucleic Acids Res. 19: 3351–3357.

Richmond, R. C. 1970. Non-Darwinian evolution: A critique. Nature 225: 1025–1028.

Richter, C., J.-W. Park, and B. N. Ames. 1988. Normal oxidative damage to mitochondrial and nuclear DNA is extensive. Proc. Natl. Acad. Sci. USA 85: 6465–6467.

Ritossa, F. M. and G. Scala. 1969. Equilibrium variations in the redundancy of rDNA in *Drosophila melanogaster*. Genetics Suppl. 61: 305–317.

Ritossa, F. M. and S. Spiegelman. 1965. Localization of DNA complementary to ribosomal RNA in the nucleolus organizer region of *Drosophila melanogaster*. Proc. Natl. Acad. Sci. USA 53: 737–745.

Ritossa, F. M., K. C. Atwood, D. L. Lindsley, and S. Spiegelman. 1966. On the chromosomal distribution of DNA complementary to ribosomal and soluble RNA. Natl. Cancer Inst. Monogr. 23: 449–472.

Rivera, M. C. and J. A. Lake. 1992. Evidence that eukaryotes and eocyte prokaryotes are immediate relatives. Science 257: 74–76.

Robertson, A. 1906. Some points in the morphology of *Phyllocladus alpinus* hook. Ann. Bot. 20: 259–265.

Rodrigo, A. G. 1993. Calibrating the bootstrap test of monophyly. Int. J. Parasitol. 23: 507–514.

Roger, A. J., P. J. Keeling, and W. F. Doolittle. 1994. Introns, the broken transposons. pp. 27–37. *Molecular Evolution of Physiological Processes*. Rockefeller University Press, New York.

Rogers, J. 1985. Exon shuffling and intron insertion in serine proteases genes. Nature 315: 458–459.

Rogers, J. 1989. How were introns inserted into nuclear genes? Trends Genet. 5: 213–216.

Rogers, J. S. 1972. Measures of genetic similarity and genetic distance. *Studies in Genetics VII* (University of Texas Publ. No. 7213) 145–153.

Rogers, S. O. and A. J. Bendich. 1985. Extraction of DNA from milligram amounts of fresh herbarium and mummified plant tissue. Plant Mol. Biol. 5: 69–76.

Rollo, F., F. M. Venanzi, and A. Amici. 1991. Nucleic acids in mummified seeds: Biochemistry and molecular genetics of pre-Columbian maize. Genet. Res. 58: 193–201.

Romans, P. and R. A. Firtel. 1985. Organization of the actin multigene family of *Dictyostelium discoideum* and analysis of variability in the protein coding regions. J. Mol. Biol. 186: 321–335.

Romer, A. S. 1966. *Vertebrate Paleontology*. University of Chicago Press, Chicago.

Rossman, M. G., A. Liljas, C.-I. Brändén, and L. J. Banaszak. 1975. Evolutionary and structural relationships among dehydrogenases. pp. 61–101. In: Boyer, P. D. (ed.) *The Enzymes*. Vol. 11, 3rd Ed. Academic Press, New York.

Roychoudhury, A. K. and M. Nei. 1988. *Human Polymorphic Genes: World Distribution*. Oxford Univ. Press, New York.

Rubin, G. 1983. Dispersed repetitive DNAs in *Drosophila*. pp. 329–361. In: Shapiro, J. A. (ed.) *Mobile Genetic Elements*. Academic Press, New York.

Ruvolo, M., T. R. Disotell, M. W. Allard, W. M. Brown, and R. Honeycutt. 1991. Resolution of the African hominoid trichotomy by use of a mitochondrial gene sequence. Proc. Natl. Acad. Sci. USA 88: 1570–1574.

Rzhetsky, A. and M. Nei. 1992. A simple method for estimating and testing minimum-evolution trees. Mol. Biol. Evol. 9: 945–967.

Rzhetsky, A. and M. Nei. 1993. Theoretical foundation of the minimum-evolution method of phylogenetic inference. Mol. Biol. Evol. 10: 1073–1095.

Saccone, S., A. De Sario, J. Wiegant, A. K. Raap, G. Della Valle, and G. Bernardi. 1993. Correlation between isochores and chromosomal bands in the human genome. Proc. Natl. Acad. Sci. USA 90: 11929–11933.

Sadler, L. A., S. H. Blanton, and S. P. Daiger. 1991. Dinucleotide repeat polymorphism at the human tissue plasminagen activator gene (PLAT). Nucleic Acids Res. 19: 6058.

Sahni, B. 1920. On certain archaic features in the seed of *Taxus baccata*, with remarks on the antiquity of Taxineae. Ann. Bot. 34: 117–133.

Saiki, R. K., S. J. Scharf, F. Faloona, K. B. Mullins, G. T. Horn, H. A. Erlich, and N. Arnheim. 1985. Enzymatic amplification of β-globin genomic sequences and restriction site analysis for diagnosis of sickle cell anemia. Science 230: 1350–1354.

Saitou, N. 1988. Property and efficiency of the maximum likelihood method for molecular phylogeny. J. Mol. Evol. 27: 261–-273.

Saitou, N. 1989. A theoretical study of the underestimation of branch lengths by the maximum parsimony principle. Syst. Zool. 38: 1–6.

Saitou, N. and T. Imanishi. 1989. Relative efficiencies of the Fitch-Margoliash, maximum-parsimony, maximum-likelihood, minimum-evolution, and neighbor-joining methods of phylogenetic tree construction in obtaining the correct tree. Mol. Biol. Evol. 6: 514–525.

Saitou, N. and M. Nei. 1987. The neighbor-joining method: a new method for reconstructing phylogenetic trees. Mol. Biol. Evol. 4: 406–425.

Sakagami, M., K. Ohshima, H. Mukoyama, H. Yasue, and N. Okada. 1994. A novel tRNA species as an origin of short interspersed repetitive elements (SINEs): Equine SINEs may have originated from tRNA$_{Ser}$. J. Mol. Biol. 239: 731–735.

Sakai, K.-I. 1955. Competition in plants and its relation to selection. Cold Spring Harbor Symp. Quant. Biol. 20: 137–157.

Sakaoka, H., K. Kurita, Y. Iida, S. Takada, K. Y.-T. Umene, C.-S. Ren, and A. J. Nahmias. 1994. Quantitative analysis of genomic polymorphism of herpes simplex virus type 1 strains from six countries: Studies of molecular evolution and molecular epidemiology of the virus. J. General Virology 75: 513–527.

Salo, W. L., A. C. Aufderheide, J. Buikstra, and T. A. Holcomb. 1994. Identification of *Mycobacterium tuberculosis* DNA in pre-Columbian Peruvian mummy. Proc. Natl. Acad. Sci. USA 91: 2091–2094.

Sankoff, D. 1975. Minimal mutation trees of sequences. SIAM J. Appl. Math. 28: 35–42.

Sankoff, D. 1990. Designer invariants for large phylogenies. Mol. Biol. Evol. 7: 255–269.

Sankoff, R. J., R. J. Cedergren, and W. McKay. 1972. A strategy for sequence phylogeny research. Nucleic Acids Res. 10: 421–431.

Sarich, V. M. and A. C. Wilson. 1966. Quantitative immunochemistry and the evolution of primate albumins: Micro-complement fixation. Science 154: 1563–1566.

Sarich, V. M. and A. C. Wilson. 1967. Immunological time scale for hominid evolution. Science 158: 1200–1203.

Sarich, V. M. and A. C. Wilson. 1973. Generation time and genomic evolution in primates. Science 179: 1144–1147.

Sattath, S. and A. Tversky. 1977. Additive similarity trees. Psychometrika 42: 319–345.

Saunders, N. C., L. G. Kessler, and J. C. Avise. 1986. Genetic variation and geographic differentiation in mitochondrial DNA of the horseshoe crab, *Limulus polyphemus*. Genetics 112: 613–627.

Sawyer, S. A. 1989. Statistical tests for detecting gene conversion. Mol. Biol. Evol. 6: 526–538.

Sawyer, S. A. and D. Hartl. 1986. Distribution of transposable elements in prokaryotes. Theor. Pop. Biol. 30: 1–16.

Sawyer, S. A. and D. L. Hartl. 1992. Population genetics of polymorphism and divergence. Genetics 132: 1161–1176.

Sawyer, S. A., D. E. Dykhuizen, R. F. DuBose, L. Green, T. Mutangadura-Mhlanga, D. F. Wolczyk, and D. L. Hartl. 1987. Distribution and abundance of insertion sequences among natural isolates of *Escherichia coli*. Genetics 115: 51–63.

Schaeffer, S. W. and C. F. Aquadro. 1987. Nucelotide sequence of the *Adh* gene region of *Drosophila pseudoobscura*: Evolutionary change and evidence for an ancient gene duplication. Genetics 117: 61–73.

Schalet, A. 1969. Exchanges at the bobbed locus of *Drosophila melanogaster*. Genetics 63: 133–153.

Schimenti, J. C. and C. H. Duncan. 1984. Ruminant globin gene structures suggest an evolutionary role for *Alu*-type repeats. Nucleic Acids Res. 12: 1641–1655.

Schlötterer, C. and D. Tautz. 1992. Slippage synthesis of simple sequence DNA. Nucleic Acids Res. 20: 211–215.

Schlötterer, C., B. Amos, and D. Tautz. 1991. Conservation of polymorphic simple sequence loci in cetacean species. Nature 354: 63–65.

Schmid, C. W. and P. L. Deininger. 1975. Sequence organization of the human genome. Cell 6: 345–358.

Schmid, C. W. and R. Maraia. 1992. Transcriptional regulation and transpositional selection of active SINE sequences. Curr. Opin. Genet. Dev. 2: 874–882.

Schmid, C. W. and C. K. J. Shen. 1985. The evolution of interspersed repetitive DNA sequences in mammals and other vertebrates. pp. 323–358. In: MacIntyre, R. J. (ed.) *Molecular Evolutionary Genetics*. Plenum Press, New York.

Schubert, F. R., K. Nieselt-Struwe, and P. Gruss. 1993. The Antennapedia-type homeobox genes have evolved from three precursors separated early in metazoan evolution. Proc. Natl. Acad. Sci. USA 90: 143–147.

Schughart, K., C. Kappen, and F. H. Ruddle. 1989. Duplication of large genomic regions during the evolution of vertebrate homeobox genes. Proc. Natl. Acad. Sci. USA 86: 7067–7071.

Schultz, A. H. 1963. *Classification and Human Evolution*. Aldine, Chicago.

Schulz, G. E. 1980. Gene duplication in glutathione reductase. J. Mol. Biol. 138: 335–347.

Schuster, W. and A. Brennicke. 1987. Plastid, nuclear and reverse transcriptase sequences in the mitochondrial genome of Oenothera: Is genetic information transferred between organelles via RNA? EMBO J. 6: 2857–2863.

Scott, A. F., P. Heath, S. Trusko, S. H. Boyer, W. Prass, M. Goodman, J. Czelusniak, L.-Y. E. Chang, and J. L. Slightom. 1984. The sequence of the gorilla fetal globin genes: Evidence for multiple gene conversions in human evolution. Mol. Biol. Evol. 1: 371–389.

Scott, M. P. 1992. Vertebrate homeobox gene nomenclature. Cell 71: 551–553.

Scott, M. P. and A. J. Weiner. 1984. Structural relationships among genes that control development: Sequence homology between the *Antennapedia*, *Ultrabithorax*, and *fushi tarazu* loci of *Drosophila*. Proc. Natl. Acad. Sci. USA 81: 4115–4119.

Segrest, J. P., R. L. Jackson, J. D. Morrisett, and A. M. Gotto Jr. 1974. A molecular theory of lipid-protein interactions in the plasma lipoproteins. FEBS Lett. 38: 247–258.

Seino, S., G. I. Bell, and W.-H. Li. 1992. Sequences of primate insulin genes support the hypothesis of a slower rate of molecular evolution in humans and apes than in monkeys. Mol. Biol. Evol. 9: 193–203.

Sellers, P. H. 1974. On the theory and computation of evolutionary distances. SIAM J. Appl. Math. 26: 787–793.

Setoyama, C., T. Joh, T. Tsuzuki, and K. Schimada. 1988. Structural organization of the mouse cytosolic malate dehydrogenase gene: Comparison with that of the mouse mitochondrial malate dehydrogenase gene. J. Mol. Biol. 202: 355–364.

Shah, D. M., R. C. Hightower, and R. B. Meagher. 1983. Genes encoding actin in higher plants: intron positions are highly conserved but the coding sequences are not. J. Mol. Appl. Genet. 2: 111–126.

Shapiro, S. G., E. A. Schon, T. M. Townes, and J. B. Lingrel. 1983. Sequence and linkage of the goat ξ^I and ξ^{II} β-globin genes. J. Mol. Biol. 169: 31–52.

Sharp, P. M. and W.-H. Li. 1986. An evolutionary perspective on synonymous codon usage in unicellular organisms. J. Mol. Evol. 24: 28–38.

Sharp, P. M. and W.-H. Li. 1989. On the rate of DNA sequence evolution in *Drosophila*. J. Mol. Evol. 28: 398–402.

Sharp, P. M. and A. T. Lloyd. 1993. Codon usage. pp. 378–397. In: Maroni, G. (ed.) *An Atlas of Drosophila Genes: Sequences and Molecular Features*. Oxford University Press, New York.

Sharp, P. M., T. M. F. Tuohy, and K. R. Mosurski. 1986. Codon usage in yeast: Cluster analysis clearly differentiates highly and lowly expressed genes. Nucleic Acids Res. 14: 5125–5143.

Sharp, P. M., E. Cowe, D. G. Higgins, D. Shields, K. H. Wolfe, and F. Wright. 1988. Codon usage patterns in *Escherichia coli*, *Bacillus subtilis*, *Saccharomyces cerevisiae*, *Schizosaccharomyces pombe*, *Drosophila melanogaster* and *Homo sapiens*: A review of the considerable within-species diversity. Nucleic Acids Res. 16: 8207–8211.

Shen, S.-H., J. L. Slightom, and O. Smithies. 1981. A history of the human fetal globin gene duplication. Cell 26: 191–203.

Shields, D. C. 1990. Switches in species-specific codon preferences: The influence of mutation biases. J. Mol. Evol. 31: 71–80.

Shields, D. C., P. M. Sharp, D. G. Higgins, and F. Wright. 1988. "Silent" sites in *Drosophila* genes are not neutral: Evidence of selection among synonymous codons. Mol. Biol. Evol. 5: 704–716.

Shields, G. F. and A. C. Wilson. 1987. Calibration of mitochondrial DNA evolution in geese. J. Mol. Evol. 24: 212–217.

Shih, M. C., P. Heinrich, and H. M. Goodman. 1988. Intron existence predated the divergence of eukaryotes and prokaryotes. Science 242: 1164–1166.

Shimmin, L. C., B. H.-J. Chang, and W.-H. Li. 1993. Male-driven evolution of DNA sequences. Nature 362: 745–747.

Shoemaker, J. S. and W. M. Fitch. 1989. Evidence from nuclear sequences that invariable sites should be considered when sequence divergence is calculated. Mol. Biol. Evol. 6: 270–289.

Shriver, M. D. 1993. Origins and evolution of VNTR loci: The apolipoprotein B 3′ VNTR. Ph. D. Thesis, University of Texas, Houston.

Shriver, M. D., G. Siest, and E. Boerwinkle. 1992. Length and sequence variation in the apolipoprotein B intron 20 *Alu* repeat. Genomics 14: 449–454.

Shriver, M. D., L. Jin, R. Chakraborty, and E. Boerwinkle. 1993. VNTR allele frequency distribution under the stepwise mutation model: A computer simulation approach. Genetics 134: 983–993.

Shyue, S.-K. 1994. Molecular and evolutionary genetics of the X-linked visual pigment genes in humans and new world monkeys. Ph. D. Thesis. The University of Texas, Houston.

Shyue, S.-K., L. Li, B. H.-J. Chang, and W.-H. Li. 1994. Intronic gene conversion in the evolution of human X-linked color vision genes. Mol. Biol. Evol. 11: 548–551.

Shyue, S.-K., D. Hewett-Emmett, H. G. Sperling, D. M. Hunt, J. K. Bowmaker, J. D. Mollon, and W.-H. Li. 1995. Adaptive evolution of color vision genes in higher primates. Science 269: 1265–1267.

Sibley, C. G. and J. E. Ahlquist. 1984. The phylogeny of the hominoid primates, as indicated by DNA-DNA hybridization. J. Mol. Evol. 20: 2–15.

Sibley, C. G. and J. E. Ahlquist. 1987. DNA hybridization evidence of hominoid phylogeny: results from an expanded data set. J. Mol. Evol. 26: 99–121.

Sibley, C. G, and J. E. Ahlquist. 1990. *Phylogeny and Classification of the Birds of the World: A Study in Molecular Evolution.* Yale University Press, New Haven, CT.

Sibley, C. G., J. A. Comstock, and J. E. Ahlquist. 1990. DNA hybridization evidence of hominoid phylogeny: A reanalysis of the data. J. Mol. Evol. 30: 202–236.

Silver, L. M. 1992. Bouncing off microsatellites. Nature Genetics 2: 8–9.

Simmons, M. J. and J. F. Crow. 1977. Mutations affecting fitness in *Drosophila* populations. Annu. Rev. Genet. 11: 49–78.

Simonsen, K. L., G. A. Churchill, and C. F. Aquadro. 1995. Properties of statistical tests of neutrality for DNA polymorphism data. Genetics 141: 413–429.

Simpson, G. G. 1944. *Tempo and Mode in Evolution.* Columbia University Press, New York.

Simpson, G. G. 1945. The principles of classification and a classification of mammals. Bull. Am. Museum Nat. Hist. 85: 1–350.

Simpson, G. G. 1961. *Principles of Animal Taxonomy.* Columbia University Press, New York.

Simpson, L. 1990. RNA editing - a novel genetic phenomenon? Science 250: 512–513.

Singer, C. E. and B. N. Ames. 1970. Sunlight ultraviolet and bacterial DNA base ratios. Science 170: 822–826.

Singer, M. F. 1982. SINEs and LINEs: Highly repeated short and long interspersed sequences in mammalian genomes. Cell 28: 433–434.

Skibinski, D. O. F. and R. D. Ward. 1982. Correlations between heterozygosity and evolutionary rate of proteins. Nature 298: 490–492.

Slagel, V., E. Flemington, V. Traina-Dorge, H. Bradshaw, and P. Deininger. 1987. Clustering and family relationships of the *Alu* family in the human genome. Mol. Biol. Evol. 4: 19–29.

Slatkin, M. 1986. Interchromosomal biased gene conversion, mutation and selection in a multigene family. Genetics 112: 681–698.

Slightom, J., A. E. Blechl, and O. Smithies. 1980. Human fetal $^G\gamma$- and $^A\gamma$-globin genes: Complete nucleotide sequences suggest that DNA can be exchanged between these duplicated genes. Cell 21: 627–638.

Smith, C. W. J., J. G. Patton, and B. Nadal-Ginard. 1989. Alternative splicing in the control of gene expression. Annu. Rev. Genet. 23: 527–577.

Smith, G. P. 1974. Unequal crossover and the evolution of multigene families. Cold Spring Harbor Symp. Quant. Biol. 38: 507–513.

Smith, G. P. 1976. Evolution of repeated DNA sequences by unequal crossover. Science 191: 528–535.

Smith, M. W. and R. F. Doolittle. 1992. Anomolous phylogeny involving the enzyme glucose-6-phosphate isomerase. J. Mol. Evol. 34: 544–545.

Smith, M. W., D. F. Feng, and R. F. Doolittle. 1992. Evolution by acquisition: The case for horizontal gene transfers. Trends Biochem. 17: 489–493.

Smith, T. F., M. S. Waterman, and W. M. Fitch. 1981. Comparative biosequence metrics. J. Mol. Evol. 18: 38–46.

Sneath, P. H. A. and R. R. Sokal. 1973. *Numerical Taxonomy.* W.H. Freeman, San Francisco.

Snell, G. D. 1968. The *H-2* locus of the mouse: Observations and speculations concerning its comparative genetics and its polymorphism. Folia Biol. 14: 335–358.

Snow, P. and L. W. Buss. 1994. HOM/Hox-type homeoboxes from *Stylaria lacustris* (Annelida: Oligochaeta). Mol. Phylogenet. Evol. 3: 360–364.

Soares, M. B., E. Schon, A. Henderson, S. K. Karathanasis, R. Cate, S. Zeitlin, J. Chirgwin, and A. Efstratiadis. 1985. RNA-mediated gene duplication: The rat preproinsulin I gene is a functional retroposon. Mol. Cell. Biol. 5: 2090–2103.

Sogin, M. L., H. J. Elwood, and J. H. Gunderson. 1986. Evolutionary diversity of eukaryotic small-subunit rRNA genes. Proc. Natl. Acad. Sci. USA 83: 1383–1387.

Sogin, M. L., J. H. Gunderson, H. J. Elwood, R. A. Alonso, and D. A. Peattie. 1989. Phylogenetic meaning of the kingdom concept: An unusual ribosomal RNA from *Giardia lamblia*. Science 243: 75–77.

Sokal, R. R. and C. D. Michener. 1958. A statistical method for evaluating systematic relationships. University of Kansas Sci. Bull. 28: 1409–1438.

Sokal, R. R. and F. J. Rohlf. 1981. *Biometry*. W.H. Freeman, San Francisco.

Soltis, P. S., D. E. Soltis, and C. J. Smiley. 1992. An *rbc*L sequence from a Miocene *Taxodium* (bald cypress). Proc. Natl. Acad. Sci. USA 89: 449–451.

Sourdis, J. and M. Nei. 1988. Relative efficiencies of the maximum parsimony and distance-matrix methods in obtaining the correct phylogenetic tree. Mol. Biol. Evol. 5: 298–311.

Sparrow, A. H. and A. F. Nauman. 1976. Evolution of genome size by DNA doublings. Science 192: 524–529.

Sparrow, A. H., H. J. Price, and A. G. Underbrink. 1972. A survey of DNA content per cell and per chromosome of prokaryotic and eukaryotic organisms: Some evolutionary considerations. Brookhaven Symp. Biol. 23: 451–494.

Sporne, K. R. 1965. *The Morphology of Gymnosperms: The Structure and Evolution of Primitive Seed Plants*. Hutchinson and Ross, London.

Spratt, B. G. 1988. Hybrid penicillin-binding proteins in penicillin-resistant strains of *Neisseria gonorrhoeae*. Nature 332: 173–176.

Spratt, B. G. 1989. Resistance to β-lactam antibiotics mediated by alterations of penicillin-binding proteins. pp. 77–100. In: Bryan, L. (ed.) *Handbook of Experimental Pharmacology*. Vol. 91, Springer-Verlag, Berlin.

Spratt, B. G., Q.-Y. Zhang, D. M. Jones, A. Hutchison, J. A. Brannigan, and C. G. Dowson. 1989. Recruitment of a penicillin-binding protein gene from *Neisseria flavescens* during the emergence of penicillin resistance in *Neisseria meningitidis*. Proc. Natl. Acad. Sci. USA 86: 988–8992.

Spratt, B. G., L. D. Bowler, Q.-Y. Zhang, J. Zhou, and J. Maynard Smith. 1992. Role of interspecies transfer of chromosomal genes in the evolution of penicillin resistance in pathogenic and commensal *Neisseria* species. J. Mol. Evol. 34: 115–125.

Stallings, R. L., A. F. Ford, D. Nelson, D. C. Torney, C. E. Hildebrand, and R. K. Moyzis. 1991. Evolution and distribution of (GT)n repetitive sequences in mammalian genomes. Genomics 10: 807–815.

Stavenhagen, J. B. and D. M. Robins. 1988. An ancient provirus has imposed androgen regulation on the adjacent mouse sex limited protein gene. Cell 55: 247–254.

Stebbins, G. L. 1971. *Chromosomal Evolution in Higher Plants*. Edward Arnold, London.

Stebbins, G. L. 1980. Polyploidy in plants: Unsolved problems and prospects. pp. 495–520. In: Lewis, W. H. (ed.) *Polyploidy: Biological Relevance*. Plenum Press, New York.

Stebbins, G. L. 1981. Coevolution of grasses and herbivores. Ann. Missouri Bot. Garden. 68: 75–86.

Steel, M., D. Penny, and P. J. Lockhart. 1993. Confidence in evolutionary trees from biological sequence data. Nature 364: 440–442.

Stephan, W. 1986. Recombination and the evolution of satellite DNA. Genet. Res. 47: 167–174.

Stephan, W. 1987. Quantitative variation and chromosomal location of satellite DNAs. Genet. Res. 50: 41–52.

Stephan, W. and S. J. Mitchell. 1992. Reduced levels of DNA polymorphism and fixed between-population differences in the centromeric region of *Drosophila ananassae*. Genetics 132: 1039–1045.

Stephan, W., T. H. E. Wiehe, and M. W. Lenz. 1992. The effect of strongly selected substitutions on neutral polymorphism: Analytical results based on diffusion theory. Theor. Pop. Biol. 41: 237–254.

Stephens, J. C. 1985. Statistical methods of DNA sequence analysis: Detection of intragenic recombination of gene conversion. Mol. Biol. Evol. 2: 539–556.

Stewart, C.-B. and A. C. Wilson. 1987. Sequence convergence and functional adaptation of stomach lysozymes from foregut fermenters. Cold Spring Harbor Symp. Quant. Biol. 52: 891–899.

Stewart, W. N. 1983. *Palaeobotany and the Evolution of Plants*. Cambridge University Press, Cambridge.

Stoltzfus, A. 1994. Origin of introns—early or late? Nature 369: 526–527.

Stoltzfus, A., D. F. Spencer, M. Zucker, J. M. Logsdon, and W. F. Doolittle. 1994. Testing the exon theory of genes: The evidence from protein structure. Science 265: 202–207.

Stoltzfus, A., D. F. Spencer, and W. F. Doolittle. 1995. Methods for evaluating exon–protein correspondences. CABIOS 11: 509–515.

Strachan, T., E. Coen, D. Webb, and G. Dover. 1982. Modes and rates of change of complex DNA families of *Drosophila*. J. Mol. Biol. 158: 37–54.

Strachan, T., D. Webb, and G. A. Dover. 1985. Transition stages of molecular drive in multiple-copy DNA families in *Drosophila*. EMBO J. 4: 1701–1708.

Strauss, S. H., J. D. Palmer, G. T. Howe, and A. H. Doerksen. 1988. Chloroplast genomes of two conifers lack a large inverted repeat and are extensively rearranged. Proc. Natl. Acad. Sci. USA 85: 3898–3902.

Strobeck, C. 1983. Expected linkage disequilibrium for a neutral locus linked to a chromosomal arrangement. Genetics 103: 545–555.

Stryer, L. 1988. *Biochemistry*, 3rd Ed. W.H. Freeman, San Francisco.

Studier, J. A. and K. J. Keppler. 1988. A note on the neighbor-joining algorithm of Saitou and Nei. Mol. Biol. Evol. 5: 729–731.

Sueoka, N. 1961. Correlation between base composition of deoxyribonucleic acid and amino acid composition of protein. Proc. Natl. Acad. Sci. USA 47: 1141–1149.

Sueoka, N. 1964. On the evolution of informational macromolecules. pp. 479–496. In: Bryson, V. and H. J. Vogel (eds.) *Evolving Genes and Proteins.* Academic Press, New York.

Sueoka, N. 1988. Directional mutation pressure and neutral molecular evolution. Proc. Natl. Acad. Sci. USA 85: 2653–2657.

Sueoka, N. 1992. Directional mutation pressure, selective contraints, and genetic equilibria. J. Mol. Evol. 34: 95–114.

Sugiura, M. 1992. The chloroplast genome. Plant Mol. Biol. 19: 149–168.

Sved, J. A. 1976. Hybrid dysgenesis in *Drosophila melanogaster*: A possible explanation in terms of spatial organization of chromosomes. Austral. J. Biol. Sci. 29: 375–388.

Sved, J. A., T. E. Reed, and W. F. Bodmer. 1967. The number of balanced polymorphisms that can be maintained in a natural population. Genetics 55: 469–481.

Swofford, D. L. 1993. *PAUP: Phylogenetic Analysis Using Parsimony.* Illinois Natural History Survey, Champaign, IL.

Swofford, D. L. and G. J. Olsen. 1990. Phylogeny reconstruction. pp. 411–501. In: Hillis, D. M. and C. Moritz (eds.) *Molecular Systematics* Sinauer Associates, Sunderland, MA.

Symington, L. S. and T. D. Petes. 1988. Expansions and contractions of the genetic map relative to the physical map of yeast chromosome III. Mol. Cell. Biol. 8: 595–604.

Syvanen, M. 1984. Conserved regions in mammalian β-globins: Could they arise by cross-species gene exchanges? J. Theor. Biol. 107: 685–696.

Syvanen, M. 1994. Horizontal gene transfer: Evidence and possible consequences. Annu. Rev. Genet. 28: 237–261.

Szalay, F. S. 1969. The Hapalodectinae and a phylogeny of the Mesonychidae (Mammalia, Condylarthra). Am. Mus. Nat. Hist. Nov. 2361: 1–26.

Szostak, J. W. and R. Wu. 1980. Unequal crossing over in the ribosomal DNA of *Saccharomyces cerevisiae*. Nature 284: 426–430.

Szostak, J. W., T. L. Orr-Weaver, R. J. Rothstein, and F. W. Stahl. 1983. The double-strand-break repair model for recombination. Cell 33: 25–35.

Taberlet, P. and J. Bouvet. 1994. Mitochondrial DNA polymorphism, phylogeography, and conservation genetics of the brown bear *Ursus arctos* in Europe. Proc. R. Soc. Lond. [B] 255: 195–200.

Tajima, F. 1983. Evolutionary relationship of DNA sequences in finite populations. Genetics 105: 437–460.

Tajima, F. 1989. Statistical method for testing the neutral mutation hypothesis by DNA polymorphism. Genetics 123: 585–595.

Tajima, F. 1992. Statistical method for estimating the standard errors of branch lengths in a phylogenetic tree reconstructed without assuming equal rates of nucleotide substitution among different lineages. Mol. Biol. Evol. 9: 168–181.

Tajima, F. 1993. Simple methods for testing the molecular evolutionary clock hypothesis. Genetics 135: 599–607.

Tajima, F. and M. Nei. 1982. Biases of the estimates of DNA divergence obtained by the restriction enzyme technique. J. Mol. Evol. 18: 115–120.

Tajima, F. and M. Nei. 1984. Estimation of evolutionary distance between nucleotide sequences. Mol. Biol. Evol. 1: 269–285.

Takahashi, Y., S. Urushiyama, T. Tani, and Y. Ohshima. 1993. An mRNA-type intron is present in the *Rhodotorula hasegawae* U2 small nuclear RNA gene. Mol. Cell Biol. 13: 5613–5619.

Takahata, N. 1988. More on the episodic clock. Genetics 118: 387–388.

Takahata, N. 1990. A simple genealogical structure of strongly balanced allelic lines and trans-species evolution of polymorphism. Proc. Natl. Acad. Sci. USA 87: 2419–2423.

Takahata, N. 1991. Statistical models of the overdispersed molecular clock. Theor. Pop. Biol. 39: 329–344.

Takahata, N. and M. Nei. 1990. Allelic geneology under overdominant and frequency-dependent selection and polymorphism of major histocompatibility compex loci. Genetics 124: 967–978.

Takahata, N. and T. Maruyama. 1979. Polymorphism and loss of duplicate gene expression: A theoretical study with application to tetraploid fish. Proc. Natl. Acad. Sci. USA 76: 4521–4525.

Takezaki, N. and M. Nei. 1994. Inconsistency of the maximum parsimony method when the rate of nucleotide substitution is constant. J. Mol. Evol. 39: 210–218.

Takhtajan, A. L. 1953. Phylogenetic principles of the system of higher plants. Bot. Rev. 19: 1–45.

Takhtajan, A. L. 1969. *Flowering Plants: Origin and Dispersal.* Oliver, Edinburgh.

Tamura, K. 1994. Model selection in the estimation of the number of nucleotide substitutions. Mol. Biol. Evol. 11: 146–149.

Tamura, K. and M. Nei. 1993. Estimation of the number of nucleotide substitutions in the control region of mitochondrial DNA in humans and chimpanzees. Mol. Biol. Evol. 10: 512–526.

Tanaka, T. and M. Nei. 1989. Positive Darwinian selection observed at the variable-region genes of immunoglobulins. Mol. Biol. Evol. 6: 447–459.

Tani, T. and Y. Ohshima. 1989. The gene for the U6 small nuclear RNA in fission yeast has an intron. Nature 337: 87–90.

Tani, T. and Y. Ohshima. 1991. mRNA-type introns in U6 small nuclear RNA genes: Implications for the catalysis in pre-mRNA splicing. Genes Devel. 5: 1022–1031.

Tateno, Y., N. Takezaki, and M. Nei. 1994. Relative efficiencies of the maximum-likelihood, neighbor-joining, and maximum-parsimony methods when the substitution rate varies with site. Mol. Biol. Evol. 11: 261–277.

Tautz, D. and M. Renz. 1984. Simple sequences are ubiquitous repetitive components of eukaryotic genomes. Nucleic Acids Res. 12: 4127–4138.

Temin, H. M. 1980. Origin of retroviruses from cellular moveable genetic elements. Cell 21: 599–600.

Temin, H. M. 1989. Retrons in bacteria. Nature 339: 254–255.

Templeton, A. R. 1983. Phylogenetic inference from restriction endonuclease cleavage site maps with particular reference to the evolution of humans and the apes. Evolution 37: 221–244.

Templeton, A. R. 1992. Human origins and analysis of mitochondrial DNA sequences. Science 255: 737.

Thewissen, J. G. M., S. T. Hussain, and M. Arif. 1994. Fossil evidence for the origin of aquatic locomotion in archaeocete whales. Science 263: 210–212.

Thomas, B. A. and R. A. Spicer. 1987. *The Evolution and Palaeobiology of Land Plants.* Croom Helm, London.

Thomas, G., P. S. Zelenka, R. A. Cuthbertson, B. L. Norman, and J. Piatigorsky. 1990. Differential expression of the two δ-crystallin/argininosuccinate lyase genes in lens, heart, and brain of chicken embryos. New Biol. 2: 903–914.

Thomas, R. H., W. Schaffner, A. C. Wilson, and S. Pääbo. 1989. DNA phylogeny of the extinct marsupial wolf. Nature 340: 465–467.

Throckmorton, L. H. 1975. The phylogeny, ecology and geography of *Drosophila*. pp. 421–469. In: King, R. (ed.) *Handbook of Genetics* Vol. III, Plenum Press, New York.

Ticher, A. and D. Graur. 1989. Nucleic acid composition, codon usage, and the rate of synonymous substitution in protein-coding genes. J. Mol. Evol. 28: 286–298.

Timmis, J. N. and N. S. Scott. 1984. Promiscuous DNA: Sequence homologies between DNA of separate organelles. Trends Biochem. Sci. 9: 271–273.

Tittiger, C., S. Whyard, and V. Walker. 1993. A novel intron site in the triosephosphate isomerase gene from the mosquito *Culex tarsalis*. Nature 361: 470–472.

Torroni, A., T. G. Schurr, M. F. Cabell, M. D. Brown, J. V. Neel, M. Larsen, D. G. Smith, C. M. Vullo, and D. C. Wallace. 1993. Asian affinities and continental radiation of the four founding native American mtDNAs. Am. J. Hum. Genet. 53: 563–590.

Troitsky, A. V., Y. F. Melekhovets, G. M. Rakhimova, V. K. Bobrova, K. M. Valiejo-Roman, and A. S. Antonov. 1991. Angiosperm origin and early stages of seed plant evolution deduced from rRNA sequence comparisons. J. Mol. Evol. 32: 255–261.

Trono, D. 1995. HIV accessory proteins: Leading roles for the supporting cast. Cell 82: 189–192.

Ullu, E. and C. Tschudi. 1984. *Alu* sequences are processed 7SL RNA genes. Nature 312: 171–172.

Ullu, E., S. Murphy, and M. Melli. 1982. Human 7S RNA consists of a 140 nucleotide middle repetitive sequence inserted in an *Alu* sequence. Cell 29: 195–202.

Uyenoyama, M. 1985. Quantitative models of hybrid dysgenesis: Rapid evolution under transposition, extrachromosomal inheritance and fertility selection. Theor. Pop. Biol. 27: 176–201.

Uzzel, T. and K. W. Corbin. 1971. Fitting discrete probability distributions to evolutionary events. Science 172: 1089–1096.

Valdes, A. M., M. Slatkin, and N. B. Freimer. 1993. Allele frequencies at microsatellite loci: The stepwise mutation model revisited. Genetics 133: 737–749.

Van Arsdell, S. W., R. A. Denison, L. B. Bernstein, A. M. Weiner, T. Manser, and R. F. Gesteland. 1981. Direct repeats flank three small nuclear RNA pseudogenes in the human genome. Cell 26: 11–17.

Van Valen, L. 1963. Haldane's dilemma, evolutionary rates, and heterosis. Am. Nat. 97: 185–190.

Van Valen, L. 1966. Deltatheridia, a new order of mammals. Bull. Am. Mus. Nat. Hist. 132: 1–126.

Van Valen, L. and V. C. Maiorana. 1980. The archaebacteria and eukaryotic origins. Nature 287: 248–250.

Vergnaud, G., D. Mariat, F. Apiou, A. Aurias, M. Lathrop, and V. Lauthier. 1991. The use of synthetic tandem repeats to isolate new VNTR loci: Cloning of a human hypermutable sequence. Genomics 11: 135–144.

Vigilant, L., M. Stoneking, H. Harpending, K. Hawkes, and A. C. Wilson. 1991. African populations and the evolution of human mitochondrial DNA. Science 253: 1503–1507.

Vogel, F. and A. G. Motulsky. 1986. *Human Genetics*, 2nd Ed. Springer-Verlag, Berlin and Heidelberg.

Wainright, P. O., G. Hinkle, M. L. Sogin, and S. K. Stickel. 1993. Monophyletic origins of the metazoa: an evolutionary link with fungi. Science 260: 340–342.

Wakeley, J. 1993. Substitution rate variation among sites in hypervariable region 1 of human mitochondrial DNA. J. Mol. Evol. 37: 613–623.

Wallace, D. C. and H. J. Morowitz. 1973. Genome size and evolution. Chromosoma 40: 121–126.

Wallace, M. R., L. B. Anderson, A. M. Saulino, P. E. Gregory, T. W. Glover, and F. S. Collins. 1991. A de novo *Alu* insertion results in neurofibromatosis type 1. Nature 353: 864–866.

Walsh, J. B. 1985. Interaction of selection and biased gene conversion in a multigene family. Proc. Natl. Acad. Sci. USA 82: 153–157.

Walsh, J. B. 1986. Selection and biased gene conversion in a multigene family: Consequences of interallelic bias and threshold selection. Genetics 112: 699–716.

Walsh, J. B. 1987. Persistence of tandem arrays: Implications for satellite and simple-sequence DNAs. Genetics 115: 553–567.

Walsh, J. B. 1988. Unusual behavior of linkage disequilibrium in two-locus gene conversion models. Genet. Res. 51: 55–58.

Walsh, J. B. 1995. How often do duplicated genes evolve new functions? Genetics 139: 421–428.

Wang, B. B., M. M. Müller-Immerglück, J. Austin, N. T. Robinson, A. Chisholm, and C. Kenyon. 1993. A homeotic gene cluster patterns the anterioposterior body axis of *C. elegans*. Cell 74: 29–42.

Warburton, D., K. C. Atwood, and A. S. Henderson. 1976. Variation in the number of genes for rRNA among human acrocentric chromosomes: correlation with frequency of satellite association. Cytogenet. Cell. Genet. 17: 221–230.

Ward, R. H., B. L. Frazier, K. Dew-Jager, and S. Pääbo. 1991. Extensive mitochondrial diversity within a single American tribe. Proc. Natl. Acad. Sci. USA 88: 8720–8724.

Waterman, M. S. 1984. General methods of sequence comparison. Math. Biol. 46: 473–500.

Waterman, M. S., T. F. Smith, and W. A. Beyer. 1976. Some biological sequence metrics. Adv. Math. 20: 367–387.

Waterman, M. S., J. Joyce, and M. Eggert. 1991. Computer alignment of sequences. pp. 59–72. In: Miyamoto, M. M. and J. Cracraft (eds.) *Phylogenetic Analysis of DNA Sequences*. Oxford University Press, New York.

Watson, J. D., N. H. Hopkins, J. W. Roberts, J. A. Steitz, and A. M. Weiner. 1987. *Molecular Biology of the Gene*, 4th Ed. Benjamin/Cummings, Menlo Park, CA.

Watterson, G. A. 1975. On the number of segregating sites in genetical models without recombination. Theor. Pop. Biol. 7: 256–276.

Watterson, G. A. 1977. Heterosis or neutrality? Genetics 85: 789–814.

Watterson, G. A. 1978. An analysis of multiallelic data. Genetics 88: 171–179.

Watterson, G. A. 1978. The homozygosity test of neutrality. Genetics 88: 405–417.

Watterson, G. A. 1983. On the time for gene silencing at duplicate loci. Genetics 105: 745–766.

Waye, J. S. and H. F. Willard. 1989. Human *β* satellite DNA: Genomic organization and sequence definition of a class of highly repetitive tandem DNA. Proc. Natl. Acad. Sci. USA 86: 6250–6254.

Wayne, R. K. and S. M. Jenks. 1991. Mitochondrial DNA analysis implying extensive hybridization of the endangered red wolf *Canis rufus*. Nature 351: 506–513.

Weatherall, D. J. and J. B. Clegg. 1979. Recent developments in the molecular genetics of human hemoglobin. Cell 16: 467–479.

Wehrhahn, C. F. 1975. The evolution of selectively similar electrophoretically detectable alleles in finite natural populations. Genetics 80: 375–394.

Weiner, A. M., P. L. Deininger, and A. Efstratiadis. 1986. Nonviral retroposons: Genes, pseudogenes, and transposable elements generated by the reverse flow of genetic information. Annu. Rev. Biochem. 55: 631–661.

Weir, B. S. 1992. Population genetics in the forensic DNA debate. Proc. Natl. Acad. Sci. USA 89: 11654–11659.

Weir, B. S. 1995. DNA statistics in the Simpson matter. Nature Genet. 11: 365–368.

Weiss, E. H., A. Mellor, L. Golden, K. Fahrner, E. Simpson, J. Hurst, and R. A. Flavell. 1983. The structure of a mutant H-2 gene suggests that the generation of polymorphism in H-2 genes may occur by gene conversion-like events. Nature 301: 671–674.

Wellauer, P. K. and I. B. Dawid. 1979. Isolation and organization of human ribosomal DNA. J. Mol. Biol. 128: 289–303.

Wellauer, P. K., R. H. Reeder, I. B. Dawid, and D. D. Brown. 1976. The arrangement of length heterogeneity in repeating units of amplified and chromosomal ribosomal DNA from *Xenopus laevis*. J. Mol. Biol. 105: 487–505.

Weller, P., A. J. Jeffreys, V. Wilson, and A. Blanchetot. 1984. Organization of the human myoglobin gene. EMBO J. 3: 439–446.

Wells, R. D., T. M. Jacob, S. A. Narang, and H. G. Khorana. 1967a. Studies on polynucleotides. J. Mol. Biol. 27: 237–263.

Wells, R. D., H. Buchi, H. Kössel, E. Ohtsuka, and H. G. Khorana. 1967b. Studies on polynucleotides. Synthetic deoxyribopolynucleotides as templates for the DNA polymerase of *Escherichia coli*: DNA-like polymers containing repeating tetranucleotide sequences. J. Mol. Biol. 27: 265–272.

Wessler, S. R. 1988. Phenotypic diversity mediated by the maize transposable elements *Ac* and *Spm*. Science 242: 399–405.

Wessler, S. R., G. Baran, and M. Varagona. 1987. The maize transposable element *Ds* is spliced from RNA. Science 237: 916–918.

White, M. J. D. 1978. *Modes of Speciation*. W. H. Freeman, San Francisco.

Widegren, B., U. Arnason, and G. Akusjarvi. 1985. Characteristics of a conserved 1,579-bp highly repetitive component in the killer whale, *Orcinus orca*. Mol. Biol. Evol. 2: 411–419.

Wiehe, T. H. E. and W. Stephan. 1993. Analysis of a genetic hitchhiking model, and its application to DNA polymorphism data from *Drosophila melanogaster*. Mol. Biol. Evol. 10: 842–854.

Willard, C., H. T. Nguyen, and C. W. Schmid. 1987. Existence of at least three distinct *Alu* subfamilies. J. Mol. Evol. 26: 180–186.

Willard, H. F. and J. S. Waye. 1987. Hierarchical order in chromosome-specific human α satellite DNA. Trends Genet. 3: 192–198.

Williams, S. A. and M. Goodman. 1989. A statistical test that supports a human/chimpanzee clade based on noncoding DNA sequence data. Mol. Biol. Evol. 6: 325–330.

Wilson, A. C., S. S. Carlson, and T. J. White. 1977. Biochemical evolution. Annu. Rev. Biochem. 46: 573–639.

Winderickx, J., L. Battisti, Y. Hibiya, A. G. Motulsky, and S. S. Deeb. 1993. Haplotype diversity in the human red and green opsin genes: Evidence for frequent sequence exchange in exon 3. Hum. Mol. Gen. 2: 1413–1421.

Wistow, G. 1993. Lens crystallins: Gene recruitment and evolutionary dynamism. Trends Biochem. Sci. 18: 301–306.

Wistow, G. J., J. W. M. Mulders, and W. W. De Jong. 1987. The enzyme lactate dehydrogenase as a structural protein in avian and crocodilian lenses. Nature 326: 622–624.

Woese, C. R. 1987. Bacterial evolution. Microbiol. Rev. 51: 221–271.

Woese, C. R. and G. E. Fox. 1977a. Phylogenetic structure of the prokaryotic domain: The primary kingdoms. Proc. Natl. Acad. Sci. USA 74: 5088–5090.

Woese, C. R. and G. E. Fox. 1977b. The concept of cellular evolution. J. Mol. Evol. 10: 1–6.

Woese, C. R., O. Kandler, and M. L. Wheelis. 1990. Toward a natural system of organisms: Proposal for the domains archaea, bacteria, and eucarya. Proc. Natl. Acad. Sci. USA 87: 4576–4579.

Wold, F. 1981. In vivo modification of proteins. Annu. Rev. Biochem. 50: 783–814.

Wolf, U., H. Ritter, N. B. Atkin, and S. Ohno. 1969. Polyploidization in the fish family Cyprinidae, order Cypriniformes. Humangenetik 7: 240–244.

Wolfe, K. H. and P. M. Sharp. 1993. Mammalian gene evolution: nucleotide sequence divergence between mouse and rat. J. Mol. Evol. 37: 441–456.

Wolfe, K. H., W.-H. Li, and P. M. Sharp. 1987. Rates of nucleotide substitution vary greatly among plant mitochondrial, chloroplast, and nuclear DNAs. Proc. Natl. Acad. Sci. USA 84: 9054–9058.

Wolfe, K. H., P. M. Sharp, and W.-H. Li. 1989a. Rates of synonymous substitution in plant nuclear genes. J. Mol. Evol. 29: 208–211.

Wolfe, K. H., M. Gouy, Y.-W. Yang, P. M. Sharp, and W.-H. Li. 1989b. Date of the monocot-dicot divergence estimated from chloroplast DNA sequence data. Proc. Natl. Acad. Sci. USA 86: 6201–6205.

Wolfe, K. H., P. M. Sharp, and W.-H. Li. 1989c. Mutation rates differ among regions of the mammalian genome. Nature 337: 283–285.

Wolff, R. K., R. Plaetke, A. J. Jeffreys, and R. White. 1989. Unequal crossing over between homologous chromosomes is not the major mechanism involved in the generation of new alleles at VNTR loci. Genomics 5: 382–384.

Wood, W. G., J. B. Clegg, and D. J. Weatherall. 1977. Developmental biology of human hemoglobins. pp. 43–90. In: Brown, X. E. B. (ed.) *Progress in Hematology*. Grune and Stratton, New York.

Woodward, S. R., N. J. Weyand, and M. Bunnell. 1994. DNA sequence from Cretaceous period bone fragments. Science 266: 1229–1232.

Worton, R. G. 1992. Duchenne muscular dystrophy: Gene and gene product; Mechanism of mutation in the gene. J. Inher. Metab. Dis. 15: 539–550.

Wright, F. 1990. The "effective number of codons" used in a gene. Gene 87: 23–29.

Wright, S. 1931. Evolution in Mendelian populations. Genetics 16: 97–159.

Wright, S. 1939. The distribution of self-sterility alleles in populations. Genetics 24: 538–552.

Wright, S. 1943. Isolation by distance. Genetics 28: 114–138.

Wright, S. 1949. Adaptation and selection. pp. 365–389. In: Jepson, G. L., G. G. Simpson, and E. Mayr (eds.) *Genetics, Paleontology, and Evolution*. Princeton University Press, Princeton, NJ.

Wright, S. 1969. *Evolution and the Genetics of Populations*. University of Chicago Press, Chicago.

Wu, C.-I. and W.-H. Li. 1985. Evidence for higher rates of nucleotide substitution in rodents than in man. Proc. Natl. Acad. Sci. USA 82: 1741–1745.

Wu, C.-I., T. W. Lyttle, M.-L. Wu, and G.-F. Lin. 1988. Association between a satellite DNA sequence and the *Responder* of *Segregation Distorter* in *Drosophila melanogaster*. Cell 54: 179–189.

Wu, C.-I., J. R. True, and N. Johnson. 1989. Fitness reduction associated with the deletion of a satellite DNA array. Nature 341: 248–251.

Wyman, A. H. and R. White. 1980. A highly polymorphic locus in human DNA. Proc. Natl. Acad. Sci. USA 77: 6754–6758.

Wyss, A. 1990. Clues to the origin of whales. Nature 347: 428–429.

Xiong, Y. and T. H. Eickbush. 1988. Similarity of reverse transcriptase-like sequences of viruses, transposable elements, and mitochondrial introns. Mol. Biol. Evol. 5: 675–690.

Xiong, Y. and T. H. Eickbush. 1990. Origin and evolution of retroelements based on their reverse transcriptase sequences. EMBO J. 9: 3353–3362.

Yamada, Y., V. E. Avvedimento, M. Mudryi, H. Ohkubo, G. Vogeli, M. Irani, I. Pastan, and B. de Crombrugghe. 1980. The collagen gene: Evidence for its evolutionary assembly by amplification of a DNA segment containing an exon of 54 bp. Cell 22: 887–892.

Yamao, F., S. Iwagami, Y. Azumi, A. Muto, and S. Osawa. 1988. Evolutionary dynamics of tryptophan tRNAs in *Mycoplasma capricolum*. Mol. Gen. Genet. 212: 364–369.

Yang, Z. 1993. Maximum-likelihood estimation of phylogeny from DNA sequences when substitution rates differ over sites. Mol. Biol. Evol. 10: 1396–1401.

Yao, M.-C., A. R. Kimmel, and M. A. Gorovsky. 1974. A small number of cistrons for ribosomal RNA in the germinal nucleus of a eukaryote, *Tetrahymena pyriformis*. Proc. Natl. Acad. Sci. USA 71: 3082–3086.

Yokoyama, S. and R. Yokoyama. 1989. Molecular evolution of human visual pigment genes. Mol. Biol. Evol. 6: 186–197.

Yuki, S. S., S. Ishimaru, S. Inouye, and K. Saigo. 1986. Identification of genes for reverse transcriptase-like enzymes in two *Drosophila* retrotransposons, *412* and *gypsy*: A rapid detection method of reverse transcriptase genes using YXDD box probe. Nucleic Acids Res. 14: 3017–3030.

Zakian, V. A. 1989. Structure and function of telomeres. Annu. Rev. Genet. 23: 579–604.

Zambryski, P. 1988. Basic processes underlying agrobacterium-mediated DNA transfer to plant cells. Annu. Rev. Genet. 22: 1–30.

Zambryski, P. 1989. Agrobacterium-plant cell DNA transfer. pp. 309–333. In: Berg, D. E. and M. M. Howe (eds.) *Mobile DNA*. American Society for Microbiology Press, Washington, DC.

Zhang, J. and M. Nei. 1996. Evolution of *Antennapedia*-class homeobox genes. Genetics 142: 295–303.

Zhang, Q.-Y. 1991. Molecular basis of penicillin resistance in *Neisseria meningitidis*. Ph. D. Dissertation, University of Sussex, England.

Zharkikh, A. 1994. Estimation of evolutionary distances between nucleotide sequences. J. Mol. Evol. 39: 315–329.

Zharkikh, A. and W.-H. Li. 1992a. Statistical properties of bootstrap estimation of phylogenetic variability from nucleotide sequences. I. Four taxa with a molecular clock. Mol. Biol. Evol. 9: 1119–1147.

Zharkikh, A. and W.-H. Li. 1992b. Statistical properties of bootstrap estimation of phylogenetic variability from nucleotide sequences. II. Four taxa without a molecular clock. J. Mol. Evol. 35: 356–366.

Zharkikh, A. and W.-H. Li. 1993. Inconsistency of the maximum-parsimony method: The case of five taxa with a molecular clock. Syst. Biol. 42: 113–125.

Zharkikh, A. and W.-H. Li. 1995. Estimation of confidence in phylogeny: The complete-and-partial bootstrap technique. Mol. Phylogenet. Evol. 4: 44–63.

Zhou, Y.-H. and W.-H. Li. 1996. Gene conversion and natural selection in the evolution of X-linked color vision genes in higher primates. Mol. Biol. Evol. 13: 780–783.

Zillig, W. 1991. Comparative biochemistry of archaea and bacteria. Curr. Opinion Genet. Devel. 1: 544–551.

Zimmer, E. A., S. L. Martin, S. M. Beverley, Y. W. Kan, and A. C. Wilson. 1980. Rapid duplication and loss of genes coding for the α chains of hemoglobin. Proc. Natl. Acad. Sci. USA 77: 2158–2162.

Zischler, H., M. Höss, O. Handt, A. Von Haeseler, A. C. Van der Kuyl, J. Goudsmit, and S. Pääbo. 1995. Detecting dinosaur DNA. Science 268: 1192–1193.

Zuckerkandl, E. 1963. Perspectives in molecular anthropology. pp. 243–272. In: Washburn, S. L. (ed.) *Classification and Human Evolution*. Aldine, Chicago.

Zuckerkandl, E. 1976. Gene control in eukaryotes and the c-value paradox: "Excess" DNA as an impediment to transcription of coding sequences. J. Mol. Evol. 9: 73–104.

Zuckerkandl, E. and L. Pauling. 1962. Molecular disease, evolution and genic heterogeneity. pp. 189–225. In: Kasha, M. and B. Pullman (eds.) *Horizons in Biochemistry*. Academic Press, New York.

Zuckerkandl, E. and L. Pauling. 1965. Evolutionary divergence and convergence in proteins. pp. 97–166. In: Bryson, V. and H. J. Vogel (eds.) *Evolving Genes and Proteins*. Academic Press, New York.

Zuckerkandl, E., R. T. Jones, and L. Pauling. 1960. A comparison of animal hemoglobins by tryptic peptide pattern analysis. Biochem. 46: 1349–1360.

Zuckerkandl. E., J. Derancourt, and H. Vogel. 1971. Mutational trends and random processes in the evolution of informational macromolecules. J. Mol. Biol. 59: 473–490.

Zuliani, G. and H. H. Hobbs. 1990. A high frequency of length polymorphism in repeated sequence adjacent to *Alu* sequences. Am. J. Hum. Genet. 46: 963–969.

Index

ABOUT THE BOOK

Editor: Andrew D. Sinauer
Project Manager: Carol J. Wigg
Copy Editor: Bobbie Lewis
Production Manager: Christopher Small
Composition and Page Layout: Precision Graphics
Illustrations: Precision Graphics
Book and Cover Design: Jefferson C. Johnson
Cover and Book Manufacture: Best Book Manufacturers